Agricultural
Development
in the
Third World

Agricultural Development in the Third World

Edited by
Carl K. Eicher & John M. Staatz

THE JOHNS HOPKINS UNIVERSITY PRESS

Baltimore and London

Originally published, 1984
Second printing (paperback), 1985
Third printing (paperback), 1988

The Johns Hopkins University Press
701 West 40th Street
Baltimore, Maryland 21211
The Johns Hopkins Press Ltd., London

LIBRARY OF CONGRESS CATALOGING IN PUBLICATION DATA

Main entry under title:

Agricultural development in the Third World.

(The Johns Hopkins studies in development)
Includes index.
1. Agriculture—Economic aspects—Developing countries
—Addresses, essays, lectures.   2. Agriculture—Develop-
ing countries—Addresses, essays, lectures.   1. Eicher,
Carl K.   II. Staatz, John M.   III. Series.
HD1417.A4483   1984   338.1'09172'4   83–19532
ISBN 0-8018-3014-1
ISBN 0-8018-3015-X (pbk.)

Contents

v

Preface

This book grew out of a graduate-level course on the economics of agricultural development that we teach at Michigan State University. In course preparation we have been struck by the cumulative knowledge that economists and agricultural scientists have gained during the past twenty-five years about the complexity and diversity of agriculture in the Third World, its interrelationships with other sectors, and its potential roles in the structural transformation of Third World economies. As a result of this new knowledge, many of the economic growth models of the 1960s seem remarkably unsophisticated today.

Despite this increased knowledge, students in both industrial and Third World countries have not had a text that pulled together different views about what has been learned theoretically and empirically about the agricultural development process since the early 1970s. Although single-author texts are invaluable in developing and testing models of agricultural development, by their very nature they cannot reflect the diversity of opinion about alternative policies and strategies to further the development of agriculture in the Third World. This book includes articles and original essays by leading scholars on the theory and practice of agricultural development. All the articles have been written since 1972, during the "growth-with-equity" era of development economics. The contribution of empirical research to policy analysis is an important theme of the book, as reflected in the strong empirical content of many of the selections.

A collection of readings on a topic as vast as agricultural development is by nature highly selective. We have chosen articles that deal with the fundamentals of agricultural development: the role of agriculture in economic growth, intersectoral linkages, mechanisms of agricultural growth, institutional reform, functioning of factor markets, choice of technique, and the generation and social impact of technical change. Most of these articles were written from a policy rather than a model-building perspective. We have not reprinted any of the classics of the 1950s and 1960s on agricultural development because these are readily available in other collections. All of the articles have been edited and some have been condensed. In addition, an effort has been made, to the extent possible, to standardize the references and spellings throughout the text.

Because of space limitations, we have not been able to include articles on several important topics in agricultural development. For example, there are no articles dealing primarily with agricultural extension, marketing, and agriculture in eastern Europe and the Soviet Union. We have, however, included extensive references in chapter 1 and in the introductions to parts II–V to the vast literature on agricultural development now available to agricultural and social scientists.

This book is designed for use as a text in beginning graduate courses in agricultural and rural development and as a supplementary text in economic development courses in both industrial and Third World countries. Most of the articles require no greater background in economics than intermediate courses in micro- and macroeconomics; therefore, the book also may be suited for upper-level undergraduate courses. Because the emphasis of the book is on policy analysis rather than on model building per se, we believe that it will be useful to agricultural scientists, planners, policy makers, and members of donor agencies.

A collection of readings is naturally a collective effort. We want to thank all the contributors, especially those who prepared original essays—chapters 7, 20, and 29—and those who revised previously published essays. We are grateful to the Department of Agricultural Economics of Michigan State University and especially to its chairman and associate chairman, Larry Connor and Lester Manderscheid, for their keen interest in Third World agriculture and for their leadership and support over the years.

Special thanks go to Vernon Ruttan, John Mellor, Alain de Janvry, and Carl Liedholm for their counsel on the overall structure of the volume. We are also indebted to numerous students who provided valuable feedback on successive drafts of the book and to numerous teachers of development courses throughout the world who provided advice on the selection of articles. We thank Michael Morris for reading and commenting on several parts of the manuscript. Our task was greatly facilitated by the able secretarial assistance of Lucy Wells (who combines the skills of a cryptographer with the patience of Job), Jeanette Barbour, and Cindy Spiegel and by the library assistance of Laura Wilson, Pat Eisele, Steve Berg, and Dale Edgington. We also appreciate Joanne Allen's careful copy editing of the manuscript and the invaluable help of Anders Richter and Barbara Lamb of The Johns Hopkins University Press in guiding the book through publication. Finally we thank our wives, Shirley Eicher and Barbara Berti, for their patience, support, and encouragement.

I

Overview

1

Agricultural Development Ideas in Historical Perspective

JOHN M. STAATZ AND CARL K. EICHER

Although economists have been concerned with growth and development since at least the time of the mercantilists, development economics has existed as a separate branch of economics only since about 1950. The history of the field can be roughly divided into two periods: the economic-growth-and-modernization era of the 1950s and 1960s, when development was defined largely in terms of growth in average per capita output; and the growth-with-equity period since around 1970, when the concern of most development economists broadened to include income distribution, employment, nutrition, and a host of other variables. The prevailing view of agriculture's role in development changed profoundly during these two periods.[1]

In this chapter we will outline the changing view of agriculture in economic development since 1950 and place the readings in this volume in historical perspective. In the first section we will briefly discuss the evolution of agricultural development theory and practice during the growth-and-modernization era of the fifties and sixties. We will examine the relatively passive role assigned to agriculture in the economic growth models of the 1950s, the increasing recognition of the interdependence between agricultural and industrial growth during the 1960s, the lessons learned from the agricultural development experience of the 1950s and 1960s, and the contribution of radical and dependency scholars to an improved understanding of the process of agricultural and rural development. In the next section we will discuss the increased emphasis given to agricultural and rural development during the growth-with-equity period beginning around 1970. During this period there was a sharp increase in microeconomic research on agricultural production and marketing, intersectoral linkages, rural factor markets, migration, and rural small-scale industry and a policy shift to integrated rural development and basic needs programs. Finally, we will outline the plan of this volume, placing the readings in the context of the theoretical and policy debates discussed in the first two sections.

JOHN M. STAATZ is assistant professor and CARL K. EICHER is professor of agricultural economics, Michigan State University.

3

THE ROLE OF AGRICULTURE IN DEVELOPMENT ECONOMICS, 1950–69

WESTERN DEVELOPMENT ECONOMISTS' PERSPECTIVES ON AGRICULTURE

Most Western development economists of the 1950s did not view agriculture as an important contributor to economic growth.[2] As I. M. D. Little comments in his survey of development economics, "It is fairly obvious from reading their works that the leading development economists of the 1950s knew little about tropical agriculture or rural life. They had no time for rural rides and there was no considerable body of empirical grassroots literature on which they could draw" (Little 1982, 106). Development was often equated with the structural transformation of the economy, that is, with the decline in agriculture's relative share of the national product and of the labor force. The role of development economics was seen as facilitating that transformation by discovering ways to transfer resources, especially labor, from traditional agriculture to industry, the presumed engine of growth. Agriculture itself was often treated as a "black box from which people, and food to feed them, and perhaps capital could be released" (ibid., 105).

Development economics throughout the 1950s and 1960s was strongly influenced by W. Arthur Lewis's 1954 article "Economic Development with Unlimited Supplies of Labour." Seldom has a single article been so instrumental in shaping the work of an entire subdiscipline of economics. In the article, Lewis presented a general equilibrium model of expansion in an economy with two sectors—a modern capitalist exchange sector and an indigenous noncapitalist sector, which was dominated by subsistence farming.[3] The distinguishing characteristics of the capitalist sector were its use of reproducible capital, its hiring of labor, and its sale of output for profit. "Capitalist" enterprises could be owned privately or by the state. The subsistence sector was pictured as the "self-employment sector," which did not hire labor or use reproducible capital. Lewis's model focused on how the transfer of labor from the subsistence sector (where the marginal productivity of a laborer approached zero as a limiting case) to the capitalist sector facilitated capitalist expansion through reinvestment of profits. The labor supply facing the capitalist sector was " 'unlimited' in the sense that when the capitalist sector offers additional employment opportunities at the existing wage rate, the numbers willing to work at the existing wage rate will be greater than the demand: the supply curve of labor is infinitely elastic at the ruling wage" (Meier 1976, 158). In Lewis's model, expansion in the capitalist sector continued until earnings in the two sectors were equated, at which point a dual-sector model was no longer relevant; growth proceeded as in a one-sector neoclassical model. Lewis's analysis was later extended by Ranis and Fei (1961, 1963, 1964) and Jorgenson (1961).

Lewis pointed out that the capitalist sector did not need to be industry (it could be mining or plantations) and that the noncapitalist sector could include handicrafts. Most analysts, however, equated the capitalist sector with industry and the noncapitalist sector with traditional agriculture and argued that "surplus" labor and other resources should be transferred from agriculture to industry in order to promote growth.[4] Many development economists concluded that since economic growth facilitated the structural transformation of the economy in the *long run,* the rapid transfer of resources (especially "surplus" labor) from agriculture to industry was an appropriate *short-run* economic development strategy.[5] But Johnston observed that "This preoccupation with 'surplus labor' often seems to have encouraged neglect of the agricultural sector as well as a tendency to assume too readily that a surplus can and should be extracted from agriculture, while neglecting the difficult requirements that must be met if agriculture is to play a positive role in facilitating overall economic growth" (Johnston 1970, 378). The propensity of development economists to give relatively little attention to agriculture's potential "positive role in facilitating overall economic growth" was based in part on the empirical observation that agriculture's share of the economy inevitably declines during the course of development for at least two reasons. First, the income elasticity of demand for unprocessed food is less than unity and declines with higher incomes; hence, the demand for raw agricultural products grows more slowly than consumption in general. Second, increasing labor productivity in agriculture means that the same farm output can be produced with fewer workers, implying a transfer of labor to other sectors of the economy. Because agriculture's share of the economy was assumed to be declining, many economists downplayed the need to invest in the agricultural sector in the short run.

The relative neglect of agriculture in the 1950s was reinforced by two other developments. In 1949 Raul Prebisch and Hans Singer independently formulated the thesis that there is a secular tendency for the terms of trade to turn against countries that export primary products and import manufactures.[6] From this they concluded that the scope for growth through agricultural and other primary exports was very limited. Prebisch and his colleagues at the United Nations Economic Commission for Latin America (ECLA) therefore advocated that priority be given to import substitution of manufactured goods rather than to production of agricultural exports.[7] The "secular-decline hypothesis" became an article of faith for some development economists and planners, and thus the tendency to downplay agriculture's potential role in development was reinforced.

The second important event affecting development economists' view of agriculture was the publication of Albert Hirschman's influential book *The Strategy of Economic Development* (1958). In this book, Hirschman introduced the concept of linkages as a tool for investigating how, during the course of development, investment in one type of economic activity induced subsequent investment in other income-generating activities. Hirschman defined the linkage effects of a given product line as the "investment-generating forces that are set in motion, through input-output relations, when productive facilities that supply inputs to that line or utilize its output are inadequate or non-existent. Backward linkages lead to new investment in input-supplying facilities and forward linkages to investment in output-using facilities" (Hirschman 1977, 72). Hirschman argued that government investment should be concentrated in activities where the linkage effects were greatest, since this would maximize indigenous investment in related, or "linked," industries. Hirschman asserted that "agriculture certainly stands convicted on the count of its lack of direct stimulus to the setting up of new activities through linkage effects—the superiority of manufacturing in this respect is crushing" (Hirschman 1958, 109–10). Therefore, Hirschman argued, investment in industry would generally lead to more rapid and more broadly based economic growth than would investment in agriculture. Hirschman's analysis thus reinforced ECLA's policy recommendation that priority be given to import substitution of manufactures.

Ironically, Lewis's two-sector growth model, which led many development economists to focus heavily on the role of industry in economic development, in the early 1960s led others to stress the interdependence between agricultural and industrial growth. In an article comparing a Lewis-type "classical" model with a neoclassical growth model, Jorgenson (1961) argued that growth in nonfarm employment depended on the rate of growth of the agricultural surplus. Jorgenson's analysis and similar analyses by Ranis and Fei (1961, 1963, 1964) and Enke (1962a, 1962b) showed that food shortages could choke off growth in the nonfarm sector by making its labor supply less than infinitely elastic. These authors therefore concluded that in order to avoid falling into a low-level equilibrium trap, in the early stages of development a country probably needed to make some net investment in agriculture to accelerate the growth of its agricultural surplus.

Many agricultural economists found it "shocking . . . that general economists such as Jorgenson and Enke . . . felt it necessary to argue the case for *some* investment in agriculture" (Johnston 1970, 378). In a seminal article entitled "The Role of Agriculture in Economic Development" (1961), Johnston and Mellor drew on insights from the Lewis model to stress the importance of agriculture as a motive force in economic growth. They argued that far from playing a passive role in development, agriculture could make five important contributions to the structural transformation of Third World economies: it could provide labor, capital, foreign exchange, and food to a growing industrial sector and could supply a market for domestically produced industrial goods. They argued further that the nature of the interrelationships between agriculture and industry at various stages of development had important implications for the types of agricultural and industrialization strategies that would be most likely to succeed.

Johnston and Mellor's article and William H. Nicholls's influential article "The Place of Agriculture in Economic Development" (1964) were instrumental in encouraging economists to view agriculture as a potential positive force in development, and they helped to stimulate debate on the interdependence of agricultural and industrial growth. This in turn led to a growing interest in the empirical measurement of intersectoral resource transfers during the course of development.[8]

The work of neoclassical agricultural economists during the 1960s stressed not only the interdependence of agriculture and industry and the potentially important role that agriculture could play in economic development but also the importance of understanding the process of agricultural growth per se if that potential was to be exploited.[9] The need for a better understanding of the process of agricultural growth was further emphasized by some of the agricultural development experiences of the 1950s and early 1960s.

The Influence of the Agricultural Development Programs of the 1950s and 1960s on Western Development Thought

Debates among Western economists about the role of agriculture in development did not take place in a vacuum; they were strongly influenced by the rural development experiences of Third World nations. Indeed, an important characteristic of the literature on agricultural development since the 1950s has been its movement from a priori theorizing towards empirical research.

Despite the emphasis that development economists placed on industrialization during the 1950s, governments of many low-income countries and donor agencies undertook a number of activities aimed at increasing agricultural output and rural incomes. The experience gained in these efforts was important in developing a better understanding of intersectoral relationships and the constraints on agricultural growth.

During the 1950s the approach of European and North American agricultural economists to development was colored by the historical experiences of their own countries and by their training in the then-current theories of development economics. For example, most Western agricultural economists working on problems of Third World agriculture during this period believed that the problem of rural surplus labor could be resolved by transferring "excess" rural workers to urban industry.[10] It was also widely assumed that Western agricultural advisers could directly transfer agricultural technology and models of agricultural extension from high-income countries to the Third World, that community development programs could help rural people overcome the shackles of traditional farming and inequitable land tenure systems, and that food aid could serve humanitarian needs and provide jobs for rural people.

Agricultural development efforts of the 1950s placed heavy emphasis on the direct transfer of agricultural technology from high-income countries to the Third World and the promotion of the American model of agricultural extension. These efforts were based on what Vernon Ruttan calls the "diffusion model" of agricultural development (see chapter 2). The diffusion model assumed that Third World farmers could substantially increase their agricultural productivity by allocating existing resources more efficiently and by adopting agricultural practices and technologies from the industrial countries.

Like the diffusion model, the community development effort of the 1950s and early 1960s (described by Lane Holdcroft in chapter 3) assumed that small farmers were often poor decision makers who required outside assistance in planning local development projects (Stevens 1977b, 5). Community development grew out of the cold-war atmosphere of the 1950s, when Western foreign assistance programs were searching for a nonrevolutionary approach to rural change. Community development advocates assumed that villagers, meeting with community development specialists, would express their "felt needs" and unite to design and implement self-help programs aimed at promoting rural development. The community development effort also implicitly assumed that rural development could be achieved through the direct transfer of Western agricultural technologies and social institutions, such as local democracy, to the rural areas of the Third World.[11]

The failure of many agricultural extension programs to achieve rapid increases in agricultural output and the inability of community development projects to solve the basic food problem in many countries (particularly India in the mid 1950s) led to a reevaluation of the diffusion model of agricultural development. Two elements were critical in this reappraisal. First, as Holdcroft points out, it became apparent that in many countries there were important structural barriers to rural development, such as highly concentrated political power and asset ownership. The research by economists such as Warriner (1955), Ladejinsky (see Walinsky 1977), Carroll (1961), Raup (1967), and Barraclough (1973) on land tenure and land reform in Asia, North Africa, and Latin America documented how institutional barriers inhibited the expansion of agricultural output. These authors argued that in some countries basic institutional reforms were prerequisites to effective agricultural extension and community development.[12]

The second element leading to a reevaluation of the diffusion model was research by scholars such as Jones (1960), Krishna (1967), and Behrman (1968) that documented the responsiveness of Third World farmers and consumers to economic incentives and helped to demolish the myth of the "tradition-bound peasant." The findings of these studies suggested that if farmers were not responsive to agricultural extension efforts, perhaps it was because extension workers had few profitable innovations to extend. This viewpoint was advanced most forcefully in T. W. Schultz's highly influential book *Transforming Traditional Agriculture* (1964).

As Ruttan points out in chapter 2, Schultz's book was iconoclastic at the time. Schultz argued that Third World farmers and herders, far from being irrational and fatalistic, were calculating economic agents who carefully weighed the marginal costs and benefits associated with different agricultural techniques. Through a long process of experimentation, these farmers had learned how to allocate the factors of production available to them efficiently, given existing technology. This implied that "no appreciable increase in agricultural production is to be had by reallocating the factors at the disposal of farmers who are bound by traditional agriculture. . . . An outside expert, however skilled he may be in farm management, will not discover any major inefficiency in the allocation of factors" (39).

Schultz's argument that despite low levels of per capita output, traditional agriculture was characterized by allocative efficiency became known as the "efficient-but-poor hypothesis."[13] Citing evidence from Guatemala (Tax [1953] 1963) and India (summarized in Hopper 1965) to support this hypothesis, Schultz argued that major increases in per capita agricultural output in the Third World would come about only if farmers were provided new, more productive factors of production (that is, new agricultural technologies) and the new skills needed to exploit them. The cause of rural poverty, in other words, lay in the lack of profitable technical packages for Third World farmers and the lack of investment in human capital needed to cope with rapidly changing agricultural technologies. Schultz later attributed the low levels of investment in agricultural research and rural education in most Third World countries to national policies that undervalued agriculture (Schultz 1978, 1981).

Transforming Traditional Agriculture called for a major shift from agricultural extension towards investment in agricultural research and human capital. The book, which appeared five years after the establishment of the International Rice Research Institute (IRRI) in the Philippines and one year after the establishment of the International Center for Maize and Wheat Improvement (CIMMYT) in Mexico, reinforced the increasing emphasis being given to agricultural research by the Rockefeller and Ford foundations and other donors in the 1960s. As a result of IRRI's and CIMMYT's success in developing high-yielding dwarf varieties of rice and wheat which were rapidly adopted in many areas of the Third World during the 1960s, the green revolution, or high-payoff input model, replaced the diffusion/community development model as the dominant agricultural development model for field practitioners.[14]

The appearance of the high-yielding grain varieties had important effects on the theory as well as on the practice of agricultural development. Several authors, such as Ohkawa (1964), Mellor (1966), and Ohkawa and Johnston (1969), noted that the new grain/fertilizer technologies were highly divisible and scale-neutral, allowing them to be incorporated into existing systems of small-scale agriculture. Therefore, these authors argued, intensification of agricultural production based on high-yielding cereal varieties offered the opportunity to provide productive employment for the rapidly growing rural labor force while at the same time it produced the wage goods needed for an expanding industrial labor force. The high-yielding varieties, it was argued, made it possible to achieve both employment and output objectives.[15]

The early enthusiasm for the green revolution was met by a barrage of criticism by Frankel (1971), Griffin (1974), and others. These authors argued that the new varieties often benefited mainly landlords and larger farmers in ecologically favored areas, while they frequently impoverished small farmers and tenants, particularly those in upland areas, by inducing lower grain prices and evictions from the land as landlords found it profitable to farm the land themselves using mechanization (see Hayami, chapter 27).

Although some authors, such as Lester Brown (1970), did tend to oversell the accomplishments of the green revolution, in Asia the impact of the new varieties was substantial. They have had a smaller effect in Latin America, however, where a high percentage of small farmers live in poor natural resource zones (Piñeiro, Trigo, and Fiorentino 1979), and they have had little impact in sub-Saharan Africa (see Eicher, chapter 31). Overall, high-yielding varieties accounted for about one-third of the area planted to wheat and rice in the Third World in the early 1980s.

In chapters 26 and 27, Scobie, Posada, and Hayami evaluate some of the income-distribution effects of the new varieties. These authors find that *within* villages there has been little difference between small farmers' and large farmers' rates of adoption of the modern varieties. Farmers in upland areas, however, may have been hurt relative to farmers in irrigated areas because the new varieties are more suited to irrigated conditions. Low-income consumers, who spend a high proportion of their income on foodgrains, have been major beneficiaries of the larger harvests and lower prices made possible by the green revolution. One of the important lessons of the 1950s and 1960s is that with rising population pressure on land throughout the Third World, technological change must be included as a central component in both the theory and the practice of agricultural and rural development.

RADICAL POLITICAL ECONOMY AND DEPENDENCY PERSPECTIVES ON AGRICULTURE

Western development economics was challenged in the 1960s and 1970s by the emergence and rapid growth of radical political economy and dependency models of development and underdevelopment. The radical political economy models have their roots in the writings of Lenin on imperialism and Kautsky on agriculture and in the post–World War II writings of Paul Baran and other Marxist economists. Baran, in an important article entitled "On the Political Economy of Backwardness" (1952), argued that in most low-income countries it would be impossible to bring about broad-based capitalist development without violent changes in social and political institutions. Although Baran was clearly ahead of his time in identifying institutional and structural barriers to development and the need to put effective demand at the center of development programs, he tended, as did many of the Western development economists he was criticizing, to see small-scale agriculture as incapable of making major contributions to economic growth. For example, Baran accepted the view that the marginal product of labor often approached zero in agriculture and that therefore "there is no way of employing it [labor] usefully in agriculture." Farmers "could only be provided with opportunities for productive work by transfer to industry." Baran, like many economists of the time, believed that "very few improvements that would be necessary in order to increase productivity can be carried out within the narrow confines of small-peasant holdings" and that therefore farm consolidation was necessary.

Marxist analysis of agricultural and rural development was further advanced in the 1950s and 1960s by several Latin American scholars, who often blended Marxian analyses with dependency theory.[16] The dependency interpretation of underdevelopment was first proposed in the 1950s by the Economic Commission for Latin America, under the leadership of Raul Prebisch. The basic hypothesis of this perspective is that underdevelopment is not a stage of development but the result of the expansion of the world capitalist system. Underdevelopment, in other words, is not simply the lack of development; it is a condition of impoverishment brought about by the integration of Third World economies into the world capitalist system. Although a number of different views of dependency have been put forward by scholars such as Sunkel, Furtado, Frank (1966), Galtung (1971), and others, the following definition of dependency by Dos Santos has been widely cited: "By dependency we mean a situation in which the economy of certain countries is conditioned by the development and expansion of another economy to which the former is subjected" (1970, 231).[17]

Dependency theorists implicitly argued that trade was often a zero-sum game—that low-income countries ("the periphery") were pauperized through both a process of unequal exchange with the industrialized world ("the center") and repatriation of profits from foreign-owned businesses. Capitalist growth in the periphery was not self-sustaining; it was stunted by policies favoring import substitution of luxury goods and export of agroindustrial products, often produced on large estates. These policies limited the internal market for consumer goods (including food and other agricultural products) and led to impoverishment of the mass of small farmers. Meaningful reform was blocked by an alliance of the landed elite, local bourgeoisie, and multinational firms, all of which benefited from the dependency relationship.

Various theories of unequal exchange played important roles in dependency theory. Marxist economists, such as Emmanuel (1972), argued that unequal exchange between industrialized and Third World countries resulted from the maintenance of precapitalist relations of production in the Third World, which depressed wage rates, and the use of monopoly power by industrialized nations to turn the terms of trade against the Third World (see chapter 6). Changes in the social relations of production in low-income countries and the organization of cartels by Third World exporters were therefore necessary to obtain countervailing power in international markets. Many non-Marxist economists, such as Singer and Prebisch, believed that unequal exchange was a logical consequence of the demand characteristics of the primary products exported by Third World countries and the noncompetitive labor markets in high-income countries, which tended to raise the price of manufactures. In these authors' view, import substitution and a shift to exports of manufactures represented ways of combatting unequal exchange. In contrast, Lewis (1978) developed a theory of unequal exchange based on a simple two-country, three-good Ricardian trade model in which food is produced in both countries and serves as the numeraire to establish the commodity terms of trade between countries. Lewis used this model to argue that unequal exchange resulted from the failure of Third World countries to invest adequately in domestic food production. If low-income countries attempted to shift from exporting primary products to exporting manufactures without improving domestic food production, they would simply "exchange one dependence for another" (70).

In the 1960s, dependency theory was imported into Africa from Latin America. Since the mid 1960s Samir Amin has provided leadership in developing a Marxist version of dependency theory. In *Accumulation on a World Scale* (1974) and *Unequal Development* (1976) Amin presents an analytical framework of underdevelopment in Africa based on surplus extraction and the domination of the world capitalist system. Amin has provided valuable insights into the development process, but his prescriptions for African agriculture have vacillated over time. During the 1960s Amin favored animal traction, promoted industrial crops, and argued that traditional social values were a serious constraint on development at the village level. He also argued that the transition to privately owned small farms was a precondition for socialism (Amin 1965, 210–11, 231). By the early 1970s Amin had reversed himself and recommended the collectivization of agricultural production and he abandoned his support for animal traction and industrial crops (Amin 1971, 231; 1973, 56).

These criticisms notwithstanding, the radical scholars made several important contributions to the understanding of agriculture and rural development. First, they helped demolish the myth of "a typical underdeveloped country" by stressing that each country's economic development had to be understood in the context of that country's historical experience. For example, they argued that Schultz's concept of "traditional agriculture"—a situation where farmers have settled into a low-level equilibrium after years of facing static technology and factor prices—abstracted from the historical process of integration of individual Third World economies into the world capitalist system and therefore was not a very useful analytical concept. Second, in arguing that rural poverty in the Third World resulted from the functioning of the global capitalist economy, the radical writers focused attention on the relationships between villagers and the wider economic system. Unlike Schultz, who attributed rural poverty to the lack of productive agricultural technologies and human capital, radical scholars stressed the importance of the linkages and exchange arrangements that tied villages to the rest of the economy. Third, the radical economists directly attacked what Hirschman (1981a, 3) has called the "mutual benefit claim" of development economics—the assertion that economic relations between high- and low-income countries (and among groups within low-income countries) could be shaped in a way to yield benefits for all. In disputing this claim, the radical scholars stressed that economic development was more than just a technocratic matter of determining how best to raise per capita GNP. Development involved restructuring institutional and political relationships, and the radicals urged neoclassical economists to include these political considerations explicitly in their analyses. In de Janvry's words, "Economic policy without political economy is a useless and utopian exercise" (1981, 263).

Both the radical analyses and the Western dual-sector models of the 1960s suffered from some of the same shortcomings—abstract theorizing, inadequate attention to the need for technical change in agriculture, lack of attention to the biological and location-specific nature of agricultural production processes, and lack of a solid micro foundation based on empirical research at the farm and village level. Recognition of some of these shortcomings was an important element leading to a reevaluation of the goals and approaches of development economics and of the role of agriculture in reaching those goals in the period following 1970.

THE GROWTH-WITH-EQUITY ERA SINCE 1970

THE BROADENING OF DEVELOPMENT GOALS

Around 1970 mainstream Western development economics began to give greater attention to employment and the distribution of real income, broadly defined. This shift in emphasis came about for at least three reasons. The first was ideological, a response to the radical critique of Western development economics, especially the critique of the "mutual benefit claim" discussed above. The goal of economic growth for Third World countries was seriously questioned by this critique, and some development economists may have felt the need to redefine the goals of development more broadly in order to preserve the legitimacy of their subdiscipline.

Second, from the 1960s onwards it became apparent that rapid economic growth in some countries, such as Pakistan, Nigeria, and Iran, had deleterious, and in some cases disastrous, side effects. Hirschman (1981a, 20–21) argues that development economists were forced to reevaluate the goals of their profession because

> the series of political disasters that struck a number of Third World countries from the sixties on . . . were clearly *somehow* connected with the stresses and strains accompanying development and "modernization." These development disasters, ranging from civil wars to the establishment of murderous authoritarian regimes, could not but give pause to a group of social scientists who, after all, had taken up the cultivation of development economics in the wake of World War II not as narrow specialists, but impelled by the vision of a better world. As liberals, most of them presumed that "all good things go together" and took it for granted that if only a good job could be done in raising the national income of the countries concerned, a number of beneficial effects would follow in the social, political, and cultural realms.
>
> When it turned out instead that the promotion of economic growth entailed not infrequently a sequence of events involving serious retrogression in those other areas, including the wholesale loss of civil and human rights, the easy self-confidence that our subdiscipline exuded in its early stages was impaired.

The third reason for the reevaluation of development goals was a growing awareness among development economists that even in countries where rapid economic growth had not contributed to social turmoil, the benefits of economic growth often were not trickling down to the poor and that frequently the income gap between rich and poor was widening (see, for example, Fishlow 1972; Nugent and Yotopoulos 1979; and Streeten 1979). Even where the incomes of the poor were rising, often they were rising so slowly that the poor would not be able to afford decent diets or housing for at least another generation.

Rather than simply waiting for increases in average per capita incomes to "solve" the problems of poverty and malnutrition, economists, political leaders in the Third World, and the leaders of major donor agencies argued in the early 1970s that greater explicit attention needed to be paid to employment, income distribution, and "basic needs," such as nutrition and housing. For example, Robert McNamara, then president of the World Bank, called on the Bank to redirect its activities towards helping people in the bottom 40 percent of the income distribution in low-income countries (McNamara 1973).

These growth-with-equity concerns stimulated a number of important theoretical and policy debates during the 1970s. The first debate concerned the interactions between income distribution and rates of economic growth. A number of economists during the 1970s included income distribution explicitly in their frameworks of analysis, and several examined the interdependence between income growth, income distribution, and other development goals, such as literacy and health.[18] These analyses focused on changes not only in the *size* distribution of income during the course of development (for example, Chenery et al. 1974; Adelman and Morris 1973) but in the *functional* distribution as well. For example, attention was given to the impact of economic growth on small farmers (Stevens 1977a; Fei, Ranis, and Kuo 1979) and on women (Boserup 1970; Tinker and Bramsen 1976; Spencer 1976).

A second debate centered on employment generation and the possible existence of employment-output trade-offs in industry and agriculture. Although Dovring (1959) had shown that the absolute number of people engaged in agriculture in developing countries would probably continue to grow for several decades, most economists during the 1960s still assumed that urban industry would absorb most of the new entrants to the labor force. By 1970, however, it had become apparent that urban industry in most countries could not expand quickly enough in the short run to provide employment for the expanding rural labor force. Hence, the concern of development planners shifted to finding ways to hold people in the countryside (Eicher et al. 1970).

The concern about creating rural jobs raised a number of questions in both agriculture and industry about the relative output and employment-generation capacities of large and small enterprises. In agriculture, debate centered on how much emphasis should be given to improving small farms as opposed to creating larger and more capital-intensive farms, ranches, and plantations. Empirical evidence from the late 1960s and early 1970s revealed that the economies of size in tropical agriculture were more limited than previously believed and that the improvement of small farms often resulted in greater output *and* employment per hectare than did large-scale farming. In industry, the small-versus-large debate led to a number of empirical studies of rural small-scale enterprises (these are reviewed by Chuta and Liedholm in chapter 20). In both agriculture and industry the concern with possible employment-output trade-offs also stimulated research on the choice of appropriate production techniques (see chapters 19 and 20 and the introduction to part IV of this volume).

During the 1970s economists and planners also began to give explicit consideration to the impact of development programs on nutrition. Empirical studies revealed that increases in average per capita income did not always lead to improved nutrition and that at times malnutrition actually increased with growing incomes (Berg 1973; Reutlinger and Selowsky 1976). Therefore, many analysts argued that nutrition projects, targeted to the poor and malnourished, were needed to supplement other development activities.[19]

IMPLICATIONS FOR AGRICULTURE

The change in orientation of development economics in the early 1970s implied a much greater role for agriculture in development programs. Because the majority of the poor in most Third World countries live in rural areas and because food prices are a major determinant of the real income of both the rural and the urban poor, the low productivity of Third World agriculture was seen as a major cause of poverty. Furthermore, because urban industry had generally provided few jobs for the rapidly growing labor force, development planners increasingly concentrated on ways to create productive employment in rural areas, if only as a holding action until the rate of population growth declined and urban industry could create more jobs. (Nonetheless, investment policies in many countries continued to favor urban areas [see Lipton 1977].) The need to create productive rural employment was underlined by a growing awareness of the increasing landlessness in many parts of the Third World, particularly South Asia (Singh n.d.),[20] Latin America (de Janvry 1981), and a few countries in Africa (Ghai and Radwan 1983).

It soon became apparent that if agriculture was to play a more important role in development programs, policy makers needed a more detailed understanding of rural economies than that provided by the simple two-sector models of the 1950s and early 1960s. In the late 1960s and early 1970s there was a rapid expansion of micro-level research on agricultural production and marketing, farmer decision making, the performance of rural factor markets, and rural nonfarm employment.[21] This micro-level research documented the complexity of many Third World farming and marketing systems and complemented the macro-level work begun in the 1950s on modeling agricultural growth and intersectoral relationships.

RESEARCH FINDINGS OF THE 1970s

Modeling Agricultural Growth

As policy makers looked to agriculture to provide more employment and wage goods for the rapidly expanding labor force, attempts to model the process of agricultural growth assumed increased importance. Hayami and Ruttan's induced innovation model of agricultural development (discussed in the introduction to part II of this book and in chapters 4 and 5) was a major contribution of the 1970s. Hayami and Ruttan argued that there are multiple technological paths to agricultural growth, each embodying a different mix of factors of production, and that changes in relative factor prices can guide a country's researchers to select the most "efficient" path. This implied that countries with different factor endowments would have different efficient growth paths and that the wholesale importation of agricultural technology from industrialized countries to the Third World could lead to highly inefficient patterns of growth. Hayami and Ruttan argued that relative factor prices not only affected technological development but also often played an important role in guiding the design of social institutions.[22]

Other major efforts to model the process of agricultural growth included detailed agricultural sector analyses (such as Thorbecke and Stoutjesdijk 1971; and Mantesch et al. 1971); the work of Mellor and Johnston and Kilby, discussed below; and the attempt by some radical scholars, notably de Janvry (1981), to move from a purely global, abstract explanation of rural poverty to a neo-Marxist analysis on a micro level.

Intersectoral Relationships

The 1970s also witnessed a great expansion in the theoretical and empirical research begun by economists in the 1960s on the interdependence between agricultural and nonagricultural growth. Particularly noteworthy was the work of Mellor and of Johnston and Kilby (1975).[23] Mellor argued that it was possible to design employment-oriented strategies of development based on the potential growth linkages inherent in the new high-yielding grain varieties.[24] Mellor's analysis drew heavily on empirical evidence from India. Unlike many authors who mainly stressed how the new varieties could increase total food supplies, Mellor emphasized that the new varieties could also raise the incomes of foodgrain producers, thereby generating increased effective demand for a wide variety of labor-intensive products. Indeed, Mellor saw most of the potential growth in employment that could result from the new varieties as lying outside the foodgrain sector itself, in sectors producing labor-intensive goods such as dairy products, fruit, other consumer products, and agricultural inputs. This expanded employment was made possible by the simultaneous increase in effective demand for these products and the increased supply of inexpensive wage goods in the form of foodgrains. Much of Mellor's analysis focused on the types of agricultural and industrial policies needed to exploit these growth linkages of the new grain varieties.

Johnston and Kilby analyzed "the reciprocal interactions between agricultural development and the expansion of manufacturing and other nonfarm sectors" (1975, xv). In particular, they focused on the factors affecting the rates of labor transfer between sectors and the level and composition of intersectoral commodity flows. Drawing on empirical evidence from England, the United States, Japan, Taiwan, Mexico, and the Soviet Union, Johnston and Kilby argued that the size distribution of farms was a critical determinant of the demand for industrial products in a developing economy. They showed that broad-based agricultural growth was more effective than estate production in stimulating the demand for industrial products and hence speeding the structural transformation of the economy. Johnston and Kilby's analysis strongly supported the view that concentrating agricultural development efforts on the mass of small farmers in low-income countries, rather than promoting a bimodal structure of small and large farms, would lead to faster growth rates of both aggregate economic output and employment.

Factor Markets and Employment Generation

Concern for creating jobs stimulated research during the 1970s on rural labor markets and employment. Krishna (1973) addressed the basic methodological problem of defining under- and unemployment in rural economies. Noting that most unemployment studies use definitions appropriate to industrial economies, Krishna identified four different criteria commonly used to classify people as under- or unemployed: (1) a time criterion, according to which a person is underemployed if he or she is gainfully occupied for less time than some full-employment standard; (2) an income criterion, by which an individual is under- or unemployed if he or she earns less than some desirable minimum; (3) a willingness criterion, which defines a person as under- or unemployed if he or she is willing to work longer hours at the prevailing wage; and (4) a productivity criterion, which defines a worker as unemployed if the worker's marginal product is zero. Krishna showed that different policy measures were appropriate for dealing with each type of underemployment.

During the 1970s several economists, particularly those associated with the International Labour Office (ILO), spoke of dethroning GNP as the target and indicator of development and replacing growth strategies with employment-oriented approaches (see, for example, Seers 1970). During the first half of the 1970s the ILO dispatched missions to Colombia, Sri Lanka, Kenya, the Philippines, and Sudan to draw up programs to expand employment (International Labour Office 1970, 1971, 1972, 1974, 1976b). The ILO missions "studied just about everything—population, education, income distribution, appropriate technology, multinationals" (Little 1982, 214), but they frequently lacked the detailed information needed to evaluate where and to what degree output-employment trade-offs existed in these countries. The impact of the ILO studies was further limited because research during the 1970s demonstrated that because 60–80 percent of the poor in most Third World countries were employed in some fashion, the critical policy issue was not one of creating jobs per se but one of increasing the productivity of workers already employed in small-scale agriculture and nonfarm enterprises. The ILO studies were, nonetheless, important in stimulating research on labor markets and on the impact of factor-price distortions on output and employment.

A large number of studies during the 1970s evaluating the performance of labor markets in low-income countries generally found that at peak periods of the agricultural cycle there was little unemployment in rural areas, while at other periods of the year there were labor surpluses. The studies also documented that earlier researchers frequently had overestimated the size of these surpluses because they had failed to take account of the considerable time devoted to rural nonfarm enterprises and to walking to and from fields. Studies also confirmed that labor markets in most countries were generally competitive, with wage rates, particularly in rural areas, following seasonal patterns of labor demand (see Berry and Sabot 1978).

The labor-market research also documented that when labor was misallocated, in many situations the misallocation was due not only to imperfections in labor markets but to poorly functioning markets for other factors of production as well. Overvalued exchange rates and subsidized credit, for example, often encouraged excessive substitution of capital for labor in low-wage economies. Concern about the impact of such factor-price distortions on output and employment stimulated research on the choice of technique in agricultural production and processing (see Timmer, chapter 19; and Byerlee et al. 1983) and on the functioning of rural financial markets in low-income countries (see Adams and Graham, chapter 21; and Gonzalez-Vega, chapter 22).

In the late 1960s and early 1970s the concern for employment generation led to questions about the productivity and labor-absorption capacity of large farms and ranches versus those of small farms. A large number of scholars (for example, Dorner and Kanel 1971; Barraclough 1973; Berry 1975; and Berry and Cline 1979) documented the strong economic case for land reform in many Third World countries because of the higher employment and land productivity potential of small family farms. The higher land productivity was due largely to greater use of labor (mainly family labor) per unit of land. Although there was widespread agreement among these scholars that land reform was an attractive policy instrument for raising farm output, increasing rural employment, and improving the equality of income distribution, political support for land reform waned during the 1970s, as discussed by de Janvry and Bromley in chapters 17 and 18, respectively.

Rural-to-Urban Migration

Rural-to-urban migration was a major area of research during the 1960s and 1970s because the rate of rural-to-urban migration in most Third World countries far outstripped the rate of growth of urban employment. This led to rising levels of open urban unemployment. The concern of policy makers therefore quickly shifted from trying to transfer surplus labor from agriculture to industry to trying to reduce "excessive" rates of urbanization.

Research by economists on migration in the Third World was sparked by Michael Todaro's attempt in the late 1960s to explain the apparently paradoxical phenomenon of accelerating rural-to-urban migration in the context of continuously rising urban unemployment in Kenya. Todaro (1969) proposed a model (later extended by Harris and Todaro 1970) in which a potential migrant's decision to migrate is motivated primarily by the difference between his or her *expected* (rather than the actual) urban income and the prevailing rural wage. The Harris-Todaro model implied that attempts to reduce urban unemployment by creating more urban jobs could paradoxically result in more urban unemployment rather than less. By leading potential migrants to believe that their chances of getting an urban job had increased, urban employment programs induced greater rural-to-urban migration. Harris and Todaro therefore argued that urban unemployment could best be addressed by reducing the incentives to migrate to the cities, for example, by raising rural incomes via a broad range of agricultural and rural development programs.[25]

The second approach to studying migration was spearheaded by several radical political economists who focused on the social, as opposed to the private, benefits and costs of migration. Samir Amin, for example, argued that although rural-to-urban migration might be privately profitable for the migrant, it imposed important social costs on the sending area, including the loss of future village leadership and the instability of rural families. Amin argued that these costs exceeded possible gains to the area from wage remittances to the home villages.

Although the net welfare impact of migration is obviously an important question, many of the studies by radical scholars lacked empirical data to support their conclusions. In a balanced and constructive assessment of both neoclassical and radical political economy studies of migration in southern Africa, Knight and Lenta (1980) conclude that there is not a clear picture of the net welfare impact of migration in the countries supplying labor to the mines in South Africa.

Product Market Performance

Rapid income growth and urbanization put increasing pressure on markets for agricultural products, particularly food, during the 1960s and 1970s. In response, economists undertook a number of studies to evaluate the performance of agricultural product markets and suggest improvements.[26] These studies generally found little support for allegations of widespread collusion and extraction of monopoly profits by private merchants in Third World countries. They did, however, document how insufficient infrastructure and the lack of reliable public information systems and other public goods often reduced market efficiency and lowered farmers' incentives to specialize for market production. The studies were often critical of state monopolies in the domestic food trade, citing the frequent high costs of state marketing agencies. The studies identified important roles for the state in providing public goods (better information systems, standardized weights and measures, and so on) to facilitate private trading, price stabilization, and regulation of international trade.[27] More recently, there has been discussion of ways the state can ensure adequate food supplies to the poor without disrupting normal market channels (see Timmer, chapter 8; and Lele and Candler, chapter 14).

Farming Systems Research and Farmer Decision Making

During the late 1960s and the 1970s economists increasingly investigated the factors that influenced farmers' decisions concerning whether to adopt new crop varieties and farming practices. This work eventually led to the development of farming systems research (FSR), described in chapter 25. Farming systems research attempts to incorporate farmers' constraints and objectives into agricultural research by involving farmers in problem identification, on-farm agronomic trials, and extension.

Interest in farming systems research and the new household economics also led to efforts to model the farm household as both a production and a consumption unit. Inspired by Chayanov's work on the behavior of Russian peasants in the early 1900s, the farm-household models stressed the need to understand how government policies could simultaneously affect both the production and the consumption decisions of small farmers. For example, these models showed that marketed surplus of a crop might, in some circumstances, actually decline as the crop's price was increased (even if production of the crop rose) because the price increase would raise farm family income, some of which would be spent on the good whose price had risen.[28]

Summary: Research in the 1970s

The results of microeconomic research during the 1970s contributed to an accumulation of knowledge about the behavior of farmers; constraints on the expansion of farm and nonfarm production, income, and employment; the linkages between agricultural research and extension institutions; and the complexity and location-specific nature of the agricultural development process. One of the major accomplishments of the 1970s was a large increase in knowledge about agricultural development in sub-Saharan Africa, an area often ignored by development economists during the 1950s and early 1960s (see Eicher and Baker 1982). But the increased orientation to micro-level research may have resulted in relatively less attention being paid to macroeconomic research on food policy and the role that agriculture can play in the structural transformation of Third World economies. A major challenge, therefore, is to incorporate the micro-research findings into models that examine agriculture's role in a general equilibrium (or disequilibrium) context (see part III).

DEVELOPMENT PROGRAMS OF THE 1970s: INTEGRATED RURAL DEVELOPMENT AND BASIC NEEDS

Reacting to some of the disappointments of the green revolution and the agricultural growth-oriented programs of the 1960s, donors and Third World governments turned increasing attention to integrated rural development and basic needs projects in the 1970s. Integrated rural development (IRD) attempts to combine in one project elements to increase agricultural production and improve health, education, sanitation, and a variety of other social services. Like the community development (CD) projects of the 1950s, IRD projects of the 1970s sometimes expanded social services much faster than they expanded the economic base to support them, and they often proved to be extraordinarily complex and difficult to implement and administer. Moreover, the inability of IRD projects to increase agricultural production rapidly often stemmed from the lack of appropriate technical packages. Lele (1975) reviewed seventeen IRD projects in Africa and found that most of the projects were based upon inadequate knowledge of local technical possibilities, small-farmer constraints, and local institutions. The projects also tended to have very high administrative costs, making them difficult to replicate over broader areas. By 1980 many donors, such as the World Bank and the U.S. Agency for International Development, had retreated from IRD projects or had redesigned these projects to give greater emphasis to agricultural production.[29] The rise and decline of IRD (1973–80) was in some ways very similar to the fate of CD in the 1950–57 period (see Holdcroft, chapter 3).

In the mid 1970s the basic needs approach was popularized by the ILO (1976a) and subsequently spearheaded by a group of economists within the World Bank under the leadership of Paul Streeten. The basic needs approach holds that development projects should give priority to increasing the welfare of the poor directly through projects to improve nutrition, education, housing, and so on, rather than focus mainly on increasing aggregate growth rates.[30] The basic needs advocates supported their case by citing impressive gains in life span, literacy, and nutrition in Cuba, Sri Lanka, and the People's Republic of China, countries that had emphasized basic needs. But the constraints on the basic needs approach were highlighted by Sri Lanka's inability during the mid 1970s to continue to finance the centerpiece of its program, universal free rice rations, which forced the government to shift to a more growth-oriented strategy. Likewise, as Lardy points out in chapter 29, the rising cost of food subsidies in China raises questions about China's ability to sustain its policy of cheap food for urban consumers.

Although investments in health, nutrition, education, and housing can contribute importantly to the welfare of the poor and to the rate of economic growth, the experience with the basic needs approach suggests that low-income countries also need to emphasize building the economic base to finance these investments. By the early 1980s many economists were once again giving greater emphasis to economic growth and to the sequence of different types of development activities, such as investments in irrigation and health facilities. This shift in emphasis did not imply a rejection of the growth-with-equity philosophy of the 1970s. Rather, it reflected an increasing recognition of the impossibility of achieving a decent living standard for the bulk of the rapidly growing populations in poor countries simply by redistributing existing assets. This recognition led the World Bank to shift to a more growth-oriented strategy in the early 1980s, and the basic needs approach faded into the background.[31]

The research results and development experiences of the 1970s suggest that in order to attain more rapid, more broad-based agricultural growth and rural development, the following components will have to be emphasized in the coming decades: strengthening of institutions in low-income countries for agricultural research, administration, policy analysis, and training; renewed emphasis on analyzing agricultural development issues in broader macroeconomic frameworks; reevaluation of the roles of international trade, food aid, and agricultural specialization in an increasingly interdependent world food economy; and movement towards more interdisciplinary approaches to problem solving. All these require expansion of the human-capital base of Third World countries. One of the clearest lessons of the 1960s and 1970s is that agricultural and rural development require strong local institutions and well-trained individuals. International research centers and expatriate advisers are at best complements to, not substitutes for, domestic research systems and policy analysts (see chapters 23–25 and 30–31). Because problems in the food system are typically multifaceted, there is also a need to move towards more interdisciplinary approaches to problem solving. Food policy research and farming systems research, discussed later in this book, are examples of areas where such interdisciplinary approaches are proving useful.

THE PLAN OF THIS BOOK

This book focuses on the agricultural and rural development literature produced during the growth-with-equity period since 1970. Our selection of articles was guided by three convictions. First, we believe that agricultural and economic development should be viewed as long-term processes. We need to view agricultural development in historical context, understanding the forces that led to current problems, opportunities, and policies and learning from past efforts to deal with similar situations. For example, what lessons can be drawn from the community development experience of the 1950s about how to design integrated rural development projects?

Second, we believe that many agricultural development issues need to be viewed in both a general equilibrium framework and a political economy context. Too often agriculture is pictured as a quasi-isolated sector, contributing food, labor, and perhaps capital but little else to the rest of the economy. Economic development, however, by its very nature involves increasing economic integration. Analysts and policy makers need to understand the growing interdependence between agriculture and the rest of the economy if they are to formulate intelligent policies. The types of policies that can be implemented in most countries, however, are severely constrained by political considerations. In order to be useful, agricultural policy analysis needs to be framed in a context that recognizes the importance of political constraints on policy and attempts to relax those constraints.

Third, we believe that although agricultural development issues are often best understood in a general equilibrium framework, that framework needs to be built upon a firm microeconomic data base. Careful empirical studies of farmer, merchant, and consumer behavior are needed in order to understand how rural economies function, how they are likely to react to government policy interventions, and how they interact with other sectors of the economy. A hallmark of research on agricultural development during the past twenty-five years has been the movement away from a priori theorizing towards empirical research. The important contribution of empirical research to policy analysis is a major theme of this book.

Part II of the book places the post-1970 agricultural and rural development literature in historical perspective by discussing models of agricultural growth that historically have been used in research and in development programs. These range from the induced innovation model of neoclassical economists to neo-Marxist models of rural underdevelopment based on marginalization and unequal exchange. The disagreements among the authors in part II over the usefulness of these models mirror the ferment among economists about the mechanisms of agricultural growth over the past twenty-five years.

Part III examines the questions of agricultural growth and agricultural policies in a general equilibrium framework. The articles in part III address three major agricultural policy issues that have both important general equilibrium consequences and significant effects on other development goals, such as improved nutrition: agricultural price policy and its effects on technical change in agriculture and on intersectoral resource flows; food security; and the role of agricultural trade in economic growth. The authors in part III argue that these issues are best analyzed in a broad framework that takes account of intersectoral relationships and the political-economic constraints facing decision makers.

This type of broad analysis, however, needs to be built upon a firm understanding of how rural economies function. Part IV provides some of that foundation by presenting a series of articles that analyze the forces that influence factor availability and use in agriculture, the links between agriculture and rural small-scale industry, and the generation and impact of technical change in agriculture. Part IV draws on the rich body of empirical findings generated during the 1970s and early 1980s.

Part V presents case studies of agricultural development in the People's Republic of China and in sub-Saharan Africa. The articles of Part V pull together many of the elements discussed earlier in the book, such as price policy, intersectoral resource flows, and technological change in agriculture. China and Africa will be at the center of many of the agricultural development debates in the 1980s and 1990s. China, a predominantly rural country, is home to approximately one-fourth of the world's population, while sub-Saharan Africa has the lowest average per capita income and the poorest agricultural growth record of any major region of the world during the past two decades.

NOTES

Larry Lev, Carl Liedholm, Michael Morris, and Robert Stevens offered insightful comments on an earlier draft of this paper. Johnston (1970) extensively reviewed the literature of the 1950s and 1960s on the role of agriculture in development.

1. Development economics began to emerge as a subdiscipline of economics in the post–World War II period with the work of Nurkse, Mandelbaum, Rosenstein-Rodan, Singer, Prebisch, and others. The first major text on economic development was W. Arthur Lewis's influential *The Theory of Economic Growth* (1955). Lewis's emphasis on economic growth set the tone for work in development economics during the "growth era" of the 1950s and 1960s. In the introduction Lewis wrote, "The subject matter of this book is growth of output per head of population . . . and not distribution" (p. 9). Lewis did, however, include an appendix entitled "Is Economic Growth Desirable?" For reviews of the history of development economics see Hirschman 1981a; Streeten 1979; Reynolds 1977, chap. 2; and Little 1982.

2. This section draws heavily on Johnston's excellent review of the literature through 1970 on the role of agriculture in economic development.

3. For an excellent summary of the Lewis model see Meier 1976, 157–63; see also Lewis 1972. For an analysis of Lewis's model on development economics see Gersowitz et al. 1982.

4. Lewis's statement that the marginal productivity of laborers in the noncapitalist sector approached zero as a limiting case stimulated a large number of efforts to measure the extent of surplus labor in agriculture. For a review of these efforts see Kao, Anschel, and Eicher 1964. For a critique of the concept of surplus labor in agriculture see Schultz 1964, chap. 4.

5. Nicholls (1964) was one of the first critics of the rapid transfer of surplus labor as a short-run strategy. See also Mellor's discussion of intersectoral resource transfers in chapter 9.

6. For a summary of this thesis see Prebisch 1959. For critical reviews see Kravis 1970; and Little 1982, chap. 4.

7. In recent years Prebisch has modified his views about import substitution (see Prebisch 1981). For a contrasting view of the potential role of agricultural exports in fostering economic growth see Myint's discussion of the "vent for surplus" model of growth in chapter 15 of this volume.

8. See chapter 9 by Mellor.

9. The literature of the 1960s on agricultural development is captured in the volumes edited by Eicher and Witt (1964), Southworth and Johnston (1967), and Wharton (1969).

10. See the first U.N. report on development problems in the Third World, *Measures for the Economic Development of Underdeveloped Countries* (1951), which focused heavily on ways of dealing with rural surplus labor.

11. Community development's emphasis on providing social services presaged the basic needs approach to development of the late 1970s, described later in this chapter. In this sense, the community development effort differed from other Western development efforts of the 1950s and 1960s, which focused mainly on increasing average per capita incomes.

12. See part IV of this volume for a further discussion of land reform.

13. For a critical appraisal see Shapiro 1977.

14. See Ruttan's discussion of the high-payoff input model in chapter 2 and Evenson's discussion of the development of the new grain varieties in chapter 24.

15. These arguments were most fully developed by Mellor and by Johnston and Kilby in the 1970s. See the discussion of these authors' work in the second section of this chapter.

16. French scholars also made important contributions to the Marxist analysis of agricultural development during the 1960s and 1970s (see Petit 1982).

17. For critiques of the dependency school of thought in Latin America see Cardoso and Faletto 1979; and de Janvry 1981.

18. A standard reference is the influential book *Redistribution with Growth,* by Chenery et al. (1974). See also Seers 1970; and Adelman 1975.

19. For a summary of this literature see Pinstrup-Andersen 1981.

20. Singh (n.d.) estimates that about one-fifth of all rural households in South Asia (India, Pakistan, and Bangladesh) are either landless or nearly landless.

21. See, for example, the volume edited by Stevens (1977a); and Jones 1972.

22. See chapter 4; and Hayami and Ruttan 1971. For attempts to test the induced innovation hypothesis empirically see Binswanger and Ruttan 1978.

23. See also the volume edited by Reynolds (1975).

24. Mellor's views are articulated in *The New Economics of Growth: A Strategy for India and the Developing World* (1976). See also Mellor and Lele 1973; and Mellor's contributions to this volume, chapters 9 and 10.

25. The major extensions of the Harris-Todaro model and empirical tests of it are summarized in Todaro 1980.

26. Many of these studies are reviewed by Lele (1977) and Riley and Staatz (1981).

27. For a critical review of some of these studies see Harriss 1979.

28. For a summary of the literature on family/farm modeling see the volume edited by Singh, Squire, and Strauss (n.d.).

29. For excellent reviews of IRD see Ruttan 1975; de Janvry 1981; and Johnston and Clark 1982.

30. The basic needs approach is not simply a call for increased social welfare spending, however; it is also based on recognition of the importance of investment in human capital in economic growth and of the synergism of nutrition, health, and family planning decisions. For an excellent discussion of basic needs projects and their relationships to rural development see Johnston and Clark 1982, chap. 4.

31. The World Bank's experience with basic needs is summarized by Streeten (1981).

REFERENCES

Adelman, Irma. 1975. "Development Economics—A Reassessment of Goals." *American Economic Review* 55 (2): 302–9.

Adelman, Irma, and Cynthia Taft Morris. 1973. *Economic Growth and Social Equity in Developing Countries.* Stanford: Stanford University Press.

Amin, Samir. 1965. *Trois éxperiences Africaines de développement: le Mali, la Guinée et le Ghana.* Paris: Presses Universitaires de France.

———. 1971. *L'Afrique de l'ouest bloquée: l'économie politique de la colonisation (1880–1970).* Paris: Les Editions de Minuit.

———. 1973. "Transitional Phases in Sub-Saharan Africa." *Monthly Review* 25(5): 52–57.

———. 1974. *Accumulation on a World Scale: A Critique of the Theory of Underdevelopment.* New York: Monthly Review Press.

———. 1976. *Unequal Development: An Essay on the Social Formations of Peripheral Capitalism.* New York: Monthly Review Press.

Baran, Paul A. 1952. "On the Political Economy of Backwardness." *Manchester School of Economic and Social Studies* 20:66–84.

Barraclough, Solon. 1973. *Agrarian Structure in Latin America.* Lexington, Mass.: Lexington Books.

Behrman, Jere R. 1968. *Supply Response in Underdeveloped Agriculture.* Amsterdam: North-Holland Publishing Co.

Berg, Alan. 1973. *The Nutrition Factor: Its Role in National Development.* Washington, D.C.: Brookings Institution.

Berry, A., and R. H. Sabot. 1978. "Labour Market Performance in Developing Countries: A Survey." *World Development* 6:1199–1242.

Berry, R. Albert. 1975. "Special Problems of Policy Making in a Technologically Heterogeneous Agriculture: Colombia." In *Agriculture in Development Theory,* edited by Lloyd G. Reynolds. New Haven: Yale University Press.

Berry, R. Albert, and William F. Cline. 1979. *Agrarian Structure and Productivity in Developing Countries*. Baltimore: Johns Hopkins University Press.

Binswanger, Hans P., and Vernon W. Ruttan. 1978. *Induced Innovation: Technology, Institutions and Development*. Baltimore: Johns Hopkins University Press.

Boserup, Ester. 1970. *Women's Role in Economic Development*. New York: St. Martin's Press.

Brown, Lester R. 1970. *Seeds of Change*. New York: Praeger.

Byerlee, Derek, Carl K. Eicher, Carl Liedholm, and Dunstan S. C. Spencer. 1983. "Employment-Output Conflicts, Factor Price Distortions and Choice of Technique: Empirical Results from Sierra Leone." *Economic Development and Cultural Change* 31 (2):315–36.

Cardoso, F. H., and E. Faletto. 1979. *Dependency and Development in Latin America*. Berkeley and Los Angeles: University of California Press.

Carroll, Thomas F. 1961. "The Land Reform Issue in Latin America." In *Latin American Issues: Essays and Comments*, edited by Albert O. Hirschman. New York: Twentieth Century Fund.

Chayanov, A. V. 1966. *The Theory of Peasant Economy*. Edited by D. Thorner, B. Kerblay, and R. Smith. Homewood, Ill.: Richard D. Irwin.

Chenery, H. B., M. S. Ahluwalia, C. L. G. Bell, J. H. Duloy, and R. Jolly, eds. 1974. *Redistribution with Growth*. London: Oxford University Press.

de Janvry, Alain. 1981. *The Agrarian Question and Reformism in Latin America*. Baltimore: Johns Hopkins University Press.

Dorner, Peter, and Donald Kanel. 1971. "The Economic Case for Land Reform: Employment, Income Distribution and Productivity." In *Land Reform in Latin America*, edited by Peter Dorner, 41–56. Land Economics Monograph, no. 3. Madison: University of Wisconsin Land Tenure Center.

Dos Santos, T. 1970. "The Structure of Dependence." *American Economic Review* 40(2):231–36.

Dovring, Folke. 1959. "The Share of Agriculture in a Growing Population." *Monthly Bulletin of Agricultural Economics and Statistics* 8 (August–September): 1–11. Reprinted in *Agriculture in Economic Development*, edited by Carl K. Eicher and Lawrence W. Witt, 78–98. New York: McGraw-Hill, 1964.

Eicher, Carl K., and Doyle C. Baker. 1982. *Research on Agricultural Development in Sub-Saharan Africa: A Critical Survey*. MSU International Development Paper, no. 1. East Lansing: Michigan State University, Department of Agricultural Economics.

Eicher, Carl K., and Lawrence W. Witt, eds. 1964. *Agriculture in Economic Development*. New York: McGraw-Hill.

Eicher, C. K., T. Zalla, J. Kocher, and F. Winch. 1970. *Employment Generation in African Agriculture*. East Lansing: Michigan State University, Institute of International Agriculture.

Emmanuel, A. 1972. *Unequal Exchange: A Study of the Imperialism of Trade*. New York: Monthly Review Press.

Enke, Stephen. 1962a. "Industrialization through Greater Productivity in Agriculture." *Review of Economics and Statistics* 44 (February): 88–91.

————. 1962b. "Economic Development with Limited and Unlimited Supplies of Labor." *Oxford Economic Papers* 14 (June): 158–72.

Fei, John C. H., Gustav Ranis, and Shirley W. Y. Kuo. 1979. *Growth with Equity: The Taiwan Case*. New York: Oxford University Press for the World Bank.

Fishlow, A. 1972. "Brazilian Size Distribution of Income." *American Economic Review, Proceedings* 62(May): 391–402.

Frank, A. G. 1966. "The Development of Underdevelopment." *Monthly Review* 18(4):17–31.

Frankel, Francine R. 1971. *India's Green Revolution: Economic Gains and Political Costs*. Princeton: Princeton University Press.

Galtung, J. 1971. "A Structural Theory of Imperialism." *Journal of Peace Research* 2:81–116.

Gersowitz, Mark, Carlos F. Diaz-Alejandro, Gustav Ranis, and Mark R. Rosenzweig, eds. 1982. *The Theory and Practice of Economic Development: Essays in Honor of Sir W. Arthur Lewis*. Boston: George Allen and Unwin.

Ghai, Dharam, and Samir Radwan, eds. 1983. *Agrarian Policies and Rural Poverty in Africa.* Geneva: International Labour Office.

Griffin, Keith. 1974. *The Political Economy of Agrarian Change: An Essay on the Green Revolution.* Cambridge: Harvard University Press.

Harris, John R., and Michael P. Todaro. 1970. "Migration, Unemployment and Development: A Two-Sector Analysis." *American Economic Review* 60(1):126–42.

Harriss, Barbara. 1979. "There Is a Method in My Madness: Or Is It Vice Versa? Measuring Agricultural Market Performance." *Food Research Institute Studies* 17(2):197–218.

Hayami, Yujiro, and Vernon W. Ruttan. 1971. *Agricultural Development: An International Perspective.* Baltimore: Johns Hopkins Press.

Hirschman, Albert O. 1958. *The Strategy of Economic Development.* New Haven: Yale University Press.

———. 1977. "A Generalized Linkage Approach to Development, with Special Reference to Staples." *Economic Development and Cultural Change* 25(supp.):67–98.

———. 1981a. "The Rise and Decline of Development Economics." In *Essays in Trespassing: Economics to Politics and Beyond.* New York: Cambridge University Press.

———. 1981b. *Essays in Trespassing: Economics to Politics and Beyond.* New York: Cambridge University Press.

Hopper, W. David. 1965. "Allocation Efficiency in a Traditional Indian Agriculture." *Journal of Farm Economics* 47(3):611–24.

International Labour Office. 1970. *Towards Full Employment: A Programme for Colombia.* Geneva.

———. 1971. *Matching Employment Opportunities and Expectations: A Programme of Action for Ceylon.* Geneva.

———. 1972. *Employment, Incomes and Equality: A Strategy for Increasing Productive Employment in Kenya.* Geneva.

———. 1974. *Sharing in Development: A Programme of Employment, Equity and Growth for the Philippines.* Geneva.

———. 1976a. *Employment, Growth and Basic Needs: A One-World Problem.* Geneva.

———. 1976b. *Growth, Employment and Equity: A Comprehensive Strategy for the Sudan.* Geneva.

Johnston, Bruce F. 1970. "Agriculture and Structural Transformation in Developing Countries: A Survey of Research." *Journal of Economic Literature* 3(2):369–404.

Johnston, Bruce F., and William C. Clark. 1982. *Redesigning Rural Development: A Strategic Perspective.* Baltimore: Johns Hopkins University Press.

Johnston, Bruce F., and Peter Kilby. 1975. *Agriculture and Structural Transformation: Economic Strategies in Late-Developing Countries.* New York: Oxford University Press.

Johnston, Bruce F., and John W. Mellor. 1961. "The Role of Agriculture in Economic Development." *American Economic Review* 51(4):566–93.

Jones, W. O. 1960. "Economic Man in Africa." *Food Research Institute Studies* 1:107–34.

———. 1972. *Marketing Staple Foods in Tropical Africa.* Ithaca: Cornell University Press.

Jorgenson, D. W. 1961. "The Development of a Dual Economy." *Economic Journal* 71 (June):309–34.

Kao, Charles H. C., Kurt R. Anschel, and Carl K. Eicher. 1964. "Disguised Unemployment in Agriculture: A Survey." In *Agriculture in Economic Development*, edited by Carl K. Eicher and Lawrence W. Witt, 129–44. New York: McGraw-Hill.

Knight, J. B., and G. Lenta. 1980. "Has Capitalism Underdeveloped the Labour Reserves of South Africa?" *Oxford Bulletin of Economics and Statistics* 42(3):157–201.

Kravis, I. B. 1970. "Trade as a Handmaiden of Growth: Similarities between the Nineteenth and Twentieth Centuries." *Economic Journal* 80(320):850–72.

Krishna, Raj. 1967. "Agricultural Price Policy and Economic Development." In *Agricultural Development and Economic Growth*, edited by Herman M. Southworth and Bruce F. Johnston, 497–540. Ithaca: Cornell University Press.

———. 1973. "Unemployment in India." *Indian Journal of Agricultural Economics* 28(1):1–23.

Lele, Uma. 1975. *The Design of Rural Development: Lessons from Africa.* Baltimore: Johns Hopkins University Press for the World Bank.

———. 1977. "Considerations Related to Optimum Pricing and Marketing Strategies in Rural Development." In *Decision Making and Agriculture,* edited by T. Dams and K. Hunt. Lincoln: University of Nebraska Press.

Lewis, W. Arthur. 1954. "Economic Development with Unlimited Supplies of Labour." *Manchester School of Economic and Social Studies* 22(2):139–91.

———. 1955. *The Theory of Economic Growth.* London: George Allen and Unwin.

———. 1972. "Reflections on Unlimited Labor." In *International Economics and Development: Essays in Honor of Raul Prebisch,* edited by Luis Eugenio Di Marco, 75–96. New York: Academic Press.

———. 1978. *The Evolution of the International Economic Order.* Princeton: Princeton University Press.

Lipton, Michael. 1977. *Why Poor People Stay Poor: A Study of Urban Bias in World Development.* London: Temple-Smith.

Little, I. M. D. 1982. *Economic Development: Theory, Policy and International Relations.* New York: Basic Books.

McNamara, Robert S. 1973. *Address to the Board of Governors. Nairobi, Kenya, September 24, 1972.* Washington, D.C.: International Bank for Reconstruction and Development.

Mantesch, Thomas J., et al. 1971. *A Generalized Simulation Approach to Agricultural Sector Analysis with Special Reference to Nigeria.* East Lansing: Michigan State University.

Meier, Gerald. 1976. *Leading Issues in Economic Development.* 3d ed. New York: Oxford University Press.

Mellor, John W. 1966. *The Economics of Agricultural Development.* Ithaca: Cornell University Press.

———. 1976. *The New Economics of Growth: A Strategy for India and the Developing World.* Ithaca: Cornell University Press.

Mellor, John W., and Uma Lele. 1973. "Growth Linkages of the New Foodgrain Technologies." *Indian Journal of Agricultural Economics* 28:(1):35–55.

Nicholls, William H. 1964. "The Place of Agriculture in Economic Development." In *Agriculture in Economic Development,* edited by Carl K. Eicher and Lawrence W. Witt, 11–44. New York: McGraw-Hill.

Nugent, Jeffrey B., and Pan A. Yotopoulos. 1979. "What Has Orthodox Development Economics Learned from Recent Experience?" *World Development* 7:541–54.

Ohkawa, Kazushi. 1964. "Concurrent Growth of Agriculture with Industry: A Study of the Japanese Case." In *International Explorations of Agricultural Economics,* edited by Roger N. Dixey, 201–12. Ames: Iowa State University Press.

Ohkawa, Kazushi, and Bruce F. Johnston. 1969. "The Transferability of the Japanese Pattern of Modernizing Traditional Agriculture." In *The Role of Agriculture in Economic Development,* edited by Erik Thorbecke, 277–303. New York: National Bureau of Economic Research.

Petit, Michel. 1982. "Is There a French School of Agricultural Economics?" *Journal of Agricultural Economics* 33(3):325–37.

Piñeiro, Martin, Eduardo Trigo, and Raul Fiorentino. 1979. "Technical Change in Latin American Agriculture: A Conceptual Framework for Its Evaluation." *Food Policy* 4(3):169–77.

Pinstrup-Andersen, Per. 1981. *Nutritional Consequences of Agricultural Projects: Conceptual Relationships and Assessment Approaches.* World Bank Staff Working Paper no. 456. Washington, D.C.

Prebisch, Raul. 1959. "Commercial Policy in the Underdeveloped Countries." *American Economic Review* 64 (May):251–73.

———. 1981. "The Latin American Periphery in the Global System of Capitalism." *CEPAL Reveiw,* April, 143–50. Reprinted in *International Economic Policies and Their Theoretical Foundations: A Source Book,* edited by John M. Letiche. New York: Academic Press, 1982.

Ranis, Gustav, and John C. H. Fei. 1961. "A Theory of Economic Development." *American Economic Review* 51(4):533–65.
———. 1963. "Innovation, Capital Accumulation, and Economic Development." *American Economic Review,* 53(3):283–313.
———. 1964. *Development of the Labor Surplus Economy: Theory and Policy.* Homewood, Ill.: Richard D. Irwin.
Raup, Philip M. 1967. "Land Reform and Agricultural Development." In *Agricultural Development and Economic Growth,* edited by Herman M. Southworth and Bruce F. Johnston, 267–314. Ithaca: Cornell University Press.
Reutlinger, Shlomo, and Marcelo Selowsky. 1976. *Malnutrition and Poverty: Magnitude and Policy Options.* World Bank Staff Occasional Paper, no. 23. Baltimore: Johns Hopkins University Press for the World Bank.
Reynolds, Lloyd G. 1977. *Image and Reality in Economic Development.* New Haven: Yale University Press.
———. ed. 1975. *Agriculture in Development Theory.* New Haven: Yale University Press.
Riley, Harold, and John Staatz. 1981. "Food System Organization Problems in Developing Countries." *A/D/C Report,* no. 23. New York: Agricultural Development Council.
Ruttan, Vernon W. 1975. "Integrated Rural Development: A Skeptical Perspective." *International Development Review* 17(4):9–16.
Schultz, Theodore W. 1964. *Transforming Traditional Agriculture.* New Haven: Yale University Press.
———. 1978. "On the Economics and Politics of Agriculture." In *Distortions of Agricultural Incentives,* edited by Theodore W. Schultz, 3–23. Bloomington: Indiana University Press.
———. 1981. *Investing in People: The Economics of Population Quality.* Berkeley and Los Angeles: University of California Press.
Seers, Dudley. 1970. *The Meaning of Development.* ADC Reprint. New York: Agricultural Development Council.
Shapiro, Kenneth H. 1977. "Efficiency Differentials in Peasant Agriculture and Their Implications for Development Policies." *Contributed Papers Read at the 16th International Conference of Agricultural Economists,* 87–98. Oxford: University of Oxford Institute of Agricultural Economics for the International Association of Agricultural Economists.
Singh, Inderjit J. n.d. *Small Farmers and the Landless in South Asia.* Washington, D.C.: World Bank. Forthcoming.
Singh, I. J., L. Squire, and J. Strauss, eds. n.d. *Agricultural Household Models: Extensions, Applications and Policy.* Washington, D.C.: World Bank. Forthcoming.
Southworth, Herman M., and Bruce F. Johnston, eds. 1967. *Agricultural Development and Economic Growth.* Ithaca: Cornell University Press.
Spencer, Dunstan S. C. 1976. "African Women in Agricultural Development: A Case Study in Sierra Leone." Occasional Paper, no. 9. Washington, D.C.: Overseas Liason Committee, American Council on Education.
Stevens, Robert D., ed. 1977a. *Tradition and Dynamics in Small-Farm Agriculture.* Ames: Iowa State University Press.
———. 1977b. "Transformation of Traditional Agriculture: Theory and Empirical Findings." In *Tradition and Dynamics in Small-Farm Agriculture,* edited by Robert D. Stevens. Ames: Iowa State University Press.
Streeten, Paul. 1979. "Development Ideas in Historical Perspective." In *Toward a New Strategy for Development,* 21–52. Rothko Chapel Colloquium. New York: Pergamon Press. A shorter version of this paper was published in *Economic Growth and Resources,* vol. 4, edited by Irma Adelman, 56–67. New York: St. Martin's Press, 1979.
Streeten, Paul, with Shadid Burke, Mahbub Haq, Norman Hicks, and Frances Stewart. 1981. *First Things First: Meeting Basic Needs in Developing Countries.* New York: Oxford University Press.
Tax, Sol. [1953], 1963. *Penny Capitalism.* Smithsonian Institution, Institute of Social Anthropology

Publication no. 16. Washington, D.C.: U. S. Government Printing Office. Chicago: University of Chicago Press.

Thorbecke, E., and E. Stoutjesdijk. 1971. *Employment and Output—A Methodology Applied to Peru and Guatemala.* Development Center Studies, Employment Series, no. 2. Paris: Development Center of Organization for Economic Cooperation and Development.

Tinker, Irene, and Michele Bo Bramsen, eds. 1976. *Women and World Development.* Washington, D.C.: Overseas Development Council.

Todaro, M. P. 1969. "A Model of Labor Migration and Urban Unemployment in Less Developed Countries." *American Economic Review* 59:138–48.

———. 1980. "International Migration in Developing Countries: A Survey." In *Population and Economic Change in Developing Countries,* edited by R. A. Easterlin, 361–402. Chicago: University of Chicago Press.

United Nations, Department of Economic and Social Affairs. 1951. *Measures for the Economic Development of Underdeveloped Countries.* New York.

Walinsky, Louis J., ed. 1977. *The Selected Papers of Wolf Ladejinsky: Agrarian Reform as Unfinished Business.* New York: Oxford University Press for the World Bank.

Warriner, Doreen. 1955. "Land Reform in Economic Development." National Bank of Egypt Fiftieth Anniversary Commemoration Lectures, Cairo. Reprinted in *Agriculture in Economic Development,* edited by Carl K. Eicher and Lawrence W. Witt, 272–98. New York: McGraw-Hill, 1964.

Wharton, Clifton R., ed. 1969. *Subsistence Agriculture and Economic Development.* Chicago: Aldine Publishing Co.

II
Models of Agricultural Development

Introduction

Economists traditionally have analyzed agricultural development in terms of its relationship to the growth of the overall economy. The physiocrats, for example, viewed agriculture as the engine of economic growth, arguing that agriculture was the only activity capable of generating a surplus large enough to stimulate growth in other sectors of the economy. The classical economists, on the other hand, believed that diminishing marginal returns to agricultural land would eventually lead to overall economic stagnation, or the "steady state."[1]

When development economists in the early 1950s turned their attention to agriculture in the Third World, they initially focused on the contribution of agriculture to overall economic growth instead of analyzing the process of agricultural growth per se. Two-sector models, such as those of Lewis (1954), Ranis and Fei (1961), and Jorgenson (1961), stressed intersectoral resource transfers—particularly of labor—from the traditional ("agricultural") sector to the modern ("industrial") sector. Most development economists of the 1950s viewed traditional agriculture as a passive sector that would decline in importance as industrial growth absorbed an increasing share of production and employment.

In the early 1960s Johnston, Mellor, and several other agricultural economists stressed the fundamental role that agriculture potentially could play in economic development and the importance of understanding the process of agricultural growth per se if that potential was to be exploited. There were numerous attempts in the fifties and sixties to develop increasingly sophisticated specifications of the agricultural sector in economic growth models and to model more carefully the dynamics of growth within the agricultural sector itself.[2] Several of these models of agricultural development are described in part II, while some of the links between agricultural policies and overall economic performance are explored in part III.

The articles presented in part II summarize many of the models that have helped guide agricultural development efforts over the past thirty years. Understanding these models of agricultural development not only is essential to understanding current policy debates but is a prerequisite to building and using more sophisticated models of the role of agriculture in overall economic development.

Most recent models of agricultural development can be classified into two broad groups: those having their origins in neoclassical theory (for example, the work of Schultz, Ruttan and Hayami, Johnston, Mellor, and Schuh) and those growing out of the tradition of radical political economics (such as the work of Beckford, de Janvry, and Amin). In part II we present examples from both schools of thought. Chapter 4 sets forth Ruttan and Hayami's induced innovation model of agricultural development. This is followed by Beckford's critique, in which Beckford questions the basic neoclassical assumptions of the model. In chapters 6 and 7 the procedure is repeated, but in reverse order, with de Janvry presenting a neo-Marxist model of rural development in Latin America, followed by Schuh, who draws on neoclassical theory to question some of de Janvry's arguments. We hope that the reader will not conclude that either approach is invariably right or wrong but will select ideas from both viewpoints when facing specific problems in particular countries or regions.

Vernon Ruttan's article "Models of Agricultural Development," which appears as chapter 2, outlines six of the most common models of agricultural development: the frontier model, the conservation model, the urban-industrial impact model, the diffusion model, the high-payoff input model, and the induced innovation model. Of these, the diffusion model, the high-payoff input model, and the induced innovation model have been the most influential in guiding Western agricultural development efforts in the Third World in the post–World War II era.

As discussed in chapter 1, the diffusion model served as the intellectual justification for the heavy emphasis by foreign assistance planners on agricultural extension during the 1950s and also had close affinities with the community development (CD) effort of that time. In chapter 3 Lane Holdcroft discusses the reasons why CD projects often failed to live up to expectations and draws lessons from the CD experience for the design of integrated rural development projects.[3]

Because agricultural extension programs and community development efforts of the 1950s often failed to increase agricultural production rapidly and solve the basic food problem in many countries, the diffusion model was deemphasized during the 1960s, and increasing attention was given to agricultural research. As discussed in chapter 1, T. W. Schultz was instrumental in convincing international donors and policy makers to devote more resources to the development of new, productive inputs for Third World farmers, such as high-yielding fertilizer-responsive grain varieties. The high-payoff input, or green revolution, model, stressing agricultural research and human capital formation, became the dominant agricultural development strategy of the 1960s.

The high-payoff input model, however, provides an incomplete explanation of why agricultural growth does or does not occur in a given country. As Ruttan and Hayami point out in chapter 4, both technical and institutional change are exogenous to the model. Although technical change is the engine of agricultural growth in the high-payoff input model, the model itself does not predict whether or what types of technical change will occur in a given country's agriculture.

Ruttan and Hayami present an induced innovation model of agricultural development, in which technical and institutional change are endogenous.[4] Central to the induced innovation model are the notions that there exist alternative paths of technical change and productivity growth in agriculture and that changes in relative factor prices, reflecting changes in relative factor scarcities, can play a determining role in guiding the search for new agricultural technologies and institutions. Arguing that differences in relative factor prices, reflecting different man-land ratios, led Japan and the United States to follow alternative paths of agricultural development, Ruttan and Hayami state that ''the classical problem of resource allocation, which was rejected as an adequate basis for agricultural productivity and output growth in the high-payoff input model, is, in this context, treated as central to the agricultural development process. Under conditions of static technology, improvements in resource allocation represent a weak source of economic growth. The efficient allocation of resources to open up new sources of growth is, however, essential to the agricultural development process.''[5]

For the induced innovation mechanism to guide technological change along an efficient path, a number of conditions must be met: changes in factor prices must reflect changes in relative factor scarcities, researchers in both the private and public sectors must adjust their research programs in response to changes in factor prices, and so on. In his comment on the induced innovation model (chapter 5), George Beckford questions whether these conditions are met in most low-income countries. He further argues that the Ruttan-Hayami model deals with agricultural growth rather than development because it pays little attention to who captures the increases in agricultural output. Beckford maintains that the process of agricultural development can be understood only in the context of the specific institutions of each country, and he argues for a political economy approach to understanding agricultural development because ''it is the specific social order that determines the institutional arrangements that influence the interplay of the proximate economic variables which are central to the Ruttan-Hayami model. So if we are to understand the development process we need to probe far beyond the proximate economic variables.''

Alain de Janvry argues in chapter 6 that rural development in Latin America must be understood not only in terms of specific Latin American institutions but also in terms of the relationship between rural Latin America and the rest of the world economy. De Janvry argues that underdevelopment and poverty cannot meaningfully be analyzed using the concept of ''traditional agriculture,'' a concept that abstracts from the historical process of integration of Third World economies into the world capitalist system. Rather, the processes of rural development and underdevelopment can be understood only in a general equilibrium framework that takes account of how small farmers are tied to the world economy. Attempts to model rural development at the village or sectoral level (and the attendant conclusion that poverty can be alleviated by providing villagers with new agricultural technologies and education) are bound to be misleading, according to de Janvry, because they ignore the mechanisms by which surpluses are extracted from the rural poor and siphoned off to fuel the development of metropolitan centers. ''Underdevelopment cannot be treated apart from development if backward areas or countries are related by the market to the advanced areas or countries. In fact, within the world capitalist system, a theory of development needs to be a theory of economic space which explains how the contradictions of development in certain areas transform, in other areas, traditional societies into underdeveloped ones.''

De Janvry uses the concepts of marginalization, unequal exchange, and the development of peripheral capitalism in his model to draw inferences about the political economy of rural development projects in Latin America. De Janvry's article and his subsequent book, *The Agrarian Question and Reformism in Latin America* (1981), are provocative and challenging to the agricultural scientist who views rural poverty as resulting from a lack of new technology and to the many neoclassical economists who view "getting prices right" as the key to rural progress. But, as Schuh points out in chapter 7, de Janvry lacks empirical support for some of his propositions.

Schuh criticizes not only specific aspects of de Janvry's article but also the neo-Marxist model of underdevelopment upon which it is based. Arguing that trade is by definition mutually beneficial, Schuh rejects de Janvry's proposition that rural poverty results from unequal exchange and surplus extraction.[6] The real culprits, according to Schuh, are inappropriate import-substitution policies and a lack of investment in agricultural research and rural infrastructure. The key to understanding rural poverty, for Schuh, lies in understanding "why economic policy is what it is.''

Both Schuh and de Janvry call for a political economy approach to understanding rural poverty and agricultural development, but their approaches to political economy are different. Schuh focuses on the policy process and on what he perceives as mistakes by policy makers who lack an understanding of the processes of agricultural and economic development. De Janvry gives greater emphasis to class conflicts and the role of the state as a social institution.

A major challenge for economists is to draw from both the neoclassical and the radical traditions in order to build a political economy approach that will lead to a better understanding of rural development and agricultural growth. The concept of induced institutional innovation, if applied broadly enough to include the two-way interaction between changes in factor prices and the evolution of social institutions, holds promise. De Janvry's attempt (1981) to move from a global and rather abstract explanation of rural poverty to a radical political economy analysis of specific agricultural programs and projects also appears to be a step in this direction. Greater application of both the ''old'' and ''new'' institutional economics (such as Commons 1934 and Roumasset 1978) to questions of agricultural development is a third area of potential promise. Many economists have been working towards this synthesis (see, for example, Timmer, Falcon, and Pearson, 1983), but, as Schuh states, ''the challenge is before us.''

NOTES

1. Ricardo argued that the corn yield per acre (taken as a proxy for agricultural productivity) ultimately determined the rate of return to capital invested in all sectors of the economy. Diminishing

marginal returns in agriculture therefore led to falling profit rates and economic stagnation. For a discussion of the physiocrats' and classical economists' views of agriculture see Deane 1978, chaps. 1 and 3.

2. See Johnston 1970 and the volume edited by Reynolds (1975).

3. Several ingredients of CD programs—local initiative, participation, matching grants—were common to integrated rural development (IRD) projects of the 1970s. The rise and decline of IRD in the early 1980s was foreseen by Vernon Ruttan in his perceptive article "Integrated Rural Development: A Skeptical Perspective" (1975).

4. For more details on the induced innovation model see Hayami and Ruttan 1971, and Binswanger and Ruttan 1978.

5. See Raj Krishna's comments on the limits of the induced innovation model's explanation of technical change in agriculture (chapter 11 in this volume).

6. In arguing that trade is never entered into coercively, Schuh implicitly uses a narrow definition of coercion. For a discussion of the relationships between coercion, property rights, and economic exchange see Schmid 1978, 8–9.

REFERENCES

Binswanger, Hans P., and Vernon W. Ruttan. 1978. *Induced Innovation: Technology, Institutions, and Development*. Baltimore: Johns Hopkins University Press.

Commons, John R. 1934. *Institutional Economics*. New York: Macmillan.

Deane, Phyllis. 1978. *The Evolution of Economic Ideas*. New York: Cambridge University Press.

de Janvry, Alain. 1981. *The Agrarian Question and Reformism in Latin America*. Baltimore: Johns Hopkins University Press.

Hayami, Yujiro, and Vernon W. Ruttan. 1971. *Agricultural Development: An International Perspective*. Baltimore: Johns Hopkins Press.

Johnston, Bruce F. 1970. "Agriculture and Structural Transformation in Developing Countries: A Survey of Research." *Journal of Economic Literature* 3(2):369–404.

Johnston, Bruce F., and John W. Mellor. 1961. "The Role of Agriculture in Economic Development." *American Economic Review* 51(4):566–93.

Jorgenson, D. W. 1961. "The Development of a Dual Economy." *Economic Journal* 71 (June):309–34.

Lewis, W. Arthur. 1954. "Economic Development with Unlimited Supplies of Labour." *Manchester School of Economic and Social Studies* 22(2):139–91.

Ranis, Gustav, and John C. H. Fei. 1961. "A Theory of Economic Development." *American Economic Review* 51(4):533–46.

Reynolds, Lloyd G., ed. 1975. *Agriculture in Development Theory*. New Haven: Yale University Press.

Roumasset, James A. 1978. "The New Institutional Economics and Agricultural Organization." *Philippines Economic Journal*, Number Thirty Seven, 17(3):331–48.

Ruttan, Vernon W. 1975. "Integrated Rural Development: A Skeptical Perspective." *International Development Review*, no. 4:9–16.

Schmid, A. Allan. 1978. *Property, Power and Public Choice: An Inquiry into Law and Economics*. New York: Praeger.

Schultz, Theodore W. 1964. *Transforming Traditional Agriculture*. New Haven: Yale University Press.

Timmer, C. Peter, Walter P. Falcon, and Scott R. Pearson. 1983. *Food Policy Analysis*. Baltimore: Johns Hopkins University Press, for the World Bank.

2

Models of Agricultural Development

VERNON W. RUTTAN

Prior to this century, almost all increase in food production was obtained by bringing new land into production. There were only a few exceptions to this generalization—in limited areas of East Asia, in the Middle East, and in Western Europe. By the end of this century almost all of the increase in world food production must come from higher yields—from increased output per hectare. In most of the world the transition from a resource-based to a science-based system of agriculture is occurring within a single century. In a few countries this transition began in the nineteenth century. In most of the presently developed countries it did not begin until the first half of this century. Most of the countries of the developing world have been caught up in the transition only since mid-century. The technology associated with this transition, particularly the new seed-fertilizer technology, has been referred to as the "green revolution."

During the remaining years of the twentieth century, it is imperative that the poor countries design and implement more effective agricultural development strategies than in the past. A useful first step in this effort is to review the approaches to agricultural development that have been employed in the past and will remain part of our intellectual equipment. The literature on agricultural development can be characterized according to the following models: (1) the frontier, (2) the conservation, (3) the urban-industrial impact, (4) the diffusion, (5) the high-payoff input, and (6) the induced innovation.

FRONTIER MODEL

Throughout most of history, expansion of the area cultivated or grazed has represented the dominant source of increase in agricultural production. The most

VERNON W. RUTTAN is professor of economics and of agricultural and applied economics, University of Minnesota.

Originally titled "How the World Feeds Itself." Published by permission of Transaction, Inc., from *Society* 17, no. 6 (1980). Copyright © 1980 by Transaction, Inc. Published with omissions and minor editorial revisions and with the permission of the author.

dramatic example in Western history was the opening up of the new continents—North and South America and Australia—to European settlement during the eighteenth and nineteenth centuries. With the advent of cheap transport during the latter half of the nineteenth century, the countries of the new continents became increasingly important sources of food and agricultural raw materials for the metropolitan countries of Western Europe.

Similar processes had occurred earlier, though at a less dramatic pace, in the peasant and village economies of Europe, Asia, and Africa. The first millennium A.D. saw the agricultural colonization of Europe north of the Alps, the Chinese settlement of the lands south of the Yangtze, and the Bantu occupation of Africa south of the tropical forest belts. Intensification of land use in existing villages was followed by pioneer settlement, the establishment of new villages, and the opening up of forest or jungle land to cultivation. In Western Europe there was a series of successive changes from neolithic forest fallow to systems of shifting cultivation of bush and grass land followed first by short fallow systems and later by annual cropping.

Where soil conditions were favorable, as in the great river basins and plains, the new villages gradually intensified their system of cultivation. Where soil resources were poor, as in many of the hill and upland regions, new areas were opened up to shifting cultivation or nomadic grazing. Under conditions of rapid population growth, the limits to the frontier model were often quickly realized. Crop yields were typically low—measured in terms of output per unit of seed rather than per unit of crop area. Output per hectare and per man-hour tended to decline—except in delta areas, such as Egypt and South Asia, and the wet rice areas of East Asia. In many areas the result was increasing immiserization of the peasantry.

There are relatively few remaining areas of the world where development along the lines of the frontier model will represent an efficient source of growth during the last two decades of the twentieth century. The 1960s saw the "closing of the frontier" in most areas of Southeast Asia. In Latin America and Africa the opening up of new lands awaits development of technologies for the control of pests and diseases (such as the tsetse fly in Africa) or for the release and maintenance of productivity of problem soils.

CONSERVATION MODEL

The conservation model of agricultural development evolved from the advances in crop and livestock husbandry associated with the English agricultural revolution and the notions of soil exhaustion suggested by the early German chemists and soil scientists. It was reinforced by the application to land of the concept, developed in the English classical school of economics, of diminishing returns to labor and capital. The conservation model emphasized the evolution of a sequence of increasingly complex land- and labor-intensive cropping systems, the production and use of organic manures, and labor-intensive capital formation in the form of drainage, irrigation, and other physical facilities to more effectively utilize land and water resources.

Until well into the twentieth century the conservation model of agricultural development was the only approach to intensification of agricultural production available to most of the world's farmers. Its application is effectively illustrated by development of the wet-rice culture systems that emerged in East and Southeast Asia and by the labor- and land-intensive systems of integrated crop-livestock husbandry which increasingly characterized European agriculture during the eighteenth and nineteenth centuries.

During the English agricultural revolution more intensive crop-rotation systems replaced the open-three-field system in which arable land was allocated between permanent crop land and permanent pasture. This involved the introduction and more intensive use of new forage and green manure crops and an increase in the availability and use of animal manures. This "new husbandry" permitted the intensification of crop-livestock production through the recycling of plant nutrients, in the form of animal manures, to maintain soil fertility. The inputs used in this conservation system of farming—the plant nutrients, animal power, land improvements, physical capital, and agricultural labor force—were largely produced or supplied by the agricultural sector itself.

Agricultural development, within the framework of the conservation model, clearly was capable in many parts of the world of sustaining rates of growth in agricultural production in the range of 1.0 percent per year over relatively long periods of time. The most serious recent effort to develop agriculture within this framework was made by the People's Republic of China in the late 1950s and early 1960s. It became readily apparent, however, that the feasible growth rates, even with a rigorous recycling effort, were not compatible with modern rates of growth in the demand for agricultural output—which typically fall in the 3–5 percent range in the less developed countries (LDCs). The conservation model remains an important source of productivity growth in most poor countries and an inspiration to agrarian fundamentalists and the organic farming movement in the developed countries.

URBAN-INDUSTRIAL IMPACT MODEL

In the conservation model, locational variations in agricultural development are related primarily to differences in environmental factors. It stands in sharp contrast to models that interpret geographic differences in the level and rate of economic development primarily in terms of the level and rate of urban-industrial development.

Initially, the urban-industrial impact model was formulated in Germany by J. H. von Thunen to explain geographic variations in the intensity of farming systems and the productivity of labor in an industrializing society. In the United States it was extended to explain the more effective performance of the input and product markets linking the agricultural and nonagricultural sectors in regions characterized by rapid urban-industrial development than in regions where the urban economy had not made a transition to the industrial stage. In the 1950s, interest in the urban-industrial impact model reflected concern with the failure of agricultural resource development and price policies, adopted in the 1930s, to remove the persistent regional disparities in agricultural productivity and rural incomes in the United States.

The rationale for this model was developed in terms of more effective input and product markets in areas of rapid urban-industrial development. Industrial development stimulated agricultural development by expanding the demand for farm products, supplying the industrial inputs needed to improve agricultural productivity, and drawing away surplus labor from agriculture. The empirical tests of the urban-industrial impact model have repeatedly confirmed that a strong nonfarm labor market is an essential prerequisite for labor productivity in agriculture and improved incomes for rural people.

The policy implications of the urban-industrial impact model appear to be most relevant for less developed regions of highly industrialized countries or lagging regions of the more rapidly growing LDCs. Agricultural development policies based on this model appear to be particularly inappropriate in those countries where the ''pathological'' growth of urban centers is a result of population pressures in rural areas running ahead of employment growth in urban areas.

DIFFUSION MODEL

The diffusion of better husbandry practices was a major source of productivity growth even in premodern societies. The diffusion of crops and animals from the new world to the old—potatoes, maize, cassava, rubber—and from the old world to the new—sugar, wheat, and domestic livestock—was an important by-product of the voyages of discovery and trade from the fifteenth to the nineteenth centuries. The diffusion approach rests on the empirical observation of substantial differences in land and labor productivity among farmers and regions. The route to agricultural development, in this view, is through more effective dissemination of technical knowledge and a narrowing of the productivity differences among farmers and among regions.

The diffusion model has provided the major intellectual foundation of much of the research and extension effort in farm management and production economics since the emergence, in the latter years of the nineteenth century, of agricultural economics and rural sociology as separate subdisciplines linking the agricultural and the social sciences. Developments leading to establishment of active programs of farm management research and extension occurred at a time when experiment station research was making only a modest contribution to agricultural productivity growth. A further contribution to the effective diffusion of known technology was provided by rural sociologists' research on the diffusion process. Models were developed emphasizing the relationship between diffusion rates and the personality characteristics and educational accomplishments of farm operators.

Insights into the dynamics of the diffusion process, when coupled with the observation of wide agricultural productivity gaps among developed and less developed countries and a presumption of inefficient resource allocation among "irrational tradition-bound" peasants, produced an extension or diffusion bias in the choice of agricultural development strategy in many LDCs during the 1950s. During the 1960s the limitations of the diffusion model as a foundation for the design of agricultural development policies became increasingly apparent as technical assistance and rural development programs, based explicitly or implicitly on this model, failed to generate either rapid modernization of traditional farms and communities or rapid growth in agricultural output.

HIGH-PAYOFF INPUT MODEL

The inadequacy of policies based on the conservation, urban-industrial impact, and diffusion models led, in the 1960s, to a new perspective—the key to transforming a traditional agricultural sector into a productive source of economic growth was investment designed to make modern, high-payoff inputs available to farmers in poor countries. Peasants in traditional agricultural systems were viewed as rational, efficient resource allocators. This iconoclastic view was developed most vigorously by T. W. Schultz in his controversial book *Transforming Traditional Agriculture*. He insisted that peasants in traditional societies remained poor because, in most poor countries, there were only limited technical and economic opportunities to which they could respond. The new, high-payoff inputs were classified according to three categories: (1) the capacity of public and private sector research institutions to produce new technical knowledge; (2) the capacity of the industrial sector to develop, produce, and market new technical inputs; and (3) the capacity of farmers to acquire new knowledge and use new inputs effectively.

The enthusiasm with which the high-payoff input model has been accepted and translated into economic doctrine has been due in part to the proliferation of studies reporting high rates of return to public investment in agricultural research.[1] It was also due to the success of efforts to develop new, high-productivity grain varieties suitable for the tropics. New high-yielding wheat varieties were developed in Mexico beginning in the 1950s, and new high-yielding rice varieties were developed in the Philippines in the 1960s. These varieties were highly responsive to industrial inputs, such as fertilizer and other chemicals, and to more effective soil and water management. The high returns associated with the adoption of the new varieties and the associated technical inputs and management practices have led to rapid diffusion of the new varieties among farmers in a number of countries in Asia, Africa, and Latin America.

INDUCED INNOVATION MODEL

The high-payoff input model remains incomplete as a theory of agricultural development. Typically, education and research are public goods not traded through the marketplace. The mechanism by which resources are allocated among education, research, and other public and private sector economic activities was not fully incorporated into the model. It does not explain how economic conditions induce the development and adoption of an efficient set of technologies for a particular society. Nor does it attempt to specify the processes by which input and product price relationships induce investment in research in a direction consistent with a nation's particular resource endowments.

These limitations in the high-payoff input model led to efforts by Yujiro Hayami and myself to develop a model of agricultural development in which technical change is treated as endogenous to the development process, rather than as an exogenous factor operating independently of other development processes. The induced innovation perspective was stimulated by historical evidence that different countries had followed alternative paths of technical change in the process of agricultural development.[2]

TECHNICAL INNOVATION

The levels achieved in each productivity grouping by farmers in the most advanced countries can be viewed as arranged along a productivity frontier. This frontier reflects the level of technical progress achieved by the most advanced countries in each resource endowment classification. These productivity levels are not immediately available to farmers in most low-productivity countries. They can only be made available by undertaking investment in the agricultural research capacity needed to develop technologies appropriate to the countries' natural and institutional environments and investment in the physical and institutional infrastructure needed to realize the new production potential opened up by technological advances.

There is clear historical evidence that technology has been developed to facilitate the substitution of relatively abundant (hence cheap) factors for relatively scarce (hence expensive) factors of production. The constraints imposed on agricultural development by an inelastic supply of land have, in economies such as Japan and Taiwan, been offset by the development of high-yielding crop varieties designed to facilitate the substitution of fertilizer for land. The constraints imposed by an inelastic supply of labor, in countries such as the United States, Canada, and Australia, have been offset by technical advances leading to the substitution of animal and mechanical power for manpower. In some cases the new technologies—embodied in new crop varieties, new equipment, or new production practices—may not always be substitutes per se for land or labor. Rather, they are catalysts which facilitate the substitution of relatively abundant factors (such as fertilizer or mineral fuels) for relatively scarce factors.

Institutional Innovation

A developing country which fails to evolve a capacity for technical and institutional innovation in agriculture consistent with its resource and cultural endowments suffers two major constraints on its development of productive agriculture. It is unable to take advantage of advances in biological and chemical technologies suited to labor-intensive agricultural systems. And the mechanical technology it does import from more developed countries will be productive only under conditions of large-scale agricultural organization. It will contribute to the emergence of a "bimodal" rather than a "unimodal" organization structure.

During the last two decades a number of developing countries have begun to establish the institutional capacity to generate technical changes adapted to national and regional resource endowments. More recently these emerging national systems have been buttressed by a new system of international crop and animal research institutes. These new institutes have become both important sources of new knowledge and technology and increasingly effective communication links among the developing national research systems.

The lag in shifting from a natural-resource-based to a science-based system of agriculture continues to be a source of national differences in land and labor productivity. Lags in the development and application of knowledge are also important sources of regional productivity differences within countries. In countries such as Mexico and India differential rates of technical change have been an important source of the widening disparities in the rate of growth of total agricultural output, in labor and land productivity, and in incomes and wage rates among regions.

Productivity differences in agriculture are increasingly a function of investments in scientific and industrial capacity and in the education of rural people rather than of natural resource endowments. The effects of education on productivity are particularly important during periods in which a nation's agricultural research system begins to introduce new technology. In an agricultural system characterized by static technology there are few gains to be realized from education in rural areas. Rural people who have lived for generations with essentially the same resources and the same technology have learned from long experience what their efforts can get out of the resources available to them. Children acquire from their parents the skills that are worthwhile. Formal schooling has little economic value in agricultural production.

As soon as new technical opportunities become available, this situation changes. Technical change requires the acquisition of new husbandry skills; acquisition from nontraditional sources of additional resources such as new seeds, new chemicals, and new equipment; and development of new skills in dealing with both natural resources and input and product market institutions linking agriculture with the nonagricultural sector.

The processes by which new knowledge can be applied to alter the rate and direction of technical change in agriculture, are, however, substantially greater than our knowledge of the processes by which resources are brought to bear on the process of institutional innovation and transfer. Yet the need for viable institutions capable of supporting more rapid agricultural growth and rural development is even more compelling today than a decade ago.

NOTES

1. See Evenson, chapter 24 in this volume.—ED.
2. For more details on the induced innovation model see chapter 4.—ED.

3

The Rise and Fall of Community Development, 1950–65: A Critical Assessment

LANE E. HOLDCROFT

INTRODUCTION

The worldwide community development (CD) movement of the 1950s faded away amid the euphoria of the "green revolution" in the early 1960s. There are numerous insights and lessons that can be drawn from the CD experience. Community development had great appeal to leaders of developing nations and external donor officials because it provided a nonrevolutionary approach to the development of agrarian societies. It is now apparent that these decision makers were rather naive.

The failure of CD and the shortcomings of the "green revolution" shifted the focus of planning and development assistance in the 1970s to integrated rural development (IRD). Some CD veterans believe that the new IRD is in fact a revival of old CD. Although the sponsors of IRD themselves would rather emphasize the differences, there are sufficient similarities to uphold the revivalist view. A question then may well be asked: Are there any major implications of the rise and fall of the CD movement for the new IRD?

While broad generalizations are often unwarranted, it may be useful, with the advantage of hindsight, to understand fully the shortcomings of CD. As a starting point, we should remember that CD was a product of the cold-war era of the late 1940s and the 1950s. Its principles were derived, consciously or unconsciously, from theories directly opposed to revolutionary doctrines. In that period, the threat of subversion was taken very seriously. Community development was designed to remove this threat. By bringing people together, inviting them into harmonious communities, and mobilizing them for common endeavors, CD promised to generate permanent political peace and quick economic growth.

LANE E. HOLDCROFT is director of the Office for Technical Resources, Bureau for Africa, U.S. Agency for International Development, Washington, D.C.

Reprinted from MSU Rural Development Paper No. 2, 1978, with omissions and revisions, by permission of the Department of Agricultural Economics, Michigan State University, and the author.

46

After a decade of experience, it became evident that neither promise could be fulfilled, except in rare and isolated cases.

The CD movement experienced phenomenal growth in the 1950s, primarily as a result of promotion and financial support by the United States. By 1960 the United Nations estimated that over sixty countries in Asia, Africa, and Latin America had CD programs in operation. About half of these were national in scope, and the remainder were regional programs of lesser importance. But even by 1960 some CD programs were faltering, and by 1965 most had been terminated or drastically reduced in scope, to the extent that they were no longer considered by national leaders to be major national development efforts. By the late 1950s, donors, including U.N. agencies and those of the United States, appeared disillusioned and shifted their resources in support of new initiatives such as the "green revolution."

The purpose of this paper is to trace the rise and fall of CD and to draw lessons for developing countries and donors interested in helping the rural poor.

ORIGINS OF COMMUNITY DEVELOPMENT

In 1948 the term "community development" was first used officially at the British Colonial Office's Cambridge Conference on the Development of African Initiative. Community development was proposed to help the British colonies in Africa prepare for independence by improving local government and developing their economies. Shortly thereafter, the term and concept spread rapidly to various external donor agencies, as well as to many national governments. A number of modest national CD efforts were launched, primarily in British Africa around 1950. The first major CD program was initiated in India in 1952 with support from the Ford Foundation and the U.S. foreign economic assistance agency. Soon thereafter, national programs were established in the Philippines, Indonesia, Iran, and Pakistan. The CD approach in the developing world in the 1950s had its early roots in (*a*) experiments by the British Colonial Service, primarily in Asia; (*b*) U.S. and European voluntary agency activities abroad; and (*c*) U.S. and British domestic programs in adult education, community development services, and social welfare.

Both the United States and the United Nations drew heavily upon the synthesis of rural reconstruction efforts in India in the 1930s and 1940s. India had more well-documented experience with rural reconstruction and community development than any other single country in the world. Gandhi and Tagore were influential personalities in spearheading rural development activities in India and in influencing how the United States and the United Nations approached community development. Also F. L. Brayne's (1946) experiments and writings on rural development in the Punjab provided important lessons, as did the work of agricultural missionaries at various locations in India and elsewhere. These experiments provided ample evidence that rural people would respond and take the initiative when they realized that they would benefit from community efforts. Post-Independence projects in India, including Etawah,[1] Nilokheri, and Faridabad, were influential prototypes for India's CD program, which was launched in 1952, as well as other early national CD programs in the developing world (Dayal 1960).

The second source of related experiences grew out of American and European voluntary-agency efforts in the developing world. These included the work of missionary groups as well as of institutions such as the Near East Foundation and the Ford Foundation. The Near East Foundation assisted in launching the Varamin Plain Project in Iran in the late 1940s, which became a prototype for the more ambitious national CD program initiated in 1952.

The third set of experiences that influenced CD were those from adult education, community services, and social welfare programs in the United States and the United Kingdom, many of which were initiated during the Depression in the 1930s. The social welfare experience in the United States and Europe also contributed to the ideology underlying the concept and approach of CD. Social welfare was, and is, rooted in relief and other charitable efforts to help the poor, but such programs historically have focused primarily on the urban poor.

Community development was defined as a process, method, program, institution, and/or movement which (a) involves people on a community basis in the solution of their common problems, (b) teaches and insists upon the use of democratic processes in the joint solution of community problems, and (c) activates and/or facilitates the transfer of technology to the people of a community for more effective solution of their common problems. Joint efforts to solve common problems democratically and scientifically on a community basis were seen as the essential elements of CD.

Community development was described as rooted in the concept of the worth of the individual as a responsible, participating member of society and, as such, was concerned with human organization and the political process. Its keystones were seen as community organization, community education, and social action. It was designed to encourage self-help efforts to raise standards of living and to create stable, self-reliant communities with an assured sense of social and political responsibility commensurate with basic free-world objectives. Community development was seen as dealing with a complex unit, the total community, and using a flexible, dynamic approach adapted to local circumstances. Precise definitions were believed to be neither realistically possible nor desirable. Rigid definition was seen as producing rigid, ritualized, and standardized programs that would be self-defeating.

The U.N., U.S., and British approaches to CD[2] focused on the initiation of comprehensive development schemes in individual villages on the basis of what village people perceived to be their "felt needs." Community development activities were customarily initiated by sending a specially trained civil servant known as a "multi-purpose village-level worker" into the village. These village-level workers were generally secondary school graduates who had received several months of preservice training in a CD institute. By living in a village and working with village people, the village-level worker was supposed to gain the villagers' confidence. He was to serve as a catalyst, one who would guide and assist villagers in identifying their felt needs, then translating these felt needs into village development plans, and finally implementing these plans—always working through the active village leaders.

The village-level worker was supposed to have some skills in a variety of subjects such as village organization and mobilization, as well as in such areas as literacy, agriculture, and health. And in areas in which he lacked special skills, technicians from specialized government agencies were supposed to support him. Usually the village-level worker administered "matching" grants to villagers in which the villagers' labor and some locally available materials would be combined with grants-in-kind from the national CD organization in order to carry out village projects. However, the products of successful CD were seen as not only the building of such community facilities as wells, roads, and schools and the creation of new crops but also the development of stable, self-reliant communities with an assured sense of social and political responsibility.

Community development proponents likened it to an enterprise by which the government and the rural people would be brought together, thus improving the lot of the more downtrodden and less fortunate peoples. Consistent with this view of CD, however, was a broader one that saw CD as an important technique for modernizing an entire society. Where national CD efforts were being implemented, usually a large new bureaucracy was established at the national, regional, and local levels to administer the program and attempt to coordinate the rural programs of technical ministries and regional offices, for example, agriculture, education, and health. Most often, these new CD organizations were well financed, primarily by external donors, and staffed with expatriate advisers. With their large foreign and domestic training programs, they were usually able to recruit highly motivated, relatively well-educated young men and women for both headquarters and field staff positions.

Thus, it can be seen that CD was appealing to the leaders of some developing nations who were looking for an ideology and technique to improve the living conditions of rural people. Community development held forth the promise not only of building "grass-roots" democratic institutions but also of improvements in the material well-being of rural people—without revolutionary changes in the existing political and economic order. In summary, the CD approach was assumed to have nearly universal application to rural societies. The United States and other donors agreed to finance most of the costs associated with launching national and pilot CD schemes.

DECADE OF PROMINENCE: 1950s

The CD movement blossomed in the developing world during the decade of the 1950s. By 1960 over sixty nations in Asia, Africa, and Latin America had launched national or regional CD programs. In some instances small pilot projects that had been launched by the British or French governments in African[3] and Asian nations in the early post–World War II period were expanded rapidly with U.S. and/or U.N. assistance. The greatly publicized launching of India's ambitious CD program in 1952 gave the movement an added impetus. Until about 1956 the Indian program[4] served as a prototype for national programs in other Asian countries. Leaders in the Indian program served as consultants and provided training materials for these new programs, and numerous government officials from around the world visited India to observe and/or attend training courses.

A few U.S. foreign aid missions established CD offices in the early 1950s, and in 1954 a Community Development Division was established in the foreign aid agency's Washington headquarters. The Community Development Division, through its personnel and consultants, was instrumental in promoting CD around the world. A relatively small number of individuals spearheaded the U.S. foreign aid support. The proponents included sociologists and anthropologists, with a smaller number of educators, economists, agriculturalists, and political scientists.

The *modus operandi* of the American foreign aid agency in spreading the CD approach consisted basically of (*a*) sending teams of CD experts to assist interested governments in planning national and pilot CD programs; (*b*) providing long-term technical and capital assistance; (*c*) publishing a CD periodical as well as numerous other CD documents; and (*d*) holding a series of six international conferences around the world in which interested governments were invited to participate.

In countries where governments indicated an interest in initiating CD programs, the usual pattern was that of small teams of CD "experts" who would assist the host government in formulating a preliminary program proposal. Usually, this would be followed by the establishment of a host government CD agency and a community development division in the U.S. country aid mission. Then, observation trips were arranged for senior host government personnel to attend the international conferences and observe programs already launched. The next step would be to train prospective CD officers in the host country or in another developing country with an active CD program. Generally, the United States would provide technical advisers, supplies, and equipment; training for host country personnel; and most of the budgetary support needed for program implementation. In some instances, rather than providing direct U.S. government assistance, the U.S. foreign aid agency would finance assistance programs by American universities or voluntary agencies.

After the national program in India was initiated in 1952 with massive support from the Ford Foundation and the U.S. foreign assistance agency, the United States assisted in launching major programs in Iran and Pakistan in 1953, the Philippines in 1955, Jordan in 1956, Indonesia in 1957, and Korea in 1958. Smaller programs were also launched with U.S. assistance in Iraq in 1952, Afghanistan and Egypt in 1953, Lebanon in 1954, and Ceylon and Nepal in 1956. At its zenith in 1959, the American foreign assistance program assisted twenty-five nations in the implementation of CD programs, and the U.S. foreign aid agency employed 105 direct-hire and contract CD advisers. During the ten-year period ending in 1962 the United States provided directly some $50 million in support of CD programs in over thirty countries via its bilateral foreign economic assistance agency and a somewhat lesser amount via the several U.N. agencies that funded CD efforts in another thirty countries.

Under the leadership of the United Nations Department of Economic and Social Affairs, the U.N. agencies generally fostered the CD movement in much the same manner as did the U.S. foreign aid agency, albeit on a reduced scale. Technical and capital assistance were provided in launching pilot programs, and international conferences were sponsored, in addition to the preparation of numerous widely disseminated CD publications.

REASONS FOR THE DECLINE

By 1960 some CD programs, including the major Indian effort, were faltering, and by 1965 most national CD programs had been terminated or drastically reduced. The precipitous decline was due primarily to (*a*) disillusionment on the part of many political leaders in developing countries with the performance of their programs vis-á-vis stated goals and (*b*) the sharp reduction in support from the United States and other donors. These interdependent causes were mutually reinforcing. Political leaders in developing countries were disillusioned because their CD programs had not demonstrated, as promised, that the CD approach would build stable "grass-roots" democratic institutions and would improve the economic and social well-being of rural people while contributing to the attainment of national economic goals.

During the era of the 1950s and 1960s, when the "trickle-down" theory of economic development was in vogue, CD programs were not intended to, nor did they, affect the basic structural barriers to equity and growth in rural communities. Rather, they accepted the existing local power structure as a given. Usually CD village-level workers aligned themselves with the traditional village elites, thus strengthening the economic and social position of the elites. There was little attention given to assuring that benefits from CD programs accrued to the rural poor. Realizing this, the poor majority of the villagers did not respond to the CD approach. Only in those few nations, such as South Korea, with rural communities composed of relatively homogeneous farm owner-operators were CD programs relatively successful in reaching their stated objectives. In some instances, efforts were made in the early 1960s to recognize that most rural communities were divided by the different interests of the landless and nearly landless laborers, subsistence tenants and owner-operators, and commercial farmers, thus calling for changes in the local power structure if CD were to succeed. However, most political leaders of developing countries turned to programs to increase food production.

The number of developing nations receiving major U.S. support for CD dropped from twenty-five to nineteen between 1959 and 1960, and the number of American CD advisers was reduced from 105 to 68. By 1963 the U.S. foreign aid agency's Community Development Division in Washington, D.C., had been abolished along with most CD offices in field missions. By the mid-1960s only a few countries continued to receive U.S. support for their CD programs. When major U.S. assistance was reduced or terminated, CD programs were terminated, drastically redirected, or greatly reduced by host country governments.

Under the Kennedy administration, the leadership of the U.S. foreign aid agency in the early 1960s was not only concerned with the lack of host country support of CD programs but also disillusioned with the widespread internal conflict and animosity between U.S. CD and technical services personnel, particularly agriculturalists. This conflict permeated the foreign aid agency both in Washington and in field missions, and it spread to host country ministries and agencies. It was an ideological battle which pitted the generalist against the specialist, the social scientist (excluding economists) against the technologist, the pluralist against the monist. Usually these conflicts were resolved in favor of technical services personnel, who were bureaucratically more established and less abstract in their perception of the development process.

By 1963, where CD offices had not been eliminated, CD and agricultural offices in U.S. field missions were combined into rural development offices. Most host countries' CD ministries or agencies became units of the agriculture or internal affairs ministry, depending on whether the current development focus was on local government or agricultural technology.

The United Nations and a few private philanthropic organizations continued to fund some CD activities throughout the 1960s, but without American and host country government support these efforts were relatively minor and increasingly shifted from a development to a social welfare orientation. Even British government support for the University of London's *Community Development Bulletin* was terminated in 1964.

Perhaps the most universal criticism of the CD movement was that its programs were inefficient in reaching economic goals, including food production. It was assumed that man would respond rationally to economic incentives, and since underdevelopment was defined in economic terms, programs that more directly focused on economic growth were considered more deserving of support. As central-planning-agency personnel became established and influential in decision making in developing countries during the 1950s, they criticized CD programs as being "uneconomic" and a "low priority investment" of scarce domestic and external development resources. Related to this issue was the concern in many nations that CD programs were not contributing to the alleviation of food shortages and poverty.

The CD program in India was well-documented.[5] The stated objective of the Indian program was to transform the economic and social life of the villages and to alleviate poverty and the scarcity of food through popular participation of village people. A massive self-help program embracing agriculture, health, education, public works, and social welfare was implemented for over a decade. Yet program performance, measured in terms of reaching stated objectives, was poor. Poverty and food scarcity were not reduced but rather became more widespread during that decade, as did disparities of wealth between the large farmers and peasants in the rural areas. Critics pointed to the wide disparity in the distribution of benefits of the program between accessible and remote villages, between cultivators and other groups within villages, and between the wealthier and the poorer farmers. Evaluators reported that the program was not accepted by people, did not reach the poor, and was a "top-down" bureaucratic empire that ignored agricultural production.

The leaders of the Indian CD program recognized early that the program was ineffective in stimulating village-level initiative and action. There was a propensity on the part of the village-level workers to work with the traditional village elite, to ignore the poor, and to lead or direct villagers rather than develop local leadership. This basic problem of being unable to arouse popular participation plagued most CD programs.

Defenders of CD in India and elsewhere maintained that success depended on more and better training for village-level workers and improved coordination of local government services. The view most often expressed was that political leaders did not understand either the complexity of the problem or the time required to transform traditional village societies.

India also provides an example of how national CD programs evolved during the 1950s. During the initial years social welfare, public works, and changes in villagers' attitudes, rather than material results, were emphasized. Then food production became the prime focus of the program in the late fifties. In the early 1960s the focus shifted to local self-government and cooperative development, as the CD effort receded and technical agriculture came to the fore again. The evolution of the Indian program from social welfare and public works to cooperatives, local government, and technical agriculture was the general pattern in CD programs around the world.

The turning point for CD in India came in 1959 with the publication of the Ford Foundation's *Report on India's Food Crisis and Steps to Meet It* (1959). The report called for an all-out emergency food and production program, and it urged the CD ministry and technical ministries to give top priority to food production by increasing the number of technical agricultural personnel assigned to blocks and villages. The report recommended that CD village-level workers concentrate on technical agricultural tasks. The CD program was described as trying to be all things to all people and not giving adequate attention to food production. The report was critical of the block development officers for not understanding agriculture and using village-level workers as errand boys. After this report was published, the focus of the government's rural programs clearly shifted to food production, and CD declined.

IMPLICATIONS OF THE COMMUNITY DEVELOPMENT EXPERIENCE FOR RURAL DEVELOPMENT PROGRAMS OF THE 1970s AND 1980s

GENERAL CONCLUSIONS

Politically, CD was ineffective because in most developing countries basic conflicts were too deep to be resolved simply by the persuasive efforts of CD workers. Factors such as distribution of land ownership, exploitation by elites, or urban domination could be neither ignored nor by-passed. Community development's attempt to proceed smoothly without friction towards general consensus was unrealistic. The expected reconciliation and common participation for the sake of development occurred as an exception rather than as a rule.

Economically, CD displayed a double weakness. First, it enlarged social services more rapidly than it enlarged the production of rural incomes. Second, it could not significantly improve the condition of the distressed poor, the share-croppers and laborers. Both aspects of rural poverty, low production and unjust distribution, were not significantly changed by CD.

Recoiling from the elitist bias of CD (and the "green revolution"), the new IRD programs are concentrating on the rural poor. In other words, IRD programs acknowledge the presence of conflict of interest, namely, class struggle, a point of view that was studiously avoided by CD. Beyond the IRD acknowledgment, however, there remains the challenge of finding ways and means to uplift the underprivileged. Perhaps for identical reasons, the new IRD, like the old CD, does not relish the prospect of highlighting politically sensitive obstacles and so diplomatically shrouds the suggestions for removing them. Similarly, even though CD's fondness for social services and neglect of production are now well known, the new development programs of the late 1970s such as "basic needs" may fall into the same trap. To strike a balance between demands for social services and conditions for increased production is, in any case, a very difficult task.

PITFALLS OF NEW MINISTRIES OF RURAL DEVELOPMENT

In the field of administration, CD was hampered by the confrontation between the generalist and the specialist. In country after country, attempts were made to bring different departments working in the rural area under unified control. The department of agriculture, usually the most rapidly expanding entity, tenaciously resisted any kind of merger. Community development in India enjoyed a brief period of supremacy as the czar of rural development and then succumbed to the department of agriculture. The new IRD programs, which demand unified control, must be prepared for this battle of departments. Perhaps the necessary coordination can be secured more peacefully, not by imposing a superdepartment from above, but by creating autonomous institutions at lower levels nearer to the village.

The experience of numerous CD programs suggests that the problem of coordination among various government agencies cannot be resolved by establishing a single new ministry or agency, even with the strong support of the chief of state. Difficulties arise from rivalries between the technical ministries, such as agriculture, health, and education (especially extension departments in these ministries), and the rural development agency or ministry. To be effective, IRD, like CD, inevitably must affect and make demands on the technical ministries. National CD organizations in developing countries were unable to provide the mechanism for coordinating rural development efforts, and there is no evidence that a national "rural" development organization could do any better today. Local-level coordination was successful when all local technical extension personnel and CD workers were supervised by the district administrator rather than by representatives of their technical ministries or the national CD agency.

STRENGTHENING THE ECONOMIC BASE

Rural development projects should include from their inception an income-producing component, usually one that entails increasing agricultural output through the introduction of a profitable "package" of technology. With an income-producing "centerpiece," other components, such as health, sanitation, and education, can follow. Many observers were properly critical of the Indian CD program for initially investing in community buildings, schools, clinics, and in social welfare, which increased consumption and population growth, rather than stressing agricultural production from the onset of the CD program. In countries where CD programs included an agricultural or other income-producing component, these programs often became internationally known. When there was a failure in agricultural production, the causes were usually the technology employed and/or the sharecropping arrangements.

PARTICIPATION

Participation, a major goal in the CD strategy, proved to be a most difficult and elusive goal to attain. Participation by nearly all segments of rural society, including the landless and nearly landless, was rarely accomplished in any of the CD programs. In most instances village CD workers tended to identify with the traditional village elite, to whom most of the program benefits accrued. Unfortunately, there has been very little analysis of the impact of the political and social milieu on villagers' incentives to participate in CD projects. The CD experience indicates that if the rural poor are to be helped, the structural barriers to greater equity must be addressed.

While most CD programs espoused participatory democracy, self-reliance, and local initiative, in practice the village CD worker was paternalistic and directed local-level programs. The reason usually given for the villagers' lack of participation was the inherent fatalism of rural people and their general apathy towards improving their own standard of living. Yet, the experience of those relatively successful pilot CD programs suggests that villagers will participate when they perceive that the benefits of the program will accrue to them.

IMPLEMENTATION

Regardless of the apparent differences in the rhetoric, most of the new IRD programs are based on the political and economic theories that sustained CD. The affinity is even more pronounced in the implementation of IRD programs.

1. Community development relied mainly on the village-level worker. He was the "catalyst" who precipitated the formation of communities. He was the agent of change, the chief modernizing influence. Although he was asked to help establish local leaders, committees, and councils, his role, in fact, reinforced the paternalistic and centralist tradition. Ultimately, CD could not foster the growth of self-reliant local institutions. IRD relies mainly on government change agents who fulfill similar functions.
2. The CD concept of "self-help" projects, boosted by matching grants brought by the village-level worker, seemed very attractive. But, it proved a poor substitute for long-term institutional growth and mobilization. The "aided self-help" projects implemented by the village-level worker unintentionally inhibited real planning and participation. IRD also uses "aided self-help" projects implemented by the government change agent.
3. The CD worker, generally a secondary school graduate himself, was biased in favor of the rural elite and their values. Furthermore, he was directed to work with the established leaders. He felt more at home with the large farmers or youth club members than with the landless laborers. He gladly strengthened the existing power structure. He did not see himself as the champion of the weak against the strong. The IRD change agent cannot ignore the elitist leadership.

EXPANSION OF PILOT PROGRAMS

Political leaders and administrators of rural development programs must exercise restraint in expanding successful pilot programs. In many nations, including India, the CD program was expanded very rapidly as a result of efforts by politicians to spread the program to their constituencies as soon as possible. This rapid expansion necessitated the recruitment of large numbers of poorly trained personnel. Village-level workers were assigned too many responsibilities in too many villages, and the damage that resulted was often worse than if no work had been attempted. Pilot programs are usually successful when adequate resources are provided for material and human inputs. Often, plans for the expansion of these programs do not take into account the additional resources and time required to replicate the carefully nurtured pilot schemes.

DRAWING ON HISTORY

Since many of the new IRD programs employ the same organizational methods as CD (that is, government change agents, aided self-help projects, and collaboration with elitist leaders), the results achieved by IRD will probably mirror those of the CD experience in many countries. The initial popularity of CD and its quick decline provides an object lesson, but it is a lesson that is rarely studied by the IRD experts of the 1970s. Architects of new IRD programs should draw on the earlier CD experiences. Since CD programs were carried out in over sixty countries in the 1950s and 1960s, these experiences should be assessed on a country-by-country basis, and the lessons learned should be incorporated into the planning and implementation of IRD programs today.

NOTES

1. The Etawah project was one of the successful post-Independence village-level development efforts; it served as a prototype for the national CD program. The Etawah project was launched under the sponsorship of the Uttar Pradesh provincial government in 1948 in sixty-four villages, and it expanded in three years to include over three hundred villages (see Singh 1976).
2. See United Nations 1955; United States 1956; and Great Britain Colonial Office 1958.
3. See Du Sautoy 1958, and Jackson 1956.
4. See Dey 1962; Dube 1958; Mukerji 1961; Ensminger 1972; and India 1956, 1957, 1958, and 1960.
5. See Lewis 1962; and Mayer, Marriott, and Park 1958.

REFERENCES

Abueva, José V. 1959. *Focus on the Barrio: The Story behind the Birth of the Philippine Community Development Program under President Ramon Magsaysay.* Manila: Institute of Public Administration, University of the Philippines.
Brayne, F. L. 1946. *Socrates in an Indian Village.* Oxford: Oxford University Press.
Dayal, Rajeshwar. 1960. *Community Development Program in India.* Allahabad: Kitab Mahal.
Dey, S[urendra] K[umar]. 1962. *Community Development—A Chronicle, 1954–1961.* Delhi: Ministry of Information and Broadcasting, Government of India.
Dube, S. C. 1958. *India's Changing Villages—Human Factors in Community Development.* Ithaca: Cornell University Press.
Du Sautoy, Peter. 1958. *Community Development in Ghana.* London: Oxford University Press.
Ensminger, Douglas. 1972. *Rural India in Transition.* New Delhi: All India Panchayat Parishad.
Ford Foundation. Agricultural Production Team. 1959. *Report on India's Food Crisis and Steps to Meet It.* New Delhi: Ministry of Food and Agriculture and Ministry of Community Development, Government of India.
Great Britain. Colonial Office. 1958. *Community Development: A Handbook.* London: H. M. Stationery Office.
Green, James W. 1961. "Success and Failure in Technical Assistance: A Case Study." *Human Organization* 20 (1): 2–10.
India. Community Projects Administration. 1956. *Manual for Village Level Workers.* New Delhi.

———. Ministry of Community Development. 1957. *Report: 1956–57.* New Delhi.

———. Planning Commission. Programme Evaluation Organisation. 1960. *The Seventh Evaluation Report on Community Development and Some Allied Fields.* New Delhi.

———. Team for the Study of Community Projects and National Extension Service, B. G. Mehta, leader. 1958. *Report.* New Delhi: Government of India Press.

Jackson, I. C. 1956. *Advance in Africa: A Study of Community Development in Eastern Nigeria.* London: Oxford University Press.

Lewis, John P. 1962. *Quiet Crisis in India.* Washington, D.C.: Brookings Institution.

Mayer, Albert, McKim Marriott, and Richard L. Park. 1958. *Pilot Project, India.* Berkeley: University of California Press.

Mukerji, B. 1961. *Community Development in India.* Calcutta: Orient Longmans.

Singh, D. P. 1976. "The Pilot Development Project, Etawah." Paper produced for Expert Consultation on Integrated Rural Development. Rome: Food and Agriculture Organization of the United Nations.

United Nations. Economic and Social Council. 1955. *Principles of Community Development—Social Progress through Local Action.*

United States. 1956. "The Community Development Guidelines of the International Cooperation Administration." *Community Development Review,* no. 3: 3–6.

4

Induced Innovation Model of Agricultural Development

VERNON W. RUTTAN and YUJIRO HAYAMI

During the 1960s a new consensus emerged to the effect that agricultural growth is critical (if not a precondition) for industrialization and general economic growth. Nevertheless, the process of agricultural growth itself has remained outside the concern of most development economists. Both technical change and institutional evolution have been treated as exogenous to their systems.

In this paper we elaborate the concept of induced technical and institutional innovation which we have employed in our own research on the agricultural development process, and we discuss the implications of the induced innovation perspective for the design of national and regional strategies for agricultural development.

AN INDUCED DEVELOPMENT MODEL

An attempt to develop a model of agricultural development in which technical change is treated as endogenous to the development process, rather than as an exogenous factor that operates independently of other development processes, must start with the recognition that there are multiple paths of technological development.

ALTERNATIVE PATHS OF TECHNOLOGICAL DEVELOPMENT

There is clear evidence that technology can be developed to facilitate the substitution of relatively abundant (hence cheap) factors for relatively scarce (hence expensive) factors in the economy. The constraints imposed on agri-

VERNON W. RUTTAN is professor of economics and of agricultural and applied economics, University of Minnesota. YUJIRO HAYAMI is professor of economics, Tokyo Metropolitan University.
Originally titled "Strategies for Agricultural Development." Reprinted from *Food Research Institute Studies in Agricultural Economics, Trade and Development* 9, no. 2 (1972): 129–48, with omissions and minor editorial revisions, by permission of the Food Research Institute, Stanford University, and the authors.

cultural development by an inelastic supply of land have, in economies such as those of Japan and Taiwan, been offset by the development of high-yielding crop varieties designed to facilitate the substitution of fertilizer for land. The constraints imposed by an inelastic supply of labor, in countries such as the United States, Canada, and Australia, have been offset by technical advances leading to the substitution of animal and mechanical power for labor. In both cases the new technology, embodied in new crop varieties, new equipment, or new production practices, may not always be a substitute by itself for land or labor; rather it may serve as a catalyst to facilitate the substitution of the relatively abundant factors (such as fertilizer or mineral fuels) for the relatively scarce factors. It seems reasonable, following Hicks, to call techniques designed to facilitate the substitution of other inputs for labor ''labor-saving'' and those designed to facilitate the substitution of other inputs for land ''land-saving.'' In agriculture, two kinds of technology generally correspond to this taxonomy: mechanical technology to ''labor-saving'' and biological and chemical technology to ''land-saving.''[1] The former is designed to facilitate the substitution of power and machinery for labor. Typically this involves the substitution of land for labor, because higher output per worker through mechanization usually requires a larger land area cultivated per worker. The latter, which we will hereafter identify as biological technology, is designed to facilitate the substitution of labor and/or industrial inputs for land. This may occur through increased recycling of soil fertility by more labor-intensive conservation systems; through use of chemical fertilizers; and through husbandry practices, management systems, and inputs (that is, insecticides) which permit an optimum yield response.

Historically there has been a close association between advances in output per unit of land area and advances in biological technology and between advances in output per worker and advances in mechanical technology. The construction of an induced development model involves an explanation of the mechanism by which a society chooses an optimum path of technological change in agriculture.

INDUCED INNOVATION IN THE PRIVATE SECTOR

There is a substantial body of literature on the ''theory of induced innovation.''[2] Much of this literature focuses on the choice of available technology by the individual firm. There is also a substantial body of literature on how changes in factor prices over time or differences in factor prices among countries influence the nature of invention. This discussion has been conducted entirely within the framework of the theory of the firm. A major controversy has centered around the issue of the existence of a mechanism by which changes or differences in factor prices affect the inventive activity or the innovative behavior of firms.

It had generally been accepted, at least since the publication of *The Theory of Wages* by J. R. Hicks (1932, 124–25), that changes or differences in the relative prices of factors of production could influence the direction of invention or innovation.[3] There have also been arguments raised by W. E. G. Salter (1960, 43–44) and others (Ahmad 1966; Fellner 1961; Kennedy 1964; Samuelson 1965) against Hicks's theory of induced innovation. The arguments run somewhat as follows: firms are motivated to save total cost for a given output; at competitive equilibrium, each factor is being paid its marginal value product; therefore, all factors are equally expensive to firms; hence, there is no incentive for competitive firms to search for techniques to save a particular factor.

The difference between our perspective and Salter's is partly due to a difference in the definition of the production function. Salter defined the production function to embrace all possible designs conceivable by existing scientific knowledge and called the choice among these designs "factor substitution" instead of "technical change" (1960, 14–16). Salter admits, however, that "relative factor prices are in the nature of signposts representing broad influences that determine the way technological knowledge is applied to production" (ibid., 16). If we accept Salter's definition, the allocation of resources to the development of high-yielding and fertilizer-responsive rice varieties adaptable to the ecological conditions of South and Southeast Asia, which are comparable to the improved varieties developed earlier in Japan and Taiwan, cannot be considered as a technical change. Rather, it is viewed as an application of existing technological knowledge (breeding techniques, plant-type concepts, and so on) to production.

Although we do not deny the case for Salter's definition, it is clearly not very useful in attempting to understand the process by which new technical alternatives become available. We regard technical change as any change in production coefficients resulting from purposeful resource-using activity directed to the development of new knowledge embodied in designs, materials, or organizations. In terms of this definition, it is entirely rational for competitive firms to allocate funds to develop a technology that facilitates the substitution of increasingly less expensive factors for more expensive factors. Using the above definition, Syed Ahmad (1960) has shown that the Hicksian theory of market-induced innovation can be defended with a rather reasonable assumption on the possibility of alternative innovations.[4]

We illustrate Ahmad's argument with the aid of figure 1. Suppose that at a point in time a firm is operating at a competitive equilibrium, A or B, depending on the prevailing factor-price ratio, p or m, for an isoquant, u_0, producing a given output; and this firm perceives multiple alternative innovations represented by isoquants, u_1, u_1', . . . , producing the same output in such a way as to be enveloped by U, a concave innovation possibility curve or meta-production function which can be developed by the same amount of research expenditure. In order to minimize total cost for given output and given research expenditure, innovative efforts of this firm will be directed towards developing Y-saving technology (u_1) or X-saving technology (u_1'), depending on the prevailing factor-price ratio, p (parallel to PP) or m (parallel to MM and MM'). If a firm facing a price ratio m develops an X-saving technology (u_1'), it can obtain an additional

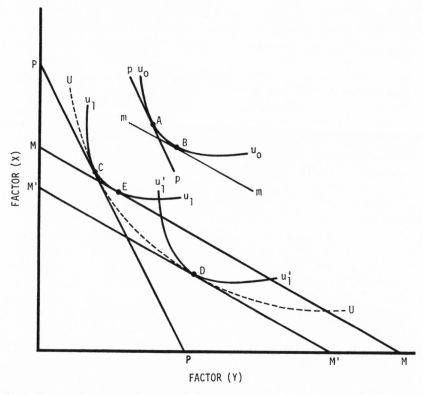

Fig.1. Factor Prices and Induced Technical Change

gain represented by the distance between M and M' compared with the case that develops a Y-saving technology (u_1). In this framework it is clear that if X becomes more expensive relative to Y over time, in any economy the innovative efforts of entrepreneurs will be directed towards developing a more X-saving and Y-using technology compared with the contrary case. Also, in a country in which X is more expensive relative to Y than in another country, innovative efforts in the country will be more directed towards X-saving and Y-using than in the other country. In this formulation the expectation of relative price change, which is central to William Fellner's theory of induced innovation, is not necessary, although expectations may work as a powerful reinforcing agent in directing technical effort.[5]

The role of changing relative factor prices in inducing a continuous sequence of non-neutral biological and mechanical innovations along the iso-product surface of a meta-production function is further illustrated in figure 2. U represents the land-labor isoquant of the meta-production function, which is the envelope of less elastic isoquants such as u_0 and u_1, corresponding to different types of

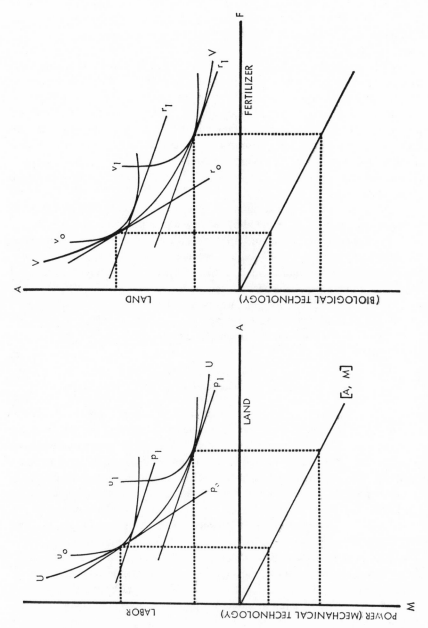

Fig.2. Factor Prices and Induced Mechanical and Biological Innovation

63

machinery or technology. A certain technology represented by u_0 (for example, a reaper) is created when a price ratio, p_0, prevails a certain length of time. When the price ratio changes from p_0 to p_1, another technology, represented by u_1 (for example, a combine), is induced in the long run, which gives the minimum cost of production for p_1.

The new technology represented by u_1, which enables enlargement of the area operated per worker, generally corresponds to higher intensity of power per worker. This implies the complementary relationship between land and power, which may be drawn as a line representing a certain combination of land and power, [A, M]. In this simplified presentation, mechanical innovation is conceived as the substitution of a combination of land and power, [A, M], for labor (L) in response to a change in wage relative to an index of labor and machinery prices, although, of course, in actual practice land and power are substitutable to some extent.

In the same context, the relation between the fertilizer-land price ratio and biological innovations represented by the development of crop varieties that are more responsive to application of fertilizers is illustrated in figure 2. V represents the land-fertilizer isoquant of the meta-production function, which is the envelope of less elastic isoquants such as v_0 and v_1, corresponding to varieties of different fertilizer responsiveness. A decline in the price of fertilizer relative to the price of land from r_0 to r_1 creates an incentive for farmers to adopt crop varieties described by isoquants to the right of v_0 and for private seed companies and public research institutions to develop and market such new fertilizer-responsive varieties.

INDUCED INNOVATION IN THE PUBLIC SECTOR

Innovative behavior in the public sector has largely been ignored in the literature on induced innovation. There is no theory of induced innovation in the public sector.[6] This is a particularly critical limitation in attempting to understand the process of scientific and technical innovation in agricultural development. In most countries that have been successful in achieving rapid rates of technical progress in agriculture, "socialization" of agricultural research has been deliberately employed as an instrument of modernization in agriculture.

Our view of the mechanism of "induced innovation" in the public sector agricultural research is similar to the Hicksian theory of induced innovation in the private sector. A major extension of the traditional argument is that we base the innovation inducement mechanism not only on the response to changes in the market prices of profit maximizing firms but also on the response of research scientists and administrators in public institutions to resource endowments and economic change.

We hypothesize that technical change is guided along an efficient path by price signals in the market, provided that the prices efficiently reflect changes in the demand and supply of products and factors and that there exists effective interaction among farmers, public research institutions, and private agricultural supply firms. If the demand for agricultural products increases, due to the growth in population and income, prices of the inputs for which the supply is inelastic will be raised relative to the prices of inputs for which the supply is elastic. Likewise, if the supply of particular inputs shifts to the right faster than that of others, the prices of these inputs will decline relative to the prices of other factors of production.

In consequence, technical innovations that save the factors characterized by an inelastic supply, or by slower shifts in supply, become relatively more profitable for agricultural producers. Farmers are induced by shifts in relative prices to search for technical alternatives that save the increasingly scarce factors of production. They press the public research institutions to develop the new technology and demand that agricultural supply firms supply modern technical inputs that substitute for the more scarce factors. Perceptive scientists and science administrators respond by making available new technical possibilities and new inputs that enable farmers profitably to substitute the increasingly abundant factors for increasingly scarce factors, thereby guiding the demand of farmers for unit cost reduction in a socially optimum direction.

The dialectic interaction among farmers and research scientists and administrators is likely to be most effective when farmers are organized into politically effective local and regional farm "bureaus" or farmers' associations. The response of the public sector research and extension programs to farmers' demand is likely to be greatest when the agricultural research system is highly decentralized, as in the United States. In the United States, for example, each of the state agricultural experiment stations has tended to view its function, at least in part, as to maintain the competitive position of agriculture in its state relative to agriculture in other states. Similarly, national policy makers may regard investment in agricultural research as an investment designed to maintain the country's competitive position in world markets or to improve the economic viability of the agricultural sector producing import substitutes. Given effective farmer organizations and a mission- or client-oriented experiment station system, the competitive model of firm behavior illustrated in figures 1 and 2 can be usefully extended to explain the response of experiment station administrators and research scientists to economic opportunities.

In this public-sector-induced innovation model, the response of research scientists and administrators represents the critical link in the inducement mechanism. The model does not imply that it is necessary for individual scientists or research administrators in public institutions to respond consciously to market prices, or directly to farmers' demands for research results, in the selection of research objectives. They may, in fact, be motivated primarily by a drive for professional achievement and recognition (Niskanen 1968). Or they may, in the Rosenberg terminology, view themselves as responding to an "obvious and compelling need'' to remove the constraints on growth of production or on factor supplies.[7] It is only necessary that there exist an effective incentive mechanism to reward the scientists or administrators, materially or by prestige, for their contributions to the solution of significant problems in the society.[8] Under these conditions, it seems reasonable to hypothesize that the scientists and administrators of public sector research programs do respond to the needs of society in an attempt to direct the results of their activity to public purpose. Furthermore, we hypothesize that secular changes in relative factor and product prices convey much of the information regarding the relative priorities that society places on the goals of research.

The response in the public research sector is not limited to the field of applied science. Scientists trying to solve practical problems often consult with or ask cooperation of those working in more basic fields. If the basic scientists respond to the requests of the applied researchers, they are in effect responding to the needs of society. It is not uncommon that major breakthroughs in basic science are created through the process of solving the problems raised by research workers in the more applied fields.[9] It appears reasonable, therefore, to hypothesize as a result of the interactions among the basic and applied sciences and the process by which public funds are allocated to research that basic research tends to be directed also towards easing the limitations on agricultural production imposed by relatively scarce factors.

We do not argue, however, that technical change in agriculture is wholly of an induced character. There is a supply (exogenous) dimension to the process as well as a demand (endogenous) dimension. Technical change in agriculture reflects, in addition to the effects of resource endowments and growth in demand, the progress of general science and technology. Progress in general science (or scientific innovation) that lowers the "cost" of technical and entrepreneurial innovations may have influences on technical change in agriculture unrelated to changes in factor proportions and product demand (Nelson 1959; Schmookler 1966). Similarly, advances in science and technology in the developed countries, in response to their own resource endowments, may result in a bias in the innovation possibility curves facing the developing countries. Even in these cases, the rate of adoption and the impact on productivity of autonomous or exogenous changes in technology will be strongly influenced by the conditions of resource supply and product demand, as these forces are reflected through factor and product markets.

Thus, the classical problem of resource allocation, which was rejected as an adequate basis for agricultural productivity and output growth in the high-payoff input model,[10] in this context is treated as central to the agricultural development process. Under conditions of static technology, improvements in resource allocation represent a weak source of economic growth. The efficient allocation of resources to open up new sources of growth is, however, essential to the agricultural development process.

INSTITUTIONAL INNOVATION

Extension of the theory of "induced innovation" to explain the behavior of public research institutions represents an essential link in the construction of a theory of induced development. In the induced development model, advances in mechanical and biological technology respond to changing relative prices of factors and to changes in the prices of factors relative to products to ease the constraints on growth imposed by inelastic supplies of land or labor. Neither this process nor its impact is confined to the agricultural sector. Changes in relative prices in any sector of the economy act to induce innovative activity, not only by private producers but also by scientists in public institutions, in order to reduce the constraints imposed by those factors of production that are relatively scarce.

We further hypothesize that the institutions that govern the use of technology or the "mode" of production can also be induced to change in order to enable both individuals and society to take fuller advantage of new technical opportunities under favorable market conditions.[11] The Second Enclosure Movement in England represents a classical illustration. The issuance of the Enclosure Bill facilitated the conversion of communal pasture and farmland into single, private farm units, thus encouraging the introduction of an integrated crop-livestock "new husbandry" system. The Enclosure Acts can be viewed as an institutional innovation designed to exploit the new technical opportunities opened up by innovations in crop rotation, utilizing the new fodder crops (turnip and clover), in response to the rising food prices.

A major source of institutional change has been an effort by society to internalize the benefits of innovative activity to provide economic incentives for productivity increase. In some cases, institutional innovations have involved the reorganization of property rights in order to internalize the higher income streams resulting from the innovations. The modernization of land tenure relationships, involving a shift from share tenure to lease tenure and owner-operator systems of cultivation in much of Western agriculture, can be explained, in part, as a shift in property rights designed to internalize the gains of entrepreneurial innovation by individual farmers.[12]

Where internalization of the gains of innovative activity are more difficult to achieve, institutional innovations involving public sector activity become essential. The socialization of much of agricultural research, particularly the research leading to advances in biological technology, represents an example of a public sector institutional innovation designed to realize for society the potential gains from advances in agricultural technology. This institutional innovation originated in Germany and was transplanted and applied on a larger scale in the United States and Japan.

Both Schultz (1968) and Kazushi Ohkawa (1969) have argued that institutional reform is appropriately viewed as a response to the new opportunities for the productive use of resources opened up by advances in technology.[13] Our view, and the view of Ohkawa and Schultz, reduces to the hypothesis that institutional innovations occur because it appears profitable for individuals or groups in society to undertake the costs. It is unlikely that institutional change will prove viable unless the benefits to society exceed the cost. Changes in market prices and technological opportunities introduce disequilibrium in existing institutional arrangements by creating profitable new opportunities for the institutional innovations.

Profitable opportunities, however, do not necessarily lead to immediate institutional innovations. Usually the gains and losses from technical and institutional change are not distributed neutrally. There are, typically, vested interests that stand to lose and that oppose change. There are limits on the extent to which group behavior can be mobilized to achieve common or group interests (Olson 1968). The process of transforming institutions in response to technical and economic opportunities generally involves time lags, social and political stress, and in some cases disruption of social and political order. Economic growth ultimately depends on the flexibility and efficiency of society in transforming itself in response to technical and economic opportunities.

AGRICULTURAL DEVELOPMENT STRATEGY

The induced innovation model outlined above does not possess formal elegance. It is partial in that it is primarily concerned with production and productivity. Yet it has added significantly to our power to interpret the process of agricultural development.

Research that we have reported elsewhere indicates that the enormous changes in factor proportions that have occurred in the process of agricultural growth in the United States and Japan are explainable very largely in terms of changes in factor-price ratios (Hayami and Ruttan 1970, 1971). When we relate the results of the statistical analysis to historical knowledge of advances in agricultural technology, we conclude that the observed changes in input mixes have occurred as the result of a process of dynamic factor substitution along a meta-production function, associated with changes in the production surface, induced primarily by changes in relative factor prices. Preliminary results of the analysis of historical patterns of technical change in German agriculture (by Adolph Weber); in Denmark, Great Britain, and France (by William Wade); and in Argentina (by Alain de Janvry) add additional support to the utility of the induced innovation model in interpreting historical patterns of technological change and agricultural development.

The question remains, however, as to whether the induced development model represents a useful guide to modern agricultural development strategy. In responding to this concern two issues seem particularly relevant.

First, we would like to make it perfectly clear that in our view the induced development model, in which technical change and institutional change are treated as endogenous to the development process, does not imply that agricultural development can be left to an "invisible hand" that directs either technology or the total development process along an "efficient" path determined by "original" resource endowments.

We do argue that the policies that a country adopts with respect to the allocation of resources to technical and institutional innovation, to the capacity to produce technical inputs for agriculture, to the linkages between the agricultural and industrial sectors in factor and product markets, and to the organization of the crop and livestock production sectors must be consistent with national (or regional) resource endowments if they are to lead to an "efficient" growth path. Conversely, failure to achieve such consistency can sharply increase the real costs, or abort the possibility, of achieving sustained economic growth in the agricultural sector.

If the induced development model is valid—if alternative paths of technical change and productivity growth are available to developing countries—the issue of how to organize and manage the development and allocation of scientific and technical resources becomes the single most critical factor in the agricultural development process. It is not sufficient simply to build new agricultural research stations. In many developing countries existing research facilities are not employed at full capacity because they are staffed with research workers with limited scientific and technical training; because of inadequate financial, logistical, and administrative support; because of isolation from the main currents of scientific and technical innovation; and because of failure to develop a research strategy that relates research activity to the potential economic value of the new knowledge it is designed to generate.

The appropriate allocation of effort between the public and private sectors also becomes of major significance in view of the extension of the induced development model to incorporate innovative activity in the public sector. It is clear that during the early stages of development the socialization of much of biological research in agriculture is essential if the potential gains from biological technology are to be realized. The potential gains from public sector investment in other areas of the institutional infrastructure characterized by substantial spillover effects are also large. One of the most important of these areas for public investment is the modernization of the marketing system through the establishment of the information and communication linkages necessary for the efficient functioning of factor and product markets.[14]

In most developing countries the market systems are relatively underdeveloped, both technically and institutionally. A major challenge facing these countries in their planning is the development of a well-articulated marketing system capable of accurately reflecting the effects of changes in supply, demand, and production relationships. An important element in the development of a more efficient marketing system is the removal of the rigidities and distortions resulting from government policy itself—including the maintenance of overvalued currencies, artificially low rates of interest, and unfavorable factor and product price policies for agriculture (Myint 1968).

The criteria specified above for public sector investment or intervention also imply a continuous reallocation of functions among public and private sector institutions. As institutions capable of internalizing a large share of the gains of innovative activity are developed, it may become possible to transfer activities—the production of new crop varieties, for example—to the private sector and to reallocate public resources to other high-payoff areas. Many governments are presently devoting substantial resources to areas of relatively low productivity—in efforts to reform the organization of credit and product markets, for example—while failing to invest the resources necessary to produce accurate and timely market information, establish meaningful market grades and standards, and establish the physical infrastructure necessary to induce technical and logistical efficiency in the performance of marketing functions (Ruttan 1969).

A second issue is whether, under modern conditions, the forces associated with the international transfer of agricultural technology are so dominant as to vitiate the induced development model as a guide to agricultural development strategy. It might be argued, for example, that the dominance of the developed countries in science and technology raises the cost of, or even precludes the possibility of the invention of, location-specific biological and mechanical technologies adapted to the resource endowments of a particular country or region.

This argument has been made primarily with reference to diffusion of mechanical technology from the developed to the developing countries. It is argued that the pattern of organization of agricultural production adopted by the more developed countries—dominated by large-scale mechanized systems of production in both the socialist and nonsocialist economies—precludes an effective role for an agricultural system based on small-scale commercial or semicommercial farm production units (Owen 1969, 1971).[15]

We find this argument unconvincing. Rapid diffusion of imported mechanical technology in areas characterized by small farms and low wages in agriculture tends to be induced by inefficient price, exchange rate, and credit policies which substantially distort the relative costs of mechanical power relative to labor and other material inputs. Nural Islam reports, for example, that as a result of such policies, the real cost of tractors in West Pakistan was substantially below the cost in the United States (1971). The preliminary findings of work by John Sanders in Latin America also stress the role of market distortions in inducing mechanization.

We are also impressed by the history of agricultural mechanization in Japan and more recently in Taiwan. Both countries have been relatively successful in following a strategy of mechanical innovation designed to adapt the size of the tractor and other farm machinery rather than to modify the size of the agricultural production unit to make it compatible with the size of imported machinery.[16]

We do insist that failure to effectively institutionalize public sector agricultural research can result in serious distortion of the pattern of technological change and resource use. The homogeneity of agricultural products and the relatively small size of the farm firm, even in the Western and socialist economies, make it difficult for the individual agricultural producer to either bear the research costs or capture a significant share of the gains from scientific or technological innovation. Mechanical technology, however, has been much more responsive than biological technology to the inducement mechanism as it operates in the private sector. In biological technology, typified by the breeding of new plant varieties or the improvement of cultural practices, it is difficult for the innovating firm to capture more than a small share of the increased income stream resulting from the innovation.

Failure to balance the effectiveness of the private sector in responding to inducements for advances in mechanical technology, and in those areas of biological technology in which advances in knowledge can be embodied in proprietary products, with institutional innovation capable of providing an equally effective response to inducements for advances in biological technology leads to a bias in the productivity growth path that is inconsistent with relative factor endowments. It seems reasonable to hypothesize that failure to invest in public sector experiment station capacity is one of the factors responsible in some developing countries for the unbalanced adoption of mechanical, relative to biological, technology. Failure to develop adequate public sector research institutions has also been partially responsible, in some countries, for the almost exclusive concentration of research expenditures on the plantation crops and for concentration on the production of certain export crops—such as sugar and bananas—in the plantation sector.

The perspective outlined in this paper can be summarized as follows: an essential condition for success in achieving sustained growth in agricultural productivity is the capacity to generate an ecologically adapted and economically viable agricultural technology in each country or development region. Successful achievement of continued productivity growth over time involves a dynamic process of adjustment to original resource endowments and to resource accumulation during the process of historical development. It also involves an adaptive response on the part of cultural, political, and economic institutions in order to realize the growth potential opened up by new technical alternatives. The "induced development model" attempts to make more explicit the process by which technical and institutional changes are induced through the responses of farmers, agribusiness entrepreneurs, scientists, and public administrators to resource endowments and to changes in the supply and demand of factors and products.

NOTES

1. The distinction made here between "mechanical" and "biological" technology has also been employed by Heady (1949). It is similar to the distinction between "laboresque" and "landesque" capital employed by Sen (1959). In a more recent article Kaneda employs the terms "mechanical-engineering" and "biological-chemical" (1969).

2. The term "innovation" employed here embraces the entire range of processes resulting in the emergence of novelty in science, technology, industrial management, and economic organization rather than the narrow Schumpeterian definition. Schumpeter insisted that innovation was economically and sociologically distinct from invention and scientific discovery. He rejected the idea that innovation is dependent on invention or advances in science. This distinction has become increasingly artificial. See, for example, Solo 1951; Ruttan 1959; and Hohenberg 1967. Our view is similar to that of Hohenberg. He defines technical effort as the product of purposive resource-using activity directed to the production of economically useful knowledge: "Technical effort is a necessary part of any firm activity, and is only in part separable from production itself. Traditionally it is part of the entrepreneur's job to provide knowledge to organize the factors of production in an optimum way, to adjust to market changes, and to seek improved methods. Technical effort is thus subsumed under entrepreneurship" (1967, 61).

3. See also the review of thought on this issue in Ahmad 1966.

4. See also discussions by Fellner (1967) and Ahmad (1967a), and by Kennedy (1967) and Ahmad (1967b).

5. The above theory is based on the restrictive assumption that there exists a concave innovation possibility curve (U) that can be perceived by entrepreneurs. This is not as strong a restrictive assumption as it may at first appear. The innovation possibility curve does not need to be of a smooth, well-behaved shape, as drawn in figure 1. The whole argument holds equally well for the case of two distinct alternatives. It seems reasonable to hypothesize that entrepreneurs can perceive alternative innovation possibilities for a given research and development expenditure through consultation with staff scientists and engineers or through the suggestions of inventors.

6. There is a growing literature on public research policy (see Nelson, Peck, and Kalachek 1967). These authors view public sector research activities as having risen from three considerations: (*a*) fields where the public interest is believed to transcend private incentives (as in health and aviation); (*b*) industries where the individual firm is too small to capture benefits from research (agriculture and housing); and (*c*) broad support for basic research and science education (pp. 151–211). For a review of thought with respect to resource allocation in agriculture see Fishel 1971.

7. Rosenberg has suggested a theory of induced technical change based on "obvious and compelling need" to overcome the constraints on growth instead of relative factor scarcity and factor relative prices (1969). The Rosenberg model is consistent with the model suggested here, since his "obvious and compelling need" is reflected in the market through relative factor prices. C. Peter Timmer has pointed out that in a linear programming sense the constraints that give rise to the "obvious and compelling need" for technical innovation in the Rosenberg model represent the "dual" of the factor prices used in our model (1970). For further discussion of the relationships between Rosenberg's approach and that outlined in this section see Hayami and Ruttan 1973.

8. Incentive is a major issue in many developing economies. In spite of limited scientific and technical manpower, many countries have not succeeded in developing a system of economic and professional rewards that permits them to have access to or make effective use of the resources of scientific and technical manpower that are potentially available.

9. The symbiotic relationship between basic and applied research can be illustrated by the relation between work at the International Rice Research Institute in (*a*) genetics and plant physiology and (*b*) plant breeding. The geneticist and the physiologist are involved in research designed to advance understanding of the physiological processes by which plant nutrients are transformed into grain yield and of the genetic mechanisms or processes involved in the transmission from parents to progenies of the physiological characteristics of the rice plant that affect grain yield. The rice breeders utilize this knowledge from genetics and plant physiology in the design of crosses and the selection of plants with the desired growth characteristics, agronomic traits, and nutritional value. The work in plant physiology and genetics is responsive to the need of the plant breeder for advances in knowledge related to the mission of breeding more productive varieties of rice.

10. For a description of the high-payoff input model see chapter 2 in this volume.—ED.

11. At this point we share the Marxian perspective on the relationship between technological

change and institutional development, though we do not accept the Marxian perspective regarding the monolithic sequences of evolution based on clear-cut class conflicts. For two recent attempts to develop broad historical generalizations regarding the relationship between insitutions and economic forces see Hicks 1969; and North and Thomas 1970.

12. For additional examples see Davis and North 1970.

13. See also North and Thomas 1970.

14. Hayami and Peterson (1972) show that the return to investment in improvements in market information is comparable to the returns that have been estimated for high-payoff research areas such as hybrid corn and poultry.

15. Owen argues that differentiation of a rural commercial sector from the rural subsistence sector is the first step towards development of relevant agricultural development policies. The "optimum sized commercial farms will comprise the maximum amount of land that can be farmed at a profit by an appropriate set of labor where the latter uses a relatively advanced level of technology for the particular farming area. . . . the optimum sized subsistence farm plot is one that comprises the minimum amount of land that is necessary to assure to the household concerned the minimum acceptable standard of subsistence living'' (1969, 107).

16. This development is reviewed in Hayami and Ruttan 1971.

REFERENCES

Ahmad, Syed. 1966. "On the Theory of Induced Invention." *Economic Journal* 76(302): 344–57.
———. 1967a. "Reply to Professor Fellner." *Economic Journal* 77(307): 664–65.
———. 1967b. "A Rejoinder to Professor Kennedy." *Economic Journal* 77(308): 960–63.
Davis, Lance, and Douglass North. 1970. "Institutional Change and American Economic Growth: A First Step Towards a Theory of Institutional Innovation." *Journal of Economic History* 30(1): 131–49.
Fellner, William. 1961. "Two Propositions in the Theory of Induced Innovations." *Economic Journal* 71(282): 305–8.
———. 1967. "Comment on the Induced Bias." *Economic Journal* 77(307): 662–64.
Fishel, W. L., ed. 1971. *Resource Allocation in Agricultural Development*. Minneapolis.
Hayami, Yujiro, and Willis Peterson. 1972. "Social Returns to Public Information Services: Statistical Reporting of U.S. Farm Commodities." *American Economic Review* 62(1): 119–30.
Hayami, Yujiro, and V. W. Ruttan. 1970. "Factor Prices and Technical Change in Agricultural Development: The United States and Japan, 1880–1960." *Journal of Political Economy* 78(5): 1115–41.
———. 1971. *Agricultural Development: An International Perspective*. Baltimore.
———. 1973. "Professor Rosenberg and the Direction of Technical Change: A Comment." *Economic Development and Cultural Change* 21(2): 352–55.
Heady, E. O. 1949. "Basic Economic and Welfare Aspects of Farm Technological Advance." *Journal of Farm Economics* 31(2): 293–316.
Hicks, J. R. 1932. *The Theory of Wages*. London.
———. 1969. *A Theory of Economic History*. London.
Hohenberg, P. M. 1967. *Chemicals in Western Europe: 1850–1914*. Chicago.
Islam, Nural. 1971. "Agricultural Growth in Pakistan: Problems and Policies." Paper presented at the Conference on Agriculture and Economic Development, Japan Economic Research Center, Tokyo, 6–10 September.
Kaneda, Hiromitsu. 1969. "Economic Implications of the 'Green Revolution' and the Strategy of Agricultural Development in West Pakistan." *Pakistan Development Review* 9(2): 111–43.
Kennedy, Charles. 1964. "Induced Bias in Innovation and the Theory of Distribution." *Economic Journal* 74(295): 541–47.

————. 1967. "On the Theory of Induced Invention—A Reply." *Economic Journal* 77(308): 958–60.

Myint, U. Hla. 1968. "Market Mechanisms and Planning—The Functional Aspect." In *The Structure and Development of Asian Economies*. Tokyo.

Nelson, R. R. 1959. "The Economics of Invention: A Survey of the Literature." *Journal of Business* 33 (April): 101–27.

Nelson, R. R., M. J. Peck, and E. D. Kalachek. 1967. *Technology, Economic Growth and Public Policy*. Washington, D.C.

Niskanen, W. A. 1968. "The Peculiar Economics of Bureaucracy." *American Economic Review* 58(2): 293–305.

North, Douglass C., and R. P. Thomas. 1970. "An Economic Theory of the Growth of the Western World." *Economic History Review*, 2d ser., 23(1): 1–17.

Ohkawa, Kazushi. 1969. "Policy Implications of the Asian Agricultural Survey—Personal Notes." In *Regional Seminar on Agriculture: Paper and Proceedings*. Makati, Philippines.

Olson, Mancur, Jr. 1968. *The Logic of Collective Action: Public Goods and the Theory of Groups*. New York.

Owen, W. F. 1969. "Structural Planning in Densely Populated Countries: An Introduction with Applications to Indonesia." *Malayan Economic Review* 14(1): 97–114.

————. 1971. *Two Rural Sectors: Their Characteristics and Roles in the Development Process*. Indiana University International Developmental Research Center Occasional Paper 1. Bloomington, Ind.

Rosenberg, Nathan. 1969. "The Direction of Technological Change: Inducement Mechanisms and Focusing Devices." *Economic Development and Cultural Change* 18(1, pt. 1): 1–24.

Ruttan, V. W. 1959. "Usher and Schumpeter on Invention, Innovation and Technological Change." *Quarterly Journal of Economics* 73(4): 596–606.

————. 1969. "Agricultural Product and Factor Markets in Southeast Asia." *Economic Development and Cultural Change* 17(4): 501–19.

Salter, W. E. G. 1960. *Productivity and Technical Change*. Cambridge.

Samuelson, P. A. 1965. "A Theory of Induced Innovation along Kennedy, Weisacker [Weizsacker] Lines." *Review of Economics and Statistics* 47(4): 343–56.

Schmookler, Jacob. 1966. *Invention and Economic Growth*. Cambridge, Mass.

Schultz, T. W. 1968. "Institutions and the Rising Economic Value of Man." *American Journal of Agricultural Economics* 50(5): 1113–22.

Sen, A. K. 1959. "The Choice of Agricultural Techniques in Underdeveloped Countries." *Economic Development and Cultural Change* 7(3, pt. 1): 279–85.

Solo, Carolyn Shaw. 1951. "Innovation in the Capitalist Process: A Critique of the Schumpeterian Theory." *Quarterly Journal of Economics* 65(3): 417–28.

Timmer, C. P. 1970. Personal communication, 9 October.

5

Induced Innovation Model of Agricultural Development: Comment

GEORGE L. BECKFORD

 In commenting on this paper by Ruttan and Hayami, I wish, first, to make some rather general remarks. Second, I will consider certain aspects of their paper which I find unsatisfactory; and, third, I wish to focus attention on certain critical questions which they largely ignore.

GENERAL REMARKS

Every contribution to the theory of agricultural development is to be welcomed, if only because this field of enquiry has not been ploughed sufficiently. As Ruttan and Hayami themselves indicate, the concern of economists has been more in the direction of examining the interaction of agriculture and overall economic growth (structural transformation)[1] than with the process of agricultural development per se. Yet we know that agricultural development is perhaps the most critical problem facing underdeveloped countries today. The bulk of the population in these countries depends on agriculture for its livelihood; so the welfare of millions of people is at stake. And, of course, we now know that overall economic advance by these countries cannot proceed without substantial expansion of agricultural output and improvements in productivity. From this general point of view, then, the Ruttan-Hayami paper can be regarded as a noteworthy contrbution.

In order adequately to assess the value of this contribution, however, we need to say something about the general usefulness of models in economic analysis.

GEORGE L. BECKFORD is professor of economics, University of the West Indies.

Originally titled "Strategies for Agricultural Development: Comment." Reprinted from *Food Research Institute Studies in Agricultural Economics, Trade, and Development* 9, no. 2 (1972): 149–54, with minor editorial revisions, by permission of the Food Research Institute, Stanford University, and the author.

All models are by definition an abstraction of what obtains in the real world. Simplifying assumptions have to be made to avoid the complexities of the real world. Ultimately, the critical factor that determines the usefulness of the model is whether or not what is left out is fundamental in understanding what goes on. We can take one of a number of approaches in assessing a particular model. One such approach is simply to check its internal consistency. Another is to see how well it explains and/or predicts what happens in the real world. This depends ultimately on whether the assumptions of the model correctly represent given situations. It is this second approach that I wish to take in the present exercise.

AN ASSESSMENT OF THE RUTTAN-HAYAMI CONTRIBUTION

The first point for us to note is that contrary to the title of their paper, Ruttan and Hayami are not concerned with agricultural development at all. They are essentially concerned with the growth of agricultural output and associated improvements in agricultural productivity. Their model is, therefore, more appropriately a model of agricultural *growth* rather than of agricultural *development*. This point is, to my mind, one of very great substance. Agricultural development is essentially a study of the process by which the material welfare of the rural population of a country is improved consistently over time. In this context, the growth of agricultural output and productivity may be a necessary, though certainly not a sufficient, condition. Indeed, there are numerous instances (in the past, as well as at present) in which substantial growth of agricultural output is accompanied by no change in the material welfare of the majority of people involved in the process of that growth.[2] In short, we must recognize that there is always a strong possibility of the phenomenon of "growth without development."

Later, I wish to return to some questions relating to development, but for now let me proceed to look at Ruttan and Hayami on their own ground. Basically, their induced development model is the conventional resource allocation model within the general framework of the traditional theory of the firm. Critical to the model is the existence of competitive conditions along with profit-maximizing behavior of decision makers. In such situations, the following endogenous sequence may be expected: resource availability determines relative factor prices, and the choice of techniques by producers is guided by the structure of factor prices. Over time, as changes in relative prices occur, technology is adjusted to maximize the use of relatively cheap factors. A further consideration, then, is the degree of technical substitutability between factors of production.

For empirical verification of the model, the authors checked the development experiences of a number of countries where the development process was played out largely in the eighteenth and nineteenth centuries. I want to suggest that the economic and social situation of underdeveloped countries today is significantly different from those that obtained for present-day advanced countries in the nineteenth century. The social order that existed in the latter countries was of a kind that permitted the emergence of economic institutions and behavioral patterns that fit the neo-classical marginalist framework of economic analysis. My contention is that such is not the case in underdeveloped countries today. These economies are for the most part characterized by imperfect market conditions and social institutional arrangements that create artificial rigidities in the flow of factor supplies and inflexibilities in the patterns of resource use.[3] Furthermore, the openness of most underdeveloped economies exposes them to exogenous influences of a kind that serves to shatter the neat links between factor endowments and factor prices and between factor prices and technological change which are central to the induced development model.

Let me quickly list some of the problems that concern me most in the Ruttan-Hayami analysis and then briefly discuss each of them.

(1) The profit maximization assumption,
(2) The association between resource endowments and the structure of relative factor prices,
(3) The aggregation problem in moving the analysis from the firm to the industry,
(4) Resource availability in the open economy,
(5) The assumption about public sector responses, and
(6) The superficiality of the model of induced institutional innovation.

Farmers in underdeveloped countries do not consistently seek to maximize profits. Profits from farm production are only one element (though a major one) in the matrix of their objectives. Considerations such as family security, social status, and risk minimization all enter into the picture, depending on the particular institutional environment.

The one-to-one association between the society's factor endowments and relative factor prices ignores two fundamental characteristics of underdeveloped agriculture. One is the marked divergencies between private and social costs and benefits that are typical of most situations; and the other is duality in the structure of some underdeveloped agricultural economies that distorts the relative factor prices faced by different producers within the same economy. The divergencies between private and social costs and benefits are very briefly organized by Ruttan and Hayami, but the question of duality entirely escapes notice. We find, for example, that in plantation economies, labor may be relatively cheap to peasants but considerably more expensive to plantations, while land may be relatively cheap to plantations but relatively expensive to peasants. In such a situation, it seems to me that there is no uniquely efficient path of technological change for the society as a whole unless of course some exogenous institutional reform to eliminate duality occurs.

On the aggregation problem of moving from the firm to the industry, what bothers me is that the Ruttan-Hayami model seems to imply that what is good for the firm is good for the industry. Let me be more specific. Given the inelastic demand for farm products, expansion of output for the individual farm firm produces different results from the expansion of output for *all* farm firms. What this implies is that there are obviously leads and lags which the induced development model does not account for in its one-to-one firm industry adjustment process.

In the modern world economy, trade is only one aspect of the characteristic of openness. Much more important is the dependence of underdeveloped countries on the capital, technology, and management resources of the economically advanced countries. In this connection, I cannot accept the cavalier manner in which Ruttan and Hayami dismiss the influence of "forces associated with the international transfer of agricultural technology." Let me take a futuristic example. Desalination of sea water is technically feasible. I suggest that its economic feasibility is likely to emerge from the research efforts of the more advanced countries. The effect of this will be to drastically alter the resources endowments of arid areas of underdeveloped countries. To be fair to Ruttan and Hayami, they admit this kind of event by saying that they do not rule out exogenous technical change. The question is whether exogenous technical change will be more important for underdeveloped agriculture in the 1970s than endogenous change. I think that given the present institutional arrangement of the world economy, exogenous factors will be more important than the endogenous for agriculture in underdeveloped countries.

A highly decentralized system of agricultural administration and the existence of strong farmers' organizations are critical for generating effective public response. But in most underdeveloped countries, local government is poorly developed and farmers' organizations are either absent or weak. In these circumstances, the kind of public sector response predicted by the induced development model will hardly be in evidence.

I am most concerned with the superficiality of the model of induced institutional innovation. It is *totally* impossible to explain institutional reform in purely economic terms, as Ruttan and Hayami have tried to do. They admit themselves that institutional change is not neutral. If that is so, as indeed it is, then we need to examine the social and political (not to mention the psychological and cultural) dimensions of the process of institutional change. And, of course, the exogenous factors are of critical importance here. We need only call to witness the "American Revolution" vs. the problem of the United States South and of black people in the United States today. Any model of induced institutional reform must explain how the existing institutional arrangements affect different groups in the society, how change will affect these groups, and the balance of power between the groups. This calls for a political, social, and psychological analysis. The simplistic Ruttan-Hayami model cannot possibly cope with these problems. A further consideration is the obvious relationship between institutional structure and technological change. Certain patterns of social organization simply do not contribute to the kind of social inputs (education and research, for example) that are critical to the process of change envisaged in the induced development model (see, for example, Nicholls 1960).

This brings me now to the question of what the model has ignored in relation to agricultural development strategies for underdeveloped countries in the 1970s.

TOWARD APPROPRIATE STRATEGIES FOR AGRICULTURAL DEVELOPMENT

Starting with the recognition that Ruttan and Hayami are not concerned with agricultural development as I have defined it earlier, I wish in conclusion to pose two basic questions.

The first is whether it is sufficient for us to concentrate simply on output growth and productivity changes in the agricultural sector. The second is whether or not our attention should be directed to institutionally specific analyses and models of agricultural development instead of seeking for a general theory. Let me say a little about each of these basic questions.

To my mind, the process of productivity change and growth of output may well be important in explaining agricultural *development* in countries like the United States. It is grossly insufficient in explaining economic adjustments in places like the United States South, the Caribbean, and elsewhere (that is, the persistence of *underdevelopment*). At least two factors need to be considered in this connection. One is the existence of duality in the agricultural sectors of underdeveloped economies and the associated question of the kind of output change. The other is the backwash effects of terms of trade adjustments in the expansion of output in *export* agriculture.

Duality assumes major proportions in the case of plantation-peasant agricultural economies. In such situations, plantations produce export output and peasants domestic output. It is the latter that is critical for the development process, for several reasons, notably its effects on structural transformation and rural welfare. Duality is an index of institutional distortions in the economic framework. So it is the institutional environment that is critical for the process of agricultural development (and underdevelopment).

The importance of the terms of trade backwash has been recently elaborated by W. A. Lewis in his 1969 Wicksell Lectures. According to Lewis, the extent to which underdeveloped countries benefit from improvements in productivity in export production depends on the relationship between export production and food production in the underdeveloped countries on the one hand; and between production of manufactures and food in the advanced countries on the other (Lewis 1969, 17–27). In an earlier presentation, Lewis verifies the point in a manner directly relevant to my reservations about the Ruttan-Hayami model of "development." It is worthwhile quoting Lewis at length on this score:

> Cane sugar production is an industry in which productivity is extremely high by any biological standard. It is also an industry in which output per acre has about trebled

over the past seventy years, a rate of growth unparalleled by any other major agricultural industry in the world—certainly not by the wheat industry. Nevertheless, workers in the cane sugar industry continue to walk barefooted, and to live in shacks, while workers in wheat enjoy among the highest living standards in the world. However vastly productive the sugar industry may become, the benefit accrues chiefly to consumers (Lewis 1955, 281).

I come, finally, to my own contribution to the evolution of thought on agricultural development. To my mind, the induced development model of Ruttan and Hayami exposes the fundamental limitations of contemporary theorizing on the nature of the process of agricultural development. If we are concerned, as I am, with the material welfare of rural people, then the problem must be approached differently from the way the authors have attempted. Basically, Ruttan and Hayami have started from the body of economic theory that we have at our disposal. That body of theory is based on the observation of economists of real situations that existed in the past. I suggest that we need to analyze the process of agricultural development from the perspective of the present. In terms of agricultural development this means developing models appropriate to the contemporary situation in Third World countries.

If we are to do this, it seems to me that we need, first, to develop a typology of underdeveloped agriculture reflecting different institutional arrangements in particular situations; and, second, to develop models appropriate to each type identified. For the most obvious lesson to be gained from the evolution on thought on this subject is that useful theories of agricultural development have been based on analyses of specific situations. It is the specific social order that determines the institutional arrangements that influence the interplay of the proximate economic variables which are central to the Ruttan-Hayami model. So if we are to understand the development process, we need to probe far beyond the proximate economic variables. And I am afraid that, as economists, we are not well equipped for that!

NOTES

1. For a summary see Johnston 1970.
2. The case of the slaves in the slave plantation economies is an outstanding example. And currently several scholars have noted that the benefits of the "green revolution" are concentrated among the larger, better-off farmers in underdeveloped countries. [See chaps. 26 and 27.—ED.]
3. I have demonstrated this in my analysis of the problem of resource allocation in plantation economies (see Beckford 1972, esp. chap. 6). A similar situation exists for the feudal-type economies of Latin America and parts of Asia, as well as for the tribal economies of Africa.

REFERENCES

Beckford, G. L. 1972. *Persistent Poverty: Underdevelopment in Plantation Economies of the Third World*. New York: Oxford University Press.

Johnston, Bruce F. 1970. "Agricultural and Structural Transformation in Developing Countries: A Survey of Research." *Journal of Economic Literature* 3(2): 369–404.

Lewis, W. A. 1955. *The Theory of Economic Growth.* Homewood, Ill.: Richard D. Irwin.

————. 1969. *Aspects of Tropical Trade, 1883–1965.* Stockholm: Almqvist and Wicksell.

Nicholls, W. H. 1960. *Southern Tradition and Regional Progress.* Chapel Hill: University of North Carolina Press.

6

The Political Economy of Rural Development in Latin America: An Interpretation

ALAIN DE JANVRY

Agricultural poverty has generally been analyzed in the context of "traditional" agriculture. Schultz (1964) has defined it as a state of economic equilibrium reached by agriculture over a long period of time and characterized by constant traditional technology and unchanging farmer preferences and motives. In this state farmers have been observed to be poor but efficient in their use of resources.

In this context the origins of agricultural poverty are dissociated from the dynamics of development in other sectors of agriculture, in other economic activities, and in the world economic system. To help farmers escape from the misery of traditional farming, recommendations are for the provision of new technological alternatives. In particular, many have looked to the application of the Green Revolution technologies in small farmer projects as a means of extricating farmers from this low-level equilibrium trap and shifting subsistence peasants to the blessed status of the commercial farmer.

To deal with rural poverty in this context is, in my view, an historical inconsistency that seriously impairs our ability to delineate and interpret the origins and dynamics of poverty and to identify the means by which it can be attacked. The alternative interpretation that is defended in this paper is that underdevelopment cannot be treated apart from development if backward areas or countries are related by the market to the advanced areas or countries. In fact, within the world capitalist system, a theory of underdevelopment and rural poverty needs to be a theory of economic space which can explain how the contradictions of development in certain areas transform, in other areas, traditional societies into underdeveloped ones.

ALAIN DE JANVRY is professor of agricultural and resource economics, University of California at Berkeley.

Reprinted from *American Journal of Agricultural Economics* 57, no. 3 (1975): 490–499, with minor editorial revisions, by permission of the American Agricultural Economics Association and the author.

For Latin America—and indeed for most less developed countries that have also been integrated into the world economic system through commercial exchange ever since the beginning of colonization—analysis of rural poverty cannot meaningfully be conducted in the framework of traditional societies. Instead, such analysis should be done against a background of historical events that is dominated by the nature of accumulation in the more developed countries, including the economic destruction of traditional societies through market forces after the first industrial revolution in England, the barriers to industrialization following the second industrial revolution in the twentieth century, the conditions of unequal commercial exchange between developed and dependent nations and the consequent need for low wages in the periphery, the exhaustion of import substitution policies to promote industrialization and the transformation of industry into an economic and social enclave in the 1960s, and the resulting reinforcement of structural dualism and marginalization of large sections of society. Because agriculture serves as a natural refuge for marginal populations, enabling them to satisfy part of their subsistence needs, rural poverty should be analyzed in the framework of marginality rather than of traditional culture.

In this paper the concept of rural marginality will first be defined. Its historical origins in Latin America and its economic rationality will be conceptualized in a consistent model of capital accumulation in the world economic system. This model will then be used to analyze the economic and social significance of alternative approaches to rural development. The purpose here is to construct a taxonomy that can be used to define and evaluate some important aspects of rural development programs. This taxonomy is based on a double dichotomy; that is, the sociopolitical structure of a country is described in terms of the nature of class alliances between traditional landed elites, national industrial capital, and foreign capital, while the sociopolitical nature of rural development programs is characterized as integrative (instruments of social control) or incorporative (instruments of social liberation).

MARGINALITY AND PERIPHERAL CAPITALIST DEVELOPMENT

The concepts of periphery and marginality are essential for the construction of a theory of underdevelopment. The periphery is that portion of economic space which is characterized by backward technology with consequent low levels of remuneration of the labor force and/or by advanced technology with little capacity to absorb the mass of the population into the modern sector.[1] These excess human masses created by the very process of economic growth are the "marginals." They can be found in all sectors of the economy and are functionally related to the modern sector that needs them to face the conditions under which growth occurs in the periphery. In agriculture they are the farmers who lose control of the means of production because they cannot withstand the competitive pressure of the modern sector, or the farmers who see their economic condition deteriorate as they retain traditional production techniques, but in both cases they cannot sufficiently proletarianize themselves to compensate for the income loss because they cannot be absorbed or fully sustained by the modern sector. Inevitably, they join the ranks of marginals as *minifundistas* and subsistence farmers, as do also many of the new entrants into the labor force who cannot find employment in the modern sector.

In contrast to the peripheral economies, the central economies have already had their industrial revolution and are characterized by an intensive rate of capital accumulation and the potential of making full use of their labor force with modern technology. Marginal populations also exist in central economies, where they consititute the *lumpen-proletariat,* but the phenomenon there is quantitatively different, as they consist principally of dispersed groups of frictionally unemployed (in the process of structural change), of temporarily unemployed (in cyclical phases of economic adjustment), or of unemployables (for physical or psychological reasons). In peripheral economies they constitute large masses who have been objectively created by the dynamics of accumulation (Quijano 1971).

Historically the transformation of Latin America into a periphery of the world capitalist system occurred first through colonization, which integrated the region into the commercial market, and later through efforts, essentially by England, to destroy the Iberian colonial empire in order to establish free trade with the rest of the world. The negative effects of free trade policies became clear during the nineteenth century as they prevented both the development of a national industry behind needed protective tariffs and the rise of a strong national bourgeoisie. Moreover, those policies encouraged the maintenance of economic systems based on the trade of primary products against manufactured consumption goods and hence provided the basis for class alliances between traditional landed elites and British capital to maintain the internal social status quo. Indeed, for the dominating classes there is economic rationality to the transformation of Latin American countries into peripheries, for such transformation has enabled them to capture part of the large surplus of the agro-exporter sector and hence to enjoy consumption patterns similar to those of developed countries while retaining all the advantages conferred by their social position in underdeveloped economies (cheap labor services, social power, and so on). Under the agro-exporter model, unequal exchange to the benefit of the center is obtained on a "voluntary" basis by co-opting the traditional elites into sharing in the surpluses extracted. Maintenance of low agricultural wages through precapitalist relations of production, imposed by the elites to tie labor to the land and alienate it from its own opportunity cost on the labor market, produces deterioration of the external terms of trade to the benefit of the center.[2]

With the second industrial revolution in the twentieth century, conditions for industrialization of peripheries have changed drastically. No longer can capital goods be produced using traditional goods, even under high tariff protection;

now their production requires modern goods, and these must be imported from the developed countries. Under these conditions, the import capacity of the economy (severely constrained by unequal exchange) has become the bottleneck to industrialization. Through severe market interventions, conscious exclusion of foreign capital, and public assistance to the development of technology, Japan (in the late 1800s) was the last country to transform itself into a developed center. Henceforth, trade protection and the rise of national entrepreneurial classes were not enough to permit development without marginality. Constrained by foreign exchange obstacles to the importation of capital goods, industrial development could not acquire the breadth and dynamics needed to generate enough employment to compensate for the destruction of traditional structures.

Early in the twentieth century, import substitution policies injected a new dynamic into the industrial sector. Industrialization aimed at capturing the existing national market, which developed in the process of exchange between exports of primary products and imports of manufactured consumption goods—first for mass consumption items and later for luxury (durable) goods. Soon, nevertheless, and particularly after the 1930s, the terms of trade in the exchange between raw materials and capital goods deteriorated "coercively" against the periphery. This deterioration was caused by systematic market distortions imposed by the central economies in taking advantage of the global monopoly which technological and capital goods dependency of the peripheral nations confers on them.[3] Given a unique world market for capital, the cost of unequal exchange had to be transferred to the only nontradable factor—unskilled labor—and continued industrial development in the periphery is now conditioned by the ability of nations to reduce their labor costs.[4]

Two means can be used to cheapen labor. One is to impose repressive labor policies—an instrument that clearly has been extensively used but is limited by the organization and insurgency potential of the working classes; the other is to lower the cost of those mass consumption items that constitute the bulk of labor's budget and hence are determinant of labor cost. Food, textiles, and popular construction are the producing sectors most negatively affected by the consequent commercial distortions that depress the internal terms of trade against wage goods. As a result, profitable investment ventures are confined to that subset of the industrial sector which is oriented to satisfy demand from the upper classes and is consequently intensive in advanced technology and imported capital goods. As the scope of import substitution expands to include intermediate goods that are more intensive in technological content and capital, the foreign exchange bottleneck becomes more acute, the employment effect is more limited, and the multiplier effect of imports on the value added to national raw materials is reduced. By the 1960s, industry had transformed itself into a large enclave of modernization that was unable to generate enough employment to absorb available labor.

Under conditions of unequal exchange, the effect of import substitution policies has been to replace the initial sectoral linkage (characteristic of peripheries) between the export sector of primary goods and the import sector of luxury-consumption items by a new linkage between the export sector of primary products and the import sector of capital goods for the production of luxury-consumption items (or, in economies that reconvert for outward growth, for the production of exportables). In contrast, in central economies the fundamental sectoral linkage is between the production sector of capital goods and the production sector of mass consumption goods. In this contrasted industrial structure of central and peripheral economies lies the economic rationality of labor incorporation in central economies and of marginality in peripheral areas.

In central economies the distribution of income between capital and labor determines the dynamics of growth. The return to capital conditions accumulation in the capital goods sector: it determines the capacity of the system to produce. The return to labor conditions the size of the market for mass consumption goods: it establishes the capacity of the system to consume. No automatic equilibrium exists between returns to capital and to labor; these are determined by relative forces in the social system.[5] Nevertheless, a necessary relation exists between the two returns for economic growth to proceed. In the continued adjustment process between capacity to consume and to produce, economic fluctuations will result with alternations of inflationary and unemployment crises, but the system will tend toward nearly full employment of its labor force and incorporation of all producing sectors in the modern nucleus of the economy.

In the central economy labor constitutes both a cost and a benefit to capital; it represents a cost as wages are subtracted from profits but also a benefit as wages serve to generate the demand that will allow further accumulation. With increasing monopolization of capital and with organization of labor at the national level, this necessary relation leads to the possibility of a "social contract" between capital and labor under the auspices of the state, which allows real wages to relate effectively to increases in labor productivity (Amin 1972). Economic rationality of this social contract is the basis of the liberal social democrat philosophy in central economies.

The rise of such a social contract is prohibited in peripheral economies by their distinct sector linkage (that between export of primary products and import of capital goods for the production of luxury items).[6] In those economies labor constitutes only a cost to capital and is not simultaneously a benefit, for industrial production is oriented not toward mass consumption but toward consumption by the upper income classes. Higher returns to labor and distributive policies in general (such as land reform) would have insignificant market effects for the existing modern industrial sector. In addition, since the capacity to produce is restricted by the foreign exchange bottleneck, the capacity to consume is rarely an effective constraint, as chronic inflationary pressures demonstrate. The dynamics of peripheral accumulation in the context of unequal exchange are based on continued dominance of capital over cheap labor. In fact, it is by this very process of labor's submission to modern capital that societies make the transition from traditional to peripheral.

Just as incorporation of labor is in the center a condition for growth, marginality is in the periphery a contradiction of growth. In the center, distributive policies do not contradict the logic of capital accumulation; in the periphery they do. In fact, especially if rates of economic growth are low, increasing the size of the domestic market for modern industry will require regressive policies to boost the purchasing power of the upper classes. In the center, traditional production structures are incorporated by the modern sector; not so in the periphery, where structural dualism will deepen, being a necessary condition for growth of the modern sector.[7]

In sum, a first major contradiction of capitalist growth in peripheral economies is the structural crisis in the balance of payments that results from capital goods dependency (a consequence of the second industrial revolution), which confers on central economies a global monopolistic privilege and with that the possibility of deterioration of the terms of trade against the periphery. Unequal exchange, in turn, implies the need for reducing labor costs, which results in unfavorable terms of trade for wage-good producing sectors and brings about a modern industrial sector confined to the production of luxury goods. Even where outward growth can confer substantial dynamics to the modern industrial enclave, reinforced structural dualism and marginality are the dominant traits of peripheral societies. This contradiction, in turn, implies contradictions within agriculture from which both agricultural stagnation and rural poverty result.

AGRICULTURAL STAGNATION AND RURAL POVERTY

In most of Latin America, the opportunities for profitable industrial investments, created in the early 1900s by import substitution policies, induced the traditional landed elites to extend their sectoral control toward industry and finance while retaining control of land in the *latifundio*. The reverse movement sometimes also occurred, whereby new industrialists sought access to the land in order to diversify portfolios and to gain access to power over the institutions of society. In all cases, agricultural interests were strongly integrated with urban interests and increasingly subject to their dominance (Stavenhagen 1968). Since industrialization requires low wages and overvalued exchange rates to cheapen capital imports, commercial and market distortions result in internal terms of trade that are dramatically unfavorable to agriculture.[8] This condition has been true, especially since the 1930s, when collapse of the international market made more pressing the need for national industrialization. Distortions have been greatest in countries like Chile, where food production does not participate vitally in the generation of foreign exchange. Except in Mexico, where land reform has eliminated dominance of the traditional elites and opened the way for ambitious infrastructure investments and the diffusion of land-saving technological change, agriculture has entered upon a long period of stagnation. Development of capitalist farming in food crops has been blocked. Only the *latifundios* have been maintained as viable economic units through institutional control by the dominating elites, who monopolize institutional services (institutional credit, technology, information, and so on) and derive from them economic compensations for the unfavorable terms of trade. Part of the cost of stagnation has been shifted to labor through miserable wages.

With the rise of rampant marginality, the social relations of production have been gradually redefined from precapitalist (where workers' subsistence plots are located within the *latifundio*) to a functional dualism between commercial sector and *minifundio* (where the subsistence sector is now external to a capitalist agricultural sector). Precapitalist relations are needed to cheapen labor by tying it to the land and alienating it from its own opportunity cost as long as it is a scarce factor, a situation characteristic of most of Latin America until the 1950s. Once marginality is widespread, labor costs can be further reduced by making use of wage labor instead of labor partially paid in land privileges, since that permits landlords to recover the land previously given to workers and to tailor the hiring of labor to fluctuating seasonal and annual needs.

In the central economy, proletarianization of labor in the process of the development of capitalism is due to two economic motives: the desire to reduce labor costs, which is accomplished by paying labor only for time actually worked (in contrast to enslaved or servile labor, whose subsistence must be covered even in periods of sickness or slack in labor needs), and the need to expand the market for consumption goods in order to permit continued accumulation. Market expansion requires destruction of the subsistence economy and redistributive policies toward labor. Because of the fundamental asymmetry between central and peripheral economies in the role of labor in market expansion, the first motive fully applies in the periphery, but the second does not. As a result, labor costs in the periphery can be further reduced by maintaining the subsistence economy, since that provides the commercial sector with labor whose subsistence is already partially covered by production in the *minifundio*. Labor wages can thus be collapsed below the subsistence needs of the worker and his family by an amount equal to the net value of production in the subsistence sector. Development of a functional dualism between subsistence agriculture and the commercial sector is fully consistent with the needs for growth in the periphery under conditions of unequal exchange. In this structure the subsistence sector produces cheap labor for the commercial sector, which, in turn, can produce cheap food for the market.

This role of the subsistence sector as a producer of cheap labor implies a specific division of labor by sex and age in the *minifundio*. While men largely do wage work outside the *minifundio,* women and children are the essential productive agents. Dramatic contrasts also exist at this level between center and periphery. In the center, women essentially have a consumption function (vividly described in Galbraith 1973, chap. 4), while in the subsistence sector of the periphery, women's function is in production. In the center, children are primarily consumption factors (a rationale that is central to the demographic analyses of the Chicago School [Leibenstein 1974, 457–79]), whereas in the subsistence sector of the periphery, they are production and protection agents. The pressure of poverty in the *minifundio* implies the need to search for additional productive resources. Strict individual economic rationality in the subsistence sector will often lead to mining the land and increasing family size in order to face poverty, and poverty thus leads to more poverty.

Generalized rural poverty in rural Latin America is hence the logical outcome of a three-level chain of exploitative relations. First is the international level between dominant centers and dependent peripheries in the context of unequal exchange between raw materials and industrial capital goods. Second is the sectoral level between modern industry, which produces commodities for the upper classes and the external market, and the sectors that produce mass consumption items in the context of the need for cheap wages and the consequent deterioration of the internal terms of trade. Third is the social level between landlords and agricultural labor and marginal populations in the context of transmission to labor of the costs of international and sectoral level unequal exchange, as well as exclusion of the mass of population from modern sector employment. Only when the food sector is itself a major component of the export sector is this chain reduced to a two-level set of relations, as unequal international exchange is brought to bear directly on the terms of trade for agriculture, which will then be generally more advantageous (as in Argentina and Uruguay) and rural poverty somewhat less acute.

To these three levels of exploitative relations correspond specific contradictions that jeopardize economic growth and social stability in the periphery. At the international level, the structural deficit in the balance of payments blocks expansion of the modern industrial enclave. At the sectoral level, agricultural stagnation raises labor costs, unleashes inflationary pressures, and worsens the balance-of-payments deficit. At the social level, miserable wages build up revolutionary pressures that are reinforced by ecological and demographic contradictions in the subsistence sector.

In the context of these fundamental contractions of growth in the periphery, it is clear that rural development programs do not have (in the rationality of capital accumulation) redistributive goals dictated by the economic need of increasing market size for the modern industrial sector. Land reform programs generally pursue two simultaneous objectives (dictated by these contradictions) whose relative importance is determined by particular historical circumstances: a production goal designed to alleviate deficits in the balance of payments and to cheapen wages and a distributive goal, which is fundamentally political, designed to promote social integration of the potentially revolutionary strata of the peasantry and their eventual incorporation into social groups that will favor maintenance of the social status quo.[9]

In Chile the land reform of the Christian Democracy aimed at increasing production in the nonreformed agricultural sector through modernizing changes and through threats of expropriation, while the *asentamiento* was used as a political instrument of social control.[10] While a small number of benefited farmers were largely incorporated in the Christian Democracy, the frustrated non-beneficiaries of the reform (whose political power had been increased by the important unionization movement that enlisted five times more members than beneficiaries of the land reform) joined the opposition forces. Simultaneously, the traditional elites, displeased by the economic cost of social reforms and the threats they represented to property in other sectors of the economy and to political stability, broke away from the Christian Democracy and presented their own presidential candidate in the 1970 elections, dangerously dividing the liberal and conservative forces. Frustrations of both excluded peasants and traditional elites with the land reform were determinant factors in the election of a socialist regime (Barraclough 1974).

The concepts of integration and incorporation are useful to characterize the scope of processes of social change (Lehmann 1971). Integration is a process that comes from above, promoted by more powerful groups in favor of less powerful ones and by which the latter are provided with new channels of relation with the central institutions of society. Integration can be obtained either by restructuring the nucleus of the incorporated so that part of their power, income, or assets can be offered to the marginals or by organization of the marginated so that they gain access to social institutions. It does not necessarily imply structural or social change but aims only at improving the relations between nonincorporated and incorporated groups. Incorporation, in contrast, is a process that develops from the bottom upward (although it may be initially induced from above) through which a social group acquires "rights of citizenship," rights which it then may use to enforce its own claims against other incorporated social groups. In different historical and social contexts, these rights will be used for different specific purposes.

It has been argued that to understand rural poverty in most of Latin America, the concept of traditional agriculture needs to be replaced by that of functional marginality. To this end, the elements of a theory of marginality as an integral part of the dynamics of growth in peripheral economies of the world capitalist system have been outlined. This theory permits identification of the variables that affect rural poverty and understanding of the political economy of alternative strategies aimed at alleviating it.

POLITICAL ECONOMY OF RURAL DEVELOPMENT PROJECTS

In addition to land reform programs, integral and partial rural development projects have been frequently used in recent years as strategies to improve the economic conditions of subsistence farmers (Mosher 1972; Adams and Coward 1972).[11] These projects are largely in the tradition of community development programs but seek new vitality in capitalizing upon the power of social change imbedded in the technology of the Green Revolution. Since, in almost all cases, subsistence farmers are marginals rather than "forgotten" or "accidental" traditional social sectors, poverty has an economic rationality within the broad economic system. Therefore, rural development projects, in both design and evaluation, will have to be consistent with this rationality.

Rural development projects, oriented at subsistence agriculture, differ in their goals from land reform programs. The production objective will be set in terms not of national necessities (improving the balance of payments and lowering food prices) but of the consumption and income levels of the marginal sector concerned. Hence, rural development projects will be of no major avail in relieving the structural deficit in the balance of payments or the aggregate output stagnation. Instead, they will have the political goal of alleviating rural poverty in order to contain the popular pressures brought about by increasing marginality and/or the economic goal of cheapening labor from the *minifundio* by increasing the share of its subsistence needs covered through subsistence agricultural production. But such projects may also be a part of broader processes of social change aimed at breaking away from the three-level chain of exploitative relations previously identified.

To describe the political economy of rural development projects, it is useful to introduce a taxonomy among social formations based on the nature of the alliances between traditional landed elites, national industrial capital, and foreign capital. The nature of these alliances permits contrasting three prevalent contexts within which rural development projects may occur in Latin America.

First is the situation in which the alliance between traditional landed elites and industrial capital and that between industrial capital and foreign capital remain intact. Belonging in this category are northeast Brazil, Argentina, and Uruguay, none of which has had any significant land reforms. In these areas land tenure is largely characterized by the *latifundio-minifundio* dualism, and traditional landed elites dominate the institutional process. Food production is stagnant because of the compound influence of unfavorable terms of trade and of the supply to agriculture of institutional services that are conditioned by the traditional elite's objective of social status quo and are, therefore, not oriented to land-saving technological change. As explained before, the internal terms of trade are severely affected by unequal international exchange, and the less agriculture participates in the generation of foreign exchange earnings, the more unfavorable these will be. Since monopolization of institutional services (credit, technology, information, and so on) by the traditional elites is essential for them to derive economic compensations from the unfavorable terms of trade and to sustain the profitability of the *latifundio*, their continued dominance over the institutions is an economic imperative. The dilemma for them is hence one of deriving maximum benefit from institutional services while simultaneously managing the institutional process in such a fashion that reproduction of institutional control is also assured. Since monopoly of the land is the basis of the social power of these elites, they will foster only those changes in technology that are not a substitute for land. This will orient the technological path toward mechanical, labor-saving, and generally non-yield-increasing technologies instead of toward biochemical land-saving technologies of the Green Revolution type (de Janvry 1973). Even if the rate of technological change is intense, the social status quo constraint imposed by the traditional elites will bias technology and largely destroy the output growth potential of technological progress.

Under prevalence of this double alliance, food output stagnation and rural poverty through transfer to labor of the cost of stagnation will result. Marginality becomes widespread because the modern industrial sector cannot absorb the mass of labor in spite of eventual high rates of growth. Rural development projects in the *latifundio-minifundio* structure are precluded when the *minifundio* is captive within the *latifundio* under precapitalist relations of production. For external *minifundios*, rural development projects probably can be only limited in scope and oriented toward integrating marginal farmers enough to alleviate social pressures but without risking incorporation. The magnitude of rural development projects will be restricted by their high economic cost for the dominating class. At this stage of social organization, land reform and other programs oriented toward giving some institutional control to the potentially productive social sectors of agriculture are probably prerequisites to rural development projects.

Once land reform programs break through the alliance between national industrial capital and traditional agrarian classes, the conditions may be set for more meaningful rural development projects, as, for example, in Mexico. While unfavorable terms of trade for food production remain as an effect of unequal exchange, institutional control has now accrued to rural and urban capitalists, and output growth can be given impetus by land-saving technological change and infrastructure investments. The social tensions in the countryside, generated by industrial and rural marginality, are transitorily frozen in a contrived dualism. But combined demographic pressures and contradictions on the distribution of income, brought about by peripheral accumulation, soon aggravate marginality and social stress.

Under these conditions, rural development projects will tend to be aimed at prolonging the social status quo of contrived dualism. Therefore, at least initially, such projects will be integrative instruments of social control over the rural marginals. But marginals can gain true access to sustained social gains only if projects evolve from integrative to incorporative. In some instances, rare because of the political risk involved, projects may be deliberately oriented toward promoting social incorporation in order to attempt capturing the political support of the newly incorporated. In other instances, the successful maturation of projects may lead them to evolve spontaneously from integrative to incorporative as a result of increasing contradictions between complexity of the production process and exogenous decision making. Creation of credit groups, cooperatives, and so on, can be effective inducement mechanisms of this transition.[12] Technological change can then become a powerful means of inducing social change rather than an end in itself. It can permit marginals to become conscious—through perception of the economic gains that can be derived from it—of the need to organize and of the possibility of thus gaining their due share of access to and control over the institutions of society. Here an analogy exists between proletarians and marginal peasants: for proletarians wage increase (income increase) is the initial perceived want, and unionization (institutional power) is the instrument toward attaining this goal; for marginal peasants technological change (income increase) is the initial perceived want, and solidarity groups (institutional power) the instrument. In both cases the long-run valid outcome is social change, based on the rise of institutional power for the working and marginal classes. Wage and technological changes are only instruments of social polarization.

The design of successful rural development projects requires a clear understanding of the process by which rural poverty is created and perpetuated. Specifically, the mechanisms by which economic surpluses are extracted from the subsistence sector and the socioeconomic reactive process to poverty within the subsistence economy need to be clearly identified, since these determine both the diffusion of program recommendations and the ultimate social distribution of welfare gains from such a program. A particular determinant is the adjustment of outside wages received by workers from the *minifundio* when resource productivity increases in the subsistence sector have been brought about by rural development projects. While productivity increases may make it possible for those *minifundistas* who have enough productive resources to progress from semi-proletarian workers to family farmers, most of them will still have to compete for outside wage work. If this competition is intense, wages may eventually collapse by an amount equal to the increase in net income in the *minifundio,* in which case the welfare gains from rural development projects would be extracted from the subsistence sector to the benefit of the employing sector. A welfare program for the rural poor could thus result in a subsidy to the commerical sector. Only by careful design of complementary institutional and structural changes will welfare gains possibly be retained by the subsistence sector.

A third situation exists when the double alliance between traditional landed elites, national industrial capital, and foreign capital has been broken down as in the ''new Latin American nationalism'' of Peru or the socialist rationality of Cuba. Here, as the influence of unequal exchange on the industrial structure is presumably eliminated, the stage is set for planned autocentered agricultural and industrial development. Industrial production can be oriented toward mass consumption items, and redistributive policies acquire logic in terms of market expansion. As in central economies, the social management of a balance between capacity of the system to produce and to consume becomes a necessary condition for growth. Terms of trade for agriculture as well as institutional processes aimed at land-saving technological change and at infrastructure investment can be managed consistently. And rural development projects can be used as powerful instruments of social incorporation. Nevertheless, potentials and achievements remain far apart in the actual transitional phases. Dependency from one ideological block tends to be replaced by dependency from another, and integrative rural development tends to be preferred over incorporative for the sake of social control.

NOTES

The author is indebted to C. Garramon for a number of ideas set forth in this paper and to C. Benito and T. Carroll for their constructive comments.

1. The center-periphery concept was first introduced by Raul Prebisch (1959) and the Economic Commission for Latin America of the United Nations. This concept has been used to develop a theory of accumulation in peripheral capitalism by Frank (1967), Cardoso (1972), Hinkelammert (1970), Dos Santos (1970), and Amin (1974). For a review of these theories see Chilcote 1974.

2. The theory of unequal exchange, based on the causality from cheap labor to deteriorated external terms of trade, was developed by Emmanuel (1969), who relaxes the Ricardian and neoclassical assumption of nontradable factors and postulates instead that capital is internationally mobile, while labor is not. Market equilibrium thus requires that the rate of return to capital equates in trading countries. Differential wage rates between center and periphery—resulting from precapitalist relations of production in the periphery and superior labor bargaining power in the center—will lead to deteriorated terms of trade against the periphery even if productivity of labor is equal in both areas.

3. The magnitude of the detrimental effects of these distortions on the peripheral nations is so great that it gave rise to the UNCTAD III meeting in 1972.

4. The theory of unequal exchange, where causality runs from coercive deterioration of the external terms of trade to low wages in the periphery, has been developed by Braun (1973) and Amin (1974).

5. This is the central proposition of the Cambridge controversy of capital, which dismisses the existence of a unique "just" return to labor and capital given as the solution of a general equilibrium economic system (Harcourt 1969).

6. Because of differences in income levels, the same durable goods that are mass consumption items in central economies are luxury consumption goods in peripheral economies. The demand for mass consumption items arises primarily from the return to labor, while that for luxuries originates mainly from the return to capital.

7. This shows that the dual economy models of both classical and neoclassical varieties simulate a dynamics of growth that characterizes central economies but is the exact opposite of peripheral development.

8. Schultz (1968, 175–84) refers to these distortions as "cheap food policies" and attributes to them major responsibility in inducing agricultural output stagnation.

9. See chapters 17 and 18 in this volume.—ED.

10. This makes it similar to the design of the Mexican land reform, where strong output growth is sought in the commerical sector and social control in the *ejido* (Gutelman 1971).

11. Mosher (1972) refers to "integrated" versus "nonintegrated" projects to contrast projects that simultaneously provide a number of services or activities for small farmers located in a specific geographic region with those that offer only a limited number of services or activities. Since the concept of integration has a different meaning in this paper, these projects shall be referred to as "integral" versus "partial" instead of according to Mosher's terminology.

12. The "solidarity groups" in Plan Puebla through which group credit is obtained are an example.

REFERENCES

Adams, Dale W., and E. Walter Coward, Jr. 1972. *Small Farmer Development Strategies: A Seminar Report.* New York: Agricultural Development Council.

Amin, Samir. 1972. "Le Modèle théorique d'accumulation et de développement dans le monde contemporain." *Tiers monde* 13.

―――. 1974. *Accumulation on a World Scale: A Critique of the Theory of Underdevelopment.* New York: Monthly Review Press.

Barraclough, Solon. 1974. "Interactions between Agrarian Structure and Public Policies in Latin America." Second International Seminar on Change in Agriculture, Reading, England.

Braun, Oscar. 1973. *Comercio internacional e imperialismo.* Mexico City: Siglo XXI.

Cardoso, Fernando. 1972. "Dependency and Development in Latin America." *New Left Review* 74: 193–226.

Chilcote, Ronald. 1974. "Dependency: A Critical Synthesis of the Literature." *Latin American Perspectives* 1: 4–29.

de Janvry, Alain. 1973. "A Socioeconomic Model of Induced Innovations for Argentine Agricultural Development." *Quarterly Journal of Economics* 87: 410–35.

Dos Santos, Theotonio. 1970. "The Structure of Dependence." *American Economic Review* 60: 231–36.

Emmanuel, Arghiri. 1969. *L 'Echange inégal.* Paris: Maspero.

Frank, Andre Gunder. 1967. *Capitalism and Underdevelopment in Latin America: Historical Studies in Chile and Brazil.* New York: Monthly Review Press.

Galbraith, John Kenneth. 1973. *Economics and the Public Purpose.* Boston: Houghton Mifflin Co.

Gutelman, Michel. 1971. *Réforme et mystification agraire en Amérique Latine: le cas du Mexique.* Paris: Maspero.

Harcourt, G. C. 1969. "Some Cambridge Controversies in the Theory of Capital." *Journal of Economic Literature* 7: 369–405.

Hinkelammert, Franz. 1970. "Dialectica del desarrollo desigual." *Cuadernos de la realidad nacional* 6.

Lehmann, David. 1971. "Political Incorporation versus Political Stability: The Case of the Chilean Agrarian Reform, 1965–70." *Journal of Development Studies* 7: 365–96.

Leibenstein, Harvey. 1974. "An Interpretation of the Economic Theory of Fertility: Promising Path or Blind Alley?" *Journal of Economic Literature* 12: 457–79.

Mosher, A. T. 1972. *Projects of Integrated Rural Development.* New York: Agricultural Development Council.

Prebisch, Raul. 1959. "Commercial Policy in Underdeveloped Countries." *American Economic Review* 44: 251–73.

Quijano, Anibal. 1971. "La Formación de un universo marginal en las ciudades de America Latina." *Espaces et societes* 3.

Schultz, Theodore W. 1964. *Transforming Traditional Agriculture.* New Haven: Yale University Press.

―――. 1968. *Economic Growth and Agriculture.* New York: McGraw-Hill.

Stavenhagen, Rodolfo. 1968. "Seven Fallacies about Latin America." In *Latin America: Reform or Revolution,* ed. J. Petras and M. Zeitlin. New York: Fawcett Publications.

7

The Political Economy of Rural Development in Latin America: Comment

G. EDWARD SCHUH

ANALYSIS OF DE JANVRY'S ARGUMENTS

Alain de Janvry has argued that generalized rural poverty in Latin America is the logical outcome of a three-level chain of exploitative relations.[1] The interpretation that de Janvry defends is that under-development cannot be treated apart from development if backward areas or countries are related by the market to the advanced areas or countries. There is an important element of truth in this argument, for two fundamental reasons. First, once there are differential rates of growth between two countries (or between the periphery and the center, if you like), a complete explanation of the *relative* position of each requires a consideration of economic conditions affecting both. Second, it is clear that growth and development in a country or group of countries, especially if they are the major components of total economic activity, clearly alter the economic conditions faced by the second group of countries. One cannot understand the development of the latter group of countries without understanding these changing economic forces.

But de Janvry's analysis is clearly something different. It assumes that the rich arrived in that state by taking away from the poor. This mercantilist idea, long since discredited, implies that the total economic pie to be divided is fixed. Its use gives de Janvry a theory of income *distribution,* not of growth or develop-ment as he implies. Moreover, his theory fails to distinguish clearly between relative and absolute poverty.

De Janvry's notion that a theory of underdevelopment and rural poverty needs to be a theory of economic space is also appealing. Economists in the neoclassi-cal tradition have been concerned with this problem also. Schultz's urban-indus-trial impact model attempts to explain spatial lags in development and has been

G. EDWARD SCHUH is professor and head of the Department of Agricultural and Applied Econom-ics, University of Minnesota.

amply tested with data from the United States and Brazil.[2] That theory, of course, is based on the assumed prevalence of market imperfections and does not appeal to exploitation or conspiracy. It has not been used to explain intercountry differences in per capita incomes, but it would lend itself to that end.

The theory of dependency pervades de Janvry's model. This has been a prominent stream of thought in the Latin American literature. Its basic premise, of course, is that economic exchanges are not entered into voluntarily, with both sides gaining; rather, one member of the exchange is dominant and therefore exploits the other. An interesting footnote to this literature is that whereas the periphery countries are assumed to be dependent on the center, the tenor of the analysis is that the center countries are dependent on the periphery for their wealth. The symmetry of this dependency, and the lack of operationality that it implies for the concept itself, is seldom recognized.

In the remainder of this discussion I will deal with nine elements of the de Janvry model.

THE TERMS OF TRADE

The supposed deterioration of the terms of trade is one of the two critical elements of the de Janvry model. Starting with Prebisch's original paper,[3] this has been a source of continuing controversy among students of Latin American development. The literature is abundant, and the issue continues to rear its ugly head despite evidence and arguments to the contrary.[4]

There are at least four key subissues in the terms-of-trade controversy. The first is whether the terms of trade have in fact declined. Given the instability in prices of primary products, whether they have declined depends critically on the point of reference and the length of the period under review. Adherents to the declining-terms-of-trade argument tend to pick a peak and measure the price relatives through the next trough. Those of an opposite disposition tend to take a longer perspective, and show that there has been little or no change. Adherents also seldom recognize that the terms of trade tend to be country specific and differ greatly according to the specific product mixes of the relevant countries.

The second subissue has to do with the measurement of the price indexes themselves. To be correct, they have to adjust for changes in the quality of the products. This is important because the quality of industrial products tends to change when new technology is imbedded in them. With the recent desire to compare prices of capital goods with the prices of primary products, this issue has come to the fore. What may appear superficially to be a rise in the price of imported goods may reflect nothing more than an improvement in quality of given products or a change in the product mix towards higher-priced industrial products that are not taken into account in measuring the price indexes.

The third issue is the cause of the change in the terms of trade. This issue is important at both the external and internal levels. Price relatives have to change if resources are to be allocated in such a way as to satisfy consumer preferences.

Understanding why they change helps in understanding the forces at work. Two points are important in this context. First, to the extent that primary products have a lower elasticity of demand than do secondary products, there is a presumption that relative prices would turn against primary products in the natural course of development unless factor markets were particularly efficient.[5] If the transfer of resources out of the primary sector should be particularly sluggish, the decline in terms of trade can be substantial, at least until the resource transfer is completed. The second point is that a shift in the terms of trade can occur because of technical progress in a particular sector. If this should be in a primary sector with a low income elasticity of demand and sluggish resource mobility, the shift in price relative can be especially severe.

These two factors can work at the international level. Hence there is no obvious reason why a decline in the terms of trade has to be *imposed* or coercive. It may be a result of the natural working of the market. To the extent that the major demand events and/or the technical change occurs in center countries, the source of required adjustments for the periphery countries may be external, of course, but that does not imply a conspiracy. Moreover, consequences of technical change in center countries could be offset in the factor-market sense (see below) in the periphery countries by appropriate investments in production technology, although resource-adjustment problems could still be severe.

Most important, the failure of countries in the periphery to respond to changing market conditions is not necessarily the fault of the center countries. Nor would it be in the best interests of the periphery countries for the center to hold back its own progress, although that often seems to be the desire of writers in the de Janvry tradition.

The fourth subissue on the terms of trade has to do with the ultimate consequences of shifts in the terms of trade. The function of relative prices in a market economy is to influence the allocation of resources. To do this, of course, they have to influence the rate of returns to resources. The extent to which factor returns are influenced, of course, will depend on resource mobility. With high mobility, resources will adjust quickly, and opportunity-cost returns will be reestablished in a short period of time.

De Janvry argues that in the international dimension the cost of a decline in the terms of trade is transferred to the only nontradeable factor—unskilled labor. This is clearly wrong, even if we accept his premise that the international capital market is perfect. Land is also untradeable among countries in the sense that he implies. But more important, within the periphery country labor can be reallocated sectorally, while for the most part land, especially that in agriculture, does not have that alternative. Hence ultimately the cost of declining terms of trade should be borne relatively more by owners of land than by unskilled labor, especially in the long run.

Finally, it should be noted that the price of the product is only one of various factors affecting the returns to resources and in turn per capita incomes.[6] And in fact it may be one of the least important. Productivity growth is one of the major factors influencing the growth of per capita income, or the level of income in any economy. Hence the level of investment in human capital will, in the final analysis, be more important than the terms of trade in determining how per capita income grows.

THE LABOR THEORY OF VALUE

De Janvry implicitly assumes a labor theory of value. This is most obvious in his argument that deterioration in the terms of trade is *produced* by maintenance of low agricultural wages, which clearly implies that labor is the only factor of production. Criticisms of the labor theory of value abound in the literature,[7] and little is to be gained from rehashing them here. What perhaps should be noted is that if the objective was to gain a decline in the terms of trade, this could have been obtained by investing in new production technology. Labor would have been released from the primary sector as a consequence and/or the terms of trade would have deteriorated. And with a gain in productivity the wages of labor could have risen while labor costs declined. Accepting de Janvry's assumption of a coalition between agricultural and industrial power groups, all could have gained from such a strategy, and the income gains resulting from such a policy would have strengthened the markets for their products. One wonders why, if these groups were so all-powerful and foreseeing, they were so lacking in intelligence and foresight in this direction.

BENEFITS FROM DEVELOPMENT

De Janvry repeatedly implies that there was something inevitable about import-substituting industrialization policies leading to self-contained enclaves and that the center countries have a vested interest in the poverty of the periphery. Neither conclusion seems valid. In the first place, most of the predicates of the infant-industry argument were ignored by the import-substituting programs. Industries were implanted with little reference to longer-term comparative advantage and the level of protection was often set inordinately high and was sustained for much longer than was required. There was nothing inevitable about these errors of policy, nor is it clear that they were imposed from abroad. Moreover, instead of favoring the upper-income groups in the society, they discriminated against them. The importation of luxury goods would have been cheaper and, at least in the short run, of higher quality.

Similarly, it is difficult to believe that the center countries had any vested interest in the poverty of the low-income countries. Trade as it has evolved in the post–World War II period has been among the center countries, not between the center and the periphery. (This in itself says something about the robustness of de Janvry's model.) The center would have gained from the growth and development of the periphery, for it would have meant growing markets and possibly gains from specialization of production. Even foreign capital coming into a periphery country would have benefited from generalized development in that country, for it would have meant a stronger internal market for its products.

At one point in his analysis de Janvry cites the Northeast of Brazil as a victim of the coalition and integration of the national industrial sector, foreign capital, and local landowners. This is factually incorrect. There has been little foreign capital in the Northeast of Brazil; nor has the integration of national industry and local landowning interests been particularly strong in that region. Where it *has* been strong is in the state of São Paulo, where both groups have been allied with foreign capital. And of course it is in that state that agricultural modernization and development have been the strongest in Brazil (and perhaps in Latin America), in part, as I have argued elsewhere,[8] *because* of the coalition. It is worth noting in this regard, moreover, that part of the investment in agricultural research in that state was made by international companies with an interest in *both* agriculture and industry.[9] There is an object lesson here for development policy: sectoral integration is to be welcomed, not condemned.

THE ROLE OF URBAN WAGE GROUPS

De Janvry identifies the landlord elite and the industrial groups as the two main sources of political power in Latin America, and he ascribes especial power to the coalition of these two groups. He totally ignores the political power of the urban wage groups, as well as the general importance of populist politics in countries such as Argentina, Brazil, Chile, Colombia, and Mexico.

Political pressures from urban wage groups have played a major role in shaping economic policies in many Latin American countries; cheap food policies in particular have arisen as much in response to that group as they have in response to any other. Peron in Argentina and Getulio Vargas and his *Estado Novo* distinctly appealed to urban wage groups.

It is also important to note that low labor absorption in most Latin American countries is in part a result of misguided labor-market policies based on the assumption that higher wages and the costs of social welfare programs could be passed on to the capitalist. Both large farms and industry in Brazil are required by law to provide education to their employees' children and to pay sizable payroll taxes to support social-welfare programs. Moreover, minimum wages have at times been set above equilibrium levels.

These policies, as much as anything else, explain the low labor-absorptive capacity of industry and of the modern sector of agriculture. The prevailing assumption behind each of these policies was that the capitalist should be forced to pay for such programs. Hence in design they were *punitive* of capital and not a subsidy to the capitalist class. If these groups had so much political power and played such a dominant role in shaping policy, why did they acquiesce in such policies?

THE POOR AND THE MODERN LABOR MARKET

A major theme of de Janvry's paper is that the marginals are functionally related to the modern sector that needs them to face the conditions under which growth occurs in the periphery. This argument demonstrates a general lack of knowledge about how rural-urban labor markets work in Latin America. The point is that modern industrial sectors require skilled labor. Because of this, there is a major market imperfection between the marginals and the labor force employable by modern industry. Consequently, industry has to train its own workers, and at considerable expense to itself. The large masses of unskilled workers are largely irrelevant to industry.

The major fallacy of de Janvry's argument in this respect, however, is that large masses of unskilled labor constitute a cheap source of labor to the industrial sector. In reality they do not. A cheap source of labor to the industrial sector would be represented by large quantities of trained or educated workers. These workers would be more productive, would be less costly per unit of labor service rendered, and would require less direct investment by the industrial sector. To the extent that marginality is a function of low investments of human capital in the worker, the industrial capital has little vested interest in it. If industrialists had so much power, why did they not impose policies that would have been more in their interest?

It is also difficult to believe that industrialists in the periphery view labor only as a cost item and not as a potential market for their product. Surely they must realize that they have to sell their product and that without a market for their product they have no existence. De Janvry fails to explain why industrialists in the periphery would be so different from those in the center in this regard.

SOURCE OF THE FOREIGN-EXCHANGE CONSTRAINT

Lack of foreign exchange has been a constraint to a more rapid rate of development in many Latin American countries at one time or another. This constraint has been largely self-imposed, however, and not imposed from abroad. A belief that import substitution would both provide an engine of growth and solve the balance-of-payments problem caused most countries to neglect their export sector. In fact, autarchy was a prevalent objective of policy as a means of cutting countries off from the international capitalist system. Exchange rates were kept overvalued as a means of cheapening imports in terms of the domestic currency; export controls and export taxes were imposed; and little attention was given to promoting or stimulating exports. Clearly exports stagnated or declined.

There was nothing inevitable about these policies or the consequences that resulted. World trade expanded in the post–World War II period faster than world GNP. And there is ample evidence to suggest that periphery countries could have participated in this growth had they themselves not withdrawn from the market. Brazil, for example, had relatively stagnant exports from 1947 through about 1965. Once quotas were removed from exports and the cruzeiro was devalued to something approaching its equilibrium level, however, exports expanded rapidly, to double and triple in a short period of time.[10] These growing exports, plus a large capital inflow due to a more receptive *ambiente* for foreign capital, eased whatever import constraint prevailed and enabled the Brazilian economy to grow at unprecedented rates from 1968 through 1974. Even in the face of the energy crisis of 1975 the Brazilian economy grew at a rate of 5 percent.

De Janvry focuses on the importance of the capital-goods constraint through the import bill. Two points seem relevant. First, the industrialization process itself did not need to be as capital-intensive as it was. This was a result of factor-price ratios that were seriously distorted in favor of capital and away from labor. Second, the particular industrial mix chosen was especially dependent on the importation of capital goods. There was nothing inevitable about this either. Moreover, policy changes in both these dimensions could have fomented a more broadly based industrialization process.

COMPETITIVENESS OF INTERNATIONAL CAPITALISM

Dependency theorists generally assume that the center countries have a monopoly on the goods they sell, and assert that systematic market distortions were imposed by the center countries to take advantage of their global monopoly. This implies, of course, that the center countries acted collusively, since despite the use of "the center," there were multiple countries involved.

In point of fact, it would appear that there were strong competitive pressures among the advanced countries for foreign markets. The United States' overvaluation of its currency can be interpreted as an attempt to exploit its position in the capital-goods industry.[11] But at the same time countries such as Germany and Japan *under*valued their currencies and competed strongly through their trade account.

The United States clearly had some technologically superior goods through part of the post–World War II period. It may have reaped some short-term economic rents to this superiority, but competition in most fields was strong. Moreover, it is worth noting that the United States did not attempt to maintain a monopoly position except in fields that were of obvious national security interest, such as computer technology. The United States has been generous in its willingness to provide technical assistance to other countries at no cost and to help develop a technological capability in countries that were willing to accept such assistance.

Perhaps more important, there were no barriers to prevent other countries from developing their own technological capability or from acquiring technical assistance. American scientists and technicians, as well as those from other countries, were highly mobile in this period. There was nothing to keep them from taking up employment in other countries. In point of fact, the technological lag in the periphery was largely self-imposed, just as was the foreign-exchange constraint.

ALTERNATIVE WAYS OF CHEAPENING LABOR

De Janvry argues that cheap labor was in the best interest of the dominant industrial and agricultural classes and that there were only two ways to obtain it:

(*a*) by imposing repressive labor policies and (*b*) by lowering the cost of wage goods by government decree. Clearly both elements of this assertion are false. Cheap labor does not necessarily mean low-wage labor. Productivity is the missing link. And de Janvry's model does not explain why this approach was not followed by the capitalist classes. Labor services can be inexpensive even if the "price" of labor is expensive so long as the productivity of labor is high. This was the lesson of the Leontief paradox, and it is surprising that this ample body of literature was ignored by de Janvry.

A key issue is why "cheap" labor was not sought by means of increasing the productivity of the labor force. Modernization of the agricultural sector, for instance, in most cases would have generated an ample product surplus. With the decline in product price that would have resulted, real wages could have risen for the urban sector even while nominal wages remained constant. The internal market for goods from the industrial sector would have been broader, labor would have been released to the nonfarm sector, and in many cases the foreign-exchange constraint would have been eased, either through increased exports or reduced imports.

The important point is that if there were in fact the interdependence and overlap between the rural aristocracy and the industrial elites that de Janvry believes existed, such a coalition would have increased the likelihood of going the productivity route.[12] One possible reason why the development process did not take this route is that the rural-industrial coalition did *not* have the political power that de Janvry postulates. In particular, Nathaniel Leff has argued in the case of Brazil[13] that the influence of both the rural aristocracy and the industrial power groups has been much exaggerated and that in most of the post–World War II period policy was made largely by a small intellectual elite. Moreover, he documents the discrimination against both rural and industrial interests by economic policy.

My own notion is that basic mistakes were made in economic policy in large part because of a misguided emulation of the advanced countries. There was a failure to understand the nature and source of growth processes in the advanced countries (we ourselves don't fully understand them yet!) and a tendency to inappropriately confuse the essence of development with the symbols of development. Unfortunately, these mistakes were reinforced by an ideology based on a conceptual model similar to de Janvry's.

The importance of honest mistakes in policy should not be neglected in attempting to understand the post–World War II development of Latin America. Economists had very little influence on policy making. There were very few of them, in any case, and the dearth of data on and knowledge of how the various economies worked was great. Under these circumstances, and in view of the desire of most countries in the region to telescope what had been a century or more of development in the advanced countries into a much shorter time period, it is not surprising that the symbols of development were confused with the essence of development. One does not have to search for conspiracy to explain the problems that arose.

It is interesting to note the changes in economic policy in Brazil in the period since 1965. Economic policy has been rationalized, investment in education has increased greatly, and a major effort has been made to strengthen the capacity for agricultural research. Although the income of the low-income groups has not increased as rapidly as that of the upper-income groups in the surge of development that has resulted, it has increased on an unprecedented scale in an absolute sense.[14] All of this has happened in a period when, most observers believe, industrial and agricultural power groups have been favored by economic policy.

What is perhaps more interesting in the case of agriculture is that between 1960 and 1970 per capita income of both the upper- and lower-income groups improved relatively, while that of middle-income groups declined.[15] Hence the development process has been much more complex than de Janvry implies, and at least on the surface there is little evidence that the poor were being exploited. Moreover, it should be noted that the general interpretation of economic policy in the late 1960s was that it favored the middle class. Clearly this was not true in the case of agriculture, which goes to show that things are not always what they appear to be on the surface.

POPULATION GROWTH

A major reason that per capita incomes in Latin America have lagged behind those in the advanced countries in the post–World War II period is that population growth rates have been quite high. Both the general economy and agricultural output have grown in the aggregate at rates comparable to those in the United States and other center countries. A major deficiency of de Janvry's model is that it gives only passing attention to population growth, which has been a major source of the lag in per capita income in the region.

Clearly, a conspiracy theory could probably be cooked up to explain the continued high rate of population growth as well. But again, the objective facts are that a demographic transition that spread over a long period of time in the center countries would have had to be compressed into a very short period of time in the periphery because the technology of reducing mortality rates was easily transferable. The economic conditions that would have reduced birth rates without a major government intervention were not forthcoming, with the result that high population growth rates are both the cause and the result of sluggish growth rates in per capita incomes.

AGRICULTURAL STAGNATION AND RURAL POVERTY

De Janvry makes a number of more specific assertions about agricultural development and rural poverty that also merit comment. In the first place, he argues that new industrialists sought access to land in order to diversify portfolios and to gain access to power over the institutions of society. The first explanation is probably a necessary and sufficient explanation for what happened, and one does not have to appeal to conspiracy. Industrialists sought access to land in large part because of the lack of viable alternative capital instruments and as a means of protecting their assets from high and unstable rates of inflation. It is interesting to note that when Brazil reformed its capital markets in the latter 1960s and the stock market boomed, there was a major shift of capital out of land and urban assets into other capital instruments. The intersectoral mobility of capital was quite high, and responded to economic incentives.

De Janvry's description of the role played by the Mexican land reform is a travesty of the facts. That land reform did little more than devise a means of fixing the worker to the land and bribing him to accept a *miseria*.[16] Rural poverty is today still associated with the *ejiditario* system and is concentrated in that region. The ambitious *infrastructure* investments and the diffusion of land-saving technological changes have occurred in other parts of Mexico, in regions where large farms and capital dominate. The lesson should be obvious.

Similarly, de Janvry argues that except in Mexico, agriculture in Latin America has entered upon a long period of stagnation and the development of capitalist farming in food crops has been blocked. In the first instance this is to ignore the incredible modernization and development of São Paulo agriculture, which through most of the post–World War II period has experienced gains in productivity growth comparable to those realized in the United States.[17] And São Paulo is not an isolated pocket. Until recently it produced some 30 percent of Brazil's agricultural output. Moreover, the modernization of its agriculture has extended into surrounding regions. This has occurred, as noted above, in a region where industrial and agricultural capital were well integrated, contrary to the predictions of the de Janvry model.

It is also incorrect to say that the production of food crops under a capitalist system has been blocked in Brazil. In terms of production value, rice is one of the most important crops in that country, and the growth in output has been well above average. Moreover, the expansion in output has been produced largely by large capitalist farms with modern equipment and modern inputs.[18] More generally, the price of grains has declined in Brazil, in part because output has been increasing faster than demand.[19] And Brazil has heavily subsidized the production of wheat, again contrary to the predictions of the de Janvry model.

De Janvry further asserts that only the *latifundios* have been maintained as viable economic units through institutional control by the dominating elites, who monopolize institutional services. It is clear that larger farmers have easier access to institutional services in Brazil, as they do in most countries. Moreover, it is clear that this is a source of continuing poverty to low-income groups. But to argue that only the *latifundios* have been maintained as economically viable is to subscribe to prevailing myths about Latin American agriculture without looking at the data. Much of São Paulo and South Brazilian agriculture is dominated by medium-sized family units.

De Janvry argues that the cost of stagnation has been shifted to labor through miserable wages. The implication is that both volition and a culprit were involved. Yet, one does not have to search for a culprit. The consequences of low productivity and stagnation are low wages. And again, as in most countries, the bulk of poverty in most Latin American countries is in agriculture. But there are at least two reasons for this. First, the natural process of development is that wages remain low in agriculture because resources have to be transferred from that sector. When barriers to migration are erected by inappropriate factor-price policy and the lack of education among the rural population, the lag in adjustment can be severe. As argued above, it is not clear, however, that these barriers were in the best interests of the industrial-agricultural classes or that those classes were responsible for imposing them.

Second, a major cause of poverty in the rural sectors of Latin America is the failure to direct appropriate development resources to the sector. This was in part due to a desire not to subsidize what was believed to be the wealthy rural aristocracy. The consequence, of course, was that the poor as well as the well-to-do in agriculture suffered, as did the economy as a whole. Equally important in explaining the underinvestment in agriculture was the simple desire to industrialize. But misordered priorities are quite different from a conspiracy to keep the poor poor.

Somewhat ironically, de Janvry argues that precapitalist relations are needed to cheapen labor by tying it to the land and alienating it from its own opportunity cost as long as it is a scarce factor. In part, this assertion is a contradiction in terms, for if the labor were scarce, its value would be high and/or rising. But more important, it was the Mexican land reform which de Janvry so admires that kept the labor tied to the land by institutional means and precluded resource adjustments that could have led to growth in productivity and increases in per capita income.

De Janvry goes on to lament the shift from sharecropping and *colono* arrangements to the use of wage labor. This argument is doubly ironic, for the sharecropper-*colono* system has been much criticized by intellectuals both inside and outside of Latin America for its supposed exploitative characteristics. Few have made the effort to understand the system as a means of wage payment or as a means of sharing risk.[20] In Brazil the sharecropper-*colono* system has declined in large part as a result of wage legislation designed to improve the lot of the rural worker. Alas, the world is often not what we think it is. Nor are the effects of policy what we expect them to be.

Finally, de Janvry returns to his theme that labor costs are reduced in the periphery countries by design. In the context of agriculture, he argues that these labor costs are reduced by maintaining the subsistence economy, since that provides the commercial sector with labor whose subsistence is already partially covered by production in the *minifundio*. This simplistic and parochial view of the world ignores the basic imperfections in the intersectoral labor market, which have caused labor to be dammed up in the agriculture of most of Latin America.

I submit that the source of these imperfections was misguided economic policies designed to obtain rapid forced-draft industrialization. The rural poverty and stagnation that resulted were a logical *consequence* of those policies; moreover, the poverty and agricultural stagnation that resulted were not in the best interest of either the agricultural landowning classes or the industrial sector. Alternative policies that would have raised productivity and in turn incomes in the agricultural sector would have served them much better.

SOME CONCLUDING COMMENTS

The issues that de Janvry raises are quite important. Although cast in the context of Latin America, the situation of rural poverty and agricultural stagnation that he attempts to understand are common around the world.

His conclusion that marginality in the periphery is a contradiction of growth is appropriate, and it is pertinent at both the international and national levels. That being the case, one wonders why those who in his view had the power in their hands were so short-sighted. The truth is that neither the center nor national power groups in the periphery had a vested interest in continued poverty, and the explanation for this problem has to be sought elsewhere.

All economic systems work imperfectly. And the nature of growth and development is such that some groups are benefited while others are either left on the sidelines or suffer downright deleterious consequences. It is a proper role of government to redress these imbalances and to see that the benefits of development are distributed in a reasonably equitable manner.

De Janvry is to be complimented for attempting to understand the problems of Latin America in an international dimension. In the interdependent world economy of the post–World War II period, developments in one country or region have obvious and important implications in other parts of the world, and the nature of development in one part of the world both opens up opportunities for and imposes constraints on other countries. But witch hunts and searches for scapegoats under the premise that a conspiracy theory of development applies can be counterproductive. They tend to shift attention from the critical issues and delay the much-desired development of the region.

Mistakes in economic policy are not rare, even in advanced countries, and their consequences can be serious. When forced-draft development policies are implemented with a limited stock of technical knowledge, limited technical capability, and sparse data, the probability of such errors is great. Unfortunately, interpretations such as de Janvry's have led to some of the most serious policy errors in Latin America, for they have given rise to the attempts at autarchic development which have led to such a dead end.

The challenge today, as in the past, is to confront alternative theories with data and thereby submit them to testing. Further, we need to understand a great deal more about how national economies grow and develop and how development is transmitted in an international system. Too little attention has been given formally to the consequences of growth in one country or another and to the international linkages among countries.

In the context of many of the issues raised by de Janvry, it is important to go back in time and analyze the alternatives before decision makers at the time that decisions were made and what influenced them. This kind of historical analysis is important, but it is seldom done. Our ignorance about how policy is made and who shapes it is vast.

The economist's task is not completed until he understands why economic policy is what it is. Understanding who benefits from particular policies and who bears the costs will be a key factor in gaining such understanding. Eventually, collaboration among economists and political scientists will be required. Out of this collaboration may eventually evolve an integration of political and economic theory into a more comprehensive body of social theory. With that in hand, perhaps we can elaborate a more relevant political economy for the rural development of Latin America. The challenge is before us.

NOTES

1. Alain de Janvry, "The Political Economy of Rural Development in Latin America: An Interpretation," *American Journal of Agricultural Economics* 57, no. 3 (1975): 490–99, reprinted as chapter 6 in this volume.

2. For a description of the urban-industrial impact model see Ruttan, chapter 2 in this volume. The urban-industrial impact model was originally published as a "retardation hypothesis" (see Theodore W. Schultz, "A Framework for Land Economics—The Long View," *Journal of Farm Economics* 33 [May 1951]: 205–15; and idem, *The Economic Organization of Agriculture* [New York: McGraw-Hill Book Co., 1953]. For a brief survey of the results obtained from the empirical work see G. Edward Schuh, "Comment," in *The Role of Agriculture in Economic Development*, ed. Erik Thorbecke [New York: National Bureau of Economic Research, 1970]).

3. *The Economic Development of Latin America and Its Principal Problems*, Economic Commission for Latin America (Lake Success, N.Y.: United Nations, Department of Economic Affairs, 1950).

4. For a critique of the declining terms-of-trade hypothesis see M. June Flanders, "Prebisch on Protectionism: An Evaluation," *Economic Journal* 74, no. 294 (1964): 305–26.

5. This is not to deny that governments in periphery countries have not turned the terms of trade against agriculture and other mass consumption items, as de Janvry notes.

6. See D. Gale Johnson, *World Agriculture in Disarray* (London: Fontana/Collins, 1973).

7. See, for example, A. C. Whitaker, *History and Criticism of the Labor Theory of Value* (New York: Columbia University Press, 1904); and J. Schumpeter, *History of Economic Analysis* (New York: Oxford University Press, 1954).

8. G. Edward Schuh, "The Modernization of Brazilian Agriculture: An Interpretation" (paper presented at the Conference on Growth, Productivity and Equity Issues in Brazilian Agriculture, The Ohio State University, 13–15 January 1975, Mimeographed).

9. It is also worth noting that recent attempts to strengthen the research and teaching in agricultural sciences were started with the collaboration of U.S. foreign aid.

10. For an analysis of the role of the exchange rate on agricultural exports and of the elasticity of export supply see Robert L. Thompson and G. Edward Schuh, "Trade Policy and Exports: The Case of Corn in Brazil" (West Lafayette, Ind.: Department of Agricultural Economics, Purdue University, 1979, Mimeographed).

11. The chronic overvaluation of currencies by the LDCs can equally as well be interpreted as an attempt to offset this price-enhancing effect of the overvalued dollar.

12. See Schuh, "The Modernization of Brazilian Agriculture," for a discussion of the benefits of such a coalition in the case of São Paulo.

13. Nathaniel H. Leff, *Economic Policy-Making and Development in Brazil, 1947–1964* (New York: Wiley, 1968).

14. See Carlos Langoni, *Distribuicão da renda e desenvolvimento economico do Brasil* (Rio de Janeiro: Editora Expressão e Cultura, 1973). For a summary of some of the pertinent data in English see G. Edward Schuh, "The Income Problem in Brazilian Agriculture" (West Lafayette, Ind.: Department of Agricultural Economics, Purdue University, 1973, Mimeographed).

15. Langoni, *Distribuicão da renda e desenvolvimento economico do Brasil*, 68.

16. For example, see Eduardo Venezian and William K. Gamble, *The Agricultural Development of Mexico* (New York: Frederick A. Praeger Publishers, 1968).

17. See *The Modernization of São Paulo Agriculture* (São Paulo: State Secretariat of Agriculture, 1973).

18. Paul J. Mandell, "The Rise of the Modern Brazilian Rice Industry: Demand Expansions in a Dynamic Economy," *Food Research Institute Studies in Agricultural Economics, Trade, and Development* 10, no. 2 (1971); and Andres Troncoso Vilas, "A Spatial Equilibrium Analysis of the Rice Economy in Brazil" (Ph.D. diss., Purdue University, 1975).

19. See G. Edward Schuh, *The Agricultural Development of Brazil* (New York: Frederick A. Praeger Publishers, 1970).

20. For a couple of important exceptions see Allen W. Johnson, *Sharecroppers of the Sertão— Economics and Dependence on a Brazilian Plantation* (Stanford: Stanford University Press, 1971); and George F. Patrick, "Efeitos de programas alternativos do governo sobre a agricultura do nordeste," *Pesquisa e Planejamento* 4, no. 1 (1974): 49–82.

III

Policy Analysis in a General Equilibrium Framework

Introduction

Most agricultural economists working on policy issues in the Third World frame their analyses in a partial equilibrium context, even though much of the literature on agricultural development stresses the interdependence between agricultural growth and overall economic development. Although a sound understanding of local microeconomic conditions is essential for project and policy analysis, many agricultural policy issues need to be analyzed in a framework that stresses the interactions between agriculture and other sectors of the economy.[1] The articles in this section examine three interrelated components of agricultural policy that require a general equilibrium framework of analysis: price policy, food security, and international trade.

Price policy, especially food price policy and its relationship to technical change in agriculture, is the first of these three components.[2] As Timmer points out (chapter 8), even in centrally planned economies many day-to-day resource allocation decisions by farmers and consumers are made in response to private incentives, which often take the form of market prices. As Third World economies become increasingly monetized and integrated, a growing number of resource allocation decisions typically are mediated through markets. Food price policy is one of the major instruments available to governments for influencing those decisions.

Food prices play two crucial roles in most economies. First, they serve as incentives to food system participants (farmers, merchants, processors) concerning what to produce, how to produce it, when to produce it and in what quantities, how to process it, where to market it, and so forth. Food prices, in other words, act as signals that guide resource allocation within the agricultural sector and among sectors. Much of the work of neoclassical economists has focused on "getting prices right" so that these signals will guide resource flows efficiently (see, for example, the volume edited by Schultz [1978]).

As Timmer (chapter 8), Mellor (chapter 10), and Falcon (chapter 12) stress, however, "getting prices right" is not simple, because food prices are a major determinant of the real income of a large proportion of the population in low-income countries. When many people spend up to 60 percent of increments to their income on food, changes in food prices substantially affect the level and distribution of real income, as Mellor discusses in chapter 10. Furthermore, because food is the major wage good in most low-income countries, food price policy can strongly influence a country's industrialization efforts (Mellor 1976) and, through effects on the government budget, a host of macroeconomic variables.

The dual role of food prices—incentives to food producers and major determinants of the real income of much of the population—complicates the process of determining an appropriate food price policy. Many governments face a dilemma: high food prices may stimulate long-term growth in agricultural output, but they may also impose severe short-term privation on consumers. Concern for the adverse effects that higher food prices would have on low-income consumers and recognition of the political power of urban consumers explain the reluctance of many governments to increase farm prices, in spite of the urgings of international agencies (see, for example, World Bank 1981). As Timmer points out, "A government that cannot raise food prices because it no longer will be the government will not raise food prices, no matter how critical that is to long-run efficiency." The situation is further complicated by the increasing complexity and interdependence of the world food economy (as discussed by Falcon in chapter 12), which makes it very costly for governments to try to insulate domestic food prices from international prices.

A major challenge for economists is devising ways to address this dilemma of the trade-off between the short-term welfare losses to consumers and the long-term gains in agricultural output that would likely accompany increases in food prices. One possibility is using food aid as a bridging mechanism during a period of transition to higher agricultural prices, but this would require long-term commitments of food aid by donors, which, as Siamwalla and Valdés point out in chapter 13, donors have been unwilling to make.[3] Increased agricultural output through cost-reducing technological change is another potential way around the dilemma.

In chapter 9, Mellor examines the relative efficacy of price policy, technological change in agriculture, and other mechanisms in promoting a net transfer of resources from agriculture to other sectors of the economy during the course of development. After reviewing the experiences of Taiwan, India, Japan, Britain, France, and the Soviet Union, he concludes that "for most contemporary low-income countries, the route of technological change in agriculture may be the only feasible one." As Mellor points out, rapid technological change in agriculture allows a government to maintain low prices in order to promote resource flows out of agriculture while still maintaining incentives for farmers to ensure an adequate food supply.

In chapter 11, Raj Krishna examines the interactions between price and technology policies, and he challenges the recommendations of "price fundamentalists," who believe that raising farm prices is the key to agricultural growth.[4] Because of low supply elasticities for aggregate agricultural output in most low-income countries, Krishna argues that an increase in farm-gate prices large enough to stimulate rapid agricultural growth would have very adverse macroeconomic effects. Instead of relying solely on price policy, low-income countries should promote technological dynamism in agriculture, within the context of a "congenial price regime" that fosters adoption of new agricultural technology. Krishna goes on to discuss how, in practical terms, a government should go about establishing the support prices necessary to institute this "congenial price regime."

Food security is the second major policy area discussed in this section of the book. Food security is related to both price policy and technological change in agriculture and is best addressed within a general equilibrium framework. Food security analysis is a process of determining an appropriate mix of domestic food production, storage, international trade, and other income-generating activities to ensure that "food-deficit countries, or regions or households within these countries . . . meet target levels of consumption" (Siamwalla and Valdés, chapter 13). A crucial element in food security planning involves trying to ensure adequate effective demand for food among the poor, as well as adequate food supplies. By recognizing that hunger often results from inadequate effective demand rather than simply inadequate total food supplies, food security analysis links hunger with poverty and the lack of productive employment.[5]

Food security research has generally been concerned with food security at two levels—the international level and the national/regional level.[6] Discussions of international food security have focused on the relative efficacy of different mechanisms for ensuring that countries have access to food during periods of low domestic food production or foreign-exchange earnings. The research of Siamwalla and Valdés has made an important contribution by documenting the high cost of trying to ensure food security through a system of international grain reserves.

Research by Siamwalla and Valdés, Goreux (1981), and Konandreas, Huddleston, and Ramangkura (1978) was instrumental in convincing the International Monetary Fund (IMF) to expand its compensatory financing facility in 1981 to include a cereal import facility. This facility allows low-income countries to finance above-trend cereal imports through borrowing from the IMF, in a manner similar to that described in the Siamwalla-Valdés paper.

Food security analysis at the national or subnational level focuses on how domestic marketing, transportation, and communication impediments may prevent the achievement of food security at the local level even if countrywide food supplies are adequate. In chapter 14, Lele and Candler contend that in East Africa the agricultural data base is inadequate to discern even basic trends in food production, let alone the deviations from trends that Siamwalla and Valdés use to define food insecurity. Lele and Candler argue that given the fragmented markets and weak transportation and communication infrastructure of many countries, food security has to be addressed primarily at the local level. In many instances, planting cassava as a drought reserve crop may be a much more effective food security measure than international grain reserves or compensatory financing schemes.

As Lele and Candler point out, much of the analysis of food security on the international level implicitly assumes that most agricultural production passes through well-functioning markets and that governments hold a significant share of total grain stocks. On the other hand, discussions of national food security sometimes focus so narrowly on local production and storage problems that they downplay the potential of interregional and international trade in assuring food security. A major challenge for analysts is to integrate more closely the work on international food security with that on food security at the national and subnational levels.

The third policy area discussed in this section of the book is one that is most generally thought of in macroeconomic terms—defining an appropriate agricultural trade policy. The rapid growth in world trade in recent years[7] and the increased interdependence it implies for growth rates in Third World and industrialized nations (Lewis 1981) have given increased salience to the debate between proponents of import substitution and advocates of export-led growth. It has also raised anew the question of the degree to which primary product exporters can benefit from commodity agreements and export cartels.[8]

In chapter 15, Myint examines theoretical arguments and empirical evidence in evaluating the debate between import substitution and export-led growth. He finds substantial support for the hypothesis that rapid export expansion leads to faster economic growth. The reasons are not primarily the static efficiency gains predicted by neoclassical theory, however, but the indirect dynamic effects of trade on the economy. These include gains due to better allocation of the flow of investible funds through reductions in product and factor-price distortions, the educative and competitive effects of an open economy on inducing greater X-efficiency (see Leibenstein 1978), and economies of scale and increasing returns from specializing for a wider export market. Myint argues that many countries in Africa and Southeast Asia also have the potential to increase growth through promoting peasant agricultural and handicraft exports, which draw into production previously unused resources.

Capturing the gains outlined by Myint may require active pro-trade intervention by government, not simply removal of existing impediments to freer trade. Greater reliance on and specialization in trade, however, expose a country to the risks as well as the benefits of international markets. Many of these markets are volatile and involve a relatively small number of large-scale commercial and state trading organizations. One way in which primary product exporters have sought to stabilize these markets and gain a measure of countervailing power has been through encouraging the creation of international commodity agreements and export cartels.

In chapter 16, Jere Behrman lays out the analytics of international commodity agreements and examines the question of who gains from commodity price stabilization. Contrary to the assertions of Johnson (1979) and other free-trade advocates, Behrman shows that the distribution of benefits from price stabilization depends on a host of empirical issues, which cannot be determined a priori. Behrman also examines the conditions under which collusive arrangements among primary product exporters are likely to enhance export revenues. The conditions turn out to be quite restrictive, which explains why few commodity groups have been as successful as OPEC in extracting monopoly rents from the industrialized world.

The three policy issues discussed in this section—price policy and its relationship to technical change in agriculture, food security, and reliance on international trade—coalesce in the area of food policy. Food policy analysis involves determining the mix of production, price, trade, technology, marketing, and other policies needed to achieve the goals a country establishes for its food system. Timmer (chapter 8) believes that most countries set at least four basic objectives for their food systems: (1) efficient growth in the food and agricultural sectors, (2) improved income distribution, (3) satisfactory nutritional status for the entire population, and (4) adequate food security. Achieving these goals involves evaluating the technical, economic, social, and political constraints facing policy makers and developing ways of addressing those constraints. This calls for a detailed understanding of the dynamics of the food system and of its interrelationships with the rest of the economy.[9] It also calls for a political economy approach that explicitly recognizes the limited degrees of freedom facing most policy makers and tries to find ways to increase those degrees of freedom.

If, as we believe, food policy will be at the center of debates about development for many years, then the role of donor agencies cannot be ignored. In chapter 12, Falcon outlines the potential role the United States, as a major donor, can play in helping to bring about the reforms necessary to alleviate hunger in the Third World. In the increasingly interdependent world food system of the 1980s, Falcon argues, food policy can no longer be made in isolation, and the United States could benefit itself as well as low-income countries by making food policy the centerpiece of its development assistance programs.

NOTES

1. See Johnston and Kilby 1975. For an excellent example of analysis conceived in a general equilibrium framework but built upon a sound microeconomic data base see Mellor 1976; see also Mellor 1978, chapter 10 in this volume. In advancing the view that many agricultural policy issues are best addressed within a general equilibrium framework, we do not mean to imply that these issues must necessarily be analyzed using a formal general equilibrium model. Often "an intuitive but sophisticated analysis of the system" is called for, provided that such analysis takes sufficient account of intersectoral linkages (Timmer, chapter 8).

2. Technical change in agriculture is discussed further in part IV, chaps. 23–27.

3. For a review of the potential roles of food aid in agricultural development see Maxwell and Singer 1979.

4. See Peterson 1979.

5. See Falcon's discussion in chapter 12 of the relationship between poverty and hunger. See also Eicher's distinction between Africa's food production gap and hunger in Africa (chapter 31).

6. See the volume edited by Valdés (1981).

7. World trade has increased much faster than output, growing at an annual rate of 6.9 percent between 1965 and 1980 (World Bank 1982, 26).

8. See, for example, Schmitz et al. 1981.

9. For an excellent discussion of ways to incorporate one of the major food policy objectives—improved nutrition—into the evaluation of agricultural projects see Pinstrup-Andersen 1981.

REFERENCES

Goreux, Louis M. 1981. "Compensatory Financing for Fluctuations in the Cost of Cereal Imports." In *Food Security for Developing Countries,* edited by Alberto Valdés, 307–33. Boulder: Westview Press.

Johnson, Harry G. 1979. "Commodities: Less Developed Countries' Demands and Developed Countries' Responses." In *The New International Economic Order and the North-South Debate,* edited by J. N. Bhagwati. Cambridge: MIT Press.

Johnston, Bruce F., and Peter Kilby. 1975. *Agriculture and Structural Transformation: Economic Strategies in Late-Developing Countries.* New York: Oxford University Press.

Konandreas, Panos; Barbara Huddleston; and Virabongsa Ramangkura. 1978. *Food Security: An Insurance Approach.* Washington, D.C.: International Food Policy Research Institute.

Leibenstein, Harvey. 1978. *General X-Efficiency Theory and Economic Development.* New York: Oxford University Press.

Lewis, W. Arthur. 1981. "Development Strategy in a Limping World Economy." In *Rural Change: The Challenge for Agricultural Economists,* edited by Glenn Johnson and Allen Maunder. Proceedings of the Seventeenth International Conference of Agricultural Economists held at Banff, Canada, 3–12 September 1979. Westmead, Farnborough, Hauts., England: Gower.

Maxwell, S. J., and H. W. Singer. 1979. "Food Aid to Developing Countries: A Survey." *World Development* 3: 225–47.

Mellor, John W. 1976. *The New Economics of Growth: A Strategy for India and the Developing World.* Ithaca: Cornell University Press.

Peterson, W. L. 1979. "International Farm Prices and the Social Cost of Cheap Food Policies," *American Journal of Agricultural Economics* 61(1): 12–21.

Pinstrup-Andersen, Per. 1981. *Nutritional Consequences of Agricultural Projects: Conceptual Relationships and Assessment Approaches.* World Bank Staff Working Paper, no. 456. Washington, D.C.

Schmitz, Andrew; Alex F. McCalla; Donald O. Mitchell; and Colin Carter. 1981. *Grain Export Cartels.* Cambridge, Mass.: Ballinger.

Schultz, Theodore W., ed. 1978. *Distortions of Agricultural Incentives.* Bloomington: Indiana University Press.

Valdés, Alberto, ed. 1981. *Food Security for Developing Countries.* Boulder: Westview Press.

World Bank. 1981. *Accelerated Development in Sub-Saharan Africa: An Agenda for Action.* Washington, D.C.

———. 1982. *World Development Report, 1982.* New York: Oxford University Press.

8

Developing a Food Strategy

C. PETER TIMMER

POLICY PERSPECTIVE

 What does it mean for a country to "develop a food strategy"? How does a country do it? What should be its objectives, where does it turn for help, what lessons can it learn from others, and what must it learn from the painful but effective techniques of trial and error?

These questions have no set answers. Each must be addressed in the context of individual country and temporal settings. Each must also be addressed analytically, and that is the purpose of this paper. The analytical perspective permits us to distinguish the set of answers—the actual building blocks of a country's food policy—from the process of identifying the right questions to ask as the first step in the design and ultimate assembly of those building blocks.

This paper is about the design process. Although the way the analytical design process is defined places some boundaries on the dimensions of the resulting food strategy, it is still possible, indeed necessary, to distinguish between what an "optimal" food policy looks like for all worlds and the analytical process that frames the right set of questions. These questions emerge from the need for a national food strategy to satisfy four basic objectives: (1) efficient growth in the food and agricultural sectors; (2) improved income distribution, primarily through efficient employment creation; (3) satisfactory nutritional status for the entire population through provision of a minimum-subsistence floor for basic needs; and (4) adequate food security to ensure against bad harvests, natural disasters, or uncertain world food supplies.

Three assumptions define the philosophical starting point for this chapter. They are as much intuitive perspective as based on verifiable reality, and their efficacy no doubt varies from country to country and from time to time. It is best,

C. PETER TIMMER is John D. Black Professor of Agriculture and Business, Graduate School of Business Administration, Harvard University.

Reprinted, with minor editorial revisions, from *Proceedings of the Conference on Food Security in a Hungry World,* International Food Policy Conference, San Francisco, 4–6 March 1981, cosponsored by the University of California, Davis, and Castle and Cooke, Inc. Published by permission of the conference organizers and the author.

however, to lay these assumptions out ahead of time rather than to discover them lurking at the core of the results.

First, the four basic objectives listed above are taken as representative of the broad set of goals most policy makers in poor and rich countries alike have for their food and agricultural sectors. Growth, jobs, a decent minimum standard of living, and security against famine or extreme food shortages capture most of what might ideally be delivered by a successful food strategy. If not all objectives are simultaneously satisfied from a given set of policies (and typically they are not), it will be necessary to understand which objectives are most important. While development plans may say all four objectives are equally important, actual budget, price, and trade policies for the food and agricultural sector usually indicate otherwise. Hence the first important starting point is a political economy perspective organized around understanding national objectives for the food and agricultural sector.

The second starting point is a concern for moving from here to there, that is, for the process of incremental change that moves a country slowly but increasingly surely toward the simultaneous achievement of all four food strategy objectives. Such a marginalist perspective is surely inappropriate in some, perhaps even many, national environments. Without radical restructuring of assets and basic production relations, little progress can be made toward any of the four objectives in such countries. This paper, however, is not about revolution but rather the search for feasible improvements in the individual components of a food strategy in those environments where the four objectives are taken seriously by some sector of the government.

What is feasible depends on what constraints the political and economic system is under. Such constraints are highly heterogeneous. They range from the political base of the national regime and the potential, for instance, to find a million workers and students in the streets when food prices rise, to the diminishing returns in additional rice production from incremental fertilizer applications when fertilizer use is already high, to the inability of national, provincial, or local bureaucracies to identify and reach the truly poor (and only the truly poor) with real resources. None of these three constraints on government policies seeking to reach the four objectives is the sort that economists normally identify as factors limiting government action. The more normal constraints—availability of foreign exchange, of domestic savings, of budgetary resources, even of talented analytical capacity—are real and important. But so too are the other, more nebulous political, economic, or technical constraints. A genuine concern for policy movement, for change in the right direction, requires that the full range of constraints be identified and incorporated into the analysis.

The third major starting point for this paper is a strong belief that understanding markets is critical to the development of an efficient and equitable food strategy. This market orientation grows out of the belief, buttressed by a now impressive range of empirical experience, that in a global economic environment undergoing rapid, almost radical change in the prices, and hence the relative scarcities, of important factors of production and commodities, the efficiency and rapidity of market decision making will generate substantially better economic progress than will strong central planning or heavy reliance on public enterprises operated according to central directives. No food system anywhere in the world is centrally planned, although some (those of the Soviet Union, China, or Cuba, for example) are more heavily influenced by planners' allocations than others. Throughout the world, however, in the fields and in the households, largely private decisions are being made in response to private incentives. How hard to work, which inputs to use, and which foods to consume are decisions that are made mostly through habit and a private calculus. Since a food strategy is designed to change many of these decisions, it is necessary that the analysis leading to the design of that food policy reveal a full understanding of the context in which the decisions are made.

Few would quarrel with this market orientation on efficiency grounds. It is the equity of the outcome, or the resulting nutritional status of the population, that is of concern. The fundamental argument of this paper is that the role of analysis and design of a food policy is to find a feasible way to use market efficiency to help provide nutritional equity. Such analysis is faced with a basic food policy dilemma: in many poor countries, especially in the heavily populated South and Southeast Asian countries, the set of incentives necessary to induce rapid and efficient growth of food production will, in the short run, simultaneously increase the hunger of the urban and rural landless poor, who are already on the nutritional margin of survival.

A food policy that deals with this basic dilemma by keeping food prices low and thus dampening agricultural incentives can partially succeed in the short run as long as food resources are available to implement the policy. The long-run consequence of such a food price policy, however, is a shortage of those essential food resources. The alternative, a food price policy designed specifically for its positive long-run productivity effects, has as a short-run consequence the significant reduction in food intake among the poor whose incomes are not linked fairly directly to food prices. This is, unfortunately, a very sizeable number of people.

Analysis that copes directly with this food price policy dilemma is at the core of the design of any successful food strategy. Recognizing this is to recognize a fundamental revolution in development thinking: macro price policy is the cutting edge of economic development; projects and programs are the cutting edge of reducing nutritional inequities. The critical bridge requires two supports: (1) understanding the nutritional consequences of that efficient macro price policy and (2) designing and implementing targeted food programs that reach the poor within reasonable budget limits.

This perspective on food policy design suggests four components to the analytical process: (1) to determine the feasible set of policies, (2) to extend the degrees of freedom for policy choice, (3) to determine what investments are needed to break the binding constraints on policy choice, and (4) to devise policies to deal with the short-run consequences of efficient long-run food sector development.

The first step is to determine the feasible set, or at least the nature of potential policies and programs that are actively discussed and debated even if considered infeasible or unlikely in the near term. Part of this process involves understanding the perceived constraints on infeasible policies and programs. Do policy makers see these constraints in budget terms, in political risks, or in unresponsive rice plants, farmers, or consumers?

The next step in this policy analysis is a search for ways to extend the degrees of freedom for policy choice. Policy makers do not live in an efficient world where all alternatives are clearly defined with costs and benefits attached, merely awaiting the nod of a person with power to decide. Many policy choices are constrained by myopic visions and faulty analysis of too aggregated a picture of the world. The first instinct of a good food policy analyst is to disaggregate, to take the data apart by producers and consumers, by income class, by commodity, by region, by urban and rural status. The poor—the ultimate focus of a successful food strategy—are different from the middle- and upper-income groups. They consume different foods, live in different places, have different kinds of jobs, and different-sized families. A full analytical understanding of these differences frequently suggests whole new policy approaches to dealing with food problems.

Assembling the basic ingredients of an effective food strategy is frequently prevented by real constraints on what committed policy makers can do. Discovering this is critical for food policy analysis. It is important to know when good policy analysis can suggest additional degrees of freedom and when the effective constraints really are binding. In the latter case the third step in the analytical process is to determine what investments are needed to break the binding constraints on policy choice. It may require investments in building the capacity to generate greater budgetary resources; investments in agricultural research to build the foundation for a locally adapted, high-yielding agriculture; investments in communications and building public trust and understanding with respect to basic policy dilemmas; or investments in analytical capacity. All of these investments no doubt have high payoffs most of the time. Good food policy analysis will identify which investments are *critical* to the design and implementation of a successful food strategy.

The fourth element of creative food policy analysis is facing squarely the short-run welfare consequences of efficient long-run food sector development strategies. Two easy approaches are possible, and both are usually wrong. The first assumes that efficient long-run policies also solve short-run welfare and nutritional problems. In this view all that is lacking is "political will." The second approach recognizes that efficient development strategies do carry severe short-run welfare costs, but the answer is to be seen in neutral fiscal transfers from the wealthy to the poor via the budget. Political will is the missing ingredient in this view of the world as well.

Both approaches beg the important question, How can good policy analysis identify the needed bridging programs that enable policy makers to make difficult choices? A government that cannot raise food prices because it will no longer be the government will not raise food prices, no matter how critical that is to long-run efficiency. A government that cannot collect income taxes cannot make neutral fiscal transfers, no matter what political will it has. Ration shops, subsidized inferior foods consumed primarily by the poor, or even direct food distribution may be feasible programs that would cushion the food consumption consequences of efficient food production strategies. It would be a wonderful world where food consumption decisions could be separated from food production decisions, but it is not the world we live in. Food prices link the two, and food policy analysis begins with this understanding.

ANALYTICAL APPROACH

An implicit hypothesis behind the search for successful food policies is that neither nutrition policies nor agricultural policies can solve the basic hunger problem. In short, "one-sided" analytical approaches, whether from the production side or from the basic needs side, fail to provide for the central role of the food system, particularly food markets and prices. Production approaches focus on the role of new technology, or, more recently, the efficiency and productivity of small farmers, but these approaches fail to integrate the demand side apart from rural farm income. Both on Java and in the United States more than half of rural income is earned from nonfarm sources,[1] and much urban poverty and hunger has no food production link at all. Similarly, despite the impressive accomplishments of nutrition planners in identifying and measuring the extent and consequences of hunger and malnutrition, their "nutritional requirements" approach to allocating resources usually runs into a contrary market economy. Nutrition simply does not have many policy levers to pull.

The alternative is to approach the hunger problem explicitly through the food sector, recognizing the linkages through that sector from the agricultural sector and to the food consumption endpoint. The advantage of such a food policy approach, apart from its central focus on food markets, is the ease with which macroeconomic influences, especially via the budget and macro price policy, and international influences can be linked to the hunger problem.

At this stage the problem sounds sufficiently complicated to require complicated models to understand it. Such models exist, and they test the ingenuity of even the brightest graduate students in their design and interpretation. Agricultural sector models nearly captured the entire agricultural development profession before the rapid changes in the 1970s revealed such models to be essentially static in nature. Not enough is known about even the agricultural growth process in the context of radically evolving macroeconomic and international environments to have much faith in the capacity of such models to capture the subtle linkages between the micro and macro sectors that now are important driving mechanisms in the modernization of agriculture.

Macro-consistency models with separate food grain sectors and even disaggregated income classes face many of the same problems. There is no question that changes in food policies in most Third World countries and many industrialized countries have important macroeconomic repercussions that few agricultural economists are trained to understand. Recognizing and roughly quantifying these macroeconomic effects are important, but two provisos limit the usefulness of the formal and complicated macroeconomic models for this purpose. First, the complexity of real-world patterns of food, services, and industrial goods production, of price formation, of income generation, and of food consumption patterns by income class of various important food commodities is such that all macro models are greatly simplified and frequently require important counterfactual assumptions in order to be "calculable." Second, most such macro models are built on Keynesian assumptions about surplus industrial capacity, closed economies or economies with foreign-exchange "constraints," and cost-plus price formation. Many, but not all, of the major countries with serious food problems more nearly resemble classical economies in the Lewis sense, where labor alone, rather than labor and industrial capacity, is in surplus supply. Such economies have important macroeconomic linkages to their food sectors, but such linkages are not always mediated by Keynesian mechanisms.

What is the alternative to such formal, complicated models if the goal remains to understand the food system and its horizontal linkages to agriculture and nutrition and its vertical linkages to the international economy and to macro price policy? Given the complexity of these linkages and of the individual components themselves, such an understanding can be generated only by an intuitive but sophisticated analysis of the system. Informal modeling of interrelationships and linkages among components of the food sector, combined with an analytical sensitivity both to the important issues and to what seems to be driving the system, is probably the best that can be achieved in the foreseeable future.

Many economists are concerned that there is more art than science to this, and certainly no two creative food policy analysts are likely to emerge with exactly the same sense of how the system works and what the policy options might be. There is, however, a reality driving the complexity. The more skilled the analysts, the more their respective visions of what is possible will correspond.

Three separate sensibilities are required of such analysts. The first is a welfare sensitivity from a macroeconomic policy perspective. This is manifested as a concern for income distribution and the food consumption consequences of budgetary, fiscal, and macro price policy changes. The concern comes from the top down.

The second sensitivity requires a macro understanding from a nutritional or food consumption perspective. The design of nutritional intervention intended to make an impact on the magnitude of basic hunger must be sensitive to budget priorities and the nature of national policy objectives and constraints. An understanding of the macroeconomic and productivity impact of food price changes must condition nutrition policy at the same time that concern for replicability and the aggregate impact of feeding interventions must condition the design of nutrition programs. This concern comes from the bottom up.

A last sensitivity is perhaps most critical of all. Food policy analysis is the search for second-best solutions in an imperfect and poorly understood world. First-best solutions are best, and the designer of food strategies will fight for them whenever they appear even reasonably attainable. But there must be a sensitivity to what is possible. A food policy analyst need not abandon ideals or principle, for these motivate the search for degrees of freedom, but the search must be in the real world and not in a model.

UNDERSTANDING THE FOOD SYSTEM

CONSUMPTION

For the food policy analyst, food consumption is the variable of proximate concern. Nutritional status is strongly influenced by water and sanitation conditions, health status, food preparation techniques, and a host of other variables that intervene between the quantities of food entering the household and how well-nourished the household members are. In most circumstances, however, it is the quantity of food consumed by the household members that determines whether hunger is a significant factor in their lives. Consequently, food policy analysis focuses major analytical attention on food consumption patterns, even to the extent of starting the analysis with this component of the food system.

Two separate analyses—one in aggregate, the other disaggregated—must be conducted and ultimately reconciled. First, it is necessary to generate an understanding of the major market demand parameters for basic foodstuffs. When average per capita GNP rises, how much is market demand for rice likely to increase? market demand for wheat? for cassava? for corn? for meat? How much meat is grain-fed, and what impact will increased meat demand have on demand for grain? How sensitive is market demand to absolute and relative food prices? If all food grain prices rise relative to nonfood prices, how much does demand drop? When rice prices rise, does wheat consumption rise while rice consumption falls?

Answering these questions is important because the resulting income, own-price, and cross-price elasticities provide necessary linkages from macro price policy and macroeconomic performance to food consumption and, through the food marketing sector, to incentives for agricultural production. Obtaining these aggregate market demand parameters with any real confidence is seldom easy. Time series data are frequently short and of dubious accuracy when domestic food grain production makes up a significant part of total consumption. Mechanical regression analysis seldom gives plausible parameters. A combination of intuitive judgment, talking to traders, evidence from household surveys, simple graphical and statistical analysis of available data, and familiarity with similar parameters in other countries is frequently the best that can be achieved. Economic theory suggests some basic consistency relationships; reality suggests that parameters can change from year to year for many reasons, including changing expectations. Aggregate demand parameters are important; they are seldom precise.

The second step in food consumption analysis is to disaggregate the first step. The motivation for this step, however, is quite different from the need to know aggregate market demand parameters to understand the macro linkages in the food sector. The disaggregated consumption understanding is needed in order to trace the effects of various price and income policies on the food intake of the poor. Indeed, it is quite possible to focus this stage of the food consumption analysis specifically on the poor. Additional insights can be gained by similar analysis of the food consumption patterns of the middle class and wealthy, especially with respect to the impact that demand for grain-fed livestock will have. If time and analytical resources are a binding constraint, however, understanding what the poor eat and why should receive first priority.

The starting point is to discover what the poor actually eat. Everyone knows that they eat less than the rich, but in virtually all societies the composition of foods the poor eat is also significantly different from the middle-income and wealthy diet. The point can usually be demonstrated and suitably quantified by preparing separate food balance sheets for three or four income classes in a society and comparing them with the aggregate food balance sheet that is published by most governments. The information needed to do this is normally available from household expenditure surveys (if not published, then in the file drawers of the statistical bureau). Any country that publishes a cost of living index has such data, although the base year may not be recent. Failing this, careful interviews with a dozen representative consumers in each income category will substitute and will not fail to provide fascinating perspectives on the variations in food habits across income classes.

Good food policy analysts should always go this far in understanding disaggregated food consumption patterns. It is critical to get the commodities and the amounts right. Whether it is then possible to disaggregate the demand parameters by income class will depend on data availability, computer facilities, and analytical capacity. With the right combination of these factors, as in Indonesia and Thailand, very powerful analysis of the determinants of food intake by income class can be conducted. Such analysis is conducted commodity by commodity, income class by income class, using extremely large household expenditure surveys conducted by statistical bureaus for other purposes. Only the analytical costs are attributable to food policy analysis.

The results of the few such analyses that have been conducted are both satisfying and exciting. Intuitive prior judgments that the poor are significantly more responsive to economic signals—both income and price signals—are strongly borne out in the analysis. The analysis has also demonstrated significant variations in food consumption levels and in parameters of change by geographic region, by age and sex, and especially by season of the year. It may not help a malnourished child to have enough food on average for the year if it is going to die during a three-month "hungry season."

PRODUCTION

Food production is also a legitimate starting point for food policy analysis. The fundamental fact is that food cannot be consumed unless it has been produced. Society cannot borrow from next year's production for this year's consumption needs (although in desperate times societies have eaten the seed for next year's production). Somehow food must be produced, and understanding that process is an important component of food policy analysis.

What should society ask of its food production sector? The question is harder to answer than it seems. Most people's immediate instinct is that the domestic food producing sector should provide the society's food consumption "needs." But then the thoughtful hedges begin. The United States should not grow all of its sugar and bananas, Japan should not grow all of its wheat, Europe need not grow all of its soybeans. In fact, the same four basic objectives for an overall food strategy can also be held out, with different weights, to the food producing sector as well.

Generation of farm income through efficient agricultural production provides the real productivity base from which all other objectives can be discussed. Without efficient income generation the entire rural sector will act as a drag on both macroeconomic performance and the ability of policy makers to deal with hunger and malnutrition. Knowing which crops, which new crop technologies, which size farms, which rural infrastructure investments, and which production techniques will generate that efficient agricultural productivity is partly the responsibility of policy analysis, public investment, and macro price policy and partly the responsibility of millions of farmers who typically make rational decisions within the constraints of their own household and resource environment.

Food production policy has three major tasks: (1) ensuring the availability of efficient agricultural technology for the various agro-climatic zones of the country; (2) providing a set of macro price policies—prices for capital, labor, foreign exchange, and food—which at the least do not actively discriminate against the rural sector and which ideally would provide more positive incentives for food production; and (3) developing a rural marketing system for inputs and outputs with equal access for all classes of farmers at a minimum and preferential access for small farmers as an ideal.

Each of these public policy roles with respect to efficient food production has an implicit or explicit income distribution consequence. The ideal policies build a concern for income distribution specifically into the policy and program design, for the evidence suggests that rural income distribution can worsen in the context of efficient growth in agricultural production without such specific attention.

Food production analysis must ask many of the same questions a food consumption analysis asks, but it is probably better to begin with the disaggregated perspective. For each of the major agro-climatic zones in a country, what yields are farmers actually obtaining? What is the seasonal pattern of land use relative to temperatures, rainfall, or water availability? Are crops grown in mono-culture stands or are they extensively intercropped? What are the substitute crop possibilities for any given season? Nothing may compete with paddy rice in the rainy season, but soybeans, peanuts, maize, and cassava may all be possibilities for the same plot immediately after the rice harvest. Little area response to changed rice prices is likely in such an environment, while extreme sensitivity to relative soybean to peanut prices may exist.

What prices are farmers receiving at the farm gate for their output, and how do such prices compare with international prices suitably discounted to local levels? The answer to this question is quite revealing about the extent of incentive or disincentive to domestic farmers for their respective crops. If farms are receiving high prices for their crops but are producing very low yields, both by world standards, then a visit to the local agricultural experiment station is in order. It is possible that the farmers simply do not know how to use available crop technology very efficiently, and then a vigorous agricultural extension program is called for. Historically, however, such circumstances have been explained by an absence of a high-yielding crop technology that worked reliably on a farmer's field with available inputs.

How these three empirical relationships—farmers' yields, farm-gate incentives, and available technology—compare with each other provides strong clues as to where government agricultural policy should focus its attention. Most developing countries have provided their farmers with very poor incentives and, until recently, with little modern biological technology. For most countries this record of neglect and discrimination must be reversed if the domestic food production sector is to play a dynamic role in the overall modernization process, if income distribution between urban and rural areas is to be more equal, and if domestic food security is to be improved without increasing resort to unstable world markets.

Three more micro aspects of agricultural production are also important to policy makers. First is the composition of commodity output. Many countries would like their farmers to produce domestically needed quantities of basic food grains at very low prices before turning to more profitable cash and export crops. China has had some, but not unlimited, success in implementing such a policy through centrally determined grain acreage allocations.[2] Countries with less control over their population, and even China at the margin (and more recently in general), have found their farmers quite recalcitrant about growing relatively unprofitable crops. If the balance of commodities produced is undesirable from a social point of view, either the technology or the incentives, and frequently both, will have to be changed.

The second micro issue is choice of technique in agricultural production.[3] How farmers combine factors of production to produce agricultural output is critical for three reasons: (1) depending on the extent of labor hiring, techniques chosen alter the income distribution within rural areas; (2) some techniques use considerably more foreign exchange than others, with resulting macroeconomic consequences; and (3) some techniques are more natural resource–intensive than others, especially in their use of liquid fuels or care with which fragile soils are cultivated. Many choices about technique are strongly influenced by macro price policy, and the choice of technique issue in agricultural production has strong linkages to macro and international economic policies.

The third micro production issue for policy makers is the responsiveness and flexibility of the agricultural system. Traditional agriculture evolved in a variable year-to-year internal environment but a static external world. There is some evidence that the weather may be becoming even more variable, but the external world is clearly no longer static. An agricultural system that is efficient in production and responsive to new signals about what is desired as output will be an essential foundation for a successful food strategy for the 1980s.

MARKETING

The need for responsiveness and flexibility in food production is transmitted from food consumers via the marketing system. If the signals do not get through efficiently, both sides of the food system are frustrated. Since foods as consumed are nearly always different in time, place, or form from foods as produced, nearly all foods must be marketed in the technical sense. Not all foods must enter the marketing system, however, since the three marketing functions are often provided by the farm household itself as it transports grain from the field, stores it, and mills it before consumption. The process of economic modernization raises the opportunity cost of such household activities, and the marketing system tends to supply an ever larger share of these services relative to household production. For food policy analysts the important questions about the marketing system involve its static efficiency with respect to time, space, and form transformations and the dynamic capacity and flexibility to handle varying and especially larger quantities.

Market efficiency refers to whether the marketing system equates costs of storage, transportation, and processing with temporal, spatial, and product form price differences. High marketing margins do not necessarily denote inefficient markets if costs are commensurately high. High costs typically reflect inadequate physical and institutional infrastructure in the marketing system (poor roads accessible to trucks only in the dry season, lack of central market places where price formation is competitive and resulting prices are publicly announced and broadcast to farmers, shortage of fertilizer go-downs close to the point of fertilizer use, inadequate milling facilities), and public policy should normally focus on high-payoff investments that will improve the physical capacity of the marketing system to handle both the inputs and output. Greater capacity usually means lower short-run marginal costs when supplies increase sharply; hence the flexibility of the system to respond to variations is also improved.

Food policy analysts must be alert to two common biases in developing country food marketing systems. First, most traditional peasant households have always sold at least a small portion of their basic grain output in order to buy a few cash goods necessities—salt, cloth, cooking ware. The traditional marketing system evolved to handle limited quantities of grain flowing out of the countryside to commercial centers, and small quantities of consumer goods flowing back to the countryside. Such a marketing system is frequently incapable of providing large quantities of agricultural inputs to the countryside or of returning a sizeable increase in the net marketed output.

The second market bias has built up in regions with heavy reliance on a single export commodity, for example, cocoa, peanuts, sugar. The marketing system then tends to be efficient at moving the export commodity out to ports and possibly at moving foodstuffs and other consumer goods back to the growing region. Even fairly strong demand for food from other parts of the country, especially the commercial and governmental centers, may not be transmitted back to potential growers. The argument is not that cocoa farmers should grow rice or peanut farmers, millet, but that the marketing system may be more efficient at transmitting both commodities and price signals for some goods than for others and in one direction rather than both directions.

The marketing system must not only transmit price signals efficiently but also serve as the arena for price formation for many agricultural commodities. Most countries have national food price policies for their basic staple grain (and frequently for many more commodities), and such policies are normally implemented by special trade arrangements at the international border with respect to imports and exports of the commodity. Even with such a macro food price policy, however, the marketing system must still transmit the desired price signals up to consumers and down to farmers. For other commodities that frequently are very important for farm income—fruits, vegetables, livestock, pulses—the marketing system must provide the price formation arena itself. The institutions by which governments encourage or discourage this activity heavily influence the net returns to farmers and the retail prices to consumers.

The income distribution consequences of inefficient price formation or transmittal of policy-determined price signals are almost uniformly bad. High marketing costs do not imply high incomes for the "middleman," but they do tend to imply high returns to those who can take the real risks of a costly marketing venture. Secondly, even in efficient and low-cost marketing systems many middlemen who are transporting and storing grain will tend to make windfall profits when food prices rise suddenly and unexpectedly. Although no one notices their losses when food prices fall unexpectedly, speculative profits on holding food grain during the food shortage seem antisocial, if not criminal, and many societies have treated the middleman as both. Honest profits are to be made in food marketing, and the role of public policy should be to see that such profits, and only such profits, are a normal component of the cost of providing food to consumers.

MACROECONOMIC ENVIRONMENT

Domestic Macro Policy

If policy is the cutting edge of development, then domestic macro policy conditions the development environment in which a food strategy is designed and implemented. Unfortunately, this macro environment is usually hostile to several of the objectives of a food policy, especially the critical one of efficient (and even equitable) growth in agricultural production. Macro policy has three essential components from the perspective of the food system: (1) fiscal, budgetary, and monetary policy; (2) macro prices for labor, capital, and land (food prices); and (3) the foreign-exchange rate.

Fiscal policy encompasses the overall tendency toward expansion or contraction of the economy; budgetary policy entails the actual sectoral allocations of the fiscal total; and monetary policy tends to be the instrument that accommodates fiscal policy when tax revenues fall short of budgetary expenditures. Monetary policy can be operated independently of fiscal policy, but in most cases the money supply tends to expand to fill fiscal deficits. The inevitable result is chronic and frequently rapid inflation. Inflation is a characteristic of the usual macro policy context in which food policies are framed.

The second essential component of macro policy is the real level of an economy's macro prices—the prices for the basic factors of production: labor, capital, land. Most countries influence land prices through the urban-rural terms of trade (although land also serves as a hedge against inflation), and these terms of trade are heavily influenced by prices for basic food staples. Thus the macro prices are wages, interest rates, and food prices as a proxy for the broader rural-urban terms of trade.

All three macro prices also tend to be biased against efficient agricultural production, at least to the extent that governments are able to implement their policy intentions. Since prices are meant to signal relative scarcity to decision makers in the economy, policy-dictated macro prices that are sharply at variance with actual scarcity values will tend not to be widely applicable as actual decision variables. Low interest rates can be maintained for a few preferential customers through a rationing system and an accommodating monetary policy, but such rates will depress savings and force informal credit rates, especially rural rates, to higher levels than otherwise.[4] High minimum wages can be enforced in large industries but reduce the demand for labor and force wages lower in the much larger informal labor market, also heavily rural. Low food prices, if enforced only by statute, will tend to generate a black market where nonpreferred customers, especially the poor, must pay much higher prices for their foods while forced market deliveries from food surplus areas depress agricultural incentives.[5] The typical macro price policy set, although perfectly understandable as a political reaction to the very difficult pressures of economic modernization, tends to place a very heavy burden on an effective food strategy.

Low food prices can also be maintained, usually more effectively, by appropriate food import or export policy controlled by the government. Restricting food exports or subsidizing imports will reduce domestic food prices. Restricting food exports usually provides a government budget surplus, as in Thailand, so it is a doubly popular policy with urban-based government officials. Since subsidizing imports requires a budget allocation, the costs of such a low food price policy are more apparent.

The third component of macro policy, however, typically hides some of these costs behind an overvalued exchange rate. The foreign-exchange rate—also a macro price because it is set and defended by the central bank—measures the price of a unit of foreign exchange, say a dollar, in terms of the number of units of domestic currency, say the Indonesian rupiah, required to purchase it. Thus the Indonesian exchange rate between August 1971 and November 1978 was Rp 415 per U.S. dollar and is now Rp 625 per U.S. dollar. From 1974 to 1978 the rupiah became increasingly overvalued as domestic inflation ran much faster than international inflation. Food prices were kept low domestically, not by large budget subsidies on imports, but by large imports valued at an overvalued exchange rate. The burden on rural producers and the benefits to urban consumers came through the overvalued exchange rate (permitted in this instance by petroleum revenues) that made imported commodities appear cheaper to the domestic economy than they should have. With the sharp devaluation in 1978 this bias was eliminated, and the costs of maintaining low food prices were transferred directly to the budget. The visibility of such costs changes the nature of food policy discussions from trying to show a bias to examining the costs and benefits of the bias. A devaluation also tends to increase sharply the terms of trade for rural commodities that are not specifically subsidized by domestic macro price and trade policy. Rural areas almost always benefit from devaluations of domestic currencies to reflect more accurately long-run scarcity.

Broad-based macroeconomic reforms encompassing all three components of macro policy—fiscal, budgetary, and monetary policy; macro prices for labor, capital, and food; and the foreign-exchange rate—are not likely to be carried out at the behest of food strategists. But first, they should be carried out anyway if the country is concerned about growth and equity, and second, food strategists should be on the right side rather than the wrong side of the macro policy debate.

Where is the side of the angels? The first point is to understand that macro policy reform will have significant short-run food consumption consequences and large political costs. Programs to deal with both should be in place before a major reform is implemented rather than assembled in a scramble after problems begin to appear. In effect, understanding the need for such programs and assisting in their design is the basic role of good food policy analysis.

With this step taken, the components of effective macro policy are fairly straightforward, even classical: (1) a reasonably balanced budget with restrained monetary growth leading to stable prices by the standards of international trading partners and (2) macro prices set to reflect long-run opportunity costs of the factors of production—determined by their marginal productivity in producing tradeables at an equilibrium exchange rate. Such opportunity costs should not be read from international prices with a narrow, short-run, or mechanical vision. The first leaves out environmental costs, the second is unduly influenced by temporary fluctations in highly leveraged international markets, especially for sugar and food grains, and the third implies that some formula can translate border prices into appropriate domestic price policy. With these provisions, however, such a macro policy reform will end up reflecting the reality of the domestic economy rather than the shadows of planners' wishes.

INTERNATIONAL ELEMENTS IN A DOMESTIC FOOD STRATEGY

Few countries have tried and even fewer have succeeded in isolating themselves from the international economy. Global interdependence brings great benefits, but it also magnifies the costs of contrary domestic policies. Long-term autarky is almost never a good policy, but a short-term hiatus from relatively free trade may be essential for some countries in the early stages of nation building. In general, however, international trade is beneficial to a young and unbalanced economy. It is fair to ask who benefits from the gains to such trade, and this is a task of food policy analysis. The burden of proof, however, falls on those who wish to show harm, and not vice versa.

This perspective suggests that the international economy serve as the standard by which domestic resource allocations are made, at least before income distribution consequences are factored into the policy decision. Tradeable goods, especially most output from the rural sector, can be imported or exported at the margin, and that marginal cost relative to domestic costs of production indicates the efficiency in income generation of a domestic project. Such an efficiency comparison need not dictate the "go–no go" decision if income distribution arguments are compelling, but the calculation should always be done.

If international opportunity costs are going to be held up as a standard of comparison to domestic production (and consumption) costs, then some understanding of the long-run functioning of major international commodity markets should be in the bag of tools carried by good food policy analysts. Such commodity markets are often said to be highly leveraged, unstable, and controlled by multinational corporations for the benefit of the industrialized world. There is enough truth in the appearance of some of these markets for the charges to have some validity, especially with respect to leverage and instability. Domestic food strategy planning would be made much easier if world food markets could be stabilized with some assurance of adequate food reserves for years of bad harvests—but planning *domestically* for such stabilization *internationally* would be foolish in today's world. Domestic planners should invest in understanding how the world grain markets actually perform, while encouraging their diplomats to search for ways to improve that performance.

World commodity markets are in fact tightly interconnected by technical and economic relationships, primarily through the livestock feeding industry and, in the not-too-distant future, possibly also through the sweetener and the alcohol fuel industries. Thus cassava, corn, wheat, and soybeans are linked through these mechanisms. Wheat and rice are linked in the world market by China through a calorie arbitrage implemented by shifting domestic food demand patterns with its urban grain rationing system. Understanding interdependence is the essence of understanding international commodity markets.[6]

What international instruments of assistance are available to domestic food strategy planners? None is important enough that it should influence the basic strategic design for the food sector, but food aid can be useful as a short-run consumption support to a bridge linking existing food price policy with greater long-run price incentives to the domestic agricultural sector. The World Bank and other assistance agencies can provide useful funding and design assistance for critical urban and rural infrastructure, and their support of the International Agricultural Research Institutes is absolutely critical for developing the long-run technological base for rapid improvement in domestic agricultural productivity. Finally, analytical assistance is available for the design of effective food strategies. This design process is extraordinarily complex and requires personnel trained to think creatively about such complexity. No country has a monopoly on such personnel or on the training institutions that produce them. Here the gains to international trade are impressive indeed.

SUMMARY: ELEMENTS OF A FOOD STRATEGY

Three important summary points come out of the above discussion. First is the critical importance of recognizing the explicit need for a reconciliation between food consumption/nutrition objectives and the policy set necessary for efficient growth in agricultural production and farm incomes. Reconciliation is not the same as compromise, at least in the sense of giving each side half. Doing so might turn out to be the worst of all possible worlds, serving neither efficiency nor equity. Rather, reconciliation involves creative disaggregation of food consumption problems in the search for targeted interventions that do not require impossible budget subsidies or bureaucratic resources.

Second, food policy analysis involves a frame of mind for finding order in complexity. It is sophisticated, intuitive, even artistic. At its best such analysis uses a set of filters accumulated with experience and training to sort out (1) what is driving the food system at any given time and (2) how policy levers might be used to affect its direction and speed.

Third, food policy analysts will have three important starting points: food consumption patterns, food production potential, and how the macro policy environment conditions both. These three sectors are obviously not independent. They are connected by the domestic food marketing sector and conditioned by the international economic environment. Food prices turn out to be the critical short-run link joining all the pieces of the system, while the distribution of increases in income provides a second important, long-run link. Successfully integrating the short-run and long-run perspectives on behavior and performance of the food system is the goal of a food strategy.

NOTES

1. See Chuta and Liedholm, chapter 20 in this volume.—ED.
2. See Tang, chapter 28, and Lardy, chapter 29, in this volume.—ED.
3. See Timmer, chapter 19 in this volume.—ED.
4. See Adams and Graham, chapter 21, and Gonzalez-Vega, chapter 22, in this volume.—ED.
5. See Mellor, chapter 10 in this volume.—ED.
6. See Falcon, chapter 12 in this volume.—ED.

9

Agricultural Development and the Intersectoral Transfer of Resources

JOHN W. MELLOR

Accelerated growth in the agricultural production of low-income countries may sharply increase the transfer of resources between agriculture and other sectors of the economy. Such changes affect relative rates of capital formation and income growth in various sectors, the structure of growth, and overall rates of growth. Recent technological breakthroughs in agriculture give current relevance to these relationships.

This paper deals with conceptual and empirical aspects of (*a*) the magnitude of resource flows between the agricultural and nonagricultural sectors under various conditions of economic growth; (*b*) the changing role of economic and institutional devices in transferring resources among sectors; and (*c*) the relationship between such resource flows and technological change in the agricultural sector. Detailed comparisons are made for Taiwan and India, while brief note is taken of the experience of Japan, Britain, and France.

There is controversy as to the timing and direction of net resource flows between agriculture and other sectors in early stages of economic development. One argument holds that net capital transfers to agriculture are needed so that agricultural production may be increased to meet the greater demand for food that accompanies industrial development. It is further argued that these capital transfers are large because of the high capital-output ratios associated with the agricultural sector—perhaps due to the diminishing returns traditionally associated with agriculture.[1]

JOHN W. MELLOR is director of the International Food Policy Research Institute, Washington, D.C.

Originally titled "Accelerated Growth in Agricultural Production and the Intersectoral Transfer of Resources." Reprinted from *Economic Development and Cultural Change* 22, no. 1 (1973): 1–16, by permission of the University of Chicago Press. Copyright © 1973 by The University of Chicago. All rights reserved. Published with omissions and minor editorial revisions by permission of the author.

A contrasting argument calls for a squeeze on agriculture, transferring resources to other sectors, presumably on the assumption that the rate of return to investment is higher in the nonagricultural than in the agricultural sectors.[2] This position is buttressed by the common assumption of diminishing returns in a technologically stagnant agriculture and rising returns through external economies in the nonagricultural sectors.

A much more complex and interesting case arises when technological change in agriculture sharply increases returns to investment in agriculture and consequently sharply reduces the capital-output ratios. In these circumstances, there will be at least a short-run net inflow of resources to agriculture unless (*a*) the incremental capital-output ratio is less than one or (*b*) consumption in agriculture declines.[3] As we shall see later, the incremental capital-output ratio may well be less than one in the case of recent agricultural technologies. Consumption in agriculture may decline if higher returns to investment associated with technological change cause a shift in the savings function or if increased production combined with inelastic demand causes a reduction in gross income.

Thus, the magnitude and direction of resource flows between agriculture and other sectors depend on the relationship between values in the two sectors for a complex of factors including (*a*) the rates of return on capital, (*b*) the capital-output ratios, (*c*) the savings rates, and (*d*) the demand for agricultural output. Each of these forces, and hence the balance among them, is likely to be substantially influenced by the nature and pace of technological change in agriculture.

Once these forces are determined, whether or not the optimal transfer of resources occurs is a function of the effectiveness of the institutional arrangements for such transfers. A number of peculiarities of the agricultural sector with respect to the way income is earned, consumed, and invested may impede optimal transfers.

A HISTORICAL VIEW OF INTERSECTORAL RESOURCE FLOWS

TAIWAN

There were large, continuous net transfers of resources from the agricultural sector of Taiwan throughout the period from 1895 to 1960. For agriculture, the period from 1911 to the mid-1920s was one of rapid expansion in irrigation investment but was otherwise stagnant technologically. From 1911 to 1920 the net resource transfer from the agricultural sector was equal to over half the value of agricultural sales and 30 percent of the value of total agricultural production. These proportions dropped to 40 percent and 25 percent for the period 1921–25.

Compared with the period 1911–26, the rate of technological change in agriculture, measured as a residual as in the Solow model, was twice as fast for the period 1926–40 and two and one-half times as rapid for 1950–60.[4] The real value of the net resource outflow increased by nearly 25 percent from 1911–15 to 1916–20. It remained at about that level throughout the technologically dynamic period in the last half of the 1920s and then increased to a 50 percent higher level in the 1930s and increased by another 27 percent for the period 1950–55.

In the technologically dynamic period of the 1920s, when, for example, the new Ponlai rice varieties were introduced and fertilizer use was increasing rapidly, purchases of commodities from nonagricultural sectors by the agricultural sector more than doubled. Nevertheless, the net outflow of resources from agriculture rose and maintained a high level. The value of net resource transfers as a percentage of production and sales declined from the earlier period, but increased production allowed larger absolute net transfers which were concurrent with agriculture's increased use of industrially produced capital and consumer goods.

In the 1950–55 period of extraordinarily rapid population growth, economic development, and technological change in agriculture, the net real resource transfer from agriculture increased to a new high. In this period, net resource transfer recovered to nearly 40 percent of agricultural sales and over 20 percent of total agricultural production. By 1956–60 the net transfer had begun to decline slightly in real absolute terms and was equal to only 15 percent of production and 24 percent of sales of agricultural products. The decline in net resource transfers from agriculture has continued subsequent to 1960.

Just as revealing as the large net transfers from agriculture are the dramatically changing roles of various transfer mechanisms. In the post–World War II period, the transfer of resources was achieved primarily by a sharp turn in the terms of trade against the agricultural sector, so that over 40 percent of the transfer was represented by invisible items in 1950–55. The most important mechanisms of this change were a barter exchange of rice for fertilizer and the compulsory purchase programs. In addition, technological change provided more than compensating production incentives in agriculture through greatly improved physical input-output relationships.

In the pre–World War II period, fiscal measures and land rent payments were vital in the transfer. In the latter part of the prewar period and in the postwar period, outflow through financial institutions was also substantial. The importance of particular methods and institutions for financing resource flows changed substantially from time to time according to economic and political factors. The choice was not necessarily the most efficient by economic criteria alone. For example, the heavy reliance in the postwar period on what was in effect a tax on fertilizer presented to farmers one of the most unfavorable fertilizer-rice price ratios in the world.[5]

OTHER COUNTRIES

It is generally agreed that Japan provided a major portion of the capital for early stages of its economic development by resource transfers from the agricultural sector.[6] The mechanisms of transfer differed from those mentioned above.

Direct investment by landlords in the nonagricultural sectors was relatively more important than in Taiwan. Also, Japan depended largely on land taxes, while Taiwan emphasized taxes on crop output (sugar cane in the prewar period and rice in the postwar period). During the 1920s, when her agriculture was technologically relatively stagnant, Japan relied on imports of rice and other agricultural commodities from the colonies to depress relative agricultural prices.[7] In post–World War II Japan, in which landlords had been largely eliminated and there were high support prices on rice and low fertilizer prices, the net flow of resources must have been channeled toward agriculture. By that time, however, the relative size of agriculture had declined sufficiently that the impact on capital formation and growth rates was less than that of compensating variables.

Scattered historical evidence for early stages of industrial growth in Great Britain suggests substantial net transfers of resources out of agriculture which were financed by rent payments to landlords, taxes, and, after repeal of the corn laws, by cheap food imports which turned the terms of trade against agriculture.[8] Economic historians suggest that the slow growth rate in the French economy during comparable periods was due in part to net transfers of resources into a low-productivity agriculture, largely through favorable terms of trade for the agricultural sector.[9]

The experience of the Soviet Union is consistent with the pattern described, although the Soviet Union relied heavily on compulsory deliveries facilitated by collectives for financing resource transfers from agriculture.[10] It is likely that the Soviet Union had unusually large transfers from agriculture relative to the amount of technological change in the agricultural sector.

The recent experience of India, in contrast to the evidence above, illustrates the special problems of development in contemporary low-income countries. Although we do not have a single comprehensive study of India comparable to Lee's work on Taiwan, we can piece together a number of parts that suggest that there were net resource flows into the agricultural sector during the first three five-year plans (1950–65).

Gandhi shows that in the period 1950–60 tax revenue attributable to agriculture was substantially less than government expenditures on agriculture and that rural people in the same income class were taxed at substantially lower rates than their urban counterparts.[11] The discrepancy was particularly large in the upper income brackets. In Taiwan, by contrast, the tax burden on agriculturalists, in any individual income class, was heavier than on those not in agriculture.[12] Simple lagged correlations of government expenditures on agriculture show little relation to growth in agricultural output in India, whereas in Taiwan government expenditures induced large complementary investment by farmers and increased output.[13]

The relative prices of agricultural and nonagricultural commodities fluctuate substantially in India according to the weather. Thus, it is not surprising that some observers of short periods erroneously note a change in the terms of trade against the agricultural sector.[14] For India in the period 1952/53 to 1964/65, there is no statistically significant evidence of a movement one way or the other of relative prices of food grains and nonagricultural commodities. If we compare prices of all agricultural commodities (including industrial raw material crops, fruits, vegetables, and livestock products), the terms of trade clearly moved toward the agricultural sector in this period and thus produced a significant invisible transfer of resources.[15] Extension of the period of analysis to the present would show an increase in this transfer.

Large fluctuations in relative prices arising from changes in weather make it possible to observe the effects of changing relative agricultural prices on growth of savings and investment in the nonagricultural sectors. The nature of the fluctuations, the complexity of the lag factors, and the short period of time for which data are relevant limit the usefulness of statistical analysis. It appears, however, that large crops and consequent relative declines in agricultural prices were associated with increased rates of domestic saving and investment. The converse relationship also seemed to occur.[16] In fact, a notable example is the sharp drop in savings, investment, and industrial growth following the disastrously poor crop years and sharply rising relative agricultural prices of 1965/66 and 1966/67. Similarly, the sharp decline in relative agricultural prices from 1953/54 to 1955/56 was associated with a sharp rise in domestic savings, which peaked in 1957. Relative agricultural prices rose for the three years ending 1958/59, and domestic savings dropped sharply to a trough in 1958. Relative agricultural prices again declined from the 1958/59 peak, and domestic savings again resumed their upward trend in 1959.[17]

These observations appear to indicate that the slight upward trend in relative agricultural prices over the past two decades has resulted in net resource transfers to agriculture and in slower rates of growth in the nonagricultural sectors. Without major technological change in agriculture, the resources transferred to agriculture were subject to diminishing physical returns and low physical response. They provided attractive economic returns only because of relative increase in agricultural prices.

It is likely that the net flow of resources on private account in India has also been to the agricultural sector. There may have been a single permanent transfer away from agriculture after the Indian land reforms of the early 1950s. Subsequently there has not been a large wealthy landowner class, as in the early stages of development of Taiwan, Japan, or Great Britain. In addition, during the two decades up to the late 1960s, there has not been an economic environment favorable to direct investment by landed classes in small-scale manufacturing. Resources have been largely directed toward large-scale industry. Savings rates in the agricultural sector have at least until recently been low, perhaps largely because of relatively unattractive investment opportunities.[18] Finally, there have been substantial remittances to the rural sector by urban wage earners, which have gone largely to consumption.[19] As will be suggested below, these relations may now be changing as a result of the new highly profitable grain varieties. In conclusion, then, we find that in the case of India all the three mechanisms—government account, price relationships, and private account—have transferred resources to the agricultural sector.

The successful cases of net transfer of resources from agriculture have had for a basis repressive extraction of a surplus from agriculture through a wealthy landowner class, a strong state operating through collectives, easily taxed major agricultural exports, a technologically progressive agriculture which generated large additional surpluses, or a combination of these. For most contemporary low-income countries, the route of technological change in agriculture may be the only feasible one.

TECHNOLOGICAL CHANGE AND INTERSECTORAL RESOURCE FLOWS

It is difficult to achieve continuous net resource transfers from a technologically stagnant agriculture if those resources are invested productively in the nonagricultural sector. Growth of the nonagricultural sector increases the demand for food. Relative agricultural prices will then rise in the face of the inelastic aggregate supply characteristic of technologically stagnant agriculture. As a result, resources will be transferred back to the agricultural sector—where they will of course be subject to diminishing, and eventually to low, returns. In these circumstances, the greater the increase in employment of low-income laborers, the greater the increase in demand for food, the greater the increase in agricultural prices, and the greater the resource transfer to agriculture. If increased food imports prevent increases in agricultural prices, growth of the nonagricultural sector may then be halted by scarcity of foreign exchange.

In order to prevent increased industrial investment from increasing the demand for food and thereby raising relative agricultural prices and transferring resources back to agriculture, either the increase in employment must be relatively small or the real wage rate must be reduced sufficiently to balance the aggregate income effect of increased employment. Both are difficult to accomplish, and both have unfavorable implications for income distribution.[20]

In contrast to the situation of a stagnant agriculture, continuous technological change in agriculture permits some expansion in demand for agricultural production to be met without higher relative agricultural prices.[21] In these circumstances, net transfers of resources from agriculture which cause expansion of employment in the nonagricultural sectors are not fully reversed by changes in the terms of trade toward the agricultural sector. In practice, technological change in agriculture also increases the elasticity of aggregate supply by increasing the relative importance of more demand-responsive inputs such as chemical fertilizers.

Technological change in agriculture directly accelerates growth in real national income. It is also likely to influence relative prices so as to encourage industrial development. Because, however, such technological change normally induces increased use of inputs from the nonagricultural sector, it may not induce increased net resource transfers to that sector. What actually happens will depend on (*a*) the capital-output ratios associated with the technological change and (*b*) the changes in consumption in the agricultural sector resulting from the technological change.

Significantly higher returns on investment associated with new agricultural technologies may shift the savings function and reduce consumption. Thus, an increase in the flow of production goods to agriculture may be balanced by a decrease in the flow of consumption goods. Alternatively, an incremental capital-output ratio less than one will allow immediate concurrent increases in agricultural production and increased resource transfers from agriculture, as was the case in Taiwan between 1920 and 1935 as well as in the period 1950–67. More generally, the lower the incremental capital-output ratio associated with technological change, the sooner net resource transfers may resume. Reduction of the rate of technological change will also shorten the period of accelerated net resource transfers—although at the cost of lower rates of growth.

IMPLICATIONS OF TECHNOLOGICAL CHANGE TO THE MEANS OF RESOURCE TRANSFER

Characteristics of technological change in agriculture suggest some specific means of resource transfer. For example, yield-increasing technological change, at least initially, raises dramatically the returns to land, thereby further strengthening the economic case for land taxes. Unfortunately, land taxes appear particularly unpalatable politically.

Relative agricultural prices tend to decline as yield increases, thereby transferring resources to other sectors. Such price declines may not discourage agricultural production due to the cost reductions accompanying new technology and due to the shape of response curves.[22] Government policy may usefully (*a*) facilitate orderly price declines and (*b*) help translate the decline in agricultural prices into accelerated growth in industrial investment and employment.

Yield-increasing technological change in agriculture is usually accompanied by increased use of purchased inputs, and greatly increases the returns to them. Under these circumstances, it is essential that input supplies be increased rapidly as the demand curves shift. Clearly, the worst possible policy, but one too often followed, is a subsidized price for inputs and inadequate supply. Technological change in agriculture offers potential for large net transfers of resources out of agriculture not because its added input requirements are small—they are in fact very large—but because the rate of return on those inputs is very high.

A tax on variable inputs complementary to technological change, such as on fertilizer, is inefficient on narrow economic grounds, as it reduces input use and causes a lower than optimal level of output. Yet, Taiwan used such a tax as a major means of drawing revenue from the argicultural sector. Similarly, a tax on fertilizer was instituted in India following introduction of high-yield crop varieties.

In the context of rapid technological change in agriculture, a tax on fertilizer is not as inefficient as at first might appear and has some features to recommend it. First, its incidence is somewhat in proportion to the benefits from research and other aspects of technological change with which fertilizer is so closely associated. Second, there is evidence that fertilizer response functions are essentially linear until they reach their maximum.[23] Thus, for a wide range of price relationships, the optimal quantity of fertilizer to use appears quite inelastic with respect to price. In early stages of adoption, diffusion of fertilizer use may be accelerated by highly favorable price relationships.[24] This factor, however, diminishes in importance with time.

Either a land tax or a relative decline in agricultural output prices is to be preferred to a tax on an input such as fertilizer. A tax on fertilizer is preferred to loss of investment opportunities which offer high rates of return. It is the political difficulty of effecting economically preferable mechanisms for resource transfer that compels taxes on variable inputs. Fortunately, in a context of technological change such devices may not be markedly inefficient.

CONCLUSION

Both in concept and in practice it is possible for the agricultural sector to make large net transfers of resources to other sectors. If these transferred resources are used productively, the rate of economic growth can be accelerated.

Net resource transfers are possible from a technologically stagnant agriculture. But such transfers are difficult to achieve without an economically and politically powerful landlord class strongly motivated to invest in the domestic non-agricultural sectors, a powerful unitary government, or major export crops. The first two conditions rarely exist in contemporary low-income countries. Low-income countries with major export crops are among the few that tax agriculture heavily.

In many areas the current technological breakthroughs in agriculture offer large increases in output at incremental capital-output ratios of less than one. This facilitates immediate and greatly accelerated net resource transfers.[25] Even if large investment in irrigation is a necessary complement to technological change, increased net resource outflows may occur shortly after rapid technological change in agriculture begins. A wide range of devices may be used to facilitate such resource transfers, including taxes of many types, lower relative agricultural prices, and direct investment outside of agriculture by wealthy agriculturalists. The change in structure of demand accompanying the increased agricultural output may well enhance these opportunities.[26]

If a low-income country is to grasp the type of opportunity so well exploited by Japan and Taiwan, it must develop the infrastructure of research and related institutions for developing, adapting, and applying suitable high-yield crop varieties. It then must ensure the ready availability of a large quantity of complementary inputs such as fertilizer. A highly elastic supply of inputs complementary to technological change is crucial to the process. The economic incentive for using additional inputs in agriculture is provided by technological change itself, which increases output per unit of input. Under these circumstances, a wide range of devices is available for transferring resources from agriculture. There remain substantial political problems of choosing a combination of these devices acceptable in the complex political and institutional framework of a modernizing agriculture.

NOTES

The data for this paper are drawn from a series of studies conducted under my direction at Cornell University as part of an AID-financed study of agricultural prices as they affect intersectoral resource flows. I am most indebted to T. H. Lee for the intersectoral study of Taiwan which I have used extensively in this paper. In addition, I have made substantial use of the work of Uma Lele, G. M. Desai, Ashok Dar, U. S. Bawa, and Sheldon Simon. I am particularly grateful to Uma Lele for suggesting major improvements in this paper.

1. This argument is developed in Maurice Dobb, "Some Reflections on the Theory of Investment Planning and Economic Growth," in *Problems of Economic Dynamics and Planning: Essays in Honour of Michal Kalecki* (Warsaw: Polish Scientific Publishers, 1964), 107–18; and in A. K. Sen, "Some Notes on the Choice of Capital-Intensity in Development Planning," *Quarterly Journal of Economics* 71 (November 1957): 561–84. It receives support from the argument that rapid growth in agricultural production requires massive inflow of resources for water and irrigation (for example, Vernon W. Ruttan, "Considerations in the Design of a Strategy for Increasing Rice Production in Southeast Asia" [Paper presented at the Pacific Science Congress Session on Modernization of Rural Areas, Tokyo, 27 August 1966]; and T. W. Schultz, *Economic Crises in World Agriculture* [Ann Arbor: University of Michigan Press, 1967]). The argument that agricultural development requires improved price incentives, which implicitly involves increased net inflow of resources through the change in terms of trade, also supports this position (for example, Edward Mason, *Economic Development in India and Pakistan,* Occasional Papers in International Affairs, vol. 13 [Cambridge: Harvard University Press, 1966]; and Schultz, *Economic Crises,* 32–33).

2. For a review of this argument see Wyn F. Owen, "The Double Development Squeeze on Agriculture," *American Economic Review* 56, no. 1 (1966): 43–70. See also Bruce F. Johnston and John W. Mellor, "The Role of Agriculture in Economic Development," *American Economic Review* 51, no. 4 (1961): 566–93.

3. Throughout this discussion of intersectoral resource flows the term "capital-output ratio" refers to the ratio of only that portion of capital represented by a flow of resources to agriculture from other sectors of the economy to increments of total agricultural output. Thus, a capital-output ratio defined in this manner may be low because a substantial portion of the capital more broadly defined arises from direct transformation of unemployed rural labor.

4. T. H. Lee, *Intersectoral Capital Flows in the Economic Development of Taiwan, 1895–1960* (Ithaca: Cornell University Press, 1971), 49.

5. For comparisons with India and a number of other countries see John W. Mellor, Thomas F. Weaver, Uma J. Lele, and Sheldon R. Simon, *Developing Rural India* (Ithaca: Cornell University Press, 1968), 106.

6. See, for example, Bruce F. Johnston, "Agricultural Productivity and Economic Development in Japan," *Journal of Political Economy* 59 (December 1951): 498–513; idem, "Agricultural Development and Economic Transformation: A Comparative Study of the Japanese Experience." *Food Research Institute Studies* 3 (November 1962): 223–76; and Kazushi Ohkawa and Henry Rosovsky, "The Role of Agriculture in Modern Japanese Economic Development," *Economic Development and Cultural Change* 9 (October 1960): 43–67.

7. Kazushi Ohkawa, "Concurrent Growth of Agriculture with Industry: A Study of the Japanese Case," in *International Explorations of Agricultural Economics,* ed. Roger N. Dixey (Ames: Iowa State University Press, 1964), 201–12; and Kazushi Ohkawa, Bruce F. Johnston, and Hiromitsu Kaneda, eds., *Agriculture and Economic Growth: Japan's Experience* (Tokyo: University of Tokyo Press, 1969).

8. T. S. Ashton, *The Industrial Revolution, 1760–1830* (London: Oxford University Press, 1962); J. D. Chambers and G. E. Mingay, *The Agricultural Revolution, 1750–1880* (London: B. T. Batsford, 1966); and Phyllis Deane, *The First Industrial Revolution* (Cambridge: Cambridge University Press, 1965).

9. See, for example, Uma J. Lele, *Agricultural Resource Transfers and Agricultural Development: A Brief Review of Experience in Japan, England, and France,* Occasional Paper 33 (Ithaca: Department of Agricultural Economics, Cornell University–USAID Prices Research Project, June 1970); Rondo E. Cameron, "Economic Growth and Stagnation in France, 1815–1914," in *The Experience of Economic Growth: Case Studies in Economic History,* ed. Barry E. Supple (New York: Random House, 1963), 328–39; J. M. Clapham, *Economic Development of France and Germany, 1815–1914* (Cambridge: Cambridge University Press, 1921); and M. J. Habakkuk, "Historical Experience of Economic Development," in *The Problems of Economic Development,* ed. E. A. G. Robinson (London: Macmillan, 1965), 112–30.

10. Alexander Erlich, "Preobrazhenski and the Economics of Soviet Industrialization," *Quarterly Journal of Economics* 64 (February 1950): 57–88.

11. Ved P. Gandhi, *Tax Burden on Indian Agriculture,* Harvard Law School International Tax Program (Cambridge: Harvard University Law School, 1966). Note that for the period from 1950/51 to 1964/65 taxes on the agricultural sector as a percentage of income rose from 3.6 percent to 5.6 percent, while they rose from 8.8 percent to 18.3 percent for the nonagricultural sectors. The ratio of additional taxes to additional income was 7.5 percent for agriculture and 44 percent for other sectors (see Ujagar S. Bawa, *The Relationship between Agricultural Production and Industrial Capital Formation in India, 1951–52 to 1964–65,* Cornell International Agricultural Development Bulletin, no. 17 [Ithaca: New York State College of Agriculture and Life Sciences at Cornell University, March 1971], 32).

12. Lee, *Intersectoral Capital Flows,* 108.

13. For the detailed data for India and Taiwan see, respectively, Bawa, *Agricultural Production and Industrial Capital Formation,* 54; and Lee, *Intersectoral Capital Flows,* 112.

14. V. M. Dandekar, "Agricultural Price Policy," *Economic and Political Weekly* 3 (16 March 1968): 454–59; Mason, *Economic Development in India and Pakistan;* and Schultz, *Economic Crises.*

15. On the basis of a nonparametric test for the period 1952–64, at the 15 percent level of significance, terms of trade between agriculture and industry turned toward agriculture. By the same test, the terms of trade between foodgrains and industry only fluctuated, with no discernible trend. The linear trend for all agricultural prices relative to industrial prices is positive, while there is no discernible trend for the ratio of foodgrain prices to industrial prices. Agricultural raw materials, including fibers and oil seeds, have a weight nearly half as large as that of foodgrains in the consumer price index. Raw materials comprise over 50 percent of the cost structure of the agricultural processing industries, which in this period provided over 50 percent of the value added in all industry in India. A too literal view of wages goods as food alone should not divert attention from these important parts of the agricultural sector (see John W. Mellor and Uma J. Lele, "Alternative Estimates of the Trend in Indian Foodgrains Production during the First Two Plans," *Economic*

Development and Cultural Change 13, no. 2 [1965]: 17; and Ashok K. Dar, *Domestic Terms of Trade and Economic Development of India, 1952–53 to 1964–65,* Cornell International Agricultural Development Bulletin, no. 12 [Ithaca: New York State College of Agriculture and Life Sciences at Cornell University, 1968], 11–14).

16. See Bawa, *Agricultural Production and Industrial Capital Formation,* 23.

17. Ibid.

18. For example, see data for India in Mellor et al., *Developing Rural India,* 329–39.

19. For example, in the period 1954–64 in the village of Senapur, which has a long history of movement to urban employment, per capita agricultural incomes held about constant, while per capita income from nonagricultural sources rose 133 percent to comprise over 40 percent of total per capita income. During this period of technologically stagnant agriculture, increases in savings were small and were used primarily for education (see ibid., 309, 329).

20. See Mellor, chapter 10 in this volume.—ED.

21. "Technological change" as used here refers to shifts in the production function at a given level of inputs. The prime example in agriculture is high-yielding seed varieties, which increase the value of gross output per unit of land and labor. There is, of course, in practice a large increase in use of certain inputs, such as fertilizer, and a consequent further increase in output.

22. For detailed analysis of this see Gunvant M. Desai, *Growth of Fertilizer Use in Indian Agriculture: Past Trends and Future Demand,* Cornell International Agricultural Development Bulletin, no. 18 (Ithaca: New York State College of Agriculture and Life Sciences at Cornell University, March 1971).

23. Ibid.

24. Ibid.

25. For these views presented as part of a larger mathematical framework see Uma J. Lele and John W. Mellor, *Technological Change and Distributive Bias in a Dual Economy,* Revised Occasional Paper, no. 43 (Ithaca: Department of Agricultural Economics, Cornell University–USAID Employment and Income Distribution Project, October 1972). Similar views are presented in a political framework in idem, "Jobs, Poverty, and the Green Revolution," *International Affairs* 48, no. 1 (1972): 20–32.

26. For a full explanation of this view see John W. Mellor and Uma J. Lele, "Growth Linkages of the New Foodgrain Technologies," *Indian Journal of Agricultural Economics* 28, no. 1 (1972): 1–22.

10

Food Price Policy and Income Distribution in Low-Income Countries

JOHN W. MELLOR

Change in relative food prices is, in the short run, one of the most important determinants of change in the relative and absolute real income of low-income people. They spend a high proportion of their income on food and depend directly or indirectly on agriculture for a high proportion of their employment and income. In the longer run, food price policy may affect shifts in the supply function for wage goods and thereby influence the extent to which total wage employment and hence income of the laboring classes can be expanded.

Thus, there are important trade-offs and conflicts among various direct short-run influences and indirect long-run effects of price policy on the real incomes of the poor.[1] Similarly, if the aggregate supply of food is fixed, there are direct income trade-offs for low-income families between change in prices and change in employment as market devices for providing the equilibrium consumption level.

Because the interrelationships among price, supply of wage goods, pattern of production, and income distribution are so complex, only a general equilibrium analysis can unequivocably determine the various effects of specific food price policies on income distribution. In contrast, the substantial literature on agricultural price policy is dominated by analysis of the partial relation between relative agricultural prices and production. Further, economists not only have concentrated their analyses of agricultural prices on production relations rather than on distributional effects but have given least attention to those aspects of production—such as marketable surplus, risk-uncertainty relationships, and shift of resources among enterprises of varying labor intensity—which are most rele-

JOHN W. MELLOR is director of the International Food Policy Research Institute, Washington, D.C.

Reprinted from *Economic Development and Cultural Change* 27, no. 1 (1978): 1–26, by permission of the University of Chicago Press. Copyright © 1978 by the University of Chicago. All rights reserved. Published with omissions and minor editorial revisions by permission of the author.

vant to the absolute and relative incomes of low-income consumers and producers.[2]

The purpose of this paper is to delineate the component parts of a general equilibrium analysis relevant to the relation of price policy to income distribution; to present data as to the relation of price change to a variety of those component parts; and to suggest the nature of the various interactions among those parts. The presentation commences with the more standard, relatively simple, and very limited questions of price influence on income distribution and provides the empirical basis for answers to those questions. It then proceeds to successively more complex questions, indicating the bases for judgments on those matters, but falling increasingly short of a definitive position.

DIRECT EFFECTS OF FOODGRAIN PRICE CHANGE ON DISTRIBUTION OF INCOME AMONG CONSUMERS

Change in foodgrain prices causes a larger percentage change in the real incomes of low-income consumers but a larger absolute change in the real incomes of high-income consumers. The absolute effect on the incomes of high-income consumers may have secondary effects on the poor through changes in consumption of other goods and services and a consequent change in employment in their production.

Thus, in India, the top 5 percent in the income distribution spends over two and a half times as much per capita on foodgrains as the lowest two deciles in the income distribution.[3] However, despite its large absolute expenditure, this upper-income class allocates only 15 percent of its total expenditure to foodgrains, compared with 54 percent for the lower-income class.

Table 1 presents data as to the income effect of an increase in foodgrain price. For high-income nations, this effect is normally considered negligible; but it is a major factor in low-income nations, where the proportion of family income spent on foodgrains is much higher. The analysis uses Indian data because of their availability, because foodgrains are dominant in Indian consumption patterns, and because of the low levels of income.

For the data presented, if foodgrain prices rise by 10 percent, the lowest two deciles in expenditure, which initially spend Rs 4.830 per capita per month on foodgrains, experience a real total expenditure decline of Rs 0.494 per capita per month (table 1). The decline in total real expenditure is, in fact, slightly greater than 10 percent of the initial expenditure on foodgrains because the proportion of total income spent on foodgrains increases in successively lower income classes. In this case, assuming all other prices as well as the proportion of income saved as constant and expressed in constant price terms, the new real expenditure will be Rs 8.44 per capita per month, or a decline of 5.5 percent.

In contrast, the top 5 percent in the expenditure distribution experiences a decline in real total expenditure of Rs 1.01, or somewhat over twice as large an absolute decline as that of the lower-income class; but at only 1.2 percent of the initial expenditure, there is less than one-quarter as large a percentage decline as that experienced by the lower-income class. The expenditures of the poor on foodgrains are, of course, far more elastic with respect to the income effects of price than are those of the rich. For the bottom two deciles in the income distribution, the elasticity of response of real expenditure (assuming no substitution effect) to change in foodgrain prices is 0.55, compared with 0.12 for the top 5 percent (table 1).

A number of other relationships important to income distribution are illuminated by this exercise. It is significant that in response to a price increase both the absolute and the percentage declines in real expenditure on foodgrains are greater the lower the income class. Thus, for a 10 percent increase in foodgrain prices, the bottom two deciles reduce their real expenditure on foodgrains by 5.9 percent, compared with a reduction of only 0.2 percent by the upper half of the tenth decile. More important, this top income class reduces its absolute real expenditure on foodgrains by only Rs 0.03, compared with over ten times as large an adjustment by the lower-income class (table 1). These data illustrate that in a market economy the bulk of adjustment to reduced supplies of foodgrains is made by low-income consumers.[4] Given the low initial level of foodgrain consumption in the lower-income deciles, the privation imposed on them by rising grain prices is very great. But any measures that seek to insulate the poor from the necessity to adjust to reduced supplies will force the need for adjustment to those whose demand is much more inelastic—thereby causing proportionately much greater, and perhaps explosive, price increases. It is clear from this analysis that it is essentially impossible to protect the poor from the major income effects of a short crop by market measures.

Further adverse consequences for the poor of a change in relative foodgrain prices may follow from the effects of such a change on the consumption of other commodities. As a result of a rise in foodgrain prices, the absolute decline in expenditure for almost all nonfoodgrain commodities is greater for higher-income classes than for lower-income classes, although of course the percentage decline in expenditure is much greater in each case for the lower-income classes. For example, in response to a 10 percent increase in foodgrain prices, the lowest two deciles in the income distribution reduce consumption of milk and milk products by 33 percent, compared with a reduction of 9 percent by the top 5 percent, though the absolute reduction by the higher-income class is over three times as great as for the lower-income class. Overall, while foodgrain consumption declines only 4 percent, consumption of most other commodities declines between 10 percent and 30 percent.

These relationships suggest, first, that an increase in foodgrain prices may lead to substantially reduced consumption by the poor of agricultural commodities of high nutritive value. It is, of course, conceivable, but neither logical nor likely, that the poor would respond to increased foodgrain prices by some substitution of higher-quality foods which, though more expensive, had not increased in price.

TABLE 1

DECLINE IN EXPENDITURE CONSEQUENT TO THE INCOME EFFECT OF A 10 PERCENT RISE IN THE PRICE OF FOODGRAINS, BY EXPENDITURE CATEGORIES AND CLASSES, INDIA, 1964/65*

(Rs)

Expenditure Category	Bottom Two Deciles	Decile 3	Deciles 4 and 5	Deciles 6, 7, and 8	Decile 9	Lower Half of Decile 10	Upper Half of Decile 10	Mean for All Classes
				Expenditure Class				
Initial total per capita monthly expenditure	8.93	13.14	17.8	24.13	30.71	41.89	85.84	24.43
Decline in monthly per capita expenditure due to the income effect of a 10% rise in food-grain price:								
Foodgrains	0.285	0.260	0.208	0.153	0.115	0.068	0.026	0.147
Milk and milk products	0.034	0.074	0.105	0.125	0.196	0.132	0.113	0.124
Meat, eggs, and fish	0.013	0.021	0.026	0.030	0.031	0.032	0.030	0.030
Other foods	0.004	0.037	0.061	0.085	0.104	0.131	0.210	0.085
Tobacco	0.008	0.010	0.011	0.012	0.012	0.012	0.011	0.012
Vanaspati	0.003	0.008	0.013	0.020	0.022	0.022	0.015	0.019
Other oils	0.020	0.033	0.037	0.034	0.030	0.023	0.010	0.034
Sweeteners	0.022	0.034	0.038	0.038	0.034	0.028	0.006	0.037
Cotton textiles	0.043	0.057	0.062	0.062	0.058	0.055	0.042	0.062
Woolen textiles	†	0.002	0.004	0.006	0.003	0.011	0.021	0.006
Other textiles	†	0.001	0.001	0.002	0.003	0.006	0.033	0.002
Footwear	†	0.006	0.007	0.008	0.009	0.008	0.008	0.008
Durables and semidurables	0.004	0.007	0.012	0.017	0.023	0.032	0.067	0.017
Conveyance	0.004	0.010	0.015	0.025	0.034	0.052	0.062	0.024
Consumer services	0.009	0.015	0.021	0.029	0.035	0.044	0.072	0.028
Education	0.005	0.009	0.015	0.025	0.035	0.054	0.135	0.025
Fuel and light	0.040	0.049	0.051	0.051	0.050	0.052	0.036	0.050
House rent	†	0.005	0.009	0.019	0.029	0.046	0.101	0.018
Absolute decline in total expenditure	0.494	0.638	0.696	0.741	0.772	0.808	1.01	0.728
Decline in total expenditure (%)	5.53	4.86	3.91	3.07	2.51	1.93	1.18	2.98

SOURCE: The data source is the National Council of Applied Economic Research, *All-India Consumer Expenditure Survey*, vol. 2 (New Delhi: NCAER, 1967). These data provide expenditure elasticities of foodgrains and milk and milk products consistent with those from the *National Sample Survey (NSS)* of 1963/64 when fitted to the function used here. R^2 for equations estimated from grouped data for different commodities varied between 0.742 and 0.981. The NCAER data provide more detailed breakdown of expenditure than the *NSS*, but for a sample biased toward higher-income groups. The mathematical functional form used for these calculations was: $\log y = a + b/x + c \log x$, where $y =$ per capita monthly expenditure on a commodity in each expenditure class and $x =$ per capita monthly total expenditure in each expenditure class. For a more complete discussion of the data see B. M. Desai, *Analysis of Consumption Expenditure Patterns in India*, Occasional Paper 54 (Ithaca: Department of Agricultural Economics, Cornell University–USAID Employment and Income Distribution Project, August 1972).

NOTE: Let the demand function for the ith commodity be

$$Q_i = Q_i(P_1, \ldots, P_i, \ldots, P_n, Y), \tag{1}$$

where $P_i =$ price of the ith commodity; $i = 1, \ldots, N$; and $Y =$ total expenditure. The effect of change in the price of ith commodity on the demand for any commodity, say jth, is given by the Slutsky equation as follows:

$$\frac{\partial Q_j}{\partial P_i} = \left(\frac{\partial Q_j}{\partial P_i}\right) \text{utility} = \text{constant} - Q_i \left(\frac{\partial Q_j}{\partial Y}\right) \text{prices} = \text{constant}. \tag{2}$$

This can be written as

$$\frac{\partial P_i Q_j}{\partial P_i} = \left(\frac{\partial P_i Q_j}{\partial P_i}\right) \text{utility} = \text{constant} - Q_i \left(\frac{\partial P_i Q_j}{\partial Y}\right) \text{prices} = \text{constant}. \tag{3}$$

In terms of elasticity, equation (3) can be written as:

$$E_{ji} = c_{ji} - b_i \eta_j, \tag{4}$$

where $E_{ji} =$ uncompensated price elasticity of demand for jth commodity with respect to the price of ith commodity, $c_{ji} =$ compensated price elasticity of demand for jth commodity with respect to the price of ith commodity, $b_i =$ budget share of ith commodity, and $\eta_j =$ expenditure elasticity of jth commodity. We further know that

$$\sum_{i=1}^{n} c_{ji} = o \text{ and } c_{ji} = c_{ij} \tag{5}$$

for all i and j. Thus, if the ith commodity is foodgrains, then the assumption that all the cross price elasticities are zero would imply that

$$c_{ji} = 0 \tag{6}$$

for all j and i whenever either the jth or ith commodity is foodgrains. Substituting equation (6) into equation (4) we get

$$E_{ji} = -b_i \eta_j \tag{7}$$

for all j. Alternatively, the change in consumer expenditure on the jth commodity as a result of a 10 percent change in foodgrain prices is given by the product of the expenditure on the jth commodity, the expenditure elasticity of the jth commodity, the proportion of expenditure on foodgrains, and the proportionate change in foodgrain prices.

*Assumes no change in the percentage saved.
†Negligible.

Second, the large absolute reduction in consumption by higher-income groups of goods and services other than foodgrains, and particularly of livestock products and vegetables (the production of which is in Asian countries generally highly labor-intensive), reduces employment. To the extent that such reduction in employment reduces incomes and hence demand by the poor, the increase in price of foodgrains will be dampened. This indirect effect of foodgrain price changes of course complicates empirical analysis: the greater the employment effect, the less will be the price increase. But either way the poor pay—in terms of lower real wages as prices rise or in reduced employment caused by the decline in real income of the upper-income classes.

The foregoing partial analysis is based on the calculation of the income effect of a relative price change. The extent to which observed behavior is consistent with it depends on the extent of the substitution effects and, of course, on the extent to which countervailing or reinforcing employment, income, and stocking effects occur. Substitution effects of relative price changes cannot be determined on an a priori basis and empirically cannot be separated from several other influences. However, useful observations can be made as an extension of the preceding analysis.

First, to the extent that foodgrains are an inferior good for the poor, the income effects will be reinforced. In assuming an inferior-good status, it should be remembered nevertheless that the poor have already reduced consumption of other goods to a very low and perhaps biologically minimal level.

Second, to the extent that higher-income groups have positive substitution effects, the income effects will, of course, be reduced by the substitution effects. However, although in the short run higher-income consumers may substitute livestock products, fruits, and vegetables for grain, the aggregate impact is likely to be small—first, because the same factors of weather and demand are likely to raise prices of those commodities also; and second, because in the longer run those commodities will compete with grain for the same production resources. It seems less likely that strong substitution effects would prevail between grain and nonfood commodities.

Finally, it should be remembered that grain is itself not a homogeneous category, so substitution of lower-priced for higher-priced varieties may allow reduced expenditures without a commensurate reduction in quantity, though with a possible reduction in nutritive value. In the short run, the supply of the inferior types of grain will be inelastic, and the search for a cheaper mix of grains will simply reduce the quality-related price disparities rather than mitigate the effects on consumption.

Substitution effects are probably less important in reducing the effect of production changes on price than are storage effects and employment effects. The market is usually insulated to some extent from declines in the current production of foodgrains by the availability of carryover stocks, unless there is a very large change in production or a succession of bad harvests. There is clear evidence that

TABLE 2
DECLINE IN CONSUMER EXPENDITURE IN DIFFERENT INCOME CLASSES AS A RESULT OF A 10 PERCENT DECLINE IN THE SUPPLY OF FOODGRAINS, BY INCOME CLASS, INDIA, 1964/65

	Bottom Two Deciles	Decile 3	Deciles 4 and 5	Deciles 6, 7, and 8	Decile 9	Lower Half of Decile 10	Upper Half of Decile 10	Mean for All Classes
Per capita monthly consumer expenditure	8.93	13.14	17.80	24.13	30.71	41.89	85.84	24.43
Expenditure on:								
Foodgrains	1.909	1.747	1.396	1.025	0.773	0.455	0.166	0.968
Milk and milk products	0.196	0.497	0.700	0.836	0.884	0.888	0.755	0.828
Meat, eggs, and fish	0.085	0.140	0.176	0.201	0.209	0.213	0.202	0.198
Other foods	0.026	0.248	0.408	0.568	0.698	0.876	1.412	0.567
Tobacco	0.051	0.065	0.075	0.077	0.079	0.079	0.070	0.078
Vanaspati	0.020	0.052	0.086	0.134	0.144	0.145	0.100	0.129
Other oils	0.136	0.220	0.247	0.230	0.200	0.152	0.070	0.226
Sweeteners	0.147	0.227	0.256	0.252	0.227	0.190	0.109	0.228
Cotton textiles	0.286	0.383	0.415	0.418	0.400	0.366	0.279	0.413
Woolen textiles	—	0.011	0.026	0.039	0.056	0.075	0.141	0.038
Other textiles	—	0.004	0.005	0.011	0.112	0.043	0.022	0.011
Footwear	—	0.040	0.050	0.054	0.109	0.058	0.050	0.056
Durables and semidurables	0.026	0.050	0.079	0.113	0.157	0.213	0.451	0.116
Conveyance	0.030	0.065	0.104	0.116	0.230	0.351	0.412	0.161
Consumer services	0.060	0.010	0.142	0.193	0.232	0.295	0.479	0.188
Education	0.031	0.060	0.100	0.165	0.237	0.360	0.922	0.163
Fuel and light	0.266	0.326	0.344	0.343	0.333	0.306	0.245	0.336
House rent	—	0.032	0.059	0.128	0.193	0.310	0.679	0.124
Total decline in real income in each class	3.27	4.17	4.67	4.90	5.27	5.38	6.77	4.89
New real expenditure	5.66	8.97	13.13	19.23	25.44	36.51	79.07	19.95
Decline in real expenditure (%)	36.6	31.7	26.7	20.3	17.2	12.8	7.9	20.0

SOURCE: See table 1.

NOTE: The percentage rise in the prices of foodgrains due to a 10 percent decline in supply is calculated, and then the adjustment made in consumption of all commodities due to the rise in the price of foodgrains is shown in this table. (1) The percentage rise in price of foodgrains = $[(dQ/Q) \times 100]/e$, where dQ/Q is a proportionate decline in the foodgrains supply and e is the price elasticity of demand for foodgrains in the mean class. From table 1, the decline in expenditure on foodgrains corresponding to mean class as a result of a 10 percent rise in the price of foodgrains is 0.147, implying a 1.52 percent decline in foodgrains consumption expenditure. Therefore, the percentage rise in the prices of foodgrains = $10/0.15 = 67$ percent. (2) This 67 percent rise in the price of foodgrains is reflected in the decline in the consumption expenditure on all commodities. See the footnote to table 1 for the basis of calculation of the decline in expenditure on each commodity.

153

farmers build stocks in good years and deplete them in poor years.[5] Thus, in India, the price response to poor crops was much more substantial in 1966/67 than in 1965/66, a much "worse" year; similarly for 1973/74 compared with 1972/73.

As pointed out above, the secondary effects on employment of an increase in foodgrain prices could be as important in depressing the real incomes of the poor as the direct effects of price on consumption.

With these important caveats in mind, table 2 is of interest in suggesting the distributive impact on consumers of a 10 percent decline in the supply of foodgrains, using the elasticities implicit in the preceding analysis and assuming no compensating decline in employment or secondary effects from the effect of the price change on producers' incomes.

In this analysis, the lowest two deciles in the income distribution of consumers suffer a 36.6 percent decline in real income, compared with a decline of 7.9 percent for the top 5 percent in the income distribution. Similarly, the lowest-income deciles experience a 39.5 percent decline in foodgrain consumption, compared with 1.3 percent for the top income group. The large reduction in consumption of nonfoodgrain commodities shows clearly that there would be large secondary effects on employment, greatly reducing the foodgrain price increases.

The foregoing data show one additional relationship between foodgrain prices and low-income consumers. The poor spend a high proportion of increments to income on foodgrains: the bottom two deciles in the income distribution spend 59 percent of increments to income on foodgrain, compared with only 2 percent for the top half of the tenth decile. Thus, the role of price on the supply function for this crucial wage good must enter importantly into a full analysis of the relation between price and income distribution, particularly given the important role of employment in determining the income of lower-income families.

DIRECT EFFECTS OF FOODGRAIN PRICE CHANGE ON DISTRIBUTION OF INCOME AMONG PRODUCERS

The effects of relative price changes on agricultural producers differ from the effects on consumers in two important respects. First, the income effect, assuming production is constant, is in the same, rather than in the opposite, direction as the price change. Second, the largest effects, both relative and absolute, fall on the producers with the largest marketings (and presumably with the higher incomes).

The effect of a price change that occurs independently of change in the volume of domestic production is easier to analyze than the effect of a price change in response to a production change, perhaps induced by variations in the weather. These two somewhat different cases will be discussed in order, followed by analysis of the effects of price change on production.

The Effect on Producers of Change in Relative Prices with Production Constant

The relationships described below refer to effects of price changes on producers when quantity produced stays constant, as could happen if the price changes were due to a foreign food-aid program. It could also happen as a result of commercial trade, although in that case the relationships are more complex because change in imports and exports of foodgrains would presumably be offset by trade in other commodities, with further price and income effects.

The effect of relative change in agricultural prices on producers' incomes depends on (1) the quantity they produce, (2) the quantity of home consumption and hence of marketings, and (3) the quantity of purchased production inputs. The effect of price changes is much greater in both absolute and percentage terms on larger farms than on smaller farms—the larger farms normally produce more, market a higher proportion of their production, and have a higher proportion of output represented by purchased inputs.

Table 3 illustrates these relationships by relating size classes of farms to level of production, home consumption, and marketings. For the smallest farmers, a price increase actually decreases income, because they are net purchasers of foodgrain, presumably paid for largely by working as laborers. It is only in the fourth and fifth deciles and above that a substantial rise in income occurs. In the upper income deciles, the price effect is more nearly proportionate to income, since the bulk of production is marketed—it is actually less than proportionate because of the fixed share of output assumed to be retained for feed, seed, and waste and because the calculation relates to gross value of output rather than to the value of output net of cash production costs.

The absolute increase in income resulting from a 10 percent price increase is about 60 percent larger for the ninth decile than the average for the sixth, seventh, and eighth deciles and over six times larger in the upper half of the tenth decile than the average for the sixth, seventh, and eighth deciles.

Thus, if a foodgrain price increase is seen as a simple transfer of income from consumers to producers, it largely takes place between high-income consumers, who expend the largest absolute amount on food, and high-income producers, who market the largest absolute quantities. In terms of a percentage change in income, the transfer causes the largest percentage decline in the income of low-income consumers and the largest percentage increase in the income of high-income producers.

The Effect on Producers of Concurrent Change in Production and Price

Changes in the relative price of foodgrains more usually occur as a result of changes in domestic output, due to either short-term weather changes or more permanent technological changes, and thus the effects of price changes on producers are usually offset wholly or in part by production changes.

TABLE 3
CHANGE IN THE GROSS VALUE OF FOODGRAINS CONSEQUENT TO VARIOUS CHANGES IN PRODUCTION AND PRICE, BY RURAL INCOME CLASS, INDIA, 1964/65

	Bottom Two Deciles	Decile 3	Deciles 4 and 5	Deciles 6, 7, and 8	Decile 9	Lower Half of Decile 10	Upper Half of Decile 10
1. Average gross acres sown per reporting holding	0.29	0.98	3.85	8.05	12.57	20.70	45.50
2. Gross value of foodgrains production*	4.06	13.72	53.91	112.72	176.01	289.85	637.11
3. Foodgrain consumption expenditure	4.83	6.84	8.31	9.58	10.45	11.37	12.80
4. Value of foodgrains marketed	-1.38	4.83	37.51	86.23	139.16	235.00	528.74
5. Effect of 10 percent rise in price of foodgrains, with no change in production, on value of foodgrains marketed	-0.14	0.48	3.75	8.62	13.92	23.50	52.87
6. Row 5 as a percentage of gross value of foodgrains production	-3.4	3.5	7.0	7.7	7.9	8.1	8.3
7. Effect of 10 percent decline in production and 5 percent rise in foodgrains' price on value of foodgrains marketed	-0.44	-0.99	-2.93	-5.75	-8.75	-14.05	-30.42
8. Row 7 as a percentage of gross value of foodgrains production	-10.8	-7.2	-5.4	-5.1	-5.0	-4.8	-4.8
9. Effect of 10 percent decline in production and 10 percent rise in foodgrains' price on value of foodgrains marketed	-0.52	-0.80	-1.29	-1.91	-2.54	-3.53	-6.69

10. Row 9 as a percentage of gross value of foodgrain production	−12.8	−5.8	−2.4	−1.7	−1.4	−1.2	−1.1
11. Effect of 10 percent decline in production and 20 percent rise in foodgrains' price on value of foodgrains marketed	−0.70	−0.44	2.01	5.75	9.88	17.52	40.77
12. Row 11 as a percentage of gross value of foodgrain production	−17.2	−3.2	3.7	5.1	5.6	6.0	6.5

SOURCES: Government of India, Cabinet Secretariat, *The National Sample Survey*, 18th Round, February 1963–January 1964, no. 142, tables with notes on consumer expenditure, 1968; and *The National Sample Survey*, 17th Round, September 1961–July 1962, no. 162, tables with notes on some features of land holdings in rural areas, 1969.

NOTE: In this analysis, expenditure on foodgrains is assumed to remain the same irrespective of changes in production and prices of foodgrains. The difference between production and consumption is due to a 15 percent allowance for feed, seed, and waste. For row 1, the cumulative percentage of rural population in various landholding classes was calculated from *National Sample Survey (NSS)* of landholdings for 1961/62. Similarly from *NSS*, 1963/64, data on consumer expenditure for rural households, the cumulative percentage distribution of rural population in various expenditure classes was calculated. The two cumulative distributions were matched to ascertain the approximate correspondence between level of expenditure and landholding. Then the average area cultivated per reporting holding in each consumer expenditure class was obtained. For row 2, the gross output of rice was obtained on the assumption that all the acreage was planted to rice. Hence, it is considered as gross output of foodgrains. Taking the all-India average yield of rice as 1,078 kg per hectare and the price of rice as Rs 1.925 per kilogram as the price of all foodgrains, the gross value of foodgrain production per annum was obtained. Further, assuming that the average farm family size was five, the gross value of foodgrain production per capita per month was obtained. For row 4, the value of foodgrain marketings per capita per month = the gross value of foodgrain production per capita per month − retentions. Retentions = seed + feed + waste + consumption. Seed, feed, and waste is taken to be 15 percent of gross value of output.

*Per capita per month for this and all subsequent categories.

While a production change in response to the weather or a technological change affects producers' incomes in direct proportion to their levels of production, the countervailing price change is more nearly proportional to marketings. Thus, if price increases by the same percentage as production declines, the incomes of all producers decline at least slightly. If the price increase is half as large in percentage terms as the production decline, producers' incomes, of course, decline more precipitously. In both cases, the percentage decline in income is greater for low-income than for high-income farmers. However, if the price increase is larger than the production decline, then farmers in the upper-income classes will experience an increase in income, while the lowest-income farmers may still experience a decline in income.

These relationships are illustrated in table 3, which depicts the effects of a 10 percent production decline combined with price increases of respectively, 5 percent, 10 percent, or 20 percent, giving the effects of alternative assumptions as to the price elasticity of demand. In general, the price change will be counter to the production change, and since demand is inelastic, the change in price will presumably be more than proportionate to the change in production. Given a 10 percent decline in production, a 10 percent increase in price is consistent with unit elasticity; a 20 percent increase in price is consistent with an elasticity of -0.5; and a 5 percent price increase is consistent with an elasticity of -2.0. The elasticity of -0.5 has often been justified on theoretical and empirical grounds.[6] The price rise of only 5 percent is included to illustrate the comparative effects of a price stabilization program.

It should be noted that price stabilization programs destabilize small producers' real incomes and stabilize consumers' incomes and large producers' incomes. Thus, it is significant that the introduction of a less unstable price (row 8 compared with row 12 of table 3) increases the decline in income in years-of-production decline for the third through the fifth deciles in the income distribution. Reduced price instability stabilizes income for the lowest-income farmers because they are net consumers, and for the highest-income producers because they market a high proportion of what they produce.

THE EFFECT OF RELATIVE CHANGES IN AGRICULTURAL PRICES ON AGRICULTURAL PRODUCTION

A slow rate of increase in food production is increasingly recognized as a major limitation to increase of employment and income of low-income families.[7] Without a commensurate increase in the supply of food, growth in paid employment and the consequent rise in demand for wage goods will cause an increase in food prices, the effects of which may substantially neutralize the benefits from increased employment and indeed force a reversal of high-employment policies. Adequate additional supplies of such wage goods can rarely be mobilized through reallocation of existing domestic supplies or through international trade. Thus, policies on employment, agricultural production, and agricultural prices must be closely linked. Change in relative prices plays at most a very limited role in increasing agricultural production in the context of traditional technology. But if used to complement technological change in agriculture, price policy may speed the growth of production significantly. In this context, too, price policy may encourage the adoption of income-increasing innovation by low-income producers through its influence on profitability as well as on risk and uncertainty.[8]

PRICES AND PRODUCTION IN THE CONTEXT OF STATIC TECHNOLOGY

With technology given, an increase in relative agricultural prices increases the supply of agricultural commodities by movement along the production function. Such an increase in production is, by definition, at an increasing real cost in resources and shifts the distribution of income against low-income consumers. Agriculture in low-income countries is commonly operating near the top of the total product curve, with low marginal returns to added inputs, and is, consequently, often a sector of highly inelastic aggregate supply.[9]

The extent to which the income effect of increased wage employment of the poor will be neutralized by increased food prices depends on (1) the incremental budget share allocated to foodgrains by those in wage employment, (2) the elasticity of supply of foodgrains, and (3) the cross-elasticities of demand for foodgrains. Income increases will be nullified by food price increases as the incremental budget share approaches one, the supply elasticity approaches zero, and the cross-elasticity approaches zero. In a low-income country for which the incremental budget share allocated to foodgrains by the laboring classes may be 0.59, the aggregate supply elasticity for basic foodgrains is as low as +0.1, and the cross-elasticity is low, the income increase to the laboring classes from employment may be largely eliminated by the consequent price increases.

The literature dealing with agricultural supply response is much greater with respect to substitutions among individual commodities than with respect to the more complex, but here more relevant, problem of aggregate supply of foodgrains and of agricultural wage goods more generally. What literature there is suggests quite inelastic aggregate supply, on the order of +0.1 to +0.2.[10]

The supply of basic agricultural wage goods is perhaps likely to be most elastic if a substantial proportion of land and other resources is devoted to annual crops consumed either domestically by high-income people or exported. A transfer of acreage may then be effected with little decline in marginal productivity. In such a case, the use of a price increase policy to encourage food-crop production, and the maintenance of high relative prices as a means of implementing that policy, may be successful.

A few countries have at times followed price policies that have depressed agricultural prices substantially relative to international relationships (for example, Thailand), with consequent low levels of input use by international standards and consequent discouragement of production for the market.[11] Particularly if the production functions are essentially linear, except near the bottom and top of the function, rectification of such policies may have dramatic effects on input use and on production.

In the face of inelastic domestic production, the supply of foodgrains may, of course, be augmented by imports, depending on the elasticities of supply and demand for exports to pay for them and on the elasticity of supply of foodgrains for import. However, the coefficients for these variables too may be quite inelastic, particularly for large countries or large aggregates of small countries. As an alternative policy, the marginal redistribution of income effected by fiscal policy is also unlikely to achieve an adequate transfer of food from high-income to low-income classes, because of the highly disparate marginal propensities to consume at the various income levels.

Thus, one may conclude that in the context of static agricultural technology the gains to the poor from expanded employment are likely to be substantially offset by increased prices of food as increased demand exerts its pressure on an inelastic supply. It is these relationships that turn attention to technological change generally and the role of price policy to technological change in agriculture more specifically.

PRICES AND PRODUCTION IN THE CONTEXT OF TECHNOLOGICAL CHANGE

Efficiency-increasing technological change in agriculture allows an increased supply of wage goods without an increase in price.[12] Thus, incomes of low-income people may increase through employment without an offsetting effect of rising wage-good prices. The benefits of scale-neutral technological change will be distributed among producers more nearly in proportion to output than to marketings and, thus, need be less skewed toward larger producers in their benefits than increased prices. Scale-neutral technological change is, of course, directly a product of development of institutions for research, education, input distribution, and so on, and not of policy for price changes. However, increased prices may play an important indirect role in the adoption of new technology. Insofar as that is the case, there may be a trade-off between short-run losses by the poor from higher agricultural prices and long-run gains from a shift of the agricultural production function, a larger supply of wage goods with favorable employment implications, and eventually lower agricultural prices as equilibrium is reached in agriculture with lower costs of production.

Despite its importance to both growth and income distribution, there is little empirical evidence as to the relation between agricultural prices and the pace of technological change. Of course, increased relative output prices increase further the disequilibrium induced by the lower cost of new technology, thereby accelerating output increase. That influence is thought in agriculture to be particularly strengthened by effects that reduce the incidence of uncertainty. Higher prices reduce the probability of loss from an innovation by increasing expected average profitability. More stable prices may also reduce uncertainty for some farmers.

The relation between price policy and variance in agricultural incomes is highly complex. High variance in net income, at the very least, has an unfavorable effect on income distribution by skewing the pattern of adoption toward already higher-income farmers—reflecting their greater risk-bearing capacity.[13]

However, in agriculture, yield variation due to weather is generally a more important source of variance than price. This is particularly true when a large proportion of output is retained for domestic consumption. And the extent to which price stabilization even increases the stability of producers' incomes is, in practice, dependent on several factors, important among which is the extent to which movements in price are inversely related to movements in production. At the micro level, that depends on the extent to which production changes in a particular area are similar to changes in national aggregates and on the degree of national integration of markets. Thus, Schluter, in a simulation of the effect of changes in price variance, using actual farm data from the Surat District, India, found that in four out of six situations price stabilization actually increased the coefficient of variation of revenue.[14]

Finally, in selecting a price policy to facilitate efficiency-increasing technological change in agriculture, it must be remembered that increased profitability and reduced variance in income may be achieved by appropriate investment in such items as research, education, and irrigation—quite possibly on a more cost-effective basis than through price policy and with lesser short-run deleterious effects on income distribution.

The new high-yielding crop varieties are noted for their generally low elasticity of employment with respect to output. Thus, the primary significance for employment of the new foodgrain technologies lies in their potential to relax the wage goods' constraint and to generate increased farm incomes, which may promote a secondary increase in employment.[15] In this section, three further aspects of the relation between agricultural prices and employment are explored: (*a*) the effect of relative agricultural prices on the labor intensity of the agricultural output mix; (*b*) the effect of relative agricultural input prices on the choice of technique in agricultural production; and (*c*) the effect of relative agricultural prices on the level and structure of nonagricultural employment.

Prices and the Agricultural Output Mix

Choice of cropping pattern is one of the most important factors influencing labor requirements in agriculture, and, thus, because of the relative magnitude of the agricultural sector, it is one of the most important determinants of overall employment in a low-income country. B. M. Desai, in a detailed analysis of farm-management data from the Surat District, India, calculates that 90 percent of the differences in income per acre among farms is due to differences in cropping pattern, which affect income largely through differences in labor input, rather than to differences in the intensity of input use in specific crops.[16] Desai and Schluter show from farm management data for the Surat District, India, that human labor use is 60 and 37 days per acre on groundnuts and cotton, respectively.[17] A transfer of one-quarter of the cotton acreage to groundnuts would add over 2.1 million man-days of employment per year in the Surat District. That is about eight times as much employment to be added in that district as by the special "crash scheme for rural employment" intended as a major source of rural employment. In this case, not only do the two crops compete for the same set of nonlabor resources and have substantial differences in labor requirements, but a portion of the domestic supplies of each commodity is also imported. Thus, the relative domestic prices, production, and aggregate employment can be influenced by import policy.

Despite the importance of agricultural output mix on employment and the substantial literature showing significant elasticities of output substitution at the farm level, there has been little policy analysis of this issue. Needed study not only would examine employment and elasticities of substitution among crops but, equally important, would analyze potentials to shift trade policy and domestic demand toward more labor-intensive commodities. Most simply, export and import policy may be used marginally to facilitate a shift in relative domestic prices and production toward more labor-intensive commodities. Temperate-zone nations have done this particularly in the case of sugar—historically a labor-intensive crop. India, in the earlier example, could shift the product mix from cotton toward groundnuts by importing more cotton and less vegetable oil, with a consequent increase in employment. Export subsidies on vegetables could have a similar effect. Such policies are particularly attractive if they compensate for the effects of the labor market and other imperfections, causing suboptimal utilization of labor on small farms.

Domestic demand structure may also be influenced toward more labor-intensive commodities through tax and subsidy schemes. Related and perhaps more important, demands for relatively labor-intensive agricultural commodities tend to be relatively elastic with respect to income, providing a significant opportunity for rising incomes to favor rising employment. Unfortunately, there is a tendency, because of the bulky, perishable nature of many such commodities, for marketing bottlenecks to result in restraint of consumption through price increases that are not transmitted to the farm level. Institutional credit gaps at the farm level for commodities with generally high working-capital requirements may also inhibit the desired expansion of demand. Thus, as for other aspects of agricultural price policy, programs must go beyond the price policies and indicators to substantive aspects of marketing and production. In this context, an employment-oriented price policy should (*a*) recognize the employment implications of alternative price relationships among agricultural commodities, (*b*) see that overt price policies do not discriminate against the relatively labor-intensive commodities, and (c) use movements in market-determined prices as indicators of marketing and production bottlenecks that may be dealt with through other policies.

RELATIVE INPUT PRICES

The dramatic displacement of labor often occasioned by farm mechanization and its frequent association with effective subsidization of farm machinery prices relative to labor has prompted a substantial literature.[18] Given that extensive treatment, attention is here brought to two features that make the statement of effective policy difficult in this area.

First, although inappropriate pricing policy is frequently deplored, there is still controversy as to whether market prices of labor adequately reflect the equivalent of perfect market supply prices.[19] It is thus difficult to know whether or not compensating taxes on machinery may be in order.

Second, and related to the first, considerable controversy exists as to the precise nature of the labor supply, particularly given the complexities of seasonal cycles in both demand and supply of labor.[20] Thus, Donovan shows with a linear programming analysis based on farm-management data for an area in Mysore, India, that the introduction of hand tractors allows a substantial increase in total employment.[21] It does so by shortening land preparation time sufficiently to allow a not otherwise possible second crop. It follows that at such time as mechanization is appropriate for breaking labor bottlenecks it may be as appropriate a recipient of price subsidy and other facilitative measures, particularly for small farmers, as any other element of production and labor-absorbing technological change. The full complexities of policy in this area are underlined by Donovan's analytic position that seasonal labor migration, which of course brings other problems, can meet the labor bottleneck as effectively as mechanization.

Finally, it should be noted that fertilizer and other chemical inputs also substitute for labor, and, hence, subsidization of their prices will reduce employment unless ancillary policies ensure an aggregate increase in output, with a set of direct and indirect employment effects as elaborated in preceding sections.

AGRICULTURAL PRICES AND NONAGRICULTURAL EMPLOYMENT

The basic relationships between agricultural price policy, technological change in agriculture, supply of wage goods, and employment were discussed above. In that argument, relatively low agricultural prices are seen as desirable in the longer run not only to bring about an immediate raising of the real incomes of the poor but also to stimulate longer-term employment growth—as long as technological change effects a continuous upward shift of the supply curve for agricultural commodities.

Relative agricultural prices are often depressed by import and foreign-exchange pricing policies specifically intended to encourage industrial growth. Clearly, such policies may directly encourage industrial employment while in the longer run discouraging production of the basic agricultural wage goods, with a consequent deleterious effect on employment potentials. Whether such policy is desirable hinges on an argument for industry essentially analogous to that for relatively higher agricultural prices intended to encourage accelerated application of efficiency-increasing technological change in agriculture. Policies that boost the domestic production of industrial consumer goods that are relatively efficient and labor-intensive, and eventually effect a reduction in the price of these goods, need have only short-run adverse effects on relative agricultural prices and agricultural production. As they expand, they generate increased demand for agricultural wage goods from increased industrial employment, tending to raise relative agricultural prices.

However, the effects may be quite different when import restrictions are introduced to protect capital-intensive, low-efficiency industrial processes. Protection of this kind generates little increase in employment or in demand for food, and so the depression of relative agricultural prices continues, potentially reducing production in the traditional agricultural sector and weakening incentives for technological change in agriculture.[22] Such an effect may continue indefinitely if those industries remain inefficient relative to foreign sources of supply.

CONCLUSION

The data presented demonstrate dramatically that the income effect on low-income people of food price changes is large and that the bulk of adjustment to reduced food supplies is made by low-income people. Conversely, of course, changes in the income of low-income people are reflected to a large degree as change in the demand for food.

Change in food prices causes a larger percentage change in the real incomes of low-income consumers but a larger absolute change in the real incomes of high-income consumers. Thus, there is potential for substantial secondary effects on employment and incomes of the poor arising from the primary effects on the consumption pattern of the more well-off.

The effects of relative price changes on agricultural producers differ from the effects on consumers in two important respects. First, the income effect, assuming production is constant, is in the same rather than in the opposite direction as the price change. Second, the largest effects, both relative and absolute, fall on the producers with the largest marketings (and presumably with the higher income).

Demand for food may be effectively brought into balance with deficit supply through reduction in employment rather than an increase in price. High-income consumers may prefer policy measures of that type. Similarly, if the increased demand for food accompanying the increased employment of low-income people is not met by an increased food supply, the employment-based increase in real income will be substantially reduced by price increases. Thus, an employment program, or an income-transfer program for the poor, will be inefficient in assisting them unless provision is made for an enlarged supply of basic food commodities. It follows that in designing income and employment programs attention needs to be paid to the material balances and not just the fiscal balances. To state the position clearly, barring a strict rationing system, increased agricultural production may be a necessary precondition for improving the incomes of the poor. It follows, of course, that a program of foreign food aid and, to a lesser extent, commercial imports can be effective in facilitating an employment increase, particularly in the short run.

The importance of an increased supply of food to the effectiveness of employment programs in raising the income of low-income families suggests the appropriate consideration in the analysis of production effects of agricultural price policy. The preponderance of the evidence shows that in the context of traditional technology aggregate food supply will normally be highly inelastic, and hence an upward shift in demand for food will result in magnified upward movements in food prices, with the consequent distribution effects shown above. However, in the context of technological change, higher prices to producers may counterbalance the added risk and uncertainty associated with such change. If experience with new technologies itself reduces risk and uncertainty, higher prices may induce a shift to a new technology, which will not be reversed if prices later decline. Of course, domestic policy may so depress relative agricultural output prices as to virtually eliminate use of certain key inputs, such as fertilizer, for market production, and in that case supply may be highly responsive to even modest changes in prices. Such a situation is often the result of government monopoly pricing. Changes in policy will be most effective if such policy has held production close to the subsistence level. The effect will be further enhanced if there is close interaction of such input use and new technology.

Research and education that dramatically reduce production costs may serve as an alternative to higher prices to induce innovation. Such measures do not have such deleterious effects on low-income consumers as higher prices, can also induce more permanent increases in productivity, and may have a less skewed distribution of net benefits to producers. Thus, price policy may be viewed as a competitor to other measures as well as a complement.

Relative agricultural prices affect employment through the labor intensity of the agricultural output mix, the labor intensity of agricultural technology, and the level and structure of nonagricultural employment. In each case, the importance of interactions of agricultural price policy with other policy measures is important.

Finally, the extent to which a change in terms of trade between agriculture and industry benefits the poor depends very much on the extent of the structural adjustments it encourages. A turn against industry will redress itself if it induces accelerated technological change in agriculture and consequent linkage effects with industry through relaxed wage-goods constraint and increased consumer demand. Conversely, a turn against agriculture will redress itself if it encourages accelerated industrial employment growth, consequent greater demand for agricultural wage goods, and increased efficiency of industrial production. A lack of the conditions for technological change in agriculture and a lack of the conditions for employment growth in industry will cause either of the respective price policies to fail.

NOTES

I am grateful to Shakuntala Desai for her assistance in developing the data and references; to Uma Lele, S. D. Tendulkar, Mohinder Mudahar, C. Ranade, B. M. Desai, and G. Doraswamy for a careful reading and set of comments; and to an anonymous reviewer for most helpful specific as well as general comments.

1. See Timmer, chapter 8 in this volume.—ED.

2. The complexity of the issues and the weight of analysis toward narrow short-run production considerations is reflected in the differences and controversy in the following: M. L. Dantwala, "Incentives and Disincentives in Indian Agriculture," *Indian Journal of Agricultural Economics* 22 (April–June 1967): 1–25; V. M. Dandekar, "Agricultural Price Policy," *Economic and Political Weekly* 3 (16 March 1968): 454–59; John W. Mellor, "The Functions of Agricultural Prices in Economic Development," *Indian Journal of Agricultural Economics* 23 (January–March 1968): 23–37; idem, "Agricultural Price Policy in the Context of Economic Development," *American Journal of Agricultural Economics* 51 (December 1969): 1413–20; Uma J. Lele, "Agricultural Price Policy," *Economic and Political Weekly* 4 (30 August 1969): 1413–19; idem, *Food Grain Marketing in India* (Ithaca: Cornell University Press, 1971); idem, "Considerations Related to Optimum Pricing and Marketing Strategies in Rural Development" (Paper presented at the Sixteenth International Conference of Agricultural Economists, Nairobi, Kenya, 26 July–4 August 1976); Raj Krishna, "Agricultural Price Policy and Economic Development," in *Agricultural Development and Economic Growth,* ed. Herman H. Southworth and Bruce F. Johnston (Ithaca: Cornell University Press, 1968); Theodore W. Schultz, *Transforming Traditional Agriculture,* Studies in Comparative Economics, no. 3 (New Haven: Yale University Press, 1964); E. S. Mason, *Economic Development in India and Pakistan* (Cambridge: Center for International Affairs, Harvard University, 1966). For an effort to look directly at the effects of agricultural price changes on income of various socioeconomic classes see Roberto Echeverria, *The Effect of Agricultural Price Policy on Intersectoral Income Transfers,* Occasional Paper 30 (Ithaca: Department of Agricultural Economics, Cornell University–USAID Employment and Income Distribution Project, June 1970).

3. In India foodgrains represent a major portion of total expenditure and exhibit sharply different marginal propensities to consume across income classes. The same analysis is relevant to countries with higher incomes than India if the subset of food items is defined sufficiently more broadly than foodgrains as to maintain these two characteristics.

4. These calculations in effect assume equating of total expenditure and total income. If the higher-income classes reduce savings, or the lower-income classes increase dissaving, in response to higher foodgrain prices, then the expenditures on various categories of goods will be even less responsive to price. If, as is usually assumed to be the case, the marginal propensity to save is higher for high-income than for low-income classes, then of course the proportion of the total adjustment that must be made by the lower-income classes is even greater. Similarly, the greater the extent to

which foodgrain expenditure of higher-income consumers is for relatively more demand-elastic services associated with the foodgrain, the more this conclusion will be reinforced.

5. For an analysis of this point and estimates of magnitudes see John W. Mellor and Ashok Dar, "Determinants and Development Implications of Foodgrains Prices, India, 1949–50 to 1963–64," *American Journal of Agricultural Economics* 50 (November 1968): 962–74.

6. See, for example, L. M. Goreux, *Demand Analysis for Agricultural Products,* Agricultural Planning Study, no. 3 (Rome: FAO, 1964); and Mellor and Dar, "Determinants and Development Implications."

7. For a full exposition of these relations see John W. Mellor, *The New Economics of Growth: A Strategy for India and the Developing World* (Ithaca: Cornell University Press, 1976). For a more theoretical presentation see Uma J. Lele and John W. Mellor, *Technological Change and Distributive Bias in a Dual Economy,* Occasional Paper 43 (Ithaca: Department of Agricultural Economics, Cornell University–USAID Employment and Income Distribution Project, June 1971). For a briefer, more policy-oriented statement see idem, "Jobs, Poverty and the 'Green Revolution,'" *International Affairs* 48 (January 1972): 20–32.

8. See also Krishna, chapter 11 in this volume.—Ed.

9. For a more complete discussion see John W. Mellor, *Economics of Agricultural Development* (Ithaca: Cornell University Press, 1966), chap. 11.

10. For an analysis of aggregate supply elasticities see Robert Herdt, "A Disaggregate Approach to Aggregate Supply," *American Journal of Agricultural Economics* 52 (November 1970): 512–20; Howard Barnum, "A Model of the Market for Foodgrains in India, 1948–64," Technical Report 23 (Berkeley: Project for the Evaluation and Optimization of Economic Growth, Institute of International Studies, University of California, 1970); Krishna, "Agricultural Price Policy"; and Mellor, *Economics of Agricultural Development,* chap. 11.

11. See Jere R. Behrman, *Supply Response in Underdeveloped Agriculture* (Amsterdam: North-Holland Publishing Co., 1968).

12. For a wide range of examples see John W. Mellor and Uma J. Lele, "Growth Linkages of the New Foodgrains Technologies," *Indian Journal of Agricultural Economics* 28, no. 1 (1973): 35–55.

13. Michael G. G. Schluter, *Interaction of Credit and Uncertainty in Determining Resource Allocation and Incomes on Small Farms, Surat District, India,* Occasional Paper 68 (Ithaca: Department of Agricultural Economics, Cornell University–USAID Employment and Income Distribution Project, February 1974).

14. The data were for rice and cotton (for each, fertilized and unfertilized), sorghum, and groundnuts. For 1962–72 the coefficient of variation of revenue (price x yield) was reduced slightly with both prices fixed at the mean and prices fixed at 50 percent of the mean for sorghum and groundnuts; the coefficient of variation of revenue was increased by a much larger margin in the other four cases (Schluter, *Interaction of Credit and Uncertainty*).

15. Mellor and Lele, "Growth Linkages."

16. B. M. Desai, *Relationship of Consumption and Production in Changing Agriculture: A Study in Surat District, India,* Occasional Paper 80 (Ithaca: Department of Agricultural Economics, Cornell University–USAID Technological Change in Agriculture Project, February 1975).

17. G. M. Desai and Michael G. G. Schluter, "Generating Employment in Rural Areas," in *Seminar on Rural Development for the Weaker Sections* (Bombay: Indian Society of Agricultural Economics, May 1974), 143–52; and Mellor, *New Economics of Growth,* 86.

18. H. Rao, "Farm Mechanisation in a Labour Abundant Economy," *Economic and Political Weekly* 7 (February 1972): 393–400. [See also chapter 27 in this volume.—Ed.]

19. See, for example, the review and analysis in Mellor, *New Economics of Growth,* chap. 4.

20. Ibid.

21. Graeme W. Donovan, *Employment Generation in Agriculture: A Study in Mandya District, S. India,* Occasional Paper 71 (Ithaca: Department of Agricultural Economics, Cornell University–USAID Employment and Income Distribution Project, June 1974).

22. For an analysis of such an effect in the Indian context see Mellor, *New Economics of Growth,* chap. 7.

11

Price and Technology Policies

RAJ KRISHNA

In the field of price policy, as in the field of growth policy, we have to reckon with the prevalence of fundamentalism. Like agricultural/industrial growth fundamentalism, there is price fundamentalism. The rational escape from the former is provided by the notion of balanced sectoral growth; likewise, a rational answer to price fundamentalism would be a balanced view of the role of price policy and nonprice (technology) policy in promoting growth. The need for balance is clearly suggested by the present state of research on farm supply response.

Although supply response is a heavily researched area, there are surprisingly few studies of the response of aggregate farm output to lagged terms of trade *inter alia*. Such studies are obviously crucial for measuring the marginal leverage of terms of trade as a means of stimulating agricultural growth. In recent survey papers of the World Bank tabulating about one hundred single-crop price elasticities of acreage/supply for developing countries, only two aggregate supply elasticities are recorded: for Argentina and the Punjab (India) (Sobhan 1977, table 1 and p. 13; Scandizzo and Bruce 1980, app. 2, table 1). For Europe and the United States, however, as many as nine of the thirty-six elasticities tabulated by D. G. Johnson (1973, 113) are aggregative. In the OECD region, of course, the (short-run) aggregative elasticities are in the same range (0.25 to 0.45) as single-major-crop elasticities.

The Argentinian (short-run) elasticity came out to be 0.21 to 0.35 in different regressions. The Punjab study yielded a significant positive elasticity (0.22) for the prewar period (1907–46) but a *negative* elasticity (−0.06) for the postwar period. The author explicitly noted that in the postwar period output expanded, while the terms of trade *declined;* and suggested that growth was primarily due to technological change (Herdt 1970).

In order to examine the effect of the terms of trade of agriculture on aggregate farm output, two new equations have been estimated for Japan and India using the available terms-of-trade series.[1]

RAJ KRISHNA is professor of economics, Delhi School of Economics.

Excerpts from "Some Aspects of Agricultural Growth, Price Policy and Equity," *Food Research Institute Studies in Agricultural Economics, Trade, and Development* 18, no. 3 (1982). Reprinted with minor editorial revisions by permission of the Food Research Institute, Stanford University, and the author.

$$Japan \ (1881-1919)$$

(J.1) $Q_t =$ constant $-132.80 \ P_{t-1} + 30.52*** \ T + 0.17 \ Q_{t-1}$
   (t) (-0.53) (4.77) (0.97)
   (e) $[-0.05]$
   $R^2 = 0.90$

$$India \ (1952/53-1974/75)$$

(I.1) $Q_t =$ constant $+ 0.14* \ P +0.49*** \ W_t + 3.08** \ Z_t + 0.45** \ Q_{t-1}$
   (t) (1.49) (4.68) (2.11) (2.44)
   (e) $[0.18]$ $[0.54]$ $[0.68]$
   $R^2 = 0.92$

Here: $Q =$ farm output, $P =$ terms of trade, $T =$ time trend, $W =$ weather index, and $Z =$ irrigation ratio. The price coefficient is only marginally significant in the Indian equation, and the price elasticity is about 0.2. In the Japanese case the price elasticity is negative, low (-0.05), and statistically zero. At best these equations (as well as the Punjab and Argentinian results cited above) would suggest a one-period aggregate price elasticity of about 0.2 and a long-run price elasticity of about 0.4.[2]

The terms-of-trade movements do seem to have a positive effect on aggregate output. And a favorable price environment must be considered indispensable for agricultural growth. But for a balanced view of the relative role of price and nonprice factors in promoting growth two implications of supply studies need to be noticed.

First, if we consider, for a moment, price policy as the sole instrument for fostering agricultural development, the order of annual terms-of-trade increases required is certainly more than a poor country can manage on macro grounds. Suppose for instance that the one-period price elasticity is 0.2, the "long run" implied by the usual lag coefficient (0.5) is about five years, and a low-income country needs 3 percent annual growth in farm output. Then the long-run elasticity being 0.4, 16 percent growth over five years would require a once-over 40 percent increase in the real terms of trade of agriculture. This is equivalent to a 7 percent annual increase over the period, which will also, of course, spread out the resulting output growth. This order of terms-of-trade increase is hardly a practical proposition, even assuming that a government can fix terms of trade. (Even if the long-run elasticity is assumed to be 0.6, that is, roughly equal to that for most of the individual cash crops, instead of 0.4, the required 27 percent one-shot increase, or a 5 percent annual increase, in the terms of trade would be infeasible.)

The second important fact relevant here is that in most supply regressions the elasticities of supply with respect to shifter variables (proxies for technological change, like the irrigation ratio) exceed the price elasticities. In the Indian function (I.1) above, the irrigation elasticity (0.68) is more than three times the price elasticity (0.18). In postwar Indian wheat functions the irrigation elasticity (0.75 to 0.80) has been found to be 1.5 times the price elasticity (about 0.5)

(Krishna and Raychaudhri 1979; Krishna and Chhibber 1980). And in an early supply-response study of eleven crops in the Punjab (India) (Krishna 1963, table 1b; 1967) it was observed that the irrigation elasticity exceeded the price elasticity in *every* equation where irrigation was included. The irrigation elasticity was in fact 1.5 to 5.5 *times* the price elasticity in the case of various crops (cotton, millets, and wheat). In the Japanese equation (J.1), the time trend is very strong, while the price coefficient is not even significant.[3] These numbers suggest that a unit percentage change in the important shifter (technology) variable will yield much greater growth than a unit percentage price shift.[4]

There are many episodes in the record of advanced countries in which the (lagged) terms of trade facing agriculture have stagnated, and yet farm productivity has grown 2–3 percent a year for considerable periods (see Johnson 1973 for OECD countries; and Kelley and Williamson 1974 and Hayami, Ruttan, and Southworth 1979 for Japan). Again, the explanation lies in technological dynamism.

The price fundamentalist would, of course, argue that technological change itself is induced by relative price movements. This proposition has a core of proven truth (Hayami and Ruttan 1971).[5] But only some aspects of innovation, in the broadest sense, can be shown to be price-induced. The price milieu determines the relative, privately perceived profitability of different techniques made available by completed applied research and hence influences the rates of their adoption (diffusion). But it cannot by itself explain the evolution of basic scientific knowledge and the level and growth of public investment in research, extension, infrastructure, and human capital in different parts of the world. The growth of basic knowledge has some irreducible nonlinearity, discontinuity, and randomness. And governments have been far less rational than peasants in making investment decisions.

The authors of the "induced innovation" hypothesis have documented the discontinuity of basic breakthroughs, such as the acquisition of the theoretical and empirical knowledge of processes of inheritance, the mastery of crossing techniques, and the development of methods of mass seed production, which added up to the "invention of a method of inventing" varieties and reproducing them (ibid., 147–48).

Public-sector investment in farm research in the United States has been shown to be economically rational since the 1920s (ibid., 151). But it would be hard to prove that the initial setting up of the farm research infrastructure in the late nineteenth century was price-induced. In Japan, too, for three hundred years of the Tokugawa period "constraints of feudalism" left a "substantial backlog of unexploited indigenous technology." And even when a rational research system existed, responses to price changes were very slow (ibid., 156, 162).

As for the currently developing countries, much evidence has been presented to demonstrate (1) the absence of any significant research prior to 1950, except in colonial crops; (2) the utter nonoptimality of the level and allocation of public investment in the production of technology; (3) the continuing neglect of research on noncolonial crops, especially on crops like cassava, coconuts, sweet potatoes, groundnuts, and chickpeas; (4) the necessity and inadequacy of research even to "adapt" imported technologies (varieties, machines, and chemicals); and (5) the failure to promote the genetic improvement of farm animals (ibid., 164–66; and Evenson 1981).[6]

If relative price movements alone were sufficient to generate high-yield technology, this technology should have emerged in the areas of recurrent drought-induced food-price inflations (in South Asia and African societies) in the nineteenth and early twentieth centuries. But it did not develop there; it developed elsewhere and is still to be indigenized and widely adopted in these scarcity-ridden societies.

The upshot seems to be that a congenial price regime is a necessary but not a sufficient condition for agricultural growth. An unbiased listing of growth factors would have to be dualistic, including favorable price movements as well as induced and autonomous technological and institutional innovations. Theoretically, one can conceive of three supply curves (responses): the response to a price increase along an unshifted supply curve (lowest); the response along a supply curve including the effect of price-induced innovations, or shifts (next highest); and the total actual supply growth *associated with* a price increase but including price-induced as well as autonomous shifts. It is tempting for the fundamentalist to attribute the whole or none of the supply shifts to price movements. The true supply response to price would most probably include a part but not the whole of the shift. But so far it has not been possible to identify this part because of the difficulty of identifying induced and autonomous innovations and because of deficiencies of data.

Technological change, by definition, increases the total factor productivity of the aggregate conventional input. Even at unchanged output and input prices, therefore, it must increase the return per unit of cost. To see this, one has only to write the return/cost ratio as the product of terms of trade and total factor productivity. Let the return/cost ratio (r) be written as PQ/pF, where Q and F are total output and total input, and P and p are output and input prices. If the terms of trade are defined as $p^* = P/p$ and total factor productivity is defined as $t^* = Q/F$, then $r = p^*t^*$. In growth rates (denoted by a dot, ˙), $\dot{r} = \dot{p}^* + \dot{t}^*$. Thus profitability r can be raised either by improving the terms of trade ($\dot{p}^*$) without innovation ($\dot{t}^* = 0$) or by improving productivity ($\dot{t}^*$) at unchanged prices ($\dot{p}^* = 0$), or both.

Successful innovation is thus an alternative, as well as a strong supplement, to an increase in the output-input price ratio, as a means of raising the return/cost ratio and thereby stimulating growth.

In this important sense, a good technology policy is equivalent to a good price policy. A balanced policy should of course include both. But there is a case for giving primacy to a technology policy, for, as we have seen, the (partial) elasticity of output with respect to indices of technical change is generally higher than that with respect to relative price indices and would in all probability remain higher even if the price elasticity were measured to include the effect of price-induced innovations. The measured social return to agricultural research and extension is also known to be very high (48–53 percent) (Hayami and Ruttan 1971, 41).[7] It would perhaps be much higher than the return to price policy alone, if the latter could be measured. National policy makers and international development bankers will therefore do well to devote at least as much attention and effort to the development of *technology, infrastructure,* and *human capital* as to the price environment.

Price policy should not, however, be negative. It is essential that the output-input price ratio for products whose output growth is to be accelerated is not allowed to fall, for the growth-inducing effect of innovation would be reduced. The relevant price ratio is of course net of taxes and subsidies. Abstracting here from tax and subsidy policy, two major issues arise with regard to direct product price policy: (1) the determination of the support-price or the government purchase-price level for any product and (2) the maintenance of appropriate interproduct price relatives in the support/purchase prices.

Despite numerous studies showing the undesirable allocative and distributive effects of price support, it is a safe assumption that support policies will continue in the OECD and developing countries alike, though the mixture of motives for these policies differs between these two sets of countries. In the OECD group it includes income support, stabilization, risk reduction, and/or the discouragement of excess production. In developing countries it covers stabilization, risk reduction, encouragement of production growth, food security, and diversification. (All these concepts carry varying meanings in different countries.)

Taking growth promotion to be the major aim in a poor country, there seems to be no alternative to the adoption of "full average cost" (including the imputed value of family resources at market prices) as the basic principle of support-price fixation for any single crop. This principle has been questioned on many theoretical grounds. First, it has been pointed out that the cost of specialized resources is demand-determined and therefore not independent of the product price. Including this cost in the administered price would involve circularity. Every time the product price rose, the "cost" of these resources would rise, and the administered price would have to be raised. Second, it has been noted that in the presence of uncertainty the cost that determines producer decisions is subjective, opportunity cost. And this cannot be measured objectively for the purpose of price fixation. And third, the variance of cost across farmer groups and regions is very high; therefore the choice of groups and regions whose cost is fully covered by the administered price would be arbitrary (Pasour 1980). But there are counter arguments favoring the full-cost principle. It is of course difficult to estimate the theoretically ideal cost as a basis of support, but for administrative purposes some less than ideal measure has to be chosen. Wherever cost data are generated by regular sample surveys the full-cost principle has proved to be administratively workable (as in India). Second, under certain circumstances the cost principle will entail a lower treasury cost than the parity principle. (It is interesting in this connection that the cost principle has been accepted recently in the

U.S. Food and Agriculture Act of 1977 as an alternative to parity.) Third, the coverage of full average cost provides downward price stability or insurance against the risk of a price decline below cost in the sense most meaningful to the farmers, particularly the small farmers in poor countries. Fourth, the cost-variance problem can be handled for practical purposes by ordering the sample deciles according to their average cost and ensuring that the average cost of a major part of output is covered.[8] And finally, the inclusion of the return to specialized (family-supplied) resources at (lagged) market-determined prices can be viewed as a way of providing a surplus for investment (a necessary incentive for growth in poor countries).

Thus, on balance the cost principle can be used as the least unsatisfactory basis for support. While the support price is a guaranteed minimum, entailing "passive" purchases by the government when the market price goes below it, many governments engage in "active" purchases of grain for running a concessional subsidized grain supply system, or building up public-sector stocks. These purchases need to be made in principle at the going market price even if it is much above the support price. If direct redistribution of income is not feasible and the low-income population of a poor country would suffer unacceptable cuts in food consumption at market-clearing prices, the operation of a concessional (subsidized) food supply system, to serve this population, and the associated dual pricing becomes a second-best necessity, though it has been criticized as a distortion. But the subsidy must be financed from the general revenues and not by forcibly reducing the price realized by (and thus taxing) farmers only.

In many countries supporting the prices of many farm products the interproduct price relatives need to be deliberately rationalized. Otherwise farmers switch resources between products (wheat and rice, cereals and pulses, fine and coarse grains, foodgrains and feedgrains, food crops and cash crops, crops and livestock, and so on) in response to "wrong" signals, generating excess demands and excess supplies in different product markets. This particular problem is often due to the practice of fixing support (purchase) prices for different products one by one, with uncoordinated formulae. The problem can be reduced by fixing a coordinated support-price package for all supported crops. The determination of this package—the consistent price set—can be guided by the solution of a farm-sector equation system for related products with exogenously projected or endogenous final demands. Operational research on such models to derive consistent administered price sets deserves priority.[9]

NOTES

1. For this paper, using series from Hayami, Ruttan, and Southworth 1979; Kelley and Williamson 1974; and Thamarajakshi 1977. Of the three terms-of-trade series available for India, the foodgrains terms of trade yielded relatively more significant coefficients. The simple Nerlovian specification is maintained to get estimates comparable with earlier ones.

Figures in the parentheses are t-values of regression coefficients. Figures in square brackets are elasticities at the means of variables. Coefficients significant at the 10 percent, 5 percent, and 1 percent levels are marked by (*), (**), and (***), respectively.

2. The possibility of these elasticities' being underestimates is discussed below. But the recent estimates of the aggregate agricultural supply elasticity ranging from 1.25 to 1.66 (Peterson 1979) would seem to be gross overestimates to the economists of developing countries. It is difficult to believe that sample observations from Japan and Pakistan, the United States and Denmark, Chile and Paraguay, Niger and Upper Volta used in the Peterson study come from the same structural universe. It is also difficult to accept the implication that in the typical developing country, say in Africa or South Asia, all that is required for a 3 percent growth of farm output is a 2.5 percent increase in the real price of output. For price increases of this order have occurred frequently and have continued for many years in these regions, while aggregate output has increased much less or stagnated.

3. In the Asian context at least, the irrigation ratio, or more inclusively, the proportion of area under water control, can be shown to be the best single proxy for the supply shifters. In many studies irrigation growth has been found to have been the critical precondition and most important determinant of the growth of area under high-yielding varieties, fertilizer consumption, and cropping intensity (for India see Jha 1980; and Sanderson and Roy 1979). The World Bank has noted that 50–60 percent of the increase in output over the past twenty years has been due to higher yields on previously or newly irrigated areas (World Bank 1982, chap. 4, p. 13).

In the cases of Meiji Japan and India, the treatment of a major part of irrigation growth as an autonomous or non–price-induced process also seems justified. For in Japan irrigation and water control had been extended to almost all paddy land in the Tokugawa period itself. This is the reason for specifying only time trend and the terms of trade as the main variables in the supply function for the Meiji period. And in India, the development of canal-irrigated area, which now accounts for more than 40 percent of total irrigated area, has been a public-sector activity. Tanks and traditional wells have existed since ancient times. But their renovation, and the recent expansion of pump irrigation, has been largely due to growing public outlays on construction, exploration, and outright grants; and the expansion of subsidized credit. About 20 percent of public irrigation investment goes for "minor irrigation," and medium-term credit from state institutions has been growing at 16 percent a year in the seventies (CMIE 1980). Surely some part of irrigation growth, and the associated technical input growth, is attributable to relative price movements. But it cannot be isolated with the available data. If this effect could be measured, the measured long-run price elasticity would be somewhat greater than 0.4.

4. Ideally, we should compare the elasticities of output with respect to the dollar costs, rather than the direct indexes, of price movements and supply shifters. But studies making such comparisons have yet to be made.

5. See also chapter 4 in this volume.—ED.

6. See Evenson, chapter 24 in this volume.—ED.

7. See Schultz, chapter 23, and Evenson, chapter 24, in this volume.—ED.

8. In practice there are three options: to cover the major part (more than 50 percent) of (*a*) output, (*b*) area, or (*c*) holdings, after ordering sample farms by average cost. (In business-management practice "bulk-line" costing sometimes covers as much as 85 percent of output.) The position taken here is that at least the cost of 50 percent of output (in sample farms ordered by average cost) should be covered. This avoids protecting the higher average cost of relatively inefficient farms. In the term "full average cost," *full* refers to the fact that the imputed value of family resources is included in cost, and *average* is applied to the cost on each farm. It is not necessarily the sample average cost and it is not marginal cost.

9. Attention may be drawn here to the recent work of IIASA on sector modeling (see Parikh and Rabar 1981).

REFERENCES

Center for Monitoring Indian Economy (CMIE). 1980. *Basic Statistics Relating to the Indian Economy.* Vol. 2, *States.* Bombay.

————. 1981. *Basic Statistics Relating to the Indian Economy.* Bombay.

Ensminger, Douglas, ed. 1977. *Food Enough or Starvation for Millions.* New Delhi: Tata McGraw-Hill.

Evenson, R. E. 1981. "Benefits and Obstacles to Appropriate Agricultural Technology." *Annals of the American Academy of Political and Social Sciences* 458 (November). Reprinted as chapter 24 in this volume.

Hayami, Y., and V. W. Ruttan, 1971. *Agricultural Development: An International Perspective.* Baltimore: Johns Hopkins Press.

Hayami, Y., V. W. Ruttan, and H. M. Southworth, eds. 1979. *Agricultural Growth in Taiwan, Korea and the Philippines.* Honolulu: University Press of Hawaii.

Herdt, Robert A. 1970. "A Disaggregated Approach to Aggregate Supply." *American Journal of Agricultural Economics* 52(4).

Jha, D. 1980. "Fertilizer Use and Its Determinants: A Review with Special Reference to Semi-Arid Tropical India." Hyderabad, India: International Crops Research Institute for the Semi-Arid Tropics. Mimeo.

Johnson, D. G. 1973. *World Agriculture in Disarray.* London: Macmillan.

Kelley, A. C., and J. G. Williamson. 1974. *Lessons from Japanese Development.* Chicago: University of Chicago Press.

Krishna, Raj. 1963. "Farm Supply Response in India-Pakistan: A Case Study of the Punjab Region." *Economic Journal,* 73(291):477–87.

————. 1967. "Agricultural Price Policy and Economic Development." In *Agricultural Development and Economic Growth,* edited by H. M. Southworth and Bruce F. Johnston. Ithaca: Cornell University Press.

Krishna, Raj, and Ajay Chhibber. 1981. "Policy Modelling of a Dual Grain Market: The Case of Wheat in India." Stanford: Food Research Institute. Mimeo.

Krishna, Raj, and G. S. Raychaudhri. 1979. "Some Aspects of Wheat Price Policy in India." *Indian Economic Review,* n.s., 14(2):101–25.

Parikh, K., and F. Rabar, eds. 1981. *Food for All in a Sustainable World.* Laxenburg, Austria: International Institute for Applied Systems Analysis.

Pasour, E. C. 1980. "Cost of Production: A Defensible Basis for Agricultural Price Supports." *American Journal of Agricultural Economics* 62(2):244–48.

Peterson, W. L. 1979. "International Farm Prices and the Social Cost of Cheap Food Policies," *American Journal of Agricultural Economics* 61(1):12–21.

Sanderson, F. H., and S. Roy. 1979. *Food Trends and Prospects in India.* Washington, D.C.: Brookings Institution.

Scandizzo, P. L., and Colin Bruce. 1980. *Methodologies for Measuring Agricultural Price Intervention Effects.* World Bank Staff Working Paper, no. 394. Washington, D.C.

Sobhan, Iqbal. 1977. "Agricultural Price Policy & Supply Response: A Review of Evidence and an Interpretation for Policy." Washington, D.C.: World Bank.

Thamarajakshi, R. 1977. "Role of Price Incentives in Stimulating Agricultural Production." In Ensminger 1977.

World Bank. 1982. *World Development Report 1982.* Washington, D.C.

12

The Role of the United States in Alleviating World Hunger

WALTER P. FALCON

DIMENSIONS OF THE HUNGER PROBLEM

With all of the previous studies on world hunger, it is truly amazing that such widely divergent views still exist on the number of people suffering moderate to severe protein-calorie malnutrition (PCM). Eberstadt (1981), for example, concludes that hunger affects as few as 100 million people, whereas Reutlinger and Selowsky (1976) conclude that more than 1 billion persons are afflicted globally. In another recent and important review of the literature, Poleman (1981) suggests that the number of undernourished individuals is likely to be less than 300 million.

These differences in opinion arise from a series of issues: the "true" magnitude of daily requirements for various nutrients, the specification of income-consumption-nutrition relationships, the extent to which behavioral or biological adjustments compensate for low nutrient intake, and the extent to which disease and food intake interact to create malnutrition. Perhaps the biggest debate, however, is whether only those suffering clinical malnutrition should be considered as seriously underfed or whether a much larger group of the subclinically malnourished also constitutes a serious problem.

The Presidential Commission on World Hunger (1980) concluded that between 500 million and 1 billion people suffer sufficiently severe PCM to constitute a difficult problem for public policy.[1] Large and disquieting as this range may be, it probably has little bearing on America's attitude or its capacity to help with solutions. Hence there seems little need, at least in this essay, to finely tune estimates, as further refinement would have little bearing on public action.

WALTER P. FALCON is Farnsworth Professor of International Agricultural Policy and director, Food Research Institute, Stanford University.

Adapted from "Reflections on the Presidential Commission on World Hunger," *American Journal of Agricultural Economics* 63, no. 5 (1981): 819–26. Published with permission of the American Agricultural Economics Association and the author.

Policy direction, however, does depend on five specific dimensions of hunger: Asia, children, calories, chronic malnutrition, and poverty.

ASIA

Of the approximately 750 million people thought to be suffering from moderate to severe undernutrition, about two-thirds are in Asia. Indeed, on a global basis, about 70 percent of all hunger is in nine countries (India, Pakistan, Bangladesh, Indonesia, the Philippines, Kampuchea, Zaire, Ethiopia, and Brazil). Any domestic or international proposals aimed at ending world hunger must deal fundamentally with these nations. The difficult formal relationships between the United States and a number of these countries underscore immediately the political dimensions of the hunger problem and the limits to which the United States can help with a solution—assuming that it wishes to do so.

VULNERABLE GROUPS

Irrespective of the exact total of individuals affected by hunger, there are special groups within populations where PCM incidence is the highest. Weanling children from ages one to four present the most serious problem. Whereas cereal-based diets are largely adequate for adults, the relatively low density of these foods means that small children cannot eat enough of them to be nourished adequately. In addition, the interactions among undernutrition, poor water quality, and other public health components are especially critical among the young. These interactions are one reason, for example, why infant mortality rates in Africa are more than six times as high as those in developed countries. Unless infant mortality rates can be reduced, it is unlikely that birth rates can be brought down significantly.

Pregnant and lactating women are also vulnerable groups, since the extra strains of childbearing place these groups seriously at risk with respect to nutrition. In addition, a generational effect deserves specific mention. There is almost no existing scientific evidence to suggest physiological relationships between mental retardation and undernutrition. An exception to this statement is that undernourished mothers fail to carry fetuses to full term more frequently, and among premature births the incidence of mental and other handicaps is substantially higher. On the other hand, there are strong correlations between moderate or severe undernutrition in children and learning motivation and behavioral patterns.

One important implication for analysts of the ''children and mothers'' component of the problem concerns household allocations of food. Many economists typically think of household consumption as the central unit of observation. Unfortunately, since the issue presents severe research and intervention difficulties, how food is allocated within families may be at the very core of the hunger problem in many situations.

178   *Walter P. Falcon*

CALORIES

A third component of the hunger problem is calories. Although this point is increasingly recognized, it also is true that for twenty years many in the nutrition profession had the world pointed in the wrong direction. Indeed, survey work by the Gallup Organization (1979) indicates that many Americans still believe that protein is the most severely limiting nutritional element. Except in a few localized regions, however, the overwhelming PCM problem is simply getting enough calories. Widespread evidence suggests that groups with sufficient food-energy resources typically have found ways to provide the necessary protein complement.

THE CHRONIC NATURE OF HUNGER

A fourth element of global hunger is its chronic dimension. Partly as a result of modern communications, the specters of war- or drought-induced famines are well known to most families in the United States. Events such as the Sahel drought or the Kampuchean and Somalian disasters are covered almost daily on television. Horrible as these situations are, they do not represent the dominant hunger problem. Clearly, the intensity of the PCM problem is worse in famine areas, but it is also true that famines occur much less frequently and severely than even fifty years ago. In addition, global support can generally be mobilized much more readily for disasters than for chronic hunger. It is easy enough for responsible persons to grasp the hunger complications caused by drought. It is almost impossible, however, for anyone to conceptualize one-sixth of the people on earth suffering from moderate to severe continuing undernutrition.

POVERTY

The fifth component, poverty, may be the most important of all. There are isolated instances of people who are undernourished because they are not making good use of local food resources in either quantity or composition. However, the overwhelming reason why people are hungry is not that they are ignorant or uneducated but that they are poor.[2] The recognition that poverty, and not food production, is the major problem is an important step forward, especially for the agriculturalists (and others) who may believe the contrary. Yet the implications are very sobering on two counts. First, a question is immediately raised as to whether those who are poor must be hungry as well. In the United States, largely through the food stamp program, it has been possible to separate these two afflictions. Given both the resource costs and the administrative problems encountered in America, however, can or should similar programs be replicated in low-income nations? Second, if it is impossible to separate hunger from poverty, what can outsiders—even well-meaning ones—do to attack the fundamental problems of income levels and income distribution?

HUNGER ALLEVIATION WITHIN THE WORLD FOOD ECONOMY OF THE 1980s

Any real attack on hunger clearly must deal with a broad range of policies that affect vulnerable groups within poor households. It must also be undertaken within the context of a world food economy whose dynamic components have undergone important changes since the mid-1970s. Recent changes in four of those components are outlined in this section.[3]

POPULATION AND INCOME PRESSURES

The world food economy embraces two widely divergent elements. One is conditioned by low-income nations and by rapid population growth. The second is driven by income growth in middle- and upper-income nations and by the demand for livestock products. A failure to separate these components and the links between them is a key reason why there exists so much confusion about global food supplies, food demands, and the reduction of undernutrition.

A part of the upward pressure on food prices experienced in the late 1970s, especially in regions such as the Middle East and Africa, came from population growth rates. In 1960, for example, world population totaled only about 3 billion, compared with about 4.5 billion in 1980. Nevertheless, by 1970 a large number of low- and middle-income countries had entered a transition era towards lower birth rates. Global population growth rates probably peaked in the mid-1970s at about 2.1 percent annually, and by 1980 the global population increase was down to about 1.7 percent per year.

The future food consequences of a global population growth rate even below 2 percent are significant for the longer run. The effects are particularly important for some thirty-five less developed countries that still have rates above 3 percent annually. But it is important also to stress that population growth was not the central element in making for rapid changes in world food markets in the 1970s, nor is it likely to be key in the 1980s. The dominant force was rising incomes in middle- and upper-income nations, which turned out to be much more important than had been anticipated. In such diverse countries as the Soviet Union, Mexico, China, Taiwan, South Korea, and Nigeria the growth of income, sometimes accompanied by a corresponding increase in the demand for meat and other livestock products, had major effects on grain markets. Between 1976/77 and 1979/80, for example, about 60 percent of the increase in world wheat trade was accounted for by middle-income countries. Similarly, the global consumption of grain by livestock doubled from about 20 percent of total production in 1960/61 to more than 40 percent in 1980/81. More grain was fed to animals in 1980/81 than was consumed by the 1.4 billion people living in countries with per capita incomes of less than $250. In short, the growing demand for feed grains, many of which were imported, was (and will be) a dominant element in the expanding world markets for grain and feed concentrates.

For the 1980s, the most important force pushing the world food system will probably be income-led demand for meat and its resulting ramifications on demand for livestock feed. Strong leverage is created by the inefficiency of converting grain to meat; roughly six to eight grain calories are required to produce one calorie of edible meat using intensive feedlot technology. A single addition to the world's population consumes about 180 kg of grain per year if it is eaten directly as food in a subsistence diet. But if that individual is combined with adequate income to consume an affluent, meat-intensive diet, then the grain demand rises to roughly 730 kg per year.

The implication is that population growth adds to food demand slowly and steadily. Since much of the population growth is in very poor countries, not all of the added food needed will show up as effective demand in world grain markets. An inherent characteristic of income growth, however, is that it can be converted directly to market demand. Income growth in societies with income levels above five hundred to six hundred dollars per capita per year tends to cause extremely rapid increases in meat demand. Despite much of the world's attention to the failures of economic development in some countries and for the very poor in many countries, a rapidly growing proportion of the world's population has in fact emerged from subsistence-level poverty and is clamoring for an improved standard of living, especially through a better-quality diet with more meat. Since more meat at the margin means more grain fed to livestock, these changes will place increasing demands on world supplies of livestock feeds. Income growth in the 1980s, especially in the middle-income countries, will become the major driving force in world food-grain markets.

AGRICULTURAL PRODUCTION CAPACITY OF NORTH AMERICA

While income growth dominated the changing world demand for grains, what happened in North America, especially in the United States, dominated the supply side. During the early 1960s the United States held some 15 percent of its cropped area idle in an effort to limit supplies and raise domestic prices. Following international events in the early 1970s, the United States brought all of its idled land back into production, with much of the consequent increase in output going into exports. In 1970/71, for example, the United States exported 41 million tons of grains, which represented about 37 percent of global trade. By 1980/81 the American export share had risen to 56 percent and to 118 million tons. A vital world-market question for the future, therefore, hinges on American capacity and willingness to expand food exports. Despite short-term surpluses arising from bumper crops in 1981 and 1982, the next decade will almost surely see a reduction in the rate of growth of export supplies from the United States.[4] Given that about half of all international trade in grains now originates in the United States and that about half of total grain production in the United States is exported, what happens to future American policy and productivity will have a profound effect on international food markets.

THE PRICE OF PETROLEUM

Rising petroleum prices constituted a third element that helped to intensify problems with the world food system during the 1970s. The success of the OPEC nations in raising oil prices had a series of direct and indirect effects on domestic agricultural systems throughout the world. Part of the effect on agriculture was directly related to costs of production. To some extent—although the correlation was far from perfect—oil prices affected fertilizer costs. Linked even more directly with changing oil prices were the costs of pumping irrigation water, the cost of fuel used for tractors, and the energy costs of food processing and distribution.

There were indirect effects of changed petroleum prices as well. Many food-deficit countries are also oil importers, and they therefore faced a double balance-of-payments problem after 1972/73. In addition, some countries' attempts to reduce the burden of imported oil are having a specific spillover onto world food markets. For example, America's attempts to use corn and Brazil's efforts to use sugar cane to produce gasohol are good illustrations of new international linkages between food and oil that may have important ramifications for the future.

EFFECTS OF NATIONAL POLICY

The 1960s and 1970s also saw an increased influence of domestic agricultural policies on international commodity markets. In general, a larger number of countries sought to achieve domestic stability in food prices and food availabilities through greater use of international markets. Such activities tended to expand the trade sector, but in a number of instances these domestic actions also created additional instability in international commodity prices.

Russia, of course, was the clearest example of a country whose domestic policy had international effects. The Soviet Union suffered very poor agricultural years in 1972/73, 1975/76, and 1979/80. From 1975/76 to 1976/77, for example, Russian grain production changed 84 million tons, a huge amount by any standard. In previous eras Russia had dealt with similar food crises by "belt tightening," by domestic price increases, and by slaughter of livestock herds. Beginning in the early 1970s Russian policy changed. Food prices were allowed to remain about constant, and livestock numbers actually were allowed to grow during most of the 1970s. Instead, the adjustment was accommodated by massive, and highly variable, Russian grain imports.

The shocks on the world system created by Russian imports (and other factors) would not have been so great if all other countries had helped with the price adjustment process. But many other countries had also isolated themselves from variations in international markets. The high and stable prices under the Common Agricultural Policy of the European Community meant that Community price signals to producers *and* consumers in those countries varied little in response to international price movements. East European countries, Japan, and numerous other importers behaved similarly. As a consequence of this wide-spread domestic segmentation from world markets, international prices were much more variable than they otherwise would have been. A further consequence was that a relatively few countries bore the major brunt of the shocks. Some low-income countries reduced imports when grain prices were high. However, the overwhelming adjustment was borne by the United States, which continued to allow an almost complete linkage between domestic and international prices. As a consequence, American consumers paid higher prices in a number of years, and livestock producers suffered badly as a result of rising grain prices. Indeed, much of the short-run adjustment to the world food variations in the 1970s was handled by American farmers feeding varying amount of grains. From 1973/74 to 1974/75, for example, the change in grain consumed by animals in the United States alone accounted for about 35 million tons.

For the 1980s, the question remains whether more or fewer countries will join the adjustment process. If there are more attempts to insulate domestic price movements from those internationally, and especially if the United States moves away from its free-trade policy on grain and soybeans, international price variability in key markets could increase even more. This increased uncertainty would obviously complicate further the efforts of low-income countries to use international commodity markets as an integral part of their food strategies.

While international specialists hold various opinions about the course of future changes in the world food economy, virtually everyone agrees on two points of slightly lesser generality. First, events of the 1970s have created a new series of commodity and country interdependencies that are likely to become even more important in the 1980s. A combination of price changes and technological breakthroughs has meant that substitutions will become more important in both production and end uses. Second, the processes of adjustment to shocks in the international food system are likely to become increasingly critical and complicated for many Third World countries. Stock adjustments, trade flows, altered livestock numbers, factor flows—even human starvation—are possible, and a better understanding of these mechanisms is vital.

This is not an essay to develop fully a prognosis for the 1980s. However, the foregoing comments imply the context in which future public policy on world hunger may be developed. Quite clearly, the hungry people of the world will not be the driving market force in the global food system of the 1980s. Yet it is equally clear that hungry people could be affected substantially by what happens in the food system of the affluent world.

PROGRAM ELEMENTS FOR THE FUTURE

Global hunger persists, and, if anything, it is likely to become a more de-stabilizing international influence in the future than it was in the past. At the risk of making very difficult issues sound superficial, or the solutions seem easy, the case for renewed American focus on hunger alleviation can be broken down into seven operational propositions.

1. *Given an increasingly interdependent food world, hunger is an appropriate focus for America's relationship with developing countries.* Of all the broad areas in which the United States could play an important leadership role, food and agriculture are preeminent. The extraordinary productivity of American agriculture, the (more than) adequate diet of the American people, and the dominance of the United States in the global food system give this country credibility in the food area as perhaps in no other. Moreover, a concern with the poor and malnourished, especially children, is very much in the American tradition.

These widely recognized points, however, may be necessary but not sufficient conditions for making hunger a central focus of development assistance. If hunger is related to poverty, as seems clearly the case, it is not a tidy area in which to involve an assistance program, nor is it the only important problem facing developing countries. Moreover, the vastness of the hunger problem may be out of balance with the size of America's aid commitment. Finally, it is abundantly clear that hunger issues go to the heart of the political economy of many nations. In some countries American concerns about hunger will go unheeded or be counterproductive, and in virtually all countries most of the resources and difficult decisions will be of a domestic nature. Nevertheless, hunger seems to be one area in which greater amounts of both public and private support can be mobilized within the United States.[5] More generally, food distribution, if not handled properly and expeditiously, could have far-reaching international consequences during the 1980s.

2. *A focus on hunger means a primary emphasis on agriculture and rural development.* If hunger alleviation is made a focal point of American development assistance, such a concentration implies a concomitant emphasis on agriculture, but not for the reason that most people believe. Increasing food output is obviously important, especially in those societies with rapidly increasing populations and incomes. In reducing hunger, however, the employment and income effects of agriculture are much more important than expanded food output. Although urban poverty and hunger may be more acutely visible, the overwhelming numbers of undernourished people are in the countryside. Many (perhaps 60 percent) of these individuals do not have direct access to land. These scattered and often forgotten groups are also among the hardest to reach with direct governmental consumption programs. In the absence of thoroughgoing agrarian reforms, the key to reducing hunger problems is additional productive jobs. The distinction between food production and income generation is an extremely important point—one often missed by agriculturalists and proponents of *Food First* (Lappé and Collins 1977). It also underscores the urgent need for choice-of-technique analyses based on social profitability rather than preconceived notions of what should be considered "modern."[6]

3. *A primary emphasis on agriculture means increased focus on relevant agri-*

cultural technology. The issue of agricultural technology is still hotly debated. Part of the controversy has to do with the problems of tractors and mechanization in labor-surplus areas, and part has to do with the failures of introducing annual crops on delicate forest soils. Issues surrounding seed technology and the appropriate use of fertilizers and pesticides also fuel the debate. At least in part because of these controversies, the United States does not yet have a well-articulated policy for technology for the Third World.

In spite of much-heralded developments in wheat and rice, involving now some 50 million acres mainly in the irrigated regions of Asia, the overall record on improved seed technology is rather poor. New developments with open-pollinated corn varieties reengineered for tropical conditions will soon be available, and new packages for sorghum and millet also offer substantial prospects. Active research under way as well for beans, cassava, and vegetables promises to be relevant. Nevertheless, in assessing technology needs and accomplishments to date, it is clear that much research is needed, especially for rain-fed agriculture. Fortunately, the general area of agricultural research is one in which the United States has a comparative advantage. Many of America's processes for developing technology are certainly relevant, even if much of existing American technology is not directly transferable. Technology is also an area where both the public and private sectors in the United States have much to contribute, as do the universities.

With a limited development-assistance program, finding an appropriate niche for American involvement is extremely important. For example, land reform may be more vital than technology in alleviating hunger in some regions, but American efforts to promote agrarian reform in other countries are almost sure to be counterproductive. By avoiding some of the mistakes in technology that have occurred in the past and by recognizing that technology cannot solve all the problems of development, it should be possible to develop a large, positive program in this field. Fortunately, many of the relevant institutions (for example, the Consultative Group on International Agricultural Research [CGIAR] and the Board for International Food and Agricultural Development [BIFAD]) are in a position to make this technological promise a reality.

4. *For agricultural technology to be effective, large investments will be required, especially in such fields as water-resource development.* Unfortunately there are no cheap fixes for food and agriculture, and the Presidential Commission on World Hunger (1980) called for a rapid tripling in the appropriations for foreign aid, much of which was to go for hunger alleviation. Particularly at a time when cuts in the American federal budget seem to be the order of the day, such a suggestion requires further comment.

Although there has always been a limited lobby on foreign aid, it seldom draws the passionate support of many other allocations. In recent years, this problem has been accentuated by an unusual combination of political forces. Many on the political right have seen foreign aid as a costly giveaway to be stopped. The left has become so enamored with the "small-is-beautiful" syndrome that they have significantly downplayed the very real investment costs that will be essential if Third World nations, with external assistance, are really to attack hunger. The net result has been an increased number of restrictions on aid allocations, such as on rural infrastructure, and deceleration of aid authorizations.

These effects can be seen in the official review of overseas development assistance published by the OECD (1980). In 1979, for example, the United States ranked fifteenth among development assistance countries (DAC) in percentage of GNP devoted to bilateral and multilateral assistance to developing countries. Indeed, with only 0.3 percent of GNP devoted to aid, the United States' share exceeded only that of Italy and Austria among the seventeen nations that make up the DAC. The specific rationale for a larger American share will be discussed later, but a case for larger sums can be made just on the basis of the foregoing data.

Two cost components deserve special comment in the context of hunger. One of the most severe problems in improving the nutrition of hungry people involves water-resource development. The Asian concentration of hunger has already been mentioned. Moreover, by the year 2000 about half of the world's population will live in areas defined by the ten largest river basins in Asia. These basins, with their problems of irrigation, erosion, flooding, salinity, and drainage, contain many of the world's poorest people. Without some improved control over the production environment, the potential for new agricultural technology is very limited. If unprecedented migration and other problems are to be avoided, substantial investments will be needed to create dynamic rural communities where productive employment and incomes can increase. Most of the resource mobilization will have to be accomplished locally, but international resource transfer is also vital. Since many of the basin problems cross borders and involve several countries, outside agencies have a particularly crucial role to play. It is regrettable, but there is no cheap way out of this investment issue. People who want to attack hunger without significantly adding to the investment totals are kidding themselves or each other.

Second, for both the technology and water-resource fields, there are important roles for bilateral and multilateral initiatives. There is urgent need, for example, for American support of the soft-loan window at the World Bank (IDA) and of the international agricultural research effort that the World Bank coordinates. Similarly, American assistance with debt restructuring could have important positive consequences, since debt rollover in many instances is identical with increased flows of untied aid. While the investment needs are large, they are not beyond the capacity of the world to manage. The International Food Policy Research Institute (IFPRI), for example, suggests that an additional $7 billion investment annually (in 1975 dollars) during the decade of the 1980s would increase the annual world cereal output by nearly 200 million tons by 1990. This sum compares with the approximately $80 billion supplied to less developed countries in 1979 from external aid and loan resources.

5. *For technology and institutions to pay off, a substantial reorientation in economic policy will be needed in less developed countries.* The recent growth in world cereal trade has been very large; it may also have begun to substitute for domestic stockholding. Whereas global cereal trade in 1980 was approximately three times as large as in 1960, ending world grain stocks in 1980 were absolutely smaller than in 1960, and only about half as large relative to annual production. For reasons alluded to earlier, some slowdown in the growth of trade can be expected, probably accompanied by rising and more variable international prices of grain over the decade of the 1980s. While cereal trade will clearly continue to be important for many "hungry" nations, the costs of an international "solution" to their food problems will likely be higher in the 1980s. This view of world trade thus provides a logic for an increased emphasis on domestic growth in agriculture to supply both food and employment.

Technology and investment provide two legs of the productivity triangle, while price and trade policy supplies the third. In general, low-income countries tend to discriminate against the agricultural sector and to provide less than international prices to their farmers. For a long-run production solution, raising prices to farmers in many countries is absolutely essential. However, it is more than sheer neglect or urban bias that keeps governments from making this change. Higher food prices also mean lower real incomes, especially for poorer people, who may spend up to 80 percent of their income on food. This basic pricing dilemma—short-run consumption losses versus long-run production gains—needs to be recognized for the very real problem that it poses, even for the most responsible government. Too many analysts have been content to deal only with the production issue. Neither the Agency for International Development (AID) nor the World Bank, for example, has been willing to provide much support for consumption/nutrition projects or for transition programs designed to put in place new food-price policies. A sympathy towards this basic consumption-production dilemma and a willingness to use food aid and other types of development assistance towards new policies is critical if the United States and other donors are to be helpful in solutions to the food problems actually faced by low-income countries.[7]

Unfortunately, American policy has been largely silent on both the price-policy dilemma and the further complications this problem creates for consumption programs. Untargeted programs, such as physical rations for everyone, have very high relative resource costs in poor societies. On the other hand, the administrative problems involved in reaching only the poorest groups, especially in rural areas, are immense. Helping to resolve this dilemma will be a task on which analysts can make an important contribution in the years ahead.

6. *It is in the economic and security interests of the United States to assist in hunger alleviation and in the creation of a more stable world food economy.* The redirection and expansion of American assistance to help fight global hunger that is being suggested here raises the obvious question of whether such changes would be worth the price to the United States. Investment and technology pro-

grams will have little chance politically unless the suggestions can be shown to be directly in America's own interest. Fortunately, the self-interest question can be answered with an unequivocal yes. The basis for this affirmative assessment is twofold. The first element stresses growing economies and trade, using rather traditional arguments. The second element, and by far the more important, stresses the national security implications of food. The fact that several of the nations with substantial hunger also have a nuclear capability is one aspect of the argument, but not the major element. More broadly, food in the 1980s could be a destabilizing force—in the manner, if not the magnitude, that oil was in the 1970s. Lest that view be casually discarded, one needs to think only of recent food crises in Poland, Russia, Egypt, Kampuchea, and Ethiopia. Consumption (though not necessarily hunger) issues were central in each case, and in several of these examples the potential for international conflict was clear-cut. This broader view of security would seem to be a natural complement to the military expenditures that have taken a heightened priority in the early 1980s.

7. *Given that hunger alleviation is in the self-interest of the United States, substantial changes will be required in American attitudes and capabilities for working with developing countries.* If the United States chooses to make hunger issues the center of its development assistance effort, more than marginal changes will be required. Additional dollars will be needed to support research and investment. AID will have to overcome its inadequacies in technical competence to deal with food and agricultural issues. The United States will need to seek new kinds of formal relationships with several key nations. Above all, the President and Congress will have to lead. Most of the leadership involves doing new things, but it sometimes involves *not* doing things as well—not attempting to use food as a political weapon, not promoting uneconomic gasohol installations, and not failing to recognize the severity of the hunger problems, even in countries whose governments the United States dislikes.

With a clearer sense of direction, the United States is now in a unique position to assist countries in helping to solve one of the world's worst problems. Without renewed efforts on the part of the United States and all countries, however, global hunger problems will become more acute and destabilizing before the end of the century.

NOTES

1. In September 1978, President Jimmy Carter appointed a twenty-member Commission on World Hunger. The commission's mandate was to identify the basic causes of hunger at home and abroad, to assess programs and policies affecting hunger, and to recommend (and publicize) specific actions to create a coherent national policy. This essay draws from the commission's work but also offers independent observations on the dimensions of global hunger and the steps that can be undertaken to alleviate undernutrition by the year 2000.

2. An exception to this statement involves infant-feeding practices, where educational efforts can sometimes make an important difference.

3. This view of the 1980s is developed much more fully in Timmer, Falcon, and Pearson 1983.

4. Record crops of wheat, corn, and rice in 1982, coupled with a global recession, put downward pressures on food prices and gave rise to short-term surpluses within the United States.

5. The Gallup Organization was employed by the Commission on World Hunger to undertake a poll of Americans about world hunger. The results indicated a widespread misunderstanding of the severity and nature of PCM problems. They also showed that, in relative terms, Americans were very concerned about world-hunger issues.

6. For example, see Timmer, chapter 19 in this volume.—ED.

7. For more details see Timmer, chapter 8 in this volume.—ED.

REFERENCES

Eberstadt, Nick. 1981. "Hunger and Ideology." *Commentary* 72(1):40–49.

Falcon, Walter P. 1981. "Reflections on the Presidential Commission on World Hunger." *American Journal of Agricultural Economics* 63(5):819–26.

Gallup Organization, The. 1979. *Tabulations Prepared for Commission on World Hunger.* Princeton.

Lappé, Francis Moore, and Joseph Collins. 1977. *Food First: Beyond the Myth of Scarcity.* Boston: Houghton Mifflin.

National Academy of Sciences. 1977. *World Food and Nutrition Study.* Washington, D.C.

Oram, Peter, Juan Zapata, George Alibaruho, and Shyamal Roy. 1979. *Investment and Input Requirements for Accelerating Food Production in Low-Income Countries by 1990.* International Food Policy Research Institute Research Report, no. 10. Washington, D.C.

Organization for Economic Cooperation and Development. 1980. *Development Cooperation: 1980 Review.* Paris.

Poleman, Thomas T. 1981. "Quantifying the Nutrition Situation in Developing Countries." *Food Research Institute Studies* 18(1):1–58.

Presidential Commission on World Hunger. 1980. *Overcoming World Hunger: The Challenge Ahead.* Washington, D.C.

Reutlinger, Shlomo, and Marcelo Selowsky. 1976. *Malnutrition and Poverty.* World Bank Occasional Paper, no. 23. Baltimore: Johns Hopkins University Press.

Timmer, C. Peter, Walter P. Falcon, and Scott R. Pearson. 1983. *Food Policy Analysis.* Baltimore: Johns Hopkins University Press, for the World Bank.

13

Food Security in Developing Countries: International Issues

AMMAR SIAMWALLA AND ALBERTO VALDÉS

In 1974 the World Food Conference in Rome was convened to discuss the world food crisis. Food security was the dominant theme. Agricultural prices had risen to record highs, carry-over stocks of grain were at precariously low levels, and concern was focused on the undernourished millions in the Third World suffering from the scarcity and high prices of food. Fears grew that the world was irrevocably moving toward chronic food shortages, attributable to unfavorable long-term climatic changes and continued high rates of population growth.

Fortunately, the current food position is much improved. International prices of cereals have fallen in real terms, grain stocks have been rebuilt, and the crisis atmosphere has abated. World food security has ceased to be a major concern for the press and for the general public. Yet the underlying causes of food crises have not disappeared. Though developing countries have themselves made some important strides in dealing with food insecurity, only limited progress has been made on the international scene to help them in these efforts.

Numerous food-security proposals have emanated from both international organizations and universities. But discussions of potential remedies have overlooked the enormous difference in the nature and magnitude of food-security problems in countries or regions in Asia, Africa, and Latin America. In these discussions only a limited number of policy instruments have been considered relevant to the problem of food security. Often the debate gives the impression that LDCs could surmount the problem if only larger grain reserves were available. Food-security proposals often underestimate the complexity of the practical problems at the country level in designing and implementing policies for stabilization of food supplies, particularly for rural areas.

AMMAR SIAMWALLA and ALBERTO VALDÉS are research fellows, International Food Policy Research Institute, Washington, D.C.

Originally titled "Food Insecurity in Developing Countries." Reprinted from *Food Policy* 5, no. 4 (1980): 258–72, published by Butterworth Scientific Ltd., P.O. Box 63, Westbury House, Bury Street, Guildford, Surry GU2 5 BH, UK. Reprinted with minor editorial revisions by permission of the publisher and the authors.

THE CONCEPT OF FOOD SECURITY

Food security may be defined as the ability of food-deficit countries, or regions or households within these countries, to meet target levels of consumption on a yearly basis. What constitutes target consumption levels and whose ability to maintain consumption is being referred to are two central issues of a country's food policy.

The analyst has some choices when considering target levels. For instance, a target level may refer to the minimum recommended level employing nutritional criteria, the average level over an agreed number of years, or the trend level of consumption. Although the most severe impact of short-term food supply instability is felt by the poor, chronic malnutrition caused by persistent poverty constitutes a long-term problem whose dimensions and solutions lie well beyond the question of food security, which we consider to be a problem of short-term variability. Thus, we adopt the trend level of consumption as the target instead of some absolute nutritional criteria.

On the second issue, we can analyze the problem in terms of the household's, the region's, or the nation's ability to attain food security. Although food supplies may be stable for a nation as a whole, there may still be large segments of the population whose food supplies are insecure. The nation must have some means to deliver food to the households that are exposed to food-insecurity risks. A complete understanding of the repercussions of food insecurity requires an analysis of the problem at all these levels.

Variability of food supply may have many causes. The focus of this article is on a cause that is inherent in agriculture: the impact of fluctuating weather on the size of harvests. Sudden acute food shortages may also be a consequence of natural disasters such as earthquakes and floods and disasters induced by political conflicts, such as those in Bangladesh in 1972–74 and Cambodia in 1979. This second kind of problem is distinct from the first in that it involves disruption of the normal channels of delivery of food to the afflicted area, in addition to production shortfall.

A resumption of the flow of food is the main priority in these circumstances. This usually demands an engineering or political solution rather than an economic one in that the quantities of food required are small, seldom more than a million tons, and the humanitarian grounds for providing it are so evident that an economic calculus is superfluous. Another reason for concentrating on weather-induced variability of supplies is that the accumulation of data (even though imperfect) enables better policy decisions to be made. In dealing with disasters, there is no way to foretell accurately when or where the problem will occur.

The focus on the impact of fluctuating weather also implies the consideration of annual flows of supplies and annual movements of prices rather than of the weekly or monthly fluctuations. This shorter-term consideration raises a distinct analytical issue. This involves an understanding of the quality and quantity of information flows on domestic supplies to policy makers and the time lags in converting information into decisions, decisions into action, and action into inflows of food. Specifically, when stocks are discussed below, we exclude consideration of working stocks, which are necessary to bridge the time lags and the unreliability of information where policies have to be devised to meet very short-term targets. However, we do not wish to belittle issues related to working stocks. We suspect that a great number of the concerns expressed by policy makers from LDCs arise from difficulties in coping with the daily operation of food markets and agencies involved in food supplies. As such, they raise management issues rather than conceptual problems, with which we are primarily concerned.

FOOD SECURITY AT THE NATIONAL LEVEL

Food insecurity is naturally associated with either fluctuations in food production or price changes. But fluctuations in nonfood production and prices also have considerable impact on food insecurity. For example, a cotton grower's consumption of wheat may be affected as much by the boll weevil which destroys his crop or by a sharp fall in cotton prices as by an increase in the price of wheat. Thus, from the analytical point of view, it is more useful to divide the causes of food insecurity into two categories: production fluctuations, regardless of whether the fluctuation is in the food or the non-food sector; and price fluctuations, regardless of whether the price of food or of nonfood is affected.

These fluctuations lead directly to the variations in real income within the community, which may affect different members of the community in different ways. To a farm household they appear simply as increases or reductions in output and therefore income; to a farm laborer they appear as fluctuations in available employment.

In addition, there is an indirect impact as the primary gain or loss in income is translated to other parts of the economy through increased or reduced spending by the affected households. This is the familiar multiplier process which magnifies the results of the original change.

These income fluctuations ultimately alter the food consumption of the farmer, the agricultural laborer, and other members of the community, with the poorer households being particularly sensitive. It follows that if fluctuations in real income could be smoothed out, food security could be attained. The conventional economist's solution is to employ some sort of capital market mechanism to transfer income intertemporally. In most cases, the mechanism needs to operate only on the households and sectors directly affected, for if their cash flow is maintained, despite the decline in income when crops are poor, this compensation will percolate to the other sectors, negating the multiplier effects of the original income loss. As will be seen below, however, it is essential that the capital market mechanism operate at all levels for the whole country as well as for individual households.

To achieve food security, the capital market mechanism chosen must involve assets that possess a high degree of liquidity, as their conversion to food must be effected rapidly. Such a mechanism may imply holding assets in the form of either food stocks themselves or monetary instruments, such as foreign-exchange reserves or cash balances, to purchase food in lean periods. The problem of which assets to hold is one of portfolio management and may be solved by using conventional economic criteria. For example, food-importing LDCs that are price takers in the world market can achieve food security at lower cost by varying the amount of imports while maintaining a relatively small buffer stock. This implies holding liquid assets in the form of foreign-exchange reserves rather than in the form of food.[1] At the household level, the same logic implies holding cash rather than storing food. But, regardless of the form in which liquid assets are held, they have lower returns than investment in plant and machinery, land, or education. Therefore, attaining food security is costly, being essentially the difference in the rate of return in illiquid versus liquid assets multiplied by the volume of diversion in investment required.[2]

It may be argued that the country or household suffering from food insecurity need not divert its own resources into liquid investments but can borrow as the need arises. Because the overall ability to borrow is limited in the long term by the wealth of the country or household, the access to a particular line of credit will tend to be at the expense of credit for longer-term investments, again implying that food security can be attained only at some cost.

At the national level, an exact measure of food insecurity would rest on real income fluctuations, but an approximate measure may be obtained by examining deviation from the trend in foreign-exchange earnings minus the excess expenditure over the trend on food imports for the current year. This is referred to as deviations in "real" export earnings.[3] An analysis of optimal foreign-exchange reserve policies could be based on this measure of real export earnings.

From this general interpretation of the results available in the main body of food-security literature, it is apparent that the arguments rest heavily on the assumption of a frictionless world. The issues that arise when such an assumption is removed need to be explored.

INADEQUACIES OF THE INTERNATIONAL MARKETS FOR CEREALS

Several writers have concluded that each country should become self-reliant in terms of food and thus build its own system of national food security. Minhas, for example, concluded that India should opt out of the world food-security system, not depend at all on food imports, raise its level of self-sufficiency, and build its own reserve stocks.[4]

The justification for such a policy is that each nation faces highly unreliable international supplies. There are several strands to this argument. In many of the discussions no distinction is made between concessional and commercial supplies. The reliability of each source differs considerably. The general perception, which we fully share, is that concessional supplies are completely unreliable. Food aid was, in fact, cut back in 1973–74. Worse still, in individual cases donors have used food aid to wring political concessions from the recipients. It seems that the only policy implication that can be drawn from these incontrovertible facts is that national food policies that rely on food aid supplies are highly risky.

It is, however, quite unjustified to move from this position to advocacy of complete independence from all forms of food imports. To justify this latter position, it has to be shown that commercial supplies are unreliable. *Unreliability* can carry two meanings. One is that world prices are highly unstable but supplies are always available for the needs of the small and medium-sized countries. The second, stronger, version is that there are periods, such as 1973–74, when supplies are unavailable at any price.

On the issue of availability of supplies, we have not seen any documentation of the charge. We do not know of any time when it was not possible to obtain supplies of wheat at some price, even though the price may have been considered exorbitant by some LDC importers. Not all cereal markets are the same as the wheat market, however. Rice and white maize are major cereal staples in Asia and parts of Africa, but their international markets are quite thin. There are no central sources of supplies. Interruptions in trade flows, by no means uncommon, can be quite costly for those dependent on the world markets. Thus, rice exports from Thailand, a major exporter, were banned in the second half of 1973. Even though the People's Republic of China then stepped in with increased supplies, the delays in the trade negotiations caused by the shift in sources led to major rice crises in many Asian cities. Similarly, there are wide areas of the world where people have strong preferences for white maize, but much of the world trade is in yellow maize. Thus Kenya, which experienced a major shortfall in its domestic production in 1979, had to undertake a major search for supplies of white maize.

The logical solution for these international market failures would seem to be to push for self-sufficiency, even in bad-weather years. Such a goal may well turn out to be unnecessarily costly. There are two alternatives that are at least as effective. The first is to recognize that while these international markets are imperfect, they do exist. Food security may then be attained by having sufficient working stocks to cover the period during which a search is going on. The second alternative is to use price policies on alternative cereals more forcefully to induce their substitution in consumption away from the more "difficult" rice and white maize.

BARRIERS TO INTERNAL TRADE

In most cases, the state has assumed the functions of the trader in foodgrains. Either through a parastatal agency or through tariff policies, governments have attempted to ensure food security by insulating the domestic price from the international price. If the problem originates in a price shift in the world market, a policy of domestic price insulation implies that the burden of real income fluctuation caused by the shift in terms of trade is borne by the public treasury. Since the public treasury has better access to capital markets (both domestic and international) than does an individual household, a policy of price insulation in this circumstance is adequate to achieve domestic food security.[5] This price insulation policy, if also backed by effective macroeconomic stabilization policies, would be sufficient to ensure stability of food consumption in the urban areas. For this reason there has been an association in the general perception of food security with stability in the price of food. However, if we take a broader view, and if the cause of food insecurity lies in production fluctuations, such a policy by itself would be insufficient, particularly for rural areas.

Another form of trade barrier is the high transport costs resulting from the geographical isolation of many areas. This includes relatively inaccessible countries and remote regions within a country. High transport costs mean that trade of foodgrains in and out of such regions is uneconomical. Thus, most isolated areas are self-sufficient in that production and consumption are the same, although this does not imply nutritional adequacy. In this case, the market for foodgrain would always clear domestically, with its price fluctuating between a lower bound set by its export price (fob) and an upper bound set by its import price (cif). Fluctuations of production away from the normal level would cause fluctuation in the real income of the particular community, which therefore needs to hold liquid assets. But, because of the high transport cost, local storage of foodgrains may be preferred to other forms of assets, however liquid the latter may be.

It is important to note that transport costs increase the desirability of storage only if they are so high that the country finds it economical to attain self-sufficiency. For countries that are consistent food importers despite high transport costs,[6] it is doubtful whether holding buffer stocks of grain is more desirable than holding other assets. Varying the volume of imports would be a less expensive means of stabilizing consumption. Storage is economically productive only when there is a severe limitation on the volume of food that can pass through a given transport channel at any given time.[7] Even in this case, however, a possibly cheaper alternative would be a better import planning machinery, to minimize the probability of sudden spurts in demands for transport that would overburden the system.

SHORTCOMINGS IN THE CAPITAL MARKET

Since food insecurity ultimately arises from real income fluctuations, for which the classical solution is a reliance on the capital market to smooth them out, it follows that the failure of the capital market lies at the root of the food-security problem. Introducing credit facilities to enable countries to tide over the bad years is a logical solution to the problem, at least at the national level. This proposal does not preclude the need for concessional aid for the poorest countries.

Within the country, for those in urban areas, food security is synonymous with price stability; this, in turn, can be achieved if the country adjusts its imports of food or the level of stocks to compensate exactly for the aggregate fluctuations in domestic production. However, a large reduction in farm output could result in the decline in sales of manufactured goods and thus indirectly reduce employment and income in urban areas. In this case, stable food prices are no longer a sufficient condition for food security.

In rural areas, where the real income of a household is directly affected as a result of fluctuations in its own production (of food or nonfood), stability of food consumption can no longer be achieved merely through domestic price stability, as this takes no account of the fall in real production and therefore the income level of the household. The problem now lies as much on the demand as on the supply side of the food market, since we can no longer be sure that the capital market will act to stabilize the effective demand for food of the household by meeting the shortfall in its income. Credit available to households (especially poor households) in LDCs is either nonexistent or extremely expensive.[8]

Any attempt on the part of the government to introduce arrangements that act as substitutes for a capital market at this level would be administratively costly. Under such circumstances, an outright transfer of income becomes unavoidable at times of shortages. Crop insurance is one means of accomplishing this, although it is administratively difficult to implement and does not address the question of the landless laborer. Public food distribution programs are another method of focusing on the poor agricultural households, and their grant element is adjusted to counter the effects of output fluctuations. Unfortunately, there are few public distribution programs that reach extensively into the rural areas, where income fluctuations arising from production fluctuations are the most severe. The scant research that exists on this problem, though useful, was designed for specific areas and institutions and does not generally apply to other countries.

An alternative route is to operate through the labor market with a "food for work" program. Such a public works program may be introduced (or intensified) at times of shortages. The rationale is that a crop failure would reduce labor demand and lead to unemployment in the producing areas. To put the unemployed on the payroll without providing for the wage good (food) at a time when it is particularly scarce would increase the local price of food; hence the combination of food and work.

Most governments seem helpless in dealing with the complexities of the food-insecurity problem in rural areas. The complexity arises not only from difficulties of transportation and other physical constraints but also from the more difficult economic problem of providing affected households with adequate purchasing power at times of crop failure. It is partly for this reason that government action on food security has historically concentrated on meeting urban demands.

QUANTITATIVE ASSESSMENT AT THE NATIONAL LEVEL

In most studies on food security, food has been identified solely with cereals. Although the share of cereals in total food consumption (measured in calorie equivalents) is high on average, it ranges from 85 percent in Afghanistan to only

16 percent in Zaire. Cereals are clearly the dominant food staple in Asia, but in Africa and Latin America the role of non-cereals in consumption is important and must be incorporated into any consumption equation. Nevertheless, cereals, particularly wheat, dominate world food trade and have traditionally been used to meet shortfalls in domestic production. In the discussion of trade we concentrate on cereal imports.

In this analysis of food consumption, instability is measured around the long-

TABLE 1
VARIABILITY IN STAPLE FOOD CONSUMPTION, 1961–76

	Staple Food Consumption Instability		Probability of Actual Consumption Falling Below 95% of Trend (%)
	Standard Deviation[a] (metric tons × 10³)	Coefficient of Variation (%)	
Asia			
Bangladesh	1,013	7.6%	26%
India	5,570	5.3	17
Indonesia	1,204	6.1	21
Korea, Republic of	531	6.5	22
Philippines	192	3.3	6
Sri Lanka	163	8.3	27
North Africa/Middle East			
Algeria	667	24.6	42
Egypt	1,164	12.6	34
Jordan	88	21.2	40
Libya	115	16.2	38
Morocco	933	19.3	40
Syria	360	18.7	39
Sub-Sahara Africa			
Ghana	134	6.1	21
Nigeria	965	5.6	19
Senegal	319	15.7	37
Tanzania	517	14.6	37
Upper Volta	126	9.5	30
Zaire	172	4.1	11
Latin America			
Brazil	1,955	5.8	20
Chile	386	14.4	36
Colombia	147	4.7	14
Guatemala	69	6.9	24
Mexico	757	5.3	17
Peru	110	3.9	10

SOURCES: Data on cereals and other major staples are from the *FAO Production Yearbook;* on net cereal imports, from the *FAO Trade Yearbook;* and on cereal stocks, from the U.S. Department of Agriculture data base. For sensitivity of results according to different sources and to the exclusion of stocks see Annex I of Alberto Valdés and Panos Konandreas, "Assessing Food Insecurity Based on National Aggregates in Developing Countries," in *Food Security for Developing Countries,* ed. Alberto Valdés (Boulder: Westview Press, 1981), 50–51.
[a]Defined as the standard deviation of the variable $C_t - \hat{C}_t$.

term trend, using the "coefficient of variation" as an indicator of variability (table 1). The observed variability in food consumption in a sample of sixty-seven LDCs ranges from a low of 3–4 percent in the Philippines and Peru to a high of 20–25 percent in Morocco and Algeria. High variability levels of 15 percent or more are concentrated in North Africa and the Middle East, where cereals' share in total food consumption is above 40 percent. Over 67 percent of the countries had an unexpected high degree of consumption variability—equal to or greater than 7 percent.[9]

Food consumption variability can also be expressed as the probability of actual consumption falling below, for example, 95 percent of trend consumption, given the level of actual imports. In fifty-one of the sixty-seven countries studied, a consumption shortfall below 95 percent of trend occurred every five years. In most Arab countries it occurred approximately once every two and one-half years.

Some may assume that shortfalls in domestic production are the basic cause of food insecurity. This need not be the case if the country concerned has the capacity to vary its food import volume to compensate for the variability of production. However, its ability to do so could be limited by sudden increases in world prices for food imports and/or decreases in export revenues. When these events occur simultaneously,[10] the ability of many LDCs to meet target consumption levels is devastated.

An alternative to relying on imports is to release stocks. However, historically, for most LDCs stock-level changes have not been sufficient to reduce consumption variability. In view of these arguments, reliance on imports is a rational strategy, except perhaps in a large country like India. Therefore, fluctuations in consumption may be considered as resulting from fluctuations in the levels of production and of imports.

PRODUCTION VARIABILITY

Production has been relatively stable in most large low-income countries. These include Bangladesh, Egypt, India, Indonesia, and the Philippines. For these countries, the coefficient of variation of production is about or below 6 percent. In contrast, in thirty-three out of sixty-seven countries analyzed this figure is 10 percent or more, and in several Arab countries it is above 20 percent. The probability of production's falling below 95 percent of trend in thirty out of the sixty-seven countries is once every three years. For the operation of a food-security system the absolute magnitude of a shortfall is critical. A country like India may have a relatively low level of instability (6.4 percent) but a high value of absolute variability (6.6 million tons); in contrast, Morocco has relatively high instability (27.2 percent) but an absolute variability only one-sixth that of India (1.2 million tons). Hence, withdrawals from an international scheme could be dominated by large countries, many of which have relatively low production variability.

VARIABILITY IN THE FOOD IMPORT BILL

In an effort to compensate for the variability in domestic production, countries may destabilize their food import bill. Fluctuations are aggravated by the world price instability of cereals. Analysis for the period 1961–76 clearly shows that, except for a few countries like Egypt, the variability of the import volume explains most of the variability of the food import bill. On average, only one-quarter of the variability of the import bill is explained by world price movements.[11]

The above analysis has two important qualifications. First, before 1972 world grain prices were relatively stable; since 1972 they have been more unstable. Second, a considerable portion of imports to developing countries has been through food aid. The quoted price of imports overestimates the true cost of food aid imports to recipient countries. To the extent that foreign aid is sometimes suddenly withdrawn or its volume curtailed, the "true" price for the recipients may fluctuate more than the above figures indicate.

Foreign-exchange availability could be the most critical factor in determining whether or not a country can import enough to stabilize food consumption. The average ratio of the actual value of food imports to total export revenues (including services) and its maximum for the period 1965–77 are presented in table 2. This ratio indicates the pressure on foreign-exchange supplies to finance actual food imports. To the extent that actual food imports are already subject to financial constraints, this ratio would underestimate the true pressure exerted by foreign-exchange shortages. The results show that, except for four out of the twenty-four cases (Bangladesh, India, Sri Lanka, and Egypt), during 1965–77 the mean ratio was less than 15 percent, which, in our opinion, does not indicate a severe constraint during normal years. However, this ratio reaches significantly higher levels in unfavorable years. In table 2 one observes that for some countries with low average ratios, such as Tanzania and Syria, exceptionally unfavorable years raise this ratio by a factor of two to three. This ratio becomes intolerably high, particularly in Asia and in Egypt, Tanzania, and Senegal. The ratio remains remarkably low even at its maximum values in a few countries such as Nigeria, Libya, and Colombia.

There is a common impression, implicit in most of the discussion about the food gap projections of developing countries, that the burden of the food import bill measured, for example, as a ratio of total export revenues is increasing. Thus, the use of an average ratio for the period 1965–77 might understate the true magnitude of the problem of financing food imports by not revealing an upward trend in this ratio.

As shown in table 2, this impression is correct for many countries if 1975–77 is compared with 1970–72, years immediately precedent to the great rise in real world prices of food. Real prices in the years used for comparison are not substantially different, so that the changes in the food import bill ratios reflect the volume effect rather than any price effect.

If a longer view is taken, however, say, compared with 1965–67, the ratios

TABLE 2
RATIO OF FOOD IMPORTS TO TOTAL EXPORT REVENUES

	1965–67 (%)	1970–72 (%)	1975–77 (%)	1965–77	
				Mean (%)	Maximum (%)
Asia					
Bangladesh	—	—	67	77[a]	123 (1975)
India	40	10	24	23[b]	45 (1966)
Indonesia	7	13	6	11	21 (1968)
Korea, Republic of	13	16	7	13	22 (1969)
Philippines	7	5	4	5	9 (1965)
Sri Lanka	25	25	36	30	50 (1975)
North Africa/Middle East					
Algeria	8	6	9	7[c]	10 (1973)
Egypt	23	12	20	17	32 (1974)
Jordan	13	17	7	12[c]	18 (1972)
Libya	2	1	2	2[c]	3 (1975)
Morocco	9	6	2	8	14 (1975)
Syria	8	14	7	9	19 (1971)
Sub-Sahara Africa					
Ghana	5	3	4	4[c]	6 (1977)
Nigeria	3	2	2	2[c]	3 (1977)
Senegal	—	10	10	10[d]	20 (1973)
Tanzania	3	3	10	6[c]	22 (1974)
Upper Volta	—	8	9	9[e]	19 (1974)
Zaire	3	3	6	3[f]	6 (1975)
Latin America					
Brazil	10	4	4	6	11 (1967)
Chile	4	4	6	6	14 (1974)
Colombia	3	3	4	3	5 (1974)
Guatemala	3	2	3	3	4 (1975)
Mexico	0.1	2	5	3[c]	10 (1975)
Peru	6	5	11	7[c]	16 (1975)

SOURCES: Export revenues, including goods and services (except Brazil for 1965–67), are from IMF, *International Financial Statistics*. Food import data, including cereals and other staples, are from the FAO tape.
NOTE: All food imports, including food aid, are at commercial prices.
   [a] 1973–77.
   [b] 1965–76.
   [c] 1967–77.
   [d] 1968–76.
   [e] 1968–75.
   [f] 1965–75.

have increased for some LDCs such as Sri Lanka, Tanzania, and Peru. But this cannot be generalized to describe the situation in LDCs, or even the majority of them. In fact, for many LDCs, export revenues have increased faster than the value of food imports.[12]

INTERNATIONAL POLICY APPROACHES

In current discussions there are two different but related approaches to the problem of food security. One addresses the question of world food security, in particular world price stabilization mechanisms. The other addresses issues of food security for a subsystem of the world—the food-deficit developing countries. For the latter, the more prices fluctuate, the less attractive is trade as an option to stabilize consumption.

Several international strategies are currently being considered that could enhance food security in LDCs. These include: greater reliance on an international grain reserve system; consumption and production adjustments in developed countries; and food aid and other financial measures to alleviate the foreign-exchange constraint. Political support for each of these initiatives is still, however, quite uncertain.

PRODUCTION AND CONSUMPTION ADJUSTMENT IN DEVELOPED COUNTRIES

Researchers have concluded that the root of the problem of world price instability lies in the lack of response of domestic prices, through policies that insulate domestic producers' prices and consumption from the effect of variations in world market conditions. Much of the international price instability is due to governmental policies and not to weather.[13] For example, in wheat, policy-induced variations in production among developed countries, which with the Soviet Union are the major importers and exporters, are large and represent a serious problem for LDC importers.

Eliminating policy-induced variations in production could more directly benefit LDCs than the various stock policies commonly discussed at international gatherings. Also, in the past, consumption stabilization policies in developed countries and the Soviet Union inadvertently exacerbated the problems arising from production variability. The response of stock changes, limited to Canada and, to a lesser extent, the United States, was not strong enough to counter their effect.[14] But neither consumption adjustments nor policy-induced production variability have ever been the subject of serious international negotiations, as they run counter to domestic objectives. Thus, as an alternative, all the price stabilization initiatives focus on grain reserves, which have to be greater than those required under a policy of more price flexibility.

INTERNATIONAL BUFFER STOCK SYSTEMS

Under the auspices of UNCTAD and the International Wheat Council (IWC) considerable effort was devoted between 1977 and 1979 to negotiations for a new Wheat Trade Convention (WTC) for the creation of an internationally coordinated system of wheat reserves. Its primary purpose was to reduce the short-term variability of the wheat price in international markets.

Undoubtedly, larger grain reserves could be managed to reduce world price variability. However, the collapse of the WTC negotiations in February 1979 confirms the suspicion that an agreement on price issues that would simultaneously consider market realities and accommodate LDC requirements is most unlikely. What implications will this failure of the WTC have for food-deficit developing countries?

For observers of these negotiations it has been hard to develop a clear position on this issue, largely due, in our opinion, to the lack of an explicit conceptual framework and relevant parameters to analyze the potential impact on price stability of the proposals for reserve accumulation and price trigger mechanisms.

A recent study by Morrow[15] concluded that even if the mechanism envisioned for the WTC had been agreed upon and functioned well, it would have had a modest impact on world price variability. There would have been little or no additional stock accumulation during the life of the agreement.[16] Especially if the price band were too broad or too low and the reserve obligations too small, the major wheat exporters could have fulfilled their WTC reserve stock obligations with stocks that they would have held anyway. Furthermore, by the time the agreement would have gone into effect, the opportunity for accumulating reserve stocks during the period of low prices might have passed.

Although its tangible effect on stocks and prices would have been questionable, a new agreement would have allowed an active periodic review of market conditions and national wheat policies by senior policy makers, thus perhaps avoiding a repeat of the 1973–74 events when world wheat prices soared. In this sense, the collapse of the wheat negotiations is a setback to LDC food security.

However, even if a new WTC were to succeed in reducing world price instability by itself, it would not preclude the need to cope with a highly unstable food import bill. As shown by Valdés and Konandreas for many individual countries and by Reutlinger on an aggregate basis for LDCs,[17] the major source of variability in food import costs is fluctuation in a country's own production and not fluctuation in the world price. This points to the need to understand how various complementary policy instruments effect food security in LDCs.

ALLEVIATING THE FOREIGN-EXCHANGE CONSTRAINT

Variable food aid programs and compensatory financing for commercial imports represent the two major types of policy that have been discussed recently.

Food Aid

The experience of the early 1970s highlighted the unresponsiveness of food aid to widespread production shortfalls aggravated by high world market prices and the inability of domestic food delivery systems to distribute large volumes of grain. In years of high world prices, some donors have not been willing to divert supplies from commercial markets to food aid. For instance, figures show that food aid was actually cut back in 1973–74. This is apart from the strictly emergency relief efforts, which involve 0.5 to 1 million tons per year.[18]

Thus, the lack of correspondence between food aid flows and import needs in LDCs clearly aggravated the burden of imports during these years of high world prices. If food aid were reoriented, by incorporating a variable component in aid flows, it would greatly enhance food security in LDCs. According to Huddleston, this could be achieved without jeopardizing the ability of donor countries to commit substantial volumes of aid for purposes other than food security, such as overall economic development. For the entire 1970–75 period, aid flows substantially exceeded the variable requirement of LDCs to achieve stability of supplies.[19] However, food aid flows would not have been sufficient to cover needs in an occasional year of widespread harvest shortfalls.

On an aggregate basis, concessional food sales have so far been dependent on erratic surpluses in donor countries and hence can hardly be considered a dependable base for food security in LDCs. Unlike the pattern for food aid in the past, the aggregate volume of food aid should increase when prices are high. But to enhance the food security of individual countries, the given quantity of food aid can be allocated among LDCs according to their individual needs in that year. This means that bilateral flows would be variable.

A variable food aid model operating through a grain insurance program was developed by Johnson.[20] The United States, alone or in cooperation with other donors, would "guarantee to each developing country that in any year in which grain production declines more than a given percentage below trend production that the shortfall in excess of that amount would be supplied." He argued that a substantial degree of internal price stability could be achieved with this proposal at low cost for each developing country. His results indicated that food security could be achieved by modifying the distribution of food aid, without significant increases in the average long-run volume of food aid, which in any case is relatively small and concentrated in a few recipient countries.[21]

Food aid programs modified along the lines of Johnson's "grain insurance program" and Huddleston's "variable component" of food aid would constitute a major change in current programs. It seems probable that such adjustments would greatly enhance food security in poor countries. However, the provision of food aid would be very closely linked to the availability of grain reserves. Consequently, the burden of food aid would fall largely on the grain-exporting countries, an aspect that must surely affect the political feasibility of variable food facilities, not to mention the loss to the current major recipients. Another politically relevant point is that donor countries would have to give up their discretionary power to determine the quantity of food aid on an annual basis.

Financial Food Facilities and Insurance Schemes

Another approach to reducing food insecurity is to enlarge the scope of existing compensatory financing schemes to include the cost of cereal imports. Existing schemes are the International Monetary Fund (IMF) facility, the STABEX scheme (administered by the European Community for some fifty countries in the African, Caribbean, and Pacific regions), and the Arab Monetary Fund.

The financial facility approach has the advantage of simplicity. It protects member countries against fluctuations in the cost of cereal imports by providing foreign exchange in years of above-trend food imports. As described by Goreux, "it would be technically possible to extend coverage of the IMF facility so as to protect members against fluctuations in the cost of imports; this could be done by subtracting the cost of imports from the value of export earnings before calculating the net shortfalls on which the drawings would be based."[22] For a sample of forty-six countries, Goreux concluded that the cost of expanding the scope of the facility to cover food imports would not raise the sum of compensation for shortfalls existing under the present scheme. Its cost would be significantly below that of establishing a separate facility dealing exclusively with actual food imports, because of the positive correlation of export earnings with cereal imports for several LDCs observed during 1963–75. Moreover, this financial food facility would benefit the lower-income countries in particular. Working with rules for a "food security level of imports" rather than the actual level of cereal imports, Valdés and Konandreas[23] concluded that the cost of expanding the IMF facility would involve somewhat larger withdrawals, illustrating the sensitivity of the expected withdrawals from the facility to the country coverage (all LDCs or only most seriously affected [MSA] countries under FAO definition) and to whether or not the facility is adjusted by export earnings.

In a variant of the above, Konandreas, Huddleston, and Ramangkura[24] analyzed an international food insurance scheme, projected for the years 1978–82. The scheme could either serve as a purely financial mechanism that would provide LDCs with funds to cover overruns in their cereal import bill, or it could be supported with a limited grain reserve deliberately restricted to years when prices are high to avoid rapid depletion of the reserve. They proposed creating a reserve of 20 million tons and a compensatory financing capacity of $5 billion for the period.

The existence of these schemes would tend to reduce the response of LDC import demand to the world price and to increase the need and the profitability of holding stocks. As yet, there is little research that examines the issue of whether the increased profitability would by itself induce sufficient private stock accumulation to match the need.[25]

A limitation common to the variable food aid scheme and the food facility scheme (to the extent that the latter has a concessional element) is that both rest on the availability of accurate information on production shortfalls. Although many claim that this is an insurmountable problem, particularly for the poorest LDCs, we share the optimistic belief that the existence of such schemes will induce greater effort to overcome the problems; after all, their solution is not technically difficult.

CONCLUDING REMARKS

The solution to the food-insecurity problem must begin at the national level, and every country can take important initiatives to reduce food insecurity. The analysis indicates that remedies will probably include large investments in food distribution systems, transport and communications, early warning systems, and a mix of stock and trade policies. Although there is considerable scope in many LDCs for larger investment in working stocks, one clear generalization, on the basis of past research, is that relying mainly on domestic grain reserves to cover yearly fluctuations is an expensive solution, where trade is a real possibility. Other than this, the circumstances of each country would have to be considered in designing food-security policies.

Given their food consumption stabilization objectives, LDCs need to minimize the cost of managing the short-term variation in real income. Three implications of these food consumption stabilization policies are: the incentive for private domestic stockholding would be reduced; the foreign-trade balance and the government budget would have to absorb the instability; and the loss in purchasing power or real income of farmers and farm laborers, due to crop failure, would have to be compensated for. Otherwise, stabilizing national food supply per se would be insufficient to offset declines in effective demand in the rural areas. This last implication, which could be resolved only at extremely high fiscal cost, has not been studied extensively.

Another piece of unfinished business is to recognize more explicitly the connections between the markets for food, nonfood products (agricultural and others), and foreign exchange. Such an understanding could lead to a wider range of policy instruments for managing real income adjustments. The manner in which a country copes with short-term instability may well influence the longer-term performance of its agricultural sector, and through that the mean levels of food consumption.

To summarize, international initiatives to foster food security include an international grain reserve system, consumption and production adjustments in developed countries, and food aid and financial approaches to alleviate the foreign-exchange constraint. Empirical analysis demonstrates that these initiatives do reduce the costs of national solutions in LDCs, although they do not replace them. Numerous opportunities exist for donor countries to help alleviate food insecurity in LDCs through both bilateral and multilateral action, by easing the foreign-exchange constraint for food imports, by providing technical and financial assistance to improve food delivery systems in LDCs, and by following agricultural policies that are less disruptive to world grain trade.

Insofar as instability in the import bill is concerned, most economists agree that compensatory financing schemes represent the most effective approach, far superior to international grain reserves. Creating a food-financing facility is highly desirable. With such an approach, the focus is where it should be—on the import bill rather than on import prices. Liberalizing the existing schemes of compensatory financing to take into account fluctuations in real export earnings, which compensates also for unexpected or excess expenditure on food imports and possibly other critical imports, appears to be not only effective but also politically feasible at the international level.

Not only is a liberalized financial facility less subject to political criteria—such as those prevailing in food aid—it is more likely to help the poorest of the developing countries, which are unlikely to cover their borrowing needs in the commercial capital market.

ADDENDUM

In May 1981 the International Monetary Fund created an integrated financial facility for food imports by amending its compensatory financing facility for export earnings to cover excess cereal import bills. *Excess* refers to years when cereal import bills are above normal for reasons beyond the control of the governments requesting assistance. The IMF food facility will not prevent emergency-induced increases in world prices. However, if such price increases occur, the facility will ease the strain for cereal-importing countries whose costs have risen sharply.[26]

NOTES

The authors gratefully acknowledge the comments of Anne del Castillo, John W. Mellor, and Grant M. Scobie on an earlier draft of this chapter.

1. Even if food is available to a country on concessional terms, to the extent that these imports have to be supplemented by commercial imports and thus represent "inframarginal" transfers, which is the case for all countries, that food aid should be used to replace current commercial imports rather than to build up reserve stocks.

2. This conclusion is arrived at by assuming that the nation or the household does not have consistently better access to market information "to play the market" and expect to win. Under the opposite assumption, they should become regular "speculators," trading in volumes not necessarily connected to their food security requirements.

3. A. Valdés and P. Konandreas, "Assessing Food Insecurity Based on National Aggregates in Developing Countries," chap. 2 in *Food Security for Developing Countries,* ed. Alberto Valdés (Boulder: Westview Press, 1980).

4. B. S. Minhas, "Toward National Food Security," *Indian Journal of Agricultural Economics* 31, no. 4 (1976):8–19.

5. This statement is strictly correct for a country considered by itself. As Josling has pointed out, a policy of price insulation has quite an adverse impact on the functioning of the world market as a whole (Timothy Josling, "Price, Stocks and Trade Policies and the Functioning of International Grain Markets," chap. 8 in Valdés, *Food Security for Developing Countries*).

6. For example, the inland Sahelian countries.

7. John McIntire, *Food Security in the Sahel: Variable Import Levy, Grain Reserves, and Foreign Exchange Assistance,* Research Report 26 (Washington, D.C.: International Food Policy Research Institute, 1981).

8. For a contrary view see David M. Morris, "What is a Famine?" *Economic and Political Weekly* 9, no. 44 (1974).

9. Valdés and Konandreas, "Assessing Food Insecurity."

10. I.e., domestic production shortfalls in years of adverse world prices, as experienced by many Asian and African countries in 1973/74.

11. Valdés and Konandreas, "Assessing Food Insecurity."

12. For a larger sample of fifty-four non-oil-exporting LDCs as a whole, between 1970–72 and 1975–77, the share of food imports in total imports increased only marginally from 4.9 percent to 5.3 percent. This ratio for the same set of countries during 1961–63 was 7.6 percent.

13. D. Gale Johnson, "World Agriculture, Commodity Policy and Price Variability," *American Journal of Agricultural Economics* 57 (1975): 823–38; S. Y. Shei and R. L. Thompson, "The Impact of Trade Restrictions on Price Stability in the World Wheat Market," ibid. 59, no. 4 (1977): 628–38; and Josling, "Price, Stocks and Trade Policies." [See also Falcon, chapter 12 in this volume.—ED.]

14. Timothy Josling, *Developed-Country Agricultural Policies and Developing-Country Supplies: The Case of Wheat*, Research Report 14 (Washington, D.C.: International Food Policy Research Institute, 1980).

15. Daniel Morrow, *The Economics of the International Stockholding of Wheat*, Research Report 18 (Washington, D.C.: International Food Policy Research Institute, 1980).

16. Abstracting from the complexities of cost-sharing (given uncertainty about the distribution of benefits of such reserves) and implementation of the Wheat Trade Convention proposals.

17. Valdés and Konandreas, "Assessing Food Insecurity"; Shlomo Reutlinger, "Food Insecurity: Magnitude and Remedies," *World Development* 6 (1978): 898–911; and Shlomo Reutlinger and K. Knapp, *Food Security in Food Deficit Countries*, World Bank Staff Working Paper (Washington, D.C., 1980).

18. A. H. Sarris and L. Taylor, "Cereal Stocks, Food Aid, and Food Security for the Poor," in *The New International Economic Order: The North-South Debate*, ed. J. Bhagwati (Cambridge: MIT Press, 1977), 286; and Barbara Huddleston, "Responsiveness of Food Aid to Variable Import Requirements," chap. 13 in Valdés, *Food Security for Developing Countries*.

19. Huddleston, "Responsiveness of Food Aid."

20. D. Gale Johnson, "Increasing Stability of Grain Supplies in Developing Countries: Optional Carry-overs and Insurance," in Bhagwati, *The New International Economic Order*.

21. During 1970 and 1977, food aid represented less than 13 percent of the total volume of cereal imports by all LDCs. Of this volume of food aid, approximately 70 percent was received by five LDCs and the balance, approximately 2.4 million tons (per annum), was received by the remaining LDCs.

22. Louis M. Goreux, "Compensatory Financing for Fluctuations in the Cost of Cereal Imports," chap. 14 in Valdés, *Food Security for Developing Countries*.

23. Valdés and Konandreas, "Assessing Food Insecurity."

24. P. Konandreas, B. Huddleston, and V. Ramangkura, *Food Security: An Insurance Approach*, Research Report 4 (Washington, D.C.: International Food Policy Research Institute, 1978).

25. This issue was examined by Peck and Gray, who argue: "There appears to be no limit to the amounts these [private American] traders are willing to carry if the market is providing full carrying costs (interest, insurance, and warehousing fees)." If the food-financing facility provides for payments to LDCs for higher world food prices, it will in effect be giving them the purchasing power to pay for the full carrying costs charged by the market. Thus, if Peck and Gray's argument applies, then there should be no physical shortage of stored grains, despite the reduced responsiveness of the LDCs' import demand (see Anne E. Peck and Roger W. Gray, "Grain Reserves—Unresolved Issues," *Food Policy* 5, no. 1 [1980]: 26–37).

26. For a recent analysis of the facility, including its potential impact and how it relates to food-security-related policies pursued by governments, see B. Huddleston, D. G. Johnson, S. Reutlinger, and A. Valdés, *International Finance for Food Security* (Baltimore: Johns Hopkins University Press, forthcoming).

14

Food Security in Developing Countries: National Issues

UMA LELE AND WILFRED CANDLER

From the perspective of those with experience in Eastern and Southern African countries, including the Sudan, Ethiopia, Kenya, Tanzania, Zambia, Malawi, and Lesotho, much of the discussion of food production and food security in the international literature seems unrealistic and nonoperational.[1] This chapter first reviews the realities of the food supply and supply reporting system in these countries and then considers the working of the various food marketing channels. These are compared with the assumptions made in the literature. The review leads us to conclude that official food trade is only the visible portion of a much larger total trade; and although the government is responsible for food security, it has no power to control major components of the food distribution system. These facts suggest that consideration of an official food reserve and/or financial reserve is probably necessary—mainly, to deal with the urban problem and thus indirectly with the rural problem. But such a consideration does not address the total problem of food insecurity adequately, as is often implied in the literature. Nor is it the most efficient way of dealing with this problem overall. Further, the proposed solutions are especially unsuited for the problems of rural areas, which require a different set of solutions. Our view of the production and marketing systems also implies a radically different basis for estimating food insecurity requirements— different than using overall production fluctuations as a criterion, as is now commonly discussed in the literature. We offer alternative ways of dealing with these various concerns.

UMA LELE is division chief, Development Strategy Division, Economics and Research Staff, The World Bank, Washington, D.C., and WILFRED CANDLER is senior economist, The World Bank.

Originally titled "Food Security: Some East African Considerations." Reprinted by permission of Westview Press from *Food Security for Developing Countries,* edited by Alberto Valdés. Copyright © 1981 by Westview Press, Boulder, Colorado. Published with omissions and revisions by permission of the authors.

UNDERDEVELOPMENT

Most discussions of food security start with the following implicit assumptions: (1) that there is an efficient, price-responsive agricultural industry; (2) that essentially all food passes through commercial marketing channels; (3) that there is a price for grain at which one can buy or sell at will; (4) that the influence of trading activity in one part of the market is quickly and effectively transmitted to all parts of the market; and (5) that reliable information on production and market performance is instantly and freely available.

Essentially, none of these implicit assumptions apply in sub-Saharan Africa. Figure 1 illustrates the marketing channels characteristic of most East African countries. Frequently, there are two production sectors: large-scale and traditional. The large-scale sector may be predominantly expatriate, as in Zambia, or may consist of state farms, as in Tanzania and more recently in Ethiopia. Typically, the large-scale sector sells only through official channels and to a monopsonistic parastatal (a semiautonomous public-sector entity that usually has a commercial function). Traditional producers, in contrast, retain some food for self-consumption, use some food for barter, and sell on the official and the unofficial markets. Although the marketing parastatal will sell to consumers and undertake official exports, in the unofficial market, sales are made to consumers or produce is exported informally, either for barter or, more usually, for differentially overvalued foreign currency.

Three features of this marketing system deserve special emphasis:

1. Frequently only the transactions of the marketing parastatal are known with any certainty (and our experience indicates that even this may be suspect). Thus, analyses that assume that production, consumption, or stocks can be ascertained simply misstate the problem.
2. Most food transactions—and hence private "food security" decisions—take place without passing through the marketing parastatal; hence they are beyond direct government control.
3. Transfers between marketing channels are frequent. There is no necessary correlation between changes in the activity of the marketing parastatal and changes in production.

This last point is well illustrated by the purchases and sales by Namboard (Zambia's marketing parastatal). During the period 1964–76 purchases grew at an annual compound rate of 8.5 percent. Between 1972 and 1975 sales grew at the rate of 14 percent compounded. It is inconceivable that changes in population or incomes accounted for the rapid rise in Namboard sales. It is more likely that during these four years Namboard's market share increased and, as a corollary, the market share of the informal channels fell. And yet, literature on food

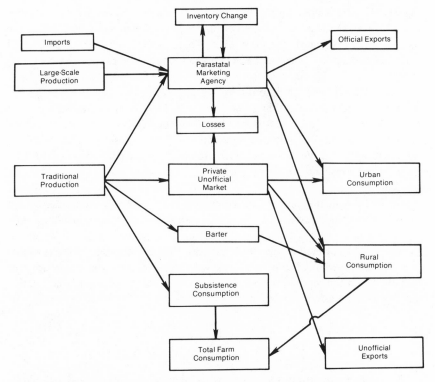

Fig.1. Conceptual Model of Foodgrain Production and Distribution

security in the international circles does not address the issue of the growth of public distribution systems, which have become national objectives in many African countries.

Because of this dual marketing structure, the parastatal tends to become over-burdened or hit with a "double whammy" during food shortages. The parastatal faces both declining inventories and additional consumers who are unable to purchase their requirements at reasonable prices from the unofficial market. Sales of milled products, that is, maize flour, wheat flour, and rice, by the National Milling Corporation in Tanzania reached nearly 300,000 tons in 1973/74, compared with 190,000 in 1970/71, and fell to 238,000 in 1977/78. Purchases were 170,000 of the same three unmilled grains in 1973/74, compared with 320,000 tons in 1970/71 and 300,000 tons in 1977/78. National food administrators must deal with the simultaneous halving of available supplies and doubling of official demand. Again, this is a problem not dealt with in the literature.

Not only does the parastatal handle only a small percentage of the crop but the parastatal's price applies for only a part of the season. After the main harvest period farmers can sell stored crop at higher prices—frequently up to twice the official price—on the unofficial market. Thus, studies that rely on the official price fail to take into account the actual situation that the farmer and consumer face.[2]

We should reiterate that information on the informal markets is imperfect and often nonexistent. We should also emphasize that the parastatal statistics may be imperfect (that is, those on stocks may refer to amounts put into storage, not the net amount after allowing for losses and pilferage). The existence of an unofficial market usually indicates a high cost structure of the parastatal where its procurement costs represent typically only half of the sale price of crops handled by the parastatal.[3]

Thus, any realistic approach to national food security policies in East Africa must be a combined exercise in "planning without facts," within the context of a complex marketing system and particular national policies. As will be shown below, we do not believe that the exercises of economists in the international agencies are inherently futile; but we do warn that such analyses should not start with the facile assumptions that all the required data are readily available and that the system is totally commercial.

DOMESTIC FOOD RESERVE

The issue of a best form for food security, that is, whether a reserve should be international, regional, or national; whether it should consist of food or be a monetary reserve; or whether it should simply be a food insurance scheme, has been debated in international circles for some time.[4] After the food crises of 1973/74, some East African countries such as Kenya and Tanzania began building a national food reserve. This chapter assumes that these national reserves exist.[5]

At any time, a domestic food reserve has three major components: (1) the official "national food reserve," (2) stocks held by the official food marketing organization for its routine purchase and sale activities, and (3) stocks held by the private sector, predominantly individual farmers. Except just prior to harvest, the private stocks will generally exceed all other stocks. Thus, to think in terms of public stocks as the only component to food security is to overlook the single most important contributor to food security: the private sector.

Many cultural practices, such as mixed cropping, planting of cassava and drought-resistant cereals, and timing of planting, are explicitly designed to offset the yield fluctuations of a monoculture in a marginal climatic environment. In addition to the stocks of harvested crop held in the farm household, many East African farmers carry an inventory of unharvested food reserves in the form of cassava.

PRODUCTION LEVELS

FOOD

Food security is frequently conceived of in terms of reducing fluctuations in domestic supply around a trend line by a certain percentage. Conceptualization of the problem in this manner is of little operational significance.

Data on domestic agriculture in most African countries are too unreliable to ascertain the level of production in any given year. Further, year-to-year production fluctuations in reported statistics are often too large to estimate a trend with any degree of confidence. Judgments about deviations from a trend by amounts as small as five or ten percentage points would be nearly impossible.

It is instructive to consider the available production estimates for Tanzania. The various series on crop production over the 1965/66–1977/78 period published by Tanzania's Ministry of Agriculture and various international agencies are presented in table 1. Maize is an important crop in Tanzania. The table indicates that USDA and FAO estimates of maize production for the year 1974/75 are 2.3 times greater than the estimate published by the Tanzanian Ministry of Agriculture. The ministry and USDA/FAO series for maize production in Tanzania have a correlation coefficient of only 0.59! The series have a higher correlation with time than they do with each other. Which series would be used to estimate food reserve requirements? For 1973/74, the USDA estimate of sorghum and millet production is 3.5 times that of the Ministry of Agriculture, whereas FAO's estimate is 88 percent of the ministry's. Incidentally, while the ministry's estimate of maize production in 1973/74—a drought year—shows a decline from the previous year's level by 27 percent, the USDA series shows an increase of 30 percent over the previous year! The ministry's maize series shows an annual growth of 19,527 tons per year over the 1966/67–1976/77 period. The USDA series for the same years shows an annual growth of 91,020 tons. Each "*b*" coefficient is highly significant. There is no a priori reason to believe that any one of these series is more accurate than another or that any rate of growth is more representative than another.[6]

Which production series should be used in the management of a food security scheme? Despite the recognition of these problems by national governments and international agencies involved in compiling statistics, only recently have small amounts of international development funds been made available to develop methods and institutions to generate reliable statistics.[7]

MARKETED PRODUCTION

Because production statistics are known to be unreliable, "marketed" production is used as the proxy for decision making about food security by many national agencies. Further, it is the production marketed through "official"

TABLE 1
VARIOUS PRODUCTION ESTIMATES OF PRINCIPAL CEREALS FOR TANZANIA, 1965/66 TO 1977/78
(IN THOUSANDS OF METRIC TONS)

Year	Maize			Paddy			Wheat			Millet and Sorghum		
	Ministry of Agriculture	USDA	FAO	Ministry of Agriculture	USDA	FAO	Ministry of Agriculture	USDA	FAO	Ministry of Agriculture	USDA	FAO
1965/66	532	—	—	—	140	—	—	39	—	265.4	—	—
1966/67	752	1,127	—	—	114	115	—	31	35	299	1,122	—
1967/68	630	549	560	125	136	136	—	44	40	254	1,145	292
1968/69	647	664	678	139	144	136	—	39	39	284.5	1,150	275
1969/70	663	525	530	184	182	182	79.8[a]	61	61	282	1,155	236
1970/71	746	637	650	193	193	193	—	84	84	322.6	1,180	245
1971/72	730	715	715	171	171	171	—	98	98	276.9	1,200	279
1972/73	856	681	863	184.9	204	152	79.8	78	80	319.1	1,220	319
1973/74	624	888	800	134.3	310	293	—	45	46	333.4	1,165	295
1974/75	623	1,446	1,446	150	385	430	46.1	56	56	189.6	950	324
1975/76	825	1,350	1,354	172	400	430	57.6	56	60	339.2	1,200	615
1976/77	897	1,619	1,619	194	400	—	70.8	50	—	—	1,200	1,190
1977/78	968	1,700	—	—	—	—	—	—	—	—	1,000	—

SOURCES: Ministry of Agriculture, *Report on Price Recommendations of the Marketing Development Bureau, 1978/79*, 53, 72, 79, 90; USDA, *Indices of Agricultural Production in Africa and the Near East, 1956–75* (the authors were informed by a USDA staff member that these were revised down considerably subsequent to our discussions with USDA headquarters staff about discrepancies among various series); FAO, *Production Yearbook*.
[a]1968–72 average.

channels, rather than total marketings, that influences policy making as it determines the supply of food to cities and major towns. Government distribution efforts are largely concentrated in urban areas. The reasons for this include (1) the desire for political and economic stability—if food prices rise, urban populations can organize to demand higher wages and/or a different government; (2) the administrative imperative—the limited manpower and management cannot be spread adequately to deal with both urban and rural distribution needs, and the civil service that makes and benefits from these decisions is largely concentrated in the cities; (3) the inadequate transport, storage, and handling facilities in rural areas, which frequently frustrate official distribution efforts; and (4) the high unit cost of public-sector distribution of grain to rural areas that require large subsidies if extensive rural distribution is undertaken.

Data are not available by sector but are recorded by administrative region. In 1977/78, a relatively good crop year, 77 percent of the Tanzanian National Milling Corporation's sales of rice, maize flour, and wheat flour were concentrated in six regions (Dar es Salaam, Morogoro, Tanga, Arusha, Dodoma, and Mwanza) that have major urban population concentrations, out of a total of twenty-two regions. Urban demand is more predictable than other variables that influence "national" food security, namely, rural demand, production, or marketed quantities. However, even urban demand is not completely met by official sources, as offtake from official channels increases in bad crop years and declines in good crop years.

The remaining 23 percent of the official distribution in Tanzania in 1977/78 was sold in sixteen mostly rural regions. Field observations suggest that the distribution occurred largely in major towns rather than in the countryside. There are always pockets of shortage in rural areas, as official distribution is thin and traditional rural markets are highly fragmented. Welfare and political considerations have prompted an improved distribution effort to the rural areas in years of extreme food shortages. Despite these improvements, the amount of rural demand satisfied by official distribution is still limited.

Although the problem of food security pervades the whole nation, government action is perceived in terms of meeting (1) urban demand plus (2) the share of rural demand that the governments feel obliged to meet, especially in bad crop years.[8] Defined in this way, the maize reserve requirements for Tanzania may be about 100,000 tons, compared with 34,000 tons and 68,000 tons if conceived in terms of offsetting production fluctuations by 5 percent and 10 percent, respectively, based on the Tanzanian official cereal production series. The figures will vary if USDA estimates are used. In Zambia the rapid market penetration by Namboard requires an extra 45,000 tons per year, quite apart from any food security considerations. No direct estimates are available of either production or consumption variability in Zambia.

The conceptualization of food security requirements has substantial implications not only for the size but also for the composition and the cost of the food reserve. Rural and urban consumption patterns may markedly influence whether the source of the supply of the reserve is imported or produced domestically. Different food security schemes will also influence domestic pricing and other agricultural production and trade policies differently. The standard international models of food security assume that a 5 percent reduction in production would lead to a corresponding percentage decline in consumption across the board. Our observation, however, suggests that the rural population, 80 percent of whose production is for subsistence, would attempt to maintain their consumption level in the face of a production decline. Thus a 5 percent decline in production is more likely to lead to a 25 percent decline in the marketed production (that is, in the 20 percent of production that is marketed).[9] This is significant in the conception of the problem and would lead to different sizes and compositions of buffer stocks than those arrived at on the basis of overall production fluctuations.

COMPOSITION OF URBAN AND RURAL DEMAND

The urban cereal consumption pattern in many African countries is relatively more homogeneous than the rural, consisting largely of maize and some wheat and rice. (In Ethiopia, teff is also important.) International food assistance is, however, frequently influenced by the particular cereal that donors can spare. This can lead to a radical shift in the urban consumption patterns away from domestically produced cereals (maize) to imported grain (as can be documented in the case of some East African countries).

Rural food consumption patterns are substantially more diverse and involve consumption of several different crops, including cassava, sorghum, millets, rice, bananas, and maize, depending on local production. This diversity aggravates the problems of ensuring rural food security through official channels, as it complicates (if it doesn't make impossible) the problem of management and distribution of stocks in different regions. Thus neither national nor international food distribution programs directly assist the rural sector. By absorbing rural surpluses for urban consumption, these distribution programs do help the rural areas indirectly.

PRODUCTION MARKETED THROUGH OFFICIAL CHANNELS

It is essential to identify the consumption requirements of the rural and urban sectors. These specific commodity requirements for national food security will have important implications for domestic food production strategy and efficient resource use in the long run. The simplistic formulation of the food security problem in the international literature, in contrast, is formulated in terms of "tonnage" and "calories." For instance, a shift in the domestic consumption pattern that tended to substitute wheat and rice for maize would increase import dependence unless domestic production of those crops increased or domestic consumption could be curbed. Increasing consumer prices and/or reducing imports are politically awkward options. Rather than control consumption, a country may attempt to increase domestic production of imported cereals to meet future urban demand. In the short run, however, an increase in the official purchases of domestic crops beyond levels of domestic needs may result in subsidized exports. A subsidy may be needed just to cover the producer prices and costs of transport and handling their extensive rural distribution. For instance, the average margin required by the National Milling Corporation in Tanzania to cover costs of official domestic distribution amounts to approximately U.S. $100 per ton. Without a subsidy, consumer prices would be too great for the effective demand in rural areas. In most of these countries the domestic market for livestock feed is very limited. Although this can be developed, it would involve a radical change in the livestock sector's practices. The scarce and poor storage facilities incur enormous losses for stocks stored longer than one year.[10] The only option left to these countries to minimize short-run costs is to subsidize exports of minor crops.

If the objective of domestic food self-sufficiency is to be achieved without sacrificing overall agricultural growth, careful attention to relative crop pricing, production, and trade policy is required. This is especially true if the crops requiring subsidized exports are not to be generated at the cost of the stagnation (or decline) of traditional export crops that contribute to the fiscal revenues.

DETERMINANTS OF "OFFICIAL" MARKETED SURPLUSES

The relationship of production to "official marketings" is not clear, although figures of 10–30 percent of the production are usually quoted. The proportion seems to vary considerably among crops (depending on their importance in home consumption) and from one year to another. A small increase in production often appears to lead to a more than proportionate increase in official marketings, implying elasticity of domestic consumption to change in production of less than one. The situation is complicated by three factors. First, marketed production may often be greater than rural consumption levels would suggest because of the tendency among farm families to sell grain during the season to meet cash requirements and to buy grain in the market for consumption in the off-season. An attractive postharvest price obviously facilitates early sales. Correspondingly, grain shortfalls or a very high price in the off-season can have an adverse impact on the food supply of rural families and hence on rural food security.

Second, apart from the level of production, the amount sold in the official channels is a function of the relative prices in the official and the "unofficial" market, the payment terms used by the official marketing agency (the official tendency being not to make prompt cash payments), and the accessibility of an official purchasing center. It is difficult to isolate the effect of policy changes and of production changes on the increase or decrease in official marketings.

Third, trading patterns in the unofficial market are not confined to the national boundaries. This exacerbates the problem of estimating the extent to which it is the production increase rather than the sealing of national borders, relative changes in exchange rates, or availability of consumer goods across national boundaries that has caused changes in unofficial exports and domestic official purchases.

A number of hypotheses can be advanced concerning levels and changes in production and marketing. But any formal analysis is made difficult by a variety of factors, including (1) the lack of reliable farm-level data, (2) the factors that affect local farming decisions, (3) the extremely diverse farming systems precluding aggregation and generalization, and (4) the low and in some cases decreasing level of monetization, especially in the case of low-income subsistence farmers, which creates difficulties in carrying out formal analysis based on the concepts of exchange and specialization.

FOOD PRICING

Although official agricultural prices are known, actual prices faced by the majority of rural producers and consumers are not known. Not only do they seem to differ from official prices but they vary from one rural area to another. These discrepancies are caused by the fragmentation of grain markets, the varying conditions that influence them, and the differing levels of prices of other goods and services that influence food pricing.[11]

The conclusion that can be drawn from this phenomenon is that little is known or predictable about the factors that influence food production and marketing, making it immensely difficult to plan a "national" food security program. The governments' strategy under these circumstances is usually twofold: (1) to meet commercial demand through a strategic reserve and (2) to increase availability of supplies in rural areas through increased production.[12]

POLICY DIRECTIONS

As indicated above, the foodgrain markets in East Africa are fragmented, with an urban sector supplied mainly by the parastatal and the rural market supplied mainly by the unofficial system. This is important to any discussion on food security policies. Increasing the supplies available to the marketing parastatal may merely tend to increase the urban supplies, with very little, if any, trickle back into the rural areas.

Rural food security is very largely a question of rural self-sufficiency. This point is well appreciated by rural producers and indeed is widely recognized as the primary consideration in the determination of cropping pattern and amounts to be marketed. This point may not be as well appreciated by urbanized policy makers.[13] Given this perspective, four policy directions suggest themselves.

1. The promotion of drought-resistant food crops to achieve self-sufficiency at the farm-household and the village or a small-area level is critical. In many cases, the more drought-resistant crops—sorghum and finger and bulrush millet—have been downgraded by policy makers in comparison with crops such as maize.

2. The provision of a market for surpluses of drought-resistant crops is another

important aspect. The major food crops, maize, wheat, and rice, have substantial commercial markets, and cassava has a minor domestic and a potential export market. However, the drought-resistant crops have essentially no market. Without government intervention, excess production of these crops in good years would be valueless and thus the incentive to produce them would be reduced. In Tanzania, a conscious policy of subsidizing these crops is being pursued as a means of encouraging self-sufficiency at the village level.[14] This may well be cheaper than providing famine-relief supplies at the village level in years of low yield. Further analysis of this issue is needed.

3. The private cash market for foodgrains in rural areas needs explicit recognition and promotion. At present the parastatals provide a market in which farmers can sell their excess foodgrain. However, to buy foodgrain, farmers have to rely on the informal market, at prices often well above the official price offered by the parastatal. Furthermore, in the rainy season it may not be possible to buy foodgrain from outside the village. This essentially forces self-sufficiency at the village level, with all of the associated losses in economic efficiency that come from a closed rather than an open economic system. Increasing the efficiency of rural markets is not easy or cheap: it involves improving transportation and communications and a tolerant attitude to the activities of the private sector.

4. Research and extension are needed on improved methods of farm-level storage. Despite some efforts in this direction, farm-level storage would seem to involve losses of 10–30 percent over a twelve-month period. (Certainly, food prices may double at the village level from the post- to preharvest periods; and physical considerations also confirm these loss estimates.) As not all grain is held for the full year, farm-level losses on average may be no more than 15–20 percent. This would appear to represent the single most promising direction for a quantum jump in available food supplies. A program of research on new farm storage methods and extension is desirable.

We consider urban food security the tail, not the dog. There are some parallels between the food security problem in East Africa and the problem as it is discussed in international economic fora. However, the East African problem differs from the assumptions underlying the discussions on international food security in several respects.

1. The variance of demand as well as that of supply poses a problem for the marketing parastatal. When the price of maize flour in the informal market rises, consumers substitute with maize flour produced by the parastatal (a product with lower consumer acceptability) and with wheat flour and rice. Thus, at fixed prices a small maize crop, with smaller supplies to the informal market, will increase urban demand for all the major foodgrains. This variation in demand is a primary source of urban food insecurity.

2. As seen in the previous paragraph, official sales fluctuate by much more than 5–10 percent. Therefore, stocks to protect against fluctuation in demand would be much greater than stocks needed to protect against production fluctuations.[15]

3. A food security program that guarantees foodgrains in essentially the pro-

portion consumers would demand at normal prices and one that guarantees the supply of a single grain have different commodity requirements. In Kenya and Tanzania strategic grain reserves are being accumulated with funding both of the grain reserve and of the necessary stores from donor countries. The intended composition of the grain reserve therefore has immediate operational significance. If the composition of the reserve is immaterial, then wheat or rice donated for the reserve can be sold on the domestic market for immediate consumption and can be replaced by domestic maize or cassava.[16] On the other hand, if the composition is important, then only wheat would be rotated with wheat.

4. There are technical limits to the size of the strategic reserve. Maize can be stored for two years before becoming unfit for human consumption. Therefore, even with 100 percent rotation of the strategic reserve, only one year's normal sales could be stored (that is, the first year after harvest the maize would be in the strategic reserve; the second year it would be sold before becoming unfit for human consumption). Thus, high rates of deterioration for stored grain may pose technical limits to the amount that can be stored. These limits are quite apart from the economic and political questions of how much should be stored. As this technical constraint was approached, the management of the food reserve would pose increasingly severe problems. Grain not milled within a year of going into strategic store would have to be downgraded to animal feed. It seems likely that substantially higher transport bills would have to be incurred with a central strategic national reserve than without it. If the reserve is a small percentage of annual marketings, its turnover can be expected to be arranged at very little extra cost. At the point where all purchases have to pass through the strategic reserve, the costs of this extra constraint could mount substantially. As a corollary to this line of thought, minor crops with little or no sale in a normal year are unsuitable candidates for the reserve. As there is no scope for turnover, there is no possibility of maintaining high-quality grain in the reserve.

CONCLUSION

This chapter describes some of the major considerations that should enter into the formulation of national food security strategies in the East African context. It contrasts these considerations with the assumptions made in the standard global models of food security. These latter are expected to be applicable not only across countries but across continents and across a range of different—and changing—policy objectives (that is, role of public distribution), institutional development (that is, the extent and efficiency of the private and public marketing systems), infrastructure, data availability, and cereal demands.

Some of the inferences of our presentation are self-evident. First, if international schemes are to be effective and responsive to the needs of the recipient, then such schemes have to be tailored to the individual countries. This requires designing schemes suited for a group of countries similar in terms of economic development, policy objectives, cereal demands, and so on, rather than one global scheme applicable to all.

Second, it should be recognized that in most East African countries a food security scheme based on public distribution will inadequately solve the problem of rural food security. Rural food security will be achieved through increased research[17] and extension on the production of drought-resistant crops, improved input supply, produce marketing, an improved communication network, and an effective farm household storage program. An effective farm household storage program will have to use local and inexpensive materials.

Third, we stress the impact of subsistence consumption maintenance as a cause of urban food supply instability. This feature leads to a markedly different perspective on the problem than does a model that implies that an across-the-board reduction in consumption is the result of a supply decrease.

The final conclusion is the vital need for additional data on total crop production and distribution in order that the basic statistics for effective national food security schemes be available.

NOTES

The views expressed in this chapter are those of the authors and do not necessarily reflect those of The World Bank.

1. See, for example, Johnson 1981; Reutlinger 1977; and Konandreas, Huddleston, and Ramangkura 1978.

2. Furthermore, the effectiveness of this price may be in question. In some regions the parastatal may shorten the procurement season, and in others payment may be delayed. The existence of an official price does not mean that farmers can necessarily sell at this price for cash whenever they wish to. Further, villagers typically have difficulty exchanging cash for consumer goods. If the parastatal really wished to increase its market share, it would take basic consumer products to the village to sell for the grain being bought.

3. It is widely recognized that per-ton transport costs are often much higher in most African countries than in most Asian countries because of poor infrastructure and the greater mileage relative to the volume of trade. The costs should thus apply equally to the public and private sector. However, in practice, the cost of public-sector distribution frequently tends to be higher due to unnecessary transport of grain, i.e., from rural areas to urban centers, before it is shipped to the distant rural areas again. This is partly a result of centralized processing facilities and management. Partly it is a result of pan-territorial official pricing, which results in relatively more being sold to and bought from the official sources in distant areas, where the official prices are competitive vis-à-vis the unofficial market prices. The reverse is the case in the areas close to the centers of urban consumption.

4. See Valdés 1981, pt. 2. See also Siamwalla and Valdés, chapter 13 in this volume.—ED.

5. Tanzania's import bill in 1974 went up to U.S. $133 million from U.S. $34 million in the previous year. To avoid such situations in the future, Tanzania plans to build a reserve of 100,000 tons. Assistance from bilateral and international agencies in the form of storage facilities and grain is expected to cover all capital costs, which are expected to be about U.S. $30 million. Management expertise is also being provided by donors. Tanzania is typical of a small country to which international assistance for national food reserves is readily forthcoming because of the small amounts involved.

The debate on the best form of food security may still be essential if all countries have not proceeded to build a domestic reserve. This is especially true if large countries such as India, Indonesia, and Bangladesh—whose demand for food in the international market is significant—do not have their own reserves. Further, even in the East African countries that are proceeding with national reserves, there is a possibility that either domestic food production may not be able to replenish the reserve or the composition of the domestic production increase may not be suitable for meeting the objectives of the reserve as seen by national policy makers (more will be said on this below). Therefore, analysis as to how to keep the national reserve operating at minimum cost to the country may be necessary. However, it is evident that a different analytical focus is needed than that noted in much of the international literature on the question of food security.

6. The three series presented in table 1 have been substantially revised by USDA, FAO, and the Ministry of Agriculture. In particular, USDA has substantially revised the series on millet and sorghum downward to about a third of what it was in table 1. These revisions also point to another crucial problem: with these changes in data of "what is," how profitable is it to analyze what might have been?

An analysis of the rate of growth of food production carried out for India has documented the other obvious fact that the growth rate obtained is highly dependent on the choice of base- and end-year production levels; that is, a poor crop in the base year and an excellent crop in the end year exaggerates the growth rate, and the reverse understates it (see, for instance, Lele and Mellor 1964).

7. Notably, the United Nations Development Programme's (UNDP) early warning, crop-yield forecasting project in Tanzania. Though useful expenditures have also been made by the Ford Foundation in Zambia and by the World Bank in several countries in a farm management context, these expenditures are not sufficient to provide a picture of the overall production situation even in a limited project area. At this stage, obtaining a national picture from these data is thus completely out of the question.

8. The recognition of the impracticality of dealing with the rural food security problem through distribution may prompt governments to promote production of drought-resistant food crops. The official policy of promoting production of cassava, sorghum, and millets since 1973/74 in Tanzania is prompted by such a concern for "national" food security.

9. The precise percentage change in the marketed surplus would of course depend on the distribution of production. For producers already on the verge of subsistence in normal years, decline in production would mean a net requirement for purchased foods. Whether this requirement is fulfilled would depend on (1) whether supplies are available in rural markets for purchases and (2) whether the rural population has the income to purchase the food.

10. Especially given the existence of a strategic reserve of 100,000 tons, which is roughly equivalent to eight months' urban consumption requirements, it is unlikely that the turnover period of "excess" domestic stocks, especially of the less preferred minor crops, would be less than one crop year.

11. J. W. Mellor, "Food Price Policy and Income Distribution in Low-Income Countries," chapter 10 in this volume.

12. It should be emphasized that an increase in average yield, with no change in the coefficient of variation, may have little effect on the level of food security (defined as the probability of having to draw down stocks to meet current consumption); that is, the variance would increase correspondingly. An extensive analysis of policy options in Zambia showed that while expected production could be substantially increased, the probability of meeting annual requirements in any one year was essentially unchanged.

13. For instance, in Lesotho, even though research results and farmers' experience shows an attractive return only at very low levels of fertilizer use, much too high dosages of fertilizer are promoted by the Ministry of Agriculture. In Tanzania, there is evidence that in the recent past, arbitrary policy decisions have included exclusive reliance on monoculture (where farmers' experience suggests that consortia of crops have more reliable yields) and indiscriminate promotion of fertilizer even in areas where its marginal productivity is very low. A less monolithic approach is now being taken, with the encouragement of drought-resistant crops such as millets and sorghum in areas

of unreliable rainfall. It is not clear, however, that at the local level sufficient appreciation exists of the adverse effects coercive efforts are likely to have on the individual farmer's efforts to assure food security for himself and his family.

14. Augmented by restriction of seed supply for other crops which farmers may wish to produce but which are not currently in demand in urban areas.

15. Because the international literature refers to fluctuations relative to total production and not official sales, the absolute level of stock requirements may be of the same general order of magnitude. But, at the same time, the parastatal might be required to hold stocks equivalent to 100 percent of its sales in order to cope with a 5 percent variation in total production.

16. Again, the acceptability of cassava, or even sorghum or millets, as a component of the strategic reserve should be the result of a conscious decision.

17. See Schultz (chapter 23), Evenson (chapter 24), and CIMMYT Economics Staff (chapter 25) in this volume.—Ed.

REFERENCES

Coopers and Lybrand Associates Limited, Management and Economics Consultants. 1977. "Grain Storage and Milling Report." London.

Johnson, D. Gale. 1981. "Grain Insurance, Reserves, and Trade: Contributions to Food Security for LDCs." In *Food Security for Developing Countries,* edited by Alberto Valdés, chap. 12. Boulder: Westview Press.

Konandreas, Panos, Barbara Huddleston, and Virabongsa Ramangkura. 1978. *Food Security: An Insurance Approach.* Research Report 4. Washington, D.C.: International Food Policy Research Institute.

Lele, Uma, and John W. Mellor. 1964. *Estimates of Change and Causes of Change in Foodgrains Production: India, 1949–50 to 1960–61.* Cornell University International Development Bulletin, no. 2. Ithaca.

Reutlinger, Shlomo. 1977. "Food Insecurity: Magnitude and Remedies." Abstract, World Bank Staff Working Paper, no. 267. Washington, D.C.

Valdés, Alberto, ed. 1981. *Food Security for Developing Countries.* Boulder: Westview Press.

15

Exports and Economic Development of Less Developed Countries

INTRODUCTION

The question of how far exports, particularly primary exports, are capable of providing the underdeveloped countries with a satisfactory basis of economic development has been extensively discussed during the last two decades and may still be regarded as something of an open question. Prima facie, the broad facts relating to the export and development experiences of these countries during the period seem to support those who advocated policies of freer trade and export expansion rather than those who advocated policies of protection and import substitution. Thus, despite the "export pessimism" of the latter, which persisted well into the 1960s, the period 1950–70 has turned out to be a period of very rapid expansion in world trade, and those underdeveloped countries that responded to the buoyant world market conditions have been able to expand their exports rapidly, typically above 5 percent per annum. This export expansion included not only the primary exports produced by the large mining and plantation enterprises but also those produced by the small peasant farmers. In addition, a smaller group of countries have expanded their exports of manufactured and semi-processed products. Furthermore, the countries that expanded their exports have also tended to enjoy rapid economic development, and significant correlations have been found between the growth of export and the growth of national income among the underdeveloped countries by cross-section studies; by time-series studies; or by a combination of both methods (Emery 1967; Maizels 1968; Kravis 1970a, 1970b;

HLA MYINT is professor of economics, London School of Economics.

Reprinted from *Economic Growth and Resources, Volume 4: National and International Policies,* edited by Irma Adelman. Copyright © International Economic Association, 1979, and reprinted by permission of St. Martin's Press, Inc., and Macmillan Press Ltd. Published with minor editorial revisions by permission of the author.

and Chenery 1971). Conversely it has been found that the countries that concentrated on import substitution tend to have lower rates of growth than those that expanded their exports (Balassa 1971).

In this paper, we shall be concerned both with the causal analysis and with the policy implications of the relationship between the exports and the economic development of the underdeveloped countries. In section I, we shall consider this relationship, not as a simple one-way causation running from exports to economic development, but as a mutual interrelationship involving other important elements of the economic system. In section II, we shall argue that it is necessary to take into account not only the "direct gains" from trade in terms of the allocative efficiency of resources but also the "indirect effects" of trade on the productive efficiency of resources in a broader sense in order to have a better understanding of why the export-oriented underdeveloped countries tend to enjoy a more rapid rate of economic growth than those that have pursued import-substitution policies. In section III, we shall argue that the expansion of peasant exports from the "traditional sector" of the underdeveloped countries tends to promote economic development, directly by a fuller and more effective utilization of their under-utilized resources in the "subsistence sector" and indirectly by extending and improving their domestic economic organization, which is incompletely developed. In section IV, we conclude that the outcome of our analysis is to support the prima facie view that freer trade and export expansion policies tend to promote the economic development of the underdeveloped countries.

I. POSSIBLE CAUSAL RELATIONSHIPS

A positive statistical association between the expansion of exports and the growth of national incomes among the underdeveloped countries does not tell us much about causal relationships. It may mean (*a*) that export expansion is the cause of economic development; or (*b*) that economic development is the cause of export expansion; or (*c*) that both are caused by some third factor. Further, the causal link running in each direction can be interpreted in a variety of ways. Without attempting to be exhaustive, we may review some of these potential elements in the mutual interrelationships between export and economic development.

(*a*) The hypothesis that the expansion of exports is the cause of economic development may be interpreted in three ways: (*a*.1) The first is the conventional free trade argument that the expansion of exports according to comparative costs will increase the direct gains from trade and thus help to promote economic development. (*a*.2) Next, we have the widely held belief that exports contribute to economic development mainly by providing the underdeveloped countries with the foreign exchange necessary for the purchase of capital goods and inputs from abroad. (*a*.3) Third, we may think of the "indirect effects" of freer trade and export expansion on domestic productive efficiency, such as the "educative effect" of an open economy, facilitating the spread of new wants and activities and new technology and new economic organization; or the gains in the productivity and the economies of scale from specialization for the export market.

It is fair to say that $(a.1)$, viz., the direct gains from trade through a more efficient allocation of resources, provided the theoretical basis on which freer trade policies were advocated for the underdeveloped countries during the last two decades. Yet the causal link running from the direct gains from trade to economic development appears to be rather weak. During the decades 1950–70, the growth rates of the underdeveloped countries with rapid export expansion have been typically between 5.5 percent and 7 percent, while those with slow or stagnant exports have been about 3–4 percent; and the difference in the growth rates between the two types of countries became more pronounced during the decade 1960–70 (cf. Chenery 1971; Balassa 1971; and Meier 1970). On the other hand, it seems to be widely held that the static gains from the removal of the distortions created by protection tend to come out as a rather "small" percentage of the aggregate national product (Harberger 1959; and Corden 1975). Even if the whole of this gain is reinvested, this would seem to be too small and short-lived to explain the difference in growth rates between the successful and the unsuccessful countries. Later, we shall see that the picture may be modified by extending the concept of the "direct" gains and losses to include those arising from the allocation of the investible funds. But even so, we shall find it necessary to go beyond the "direct gains" to the "indirect effects" of the type listed under $(a.3)$.

Leaving $(a.2)$ and $(a.3)$ for later consideration, let us now go on to the second hypothesis (b), viz., that economic development is the cause of export expansion. This may be interpreted in two ways: $(b.1)$ Economic development may lead to an expansion of exports via increasing the supply of exportable goods. This is most easily seen when we assume exports as a constant proportion of a growing national output, but this proposition will hold so long as the share of exports does not decrease sufficiently to counteract the effects of the growth in total output. $(b.2)$ Economic development may act via the increase in demand and the widening of the domestic market, having beneficial effects both upon the domestic industries and upon the export industries. Both these propositions are unexceptionable and may be included in the process of interaction between exports and economic development. The crucial question here is not whether a given increase in national output and income will tend to increase exports, but what is to be regarded as the main cause of this increase in output and income.

This leads up to the third hypothesis (c), viz., that both the expansion of exports and economic development are caused by a third factor. Here we approach the central battleground of the debates on trade and aid policies during the postwar decades, for it turns out that the choice of the "third factor" is nothing short of the choice of a theory of economic development that shapes our views about the relationship between exports and economic development. There are two main theories here.

(*c*.1) The first is deeply rooted in the thinking of the 1950s, when the supply of resources available for capital investment was regarded as the central problem of economic development. In this view, the level of investment (financed out of domestic savings and aid) is regarded as the "third factor" which will increase the total national product according to some assumed capital-output ratio and thus increase the supply of the exportable commodities. Combined with the "export pessimism" and the jaundiced attitude towards international trade which prevailed well into the 1960s, this has led to the familiar "inward-looking" policies.

(i) It was held that an underdeveloped country seeking a rapid rate of economic growth should step up the rate of capital investment oriented towards its internal investment opportunities independently of its external economic opportunities, which were assumed to be unfavorable.

(ii) Earlier on, these internal investment opportunities were supposed to be generated by a "balanced growth" between the domestic manufacturing and the domestic agricultural sectors. But given the prevailing faith in the power of modern manufacturing industry based on capital-intensive advanced technology to promote economic development, both domestic agriculture and primary exports came to be increasingly neglected in favor of the import-substituting manufacturing industry.

(iii) The expansion of the modern manufacturing industries, however, required imports of capital goods and other inputs and materials. Thus, the importance of foreign exchange as the means of acquiring the imported capital goods came to be emphasized. However, since it was assumed that the world market demand for primary exports would be extremely inelastic, it did not lead to export expansion policies (suggested by *a*.2 above), but to the restrictions on the imports of consumer goods to save foreign exchange for the imports of capital goods and to the renewed pleas for greater international aid to fill the "foreign exchange gap."

(*c*.2) The alternative view regards both the expansion of export and economic development as being caused not so much by the level of domestic investment per se as by the appropriate domestic economic policies that enable a country to allocate its resources more efficiently, taking into account both its internal and external economic opportunities. This is the position adopted by most economists who advocated policies of freer trade during the last two decades. They emphasized that an underdeveloped country would not be able to reap fully the direct gains from international trade unless it also pursued appropriate domestic policies, viz.: (1) appropriate pricing policies in both factor and product markets that remove discriminatory treatment of different lines of economic activities, both within the domestic economy and between the domestic and the foreign trade sectors; (2) appropriate macroeconomic policies that prevent overvaluation of the foreign exchange rate and avoid the need for ad hoc import restrictions and quantitative controls resulting in an unplanned protection arising from a chronic foreign exchange shortage; and (3) appropriate investment policies that reflect the country's comparative costs by valuing both the inputs and the outputs of investment projects at world market prices (cf. Johnson 1967; Little, Scitovsky, and Scott 1970; and Little and Mirrlees 1974).

It is fair to say that there are considerable empirical and theoretical grounds for supporting (*c*.2) against (*c*.1).

First, most economists would agree nowadays that the earlier approach to economic development in terms of increasing the *supply* of resources available for capital investment has turned out to be seriously inadequate and that in a fundamental sense economic development involves the raising of the productive efficiency in the use of resources. Despite the low ratios of domestic savings to their GDPs in the early 1950s, the overall rate of growth of the underdeveloped countries during the 1950–60 decade was about 4.5 percent per annum—much higher than was generally expected. This growth rate in GDP was maintained between 4.5 percent and 5 percent during the 1960–70 decade. During this time, the ratio of their domestic savings also increased to some 15 percent of the GDPs. A comparison of the savings ratio and the growth rates serves to bring out the modest contribution of capital to economic development. Thus if we follow Cairncross (1962) and assume the average rate of return on capital investment in the underdeveloped countries to be 10 percent, then the 15 percent saving ratio would have contributed no more than 1.5 percent of the total growth rate between 4.5 percent and 5 percent. If the level of domestic investment cannot explain the average growth rate of the underdeveloped countries as a whole, it is also not able to explain the difference in the growth rates between the successful and the unsuccessful countries. There does not seem to be any striking difference in the level of domestic investment between the successful countries that pursued export expansion policies and the less successful countries that pursued import-substitution policies (Bruton 1967).

Second, in the perspective of the 1950–70 decades, during which world trade expanded at an unprecedented rate, it is now difficult to maintain the extreme type of export pessimism that dominated discussions on trade and development well into the 1960s. Of course, it is anyone's guess whether world trade will resume its previous rate of expansion after the current recession. But even so, a more reasonable hypothesis would be that "under normal conditions" most underdeveloped countries could expand their exports, provided they pursued appropriate domestic economic policies. Kravis (1970a, 1970b) has taken considerable pains to dispel the fatalistic view that the underdeveloped countries' exports depend solely on the world market factors over which they have no control. He has shown that successful export performance among the underdeveloped countries during the postwar decades is determined mainly by domestic supply factors and "competitiveness." Thus the successful countries expanded their primary exports mainly by increasing their shares of the world market for their traditional exports and, to some extent, by diversifying into new exports. De Vries (1967) has also shown that some underdeveloped countries have succeeded in expanding their ''minor'' exports substantially by controlling their domestic rate of inflation with given exchange rates.

Third, the limitations of the concept of the ''foreign exchange gap,'' popularized by UNCTAD economists during the 1960s, have been extensively discussed (for example, in Findlay 1973), and here we need touch upon only one aspect of the subject. We have seen above that the once-for-all gains obtainable from the removal of the static distortions are likely to be a small percentage of the GDP and that therefore the reinvestment of these once-for-all gains would have a very small effect on economic development. But this is not to deny that losses can be great and economic development can be seriously retarded when the *annual* flow of investible funds, including the supply of foreign exchange, is directed into unproductive channels by the import-substitution policies. This suggests an important qualification of the role of foreign exchange in economic development.

In the standard case in which a country acquires imports as final consumption goods, the ''direct'' gains from trade can be readily identified with the cheaper imports leading to a greater consumers' welfare. In the case of an underdeveloped country wishing to acquire the imports, not for final consumption but as capital goods and inputs for further production, the concept of ''direct'' gains from trade has to be defined more carefully. Following Hicks (1959), we may picture the country as being faced with a problem of choice in two stages. In the first stage, the country has the choice of converting its given savings into capital goods either by producing them at home or by converting the savings into exports which can be exchanged for the imported capital goods from abroad. Having acquired the capital goods in one way or another, the country has the further choice of either using them to produce the final consumer goods at home or using them to produce exports which can be exchanged for the imported consumer goods from abroad. Comparative advantage enters into both stages of choice, and the country will not maximize its final consumer gains unless the correct choices have been made at both stages.

We can now see why we need to qualify the popular argument that foreign exchange plays a crucial role in economic development by enabling an underdeveloped country to purchase the much-needed capital goods and technical inputs that it could not produce at home except at a prohibitive cost. This is true enough as far as it goes, but the gain from the opportunity to import the capital goods is only a potential gain; it can be turned into an actual welfare gain only if the imported capital goods are *economically* suitable, that is to say, only if the technology they represent fits in with the factor proportions of the country and the final products they produce are in accordance with the country's comparative costs. If the ''wrong'' type of capital goods is imported because of the protection of the domestic manufacturing industry and if, moreover, the importation of the wrong type of capital goods is actively encouraged by the provision of cheap capital funds and foreign exchange by government policies, it is not difficult to see that the potential gain from the availability of foreign exchange may be reduced or even turned into an actual loss (cf. Johnson 1967; and Myint 1969). Bruton's study (1967) of import substitution of five Latin American countries vividly illustrates this point. According to him, during the war period 1940–45, when these countries could not obtain their supplies of capital equipment and materials from abroad, the productivity of domestic resources rose through a process of improvisation and adaptation of the existing capital equipment to fit the local market size and the product to fit the local market demand, resulting in a fuller utilization of the productive capacity. In contrast, in the post-1955 period, when import-substitution policies were actively pursued with overvalued exchange rates and subsidies on the import of capital goods and technically necessary inputs, there was virtually no increase in the "pure" productivity of resources. On the contrary, with a growing inappropriateness of input mix of production due to the overvaluation of currency and distortions in the factor markets, and with a growing inappropriateness of the composition of output due to protection and to a decline in competition, "an industrial structure has tended to emerge that is so alien to factor endowments that full utilization of existing capacity came to depend more, not less, on a constant flow of imports" (Bruton 1967, 1112–13).

In this section we have been considering the mutual interrelationship between exports and economic development in terms of its constituent elements: (*a*) the effects of export expansion on economic development; (*b*) the effects of economic development on export expansion, through both an increase in total supply and an increase in total demand; and (*c*) the effects on both export expansion and economic development of a "third factor," notably the appropriate domestic economic policies. We have found that the causal link running from the direct gains from trade to economic development (*a*.1) is rather weak but that the static theory of comparative costs can be effectively extended to explain the *negative* proposition as to why the countries that pursued import-substitution policies tended to enjoy a lower growth rate through a misdirection of the investible resources. This has also suggested an important qualification to the proposition (*a*.2), viz., exports can contribute to economic development through the foreign exchange earnings only if these are correctly reinvested, according to comparative costs. We now turn to (*a*.3), the "indirect effects" of exports on economic development through their effects on the productive efficiency of the domestic economic system.

II. INDIRECT EFFECTS OF TRADE ON PRODUCTIVE EFFICIENCY

If we adhere strictly to the assumption of the static comparative costs theory and the "perfect competition" model on which it is based, then there would be little scope for the indirect effects of trade to operate on the domestic economic system. In such a model, the country's maximum production possibility frontier is supposed to be determined in an unambiguous manner by the autonomously given resources and technology. The country is assumed to be already on the production possibility curve before trade takes place, and it responds to the opportunity to trade by reversible movements along this curve. Since the country's production possibility curve cannot be shifted except by autonomous changes in the supply of resources and technology, all that trade can do is to change the allocation of resources. The possibility that the country's productive efficiency in a broader sense might be affected, either favorably or unfavorably, by the process of adapting its capacity to the requirements of international trade is ruled out by assumption.

(1) The possible increase in productive efficiency through the "educative effect" of an open economy in facilitating the spread of new skills and technology is ruled out if we adhere strictly to the assumption of "perfect knowledge," which implies a zero cost of search for information and transmission of knowledge. Thus "given" the technology in the outside world, the producers within a country supposedly would be able to adopt this technology in an instantaneous and costless manner.

(2) The possible increase in productive efficiency through the response of the domestic entrepreneurs to the pressure of foreign competition is similarly ruled out. Each producer is already supposed to be operating on his production function representing the minimum combinations of inputs required to produce a given output. With the given technology, these production functions are assumed to be determinate and can be changed only through an autonomous technological change.

(3) The possible increase in productive efficiency through the process of specialization for a wider export market in a genuine sense, involving the adaptation in the quality of resources and investment in durable productive capacity and human capital to meet a specific international demand, is ruled out by the static assumptions. Movements along the production possibility curve are reversible, and resources of each type are assumed to be homogeneous and divisible. There are no differential rents arising out of the differences in the quality of the resources and their suitability to international demand; all that adjustments to trade can bring about are the changes in the relative scarcities of different types of resources or their scarcity rents. The assumption of perfect divisibility of resources rules out the economies of scale. If there are no gains in productivity from specialization for the export market, there are also no risks or commitments in specialization. The economic system is assumed to be able to allocate resources either to expand or to contract exports in a smooth and flexible manner.

(4) Finally, it is assumed that with a given physical endowment of resources and given technology, the production possibility curve of a country cannot move upwards through a fuller utilization of these given resources. Resources are assumed to be fully employed, given the assumptions of their perfect mobility and the flexibility of their prices. This is reinforced by the assumption of "perfect knowledge," according to which the producers are supposed to know about the availabilities of the resources within the country in the same way as they are supposed to know about the available technology in the outside world. Further,

the "perfect competition" model has an implicit assumption that is of considerable importance to our later analysis, viz., that the domestic economic organization of a trading country, including its market system and network of transport and communication, is sufficiently well developed to bring the physically available resources into full utilization. This excludes the possibility that the process of international trade might introduce improvements of the organizational framework of the domestic economic system resulting in a fuller utilization of its existing resources.

Let us now consider how far we can combine these indirect effects of trade on the broader productive efficiency of the domestic economic system with the direct static gains from trade.

(1) The "educative effect" of an open economy arises not only from free commodity trade but also from the auxiliary functions of trade in facilitating the spread of new wants and activities and new methods of production and economic organization through a greater degree of international mobility of resources and human contacts. The conditions favorable for the "educative effect" are thus not identical with the static optimum conditions of free trade. A moderate degree of tariff or indirect controls on trade (as distinct from detailed quantitative controls) need not reduce the "educative effect," provided other aspects of international economic contacts are relatively free. Similarly, some of the concepts of static welfare analysis, such as "the tariff equivalent" of quantitative controls, may be inappropriate for the analysis of the "educative effect." But on the whole there does not seem to be any inherent logical difficulty in combining it with the static gains from trade.

(2) It is not clear how far the possible increase in productive efficiency or X-efficiency arising from the response of domestic entrepreneurs to the "pressure of foreign competition" can be grafted on to the static framework of analysis. A protectionist can argue that while the sheltering from foreign competition may lead to entrepreneurial inefficiency, exposure to the pressure of competition, particularly from the producers in the advanced countries, may present the domestic producers of the underdeveloped countries with an excessive challenge which outclasses them, and that they are likely to succumb rather than respond to such a challenge. In spite of this indeterminacy, there can be no disagreement about the proposition that under the typical conditions in which import-substitution policies are pursued by means of a network of controls, the domestic entrepreneurs of the underdeveloped countries would find it more profitable to direct their energies to the task of procuring the government licenses or exploiting the loopholes in the regulations rather than to the task of raising productivity. Thus a removal of such controls and a redirection of entrepreneurial incentives would tend to raise productive efficiency in addition to the purely static gains from the correction of the distortions.

(3) We shall definitely have broken out of the static framework of the comparative costs theory when we try to broaden the argument for export expansion by incorporating the gains from "specialization" for a wider export market involving the adaptation of the quality of resources and the building up of special

skills and productive capacity to meet the specific requirements of the export market. These gains represent a nonreversible outward shift of the production possibility curve of a country in the direction of export production. In order to obtain them, a country would usually have to commit its resources to export production on a large enough scale to overcome the indivisibilities in the production process or in the auxiliary facilities such as transport and communications. This is likely to take the country beyond the static optimum point indicated by the given comparative advantage before the process of specialization takes place and is therefore likely to impose some initial sacrifice of static gains from trade and also the risks of specialization. On the other hand, there is a general presumption, dating back to Adam Smith, that a country is likely to increase its productive efficiency by taking advantage of the opportunity to exploit the economies of scale and specialization in certain selected lines of export production for a wider world market instead of matching its resource allocation to the pattern of domestic demand inside the narrow home market.

There is some empirical evidence to suggest that indirect effects of this nature are more relevant for the understanding of the higher growth rates of the export-oriented underdeveloped countries than the conventional approach in terms of the removal of the static distortions. Bhagwati and Krueger (1973) have found that some of the export-oriented countries, such as South Korea, appear "to have intervened virtually as much and as 'chaotically' on the side of promoting new exports as others have on the side of import substitution" and that their success cannot be attributed to "the presence of a neo-classically efficient allocating mechanism *in toto* in the system." Similarly, one may deduce that the need for the "free trade zones" in other export-oriented countries such as Taiwan and Mexico implies that the rest of the economy, outside these zones, is not so free. Bhagwati and Krueger suggest that the success of the export-oriented countries may be attributed to the factors emphasized by the older writers on international trade, viz., the built-in budgetary constraints preventing excessive export subsidization; the relatively greater use of indirect rather than direct interventions in export promotion; the pressure of international competition on the producers of the subsidized exports; and the economies of scale.

The protectionist case against primary exports is based entirely on the alleged unfavorable indirect effects of such exports on the long-run productive efficiency of the domestic system contrasted with the external economies and "linkages" from the manufacturing industry. Since the whole argument is based on the presumption that it is normally worthwhile for an underdeveloped country to sacrifice some of the direct gains from trade to secure these longer-run indirect benefits, the issues of the debate are not properly joined so long as the free trade argument is limited to the direct gains from trade. We have to go on to consider the adverse indirect effects attributed to the expansion of primary export production.

The belief that primary exports tend to create the colonial-type "export enclaves" and tend to "fossilize" the economic structure of the underdeveloped countries still exerts a powerful influence and requires a critical examination.

Basically, the argument envisages the primary exports as being produced by the larger mining and plantation enterprises operated by direct private foreign investment. The objections usually advanced against such exports may be summarized as follows: (1) There may be "unfair" distribution of the gains from trade because the foreign investors have been able to obtain cheap concessions to exploit the country's natural resources, paying less than the "economic rent" for the use of these resources. (2) An "export bias" may be created because too large a portion of government revenues from the exports and the external supply of capital has been devoted to the type of social overhead capital, such as transport and communications and research, which benefits the larger enterprises in the export sector rather than the small peasants in the domestic sector. (3) This tends to "fossilize" the production structure, trapping resources in special lines of primary export production in the face of a long-run decline for the world market demand for primary products. (4) Primary export production tends to result in "enclaves" with few "linkages" with the rest of the economy, in contrast to the manufacturing industry's capacity to create such linkages and external economies.

(1) Leaving aside the complications introduced by the exhaustible natural resources, the possibility of "unfair" distribution of the gains from trade is an argument for the government of an underdeveloped country to charge the full economic rent on its natural resources rather than an argument against allowing either the foreign or the domestic investors to use these resources for the production of primary exports (see Myint 1972 for a fuller discussion of the issues). It does, however, serve to bring out a weakness in the standard Heckscher-Ohlin type of trade theory, based on the assumption of homogeneous resources and concerned with the effects of trade on the relative scarcities or the scarcity rents of the different types of homogeneous resources. Typically the "economic rents" on the natural resources required for primary exports largely consist of "differential rents" arising from the qualitative differences in a country's natural resources, including their location. This means that the direct gains from trade from primary exports are likely to be much larger than the conventional gains in terms of allocative efficiency based on the assumption of homogeneous resources. If withdrawn from export production, the earnings of special types of natural resources, such as land bearing mineral deposits or suitable for particular types of tropical product, would drop sharply in their alternative uses in domestic production. Thus so long as the governments of the underdeveloped countries are able to extract the full economic rent from the investors (and they have increasingly proven their ability to do so), this argument strengthens the case for primary export production.

(2) There are two different versions of the "export bias" argument. The first is directed against the flow of external capital into the underdeveloped countries for the construction of social overhead capital to facilitate the primary exports from the mines and plantations. But since this type of capital investment is "induced" by the specific purpose of expanding these exports and would not

have been available to the country for other purposes, the country does not suffer from foregone opportunities of investment elsewhere. It is true, for example, that a railway line running straight from the seaport to the mines, without the "feeder" lines to develop the surrounding countryside, may give little stimulus to domestic economic development. But it would not have come into existence without the mines, and it does not hinder domestic development in any way. It may be easier subsequently to build the feeder lines because of the existence of the trunk routes than if they had not existed. Further, the mines should provide the government with the revenues for domestic economic development. This brings us to the second version of the "export bias" arguments, viz., that in the past the colonial governments have devoted too much of the country's revenue for the benefit of the larger mining and plantation enterprises in the export sector rather than for the benefit of the small peasants in domestic agriculture. Whatever the truth of this allegation, it would now be entirely within the control of the independent governments of the underdeveloped countries to correct this bias against the "domestic sector." Unfortunately, however, in many underdeveloped countries the "domestic sector" has come to be identified with the modern manufacturing industry in the urban centers rather than with the small peasant producers in traditional agriculture. Thus the small peasant producers in the traditional sector may have suffered doubly—under the old "dualism" because of the colonial governments' encouragement of the larger-scale mining and plantation enterprises and under the new "dualism" because of the independent governments' encouragement of the modern manufacturing industry.

(3) The argument that the underdeveloped countries would tend to "fossilize" their domestic economic structure by specialization in primary exports has two aspects. The first stems from pessimistic views of the long-term patterns of world market demand for primary exports. These have proved unfounded in the light of the experiences of the postwar decades and the present concern with raw-materials shortage and exhaustible resources. In particular, Porter (1970) has shown that while the demand for primary products may be typically very price-inelastic or very income-inelastic, they are not both price- and income-inelastic and that the advanced countries of North America and Western Europe have managed to become dominant producers of the income-elastic primary products at least partly through an ability to cut costs. The second strand of the "fossilization" argument stems from the view that it is more difficult to shift resources out of primary production than out of manufacturing industry. Here, as we have seen, any genuine process of specialization in an export production, involving specific investment in durable productive capacity and infrastructure, is bound to give rise to heavy costs of readjustment in output, not taken into account in the conventional trade model with reversible movements along the production possibility curve. The question is whether it is significantly more difficult to shift out of specialization in primary production than from specialization in manufacturing industry. Although this is widely assumed to be true, it is difficult to find cogent theoretical or historical reasons for it. Historically, it is not difficult to find

instances of successful switches from one line of primary exports to another. Thus Malaysia switched from coffee production to her highly successful rubber industry and is in the process of switching at least some of her land to palm oil. Brazil is in the process of shifting from coffee to soya beans. Conversely, the British textile industry has for long been a well-known example of "fossilization" in manufacturing industry following an early leadership in the export market (cf. Allen and Donnithorne 1957; and Kindleberger 1961. See also, however, Chenery 1976, where it is argued that few governments would have sufficient foresight to switch out of primary exports in time, as though this "foresight" in an operational sense could be obtained from broad cross-section average relationships between the level of per capita incomes and the share of manufacturing industry in GNP).

(4) Finally, we come to the widely held opinion that primary exports are inherently less capable than manufacturing industry of creating "linkages" with the rest of the economy. On the face of it, there are as many technical possibilities for primary export production to create "forward linkages"—say, by increasing the supply of locally produced raw materials for processing industries—as for manufacturing industry to create "backward linkages" by setting up demand for locally produced inputs. How far these technical possibilities in either case can be economically realizable depends on the relative costs and prices of inputs and outputs; in other words, on the comparative costs principle that operates on the pattern of vertical international specialization between different intermediate stages of production in the same way that it operates on the pattern of horizontal international specialization between different final products. Advantages in location and the saving of transport costs may help with the creation of linkages. Here, as we shall see, the expansion of primary exports by peasant producers located in the traditional sector of the underdeveloped countries may be more favorable for the generation of some kinds of linkages than manufacturing industry located in the urban centers.

III. PEASANT EXPORTS AND ECONOMIC DEVELOPMENT

So far we have been concerned with the exports from the "modern sector" of the underdeveloped countries. We may now turn to the peasant exports from the "traditional sector" and their effect on economic development. At this point, it is necessary to re-examine the two related assumptions of the conventional trade theory, viz., that the domestic economic organization of the trading country, including its market system and its internal network of transport and communications, is fully developed before trade takes place; and that its resources are fully employed, given that they are internally mobile and that their prices are flexible. These two assumptions effectively rule out the possibility that the process of "opening up" an underdeveloped country to international trade will extend and improve its domestic economic framework and thereby enable it to utilize its "given" resources more fully and effectively.

There is, however, considerable historical evidence that in the early stages of development, peasant exports from the underdeveloped countries typically expanded, not by a contraction of domestic outputs, as implied by the full employment assumption of the comparative costs theory, but by bringing in the hitherto underutilized resources of the "subsistence sector," as suggested by the "vent-for-surplus" theory (Myint 1958). On the demand side, there were the foreign merchants and entrepreneurs, who actively searched for the sources of the exportable peasant products from the underdeveloped countries and subsequently built up the channels of trade consisting of the chains of transport and communications and retail-wholesale links that connected the traditional sectors of these countries to the world market. On the supply side, there were the peasant producers, who responded vigorously to the stimulus of new or cheaper imported consumer goods by clearing more land and by devoting the labor they did not require for subsistence production to export production. This historical process started in the late nineteenth century, but its potentialities were by no means exhausted in the postwar decades, particularly for the peasant export economies of Africa and Southeast Asia (Lewis 1969; and Myint 1972). Thus the rapid rate of economic growth of some of the export-oriented countries, such as the Ivory Coast and Thailand during the decade 1960–70, may be largely accounted for in terms of the vent-for-surplus theory. The direct gains from trade obtainable by bringing in the underutilized resources from the subsistence sector into export production offer a more convincing explanation of their growth than the conventional gains in terms of an increase in the static allocative efficiency of the "given" and fully employed resources.

The protectionist reaction to this type of export expansion policy would be to say that since the extension of peasant production takes place on the basis of unchanged agricultural techniques, it would sooner or later be stopped by the limits of the cultivable hinterland and that in order to forestall this and obtain a self-sustained basis for economic growth, the proceeds from peasant exports should be used to finance domestic industrialization. There has been no shortage of underdeveloped countries that have followed this protectionist domestic industrialization policy, in spite of the fact that their resource endowments and stage of development would seem to indicate that they could still greatly benefit from the expansion of peasant exports. Thus Ghana and Burma have used the device of state agricultural marketing boards with considerable ruthlessness to extract the revenue from their peasant exports of cocoa and rice for the purposes of domestic industrialization. Their poor performance both in export and economic growth contrasts sharply with that of their respective neighbors, the Ivory Coast and Thailand.

The idea that unlike manufacturing industry, peasant export production is incapable of generating "linkages" and should therefore be treated merely as the milch cow for the domestic industrialization programs is very widespread. It arises from a failure to appreciate the strategic position that peasant exports occupy in the incompletely developed domestic economic organization of the underdeveloped countries.

First, there is the familiar "dualism" in their economic organization, characterized by the co-existence of a "modern sector," consisting of the larger-scale economic units engaged in mining, plantation, or manufacturing industry, and a "traditional sector," consisting of small economic units engaged in peasant agriculture and handicraft industries. The larger economic units in the modern sector are well provided with the auxiliary facilities such as transport, communications, power, and so on, and have access to a modern banking system and organized capital market. Even with a neutral government policy, the small economic units would have been handicapped by the incomplete development of the domestic economic organization, by the higher transport and transactions costs in retail trade, by the higher risks and costs of organizing a credit supply for small borrowers, and by the higher administrative costs of providing transport and communications and other public services to the widely dispersed small economic units in the rural areas. These natural handicaps are greatly aggravated when the government pursues a deliberate policy of encouraging domestic industrialization by discriminating in favor of the larger economic units engaged in the import-substituting manufacturing industry.

Second, although all types of peasant agriculture are handicapped compared with the economic activities in the modern sector, peasant export production is relatively favored under free market conditions compared with subsistence production and with cash production of food crops for the local or the domestic market. Thus peasant export crops are better served with marketing, transport, and credit facilities than are cash crops for the local market. Few other types of peasant products would enjoy a nationwide marketing network normally associated with the export products. Now the vent-for-surplus theory is concerned with the process whereby the surplus land and labor not required for "subsistence production" are attracted into export production in the early stages of the development of the traditional sector. At a later stage of development, it becomes necessary to consider also the relationship between the staple export crops and the development of cash crops for the domestic market and of the new subsidiary export crops. It is in this context that the full employment assumption of the conventional trade theory has to be modified.

An underdeveloped country that has apparently exhausted the more obvious possibilities of extending cultivation into the unused hinterland may nevertheless possess considerable "pockets" of underutilized resources which can be brought into fuller use by improvements in transport and communications and by widening of the local markets that extend the marketing facilities to cover some of the "gaps" in the existing network. In this setting of an incompletely developed domestic economic organization there would be possibilities for *complementary* relationships between the expansion of peasant exports and the development of cash production for the domestic market and the new peasant exports. The full employment assumption stresses the *competitive* relationship in allocating the "given" resources between one use and another. But if there were considerable "slacks" arising from the gaps in the organizational framework, the complementary relationships might be more important than the competitive relationships. Thus it has been observed that in some of the West African countries, the trade credit advanced by the export–import firms to the middlemen for the purchase of cocoa is frequently turned over two or three times to finance the local food crops before it is used for its original purpose of purchasing the export crop (Jones 1972, 251). Similarly, the "slacks" in other trading and transport facilities in the staple export crop can be used in the marketing of local food crops and subsidiary exports. Thus the effect of introducing state agriculture marketing boards into the export economies of West Africa and Southeast Asia has been not only to retard the growth of peasant exports by fixing buying prices well below the world market prices but also to repress the growth of the credit and marketing institutions in the traditional sector by replacing the private middlemen with state purchasing agencies at the local level.

Third, it is true that in the past, peasant exports have expanded on the basis of more or less unchanged methods of production, relying mainly on the possibility of extending cultivation into unused land. But this is not to say that peasant exports are inherently incapable of technical improvement. They may also indirectly contribute to the productive efficiency of the resources in the traditional sector in a number of ways. (i) First, there are the gains in productive efficiency from the division of labor and specialization through the widening of the local markets. In the initial stages, peasant producers entered into export production on a spare-time basis while continuing to produce all their subsistence requirements. At a later stage, some of them would specialize or devote all their resources to export production, so setting up a cash demand for locally produced foodstuffs. The consequent development of specialized food producers for the local markets tends to stimulate the growth of a marketable food surplus for the country as a whole, so enabling a labor-abundant underdeveloped country to take up its potential comparative advantage in labor-intensive exports of manufactured and semi-processed products. (ii) The growth of cash crops both for the export and the domestic markets should in turn facilitate the adoption of the improved cash-intensive methods of production based on purchased inputs, such as high-yielding seeds, fertilizer, and more efficient farm equipment. So far, these improved agricultural methods introduced by the "Green Revolution" have been mainly taken up by the food-deficit countries for the purpose of attaining "national self-sufficiency" in food, based on a considerable amount of agricultural protection. Freer trade and export expansion policies have an important role of counteracting this tendency towards import substitution in agriculture and in bringing out the potential export possibilities of peasant agriculture after the introduction of these new methods. (iii) Finally, the development of both the product and the factor market within the traditional sector, which is perhaps the most important indirect effect of the expansion of peasant exports, should serve to improve the information network linking up the technology available in the outside world with the local availability of resources within the traditional sector. Since agricultural technology is "location-specific," this should facilitate the adaptation of new technology to local requirements.

To sum up: the expansion of peasant exports would tend to promote the economic development of the underdeveloped countries, both directly, through the "vent-for-surplus" mechanism, and indirectly, through the extension and improvement of their domestic economic organization, particularly in relation to the traditional sector of their economy. These indirect effects will continue to be important as long as these countries are characterized by a pronounced "dualism" between the modern and the traditional sectors and so long as a considerable proportion of resources in the traditional sector still remains in subsistence production.

IV. EXPORT POLICY AND ECONOMIC DEVELOPMENT

The result of our analysis is to support the prima facie impression that the economic development of the underdeveloped countries is likely to be promoted by freer trade and export-expansion policies rather than by protection and import-substitution policies. This result has been arrived at both in terms of the "direct gains" from trade within the conventional framework of the comparative costs theory and in terms of the broader "indirect effects," which take us beyond the confines of the conventional theory.

(1) The conventional approach yields the negative proposition that the countries that pursue import-substitution policies tend to suffer from a lower rate of growth. Thus, although the re-investment of the once-for-all gains obtainable from the removal of static distortions may only have a small effect in increasing the growth rate, the misallocation of the *annual* flow of investible funds into inappropriate channels dictated by the import-substitution policies is likely to lead to serious losses and a significant retardation of economic growth.

(2) In order to put forward the positive proposition that freer trade and export expansion policies are likely to lead to a higher growth rate, it is necessary to bring in the "indirect effects" of trade. These include the "educative effect" of the open economy, the pressure of foreign competition in stimulating productive efficiency, and the economies of scale and increasing returns from specializing for a wider export market. The last type of gain involves a commitment of resources on a large enough scale to overcome indivisibilities and may require a pro-export policy instead of a strict neutrality between exports and domestic production implied by the static comparative costs theory. It also involves changes in the production structure that can be reversed only at considerable cost. But, on the other hand, genuine specialization in any line of economic activity, whether in manufacturing or primary products, would entail the commitment of resources and the risk of wrong or excessive investment. Specialization for the export market at least exposes the country to the discipline of international competition. These "indirect effects" are not theoretically demonstrable in the same way as the "direct" static gains from trade and are not susceptible of accurate measurement. But they are based on widely acceptable general presumptions that are, at least in principle, capable of empirical verification.

Our analysis of the relationship between the expansion of peasant exports and economic development gives an additional support to our general conclusion. The expansion of peasant exports tends to promote economic development, both directly, through the "vent-for-surplus" mechanism, and indirectly, by improving the domestic economic organization of an underdeveloped country.

REFERENCES

Allen, G. C., and A. Donnithorne. 1957. *Western Enterprise in Indonesia and Malaya*. London: Allen & Unwin.

Balassa, B. 1971. "Trade Policies in Developing Countries." *American Economic Review* 61(2): 178–87.

Bhagwati, J., and A. Krueger. 1973. "Exchange Control, Liberalization and Economic Development." *American Economic Review* 63(2): 419–27.

Bruton, H. J. 1967. "Productivity Growth in Latin America." *American Economic Review* 57(5): 1099–1116.

Cairncross, A. K. 1962. *Factors in Economic Development*. Chap. 4. London: Allen & Unwin.

Chenery, H. B. 1971. "Growth and Structural Change." *Finance and Development Quarterly* 8(3).

————. 1976. *Transitional Growth and World Industrialization*. Nobel Symposium on the International Allocation of Economic Activity, Stockholm.

Corden, W. M. 1975. "The Costs and Consequences of Protection. A Survey of Empirical Work." In *International Trade and Finance: Frontiers for Research*, edited by Peter B. Kenen. Cambridge: Cambridge University Press.

De Vries, B. A. 1967. *The Export Experience of Developing Countries*, World Bank Staff Occasional Papers, no. 3. Washington, D.C.

Emery, R. F. 1967. "The Relation of Exports to Economic Growth." *Kyklos* 20, fasc. 2.

Findlay, R. 1973. *International Trade and Development Theory*. Chap. 10. New York: Columbia University Press.

Harberger, A. C. 1959. "Using the Resources at Hand More Effectively." *American Economic Review, Proceedings* 49(2): 134–46.

Hicks, J. R. 1959. *Essays in World Economics*. Chap. 8. Oxford: Clarendon Press.

Johnson, H. G. 1967. "The Possibility of Income Losses from Increased Efficiency or Factor Accumulation in the Presence of Tariffs." *Economic Journal* 77(305): 151–54.

Jones, W. O. 1972. *Marketing Staple Food Crops in Tropical Africa*. Chap. 9. Ithaca: Cornell University Press.

Kindleberger, C. P. 1961. "Foreign Trade and Economic Growth: Lessons from Britain and France, 1850–1913." *Economic History Review* 14(2).

Kravis, I. B. 1970a. "Trade as a Handmaiden of Growth: Similarities between the Nineteenth and Twentieth Centuries." *Economic Journal* 80(320): 850–72.

————. 1970b. "External Demand and Internal Supply Factors in LDC Export Performance." *Banca Nazionale del Lavoro Quarterly Review* 93(June): 157–79.

Lewis, W. A. 1969. *Aspects of Tropical Trade*. Stockholm: Almqvist and Wicksell.

Little, I., T. Scitovsky, and M. Scott. 1970. *Industry and Trade in Some Developing Countries: A Comparative Study*. London: Oxford University Press for the Organisation for Economic Cooperation and Development.

Little, I., and J. Mirrlees. 1974. *Project Appraisal and Planning for Developing Countries*. London: Heinemann.

Maizels, A. 1968. *Exports and Economic Growth of Developing Countries*. Cambridge: Cambridge University Press.

Meier, G. M. 1970. *Leading Issues in Economic Development*. 2d ed. New York: Oxford University Press.

Myint, H. 1958. "The 'Classical' Theory of International Trade and the Underdeveloped Countries." *Economic Journal* 68(270): 317–37.

————. 1969. "International Trade and The Developing Countries." In *International Economic Relations,* edited by P. A. Samuelson. London: Macmillan.

————. 1972. *South East Asia's Economy: Development Policies in the 1970s.* Harmondsworth: Penguin Books.

Porter, R. 1970. "Some Implications of Post-war Primary Product Trends." *Journal of Political Economy* 78(3):586–97.

16

The Analytics of International Commodity Agreements

JERE R. BEHRMAN

INTRODUCTION

In the United Nations and in other international forums the developing countries have been calling for a "new international economic order." According to the "Declaration" and "Program of Action" adopted by the U.N. General Assembly on 1 May 1974 [U.N. General Assembly resolutions 3201 (S-V1) and 3202 (S-V1)], the new order is to "redress existing injustices" and "make it possible to eliminate the widening gap between the developed and the developing countries." The past and current dominant-subservient economic relations between these two country groups are to be replaced by more symmetrical interaction.[1]

In the ensuing debates and declarations about the "new international economic order," high priority has been given to international commodity trade. This emphasis is not surprising. Although nonpetroleum commodity exports from developing countries account for only about 29 percent of world nonpetroleum commodity exports, they account for about 55 percent of total nonpetroleum exports from the developing countries.

The United Nations Conference on Trade and Development (UNCTAD) has provided a natural forum for the development of positive programs for international commodities. In December 1974 it unveiled a new plan for an integrated commodity program (UNCTAD 1974a, 1974b, 1974c, 1974d, 1974e), which it further elaborated in 1975 (UNCTAD 1975a, 1975b, 1975c, 1975d, 1975e, 1975f).

The broad objectives of that proposal are:

JERE R. BEHRMAN is professor of economics, University of Pennsylvania.

Excerpts from Jere R. Behrman, "International Commodity Agreements: An Evaluation of the UNCTAD Integrated Commodity Programme," in *Policy Alternatives for a New International Economic Order: An Economic Analysis,* ed. William R. Cline (New York: Praeger Publishers for the Overseas Development Council, 1979). Published with minor editorial revisions by permission of the Overseas Development Council and the author.

(*a*) to improve the terms of trade of developing countries and to ensure an adequate rate of growth in the purchasing power of their aggregate earnings from their exports of primary commodities while minimizing short-term fluctuations in those earnings; and

(*b*) to encourage more orderly development of world commodity markets in the interests of both producers and consumers.

The proposal recommends focus on seventeen commodities that cover about three-quarters of the nonpetroleum commodity trade of the developing countries. These include ten "core" commodities (cocoa, coffee, copper, sugar, cotton, jute, rubber, sisal, tea, and tin) and seven "other" commodities (bananas, bauxite, beef and veal, iron ore, rice, wheat, and wool). The ten core commodities together account for about three-quarters of the export value to developing countries of all seventeen commodities; they are storable; and they are recommended for initial individual international stockpiling agreements.

THEORETICAL CONSIDERATIONS

Economic theory provides useful guidelines for considering some important issues about which there is some confusion in much of the speculation about commodity market agreements. Therefore, it is useful to discuss briefly the following four theoretical questions: What are the normative implications of market solutions to economic problems? What are the implications of price stabilization attempts for revenues? Who benefits from stabilization? Under what conditions is it probable that collusive action can raise market prices?

1. NORMATIVE IMPLICATIONS OF MARKET SOLUTIONS

Pure competition is defined to be the situation in which no single participant has the capacity to affect market prices more than infinitesimally (that is, no single participant has market power). From the point of view of individual entities in the market place, prices seem to be given parameters (not fixed over time), independent of their own behavior. Pure competition generally is considered an interesting paradigm, for reasons summarized below, but not of very general applicability in the real world. However, most of the initial producers of agricultural commodities that enter into international commodity markets (and agricultural products account for about 85 percent of total nonpetroleum commodity exports from developing countries) and most of the ultimate consumers of both the agricultural and non-agricultural internationally traded commodities sell and purchase these goods (generally in processed form), respectively, under conditions approximating pure competition. At both ends of the market chain for the relevant international commodities (but not in the middle, where marketing boards, other government agencies, and large companies dominate), therefore, the purely competitive model may have substantial applicability.

What are the advantages of pure competition? Some answers are in the area of political economy and, thus, derivative of particular value systems. Under pure competition economic problems (such as what is produced, how it is produced, and for whom it is produced) are solved in an impersonal manner, independently of personal ties or characteristics such as race or national origin.[2] The atomistic structure of buyers and sellers required for competition also decentralizes and disperses power. Moreover, if the conditions necessary for pure competition to exist do in fact prevail, freedom of entry into various industries and individual mobility both will be high.

Most economists focus on answers related to economic efficiency. In a world with the correct initial distribution of assets for a given social welfare function, with easy entry (due, for example, to a lack of legal restrictions and limited increasing returns to scale relative to the size of industries), with no externalities, with no uncertainty, and with pure competition everywhere else, pure competition in international commodity markets results in maximization of that social welfare function.

This is a strong result. But these are very strong conditions, obviously not even approximately satisfied in the real world. If the first one is dropped, the given social welfare function no longer is maximized, although economic efficiency remains in the Paretian sense. No shift in resources, and so on, exists that would improve the welfare of any one individual without reducing the welfare of at least one other individual. Attainment of this state seems desirable, *ceteris paribus,* and its virtues are emphasized (perhaps overemphasized) by many economists. But without the correct distribution of assets, it does not maximize a given social welfare function. Within a static framework, much of the conflict between the developing and developed countries may arise at this point. Even if all the other conditions given above are satisfied, so that economic efficiency is attained, the initial distribution of assets is seen by many to be so inequitable that the world is far away from a welfare maximization. Efficiency concerns may be unimportant in light of the distribution question.

But even economic efficiency goes by the wayside once the real-world applicability of the other necessary conditions is considered. Entry is not easy in many cases due to legal, natural, and technological monopolies or due to increasing returns to scale that are large relative to the industry. The resulting market power and behavior approximating profit maximization lead to economic inefficiency by driving a wedge between marginal cost and the price that the consumer uses in his or her utility maximization decision. Externalities in consumption (such as keeping up with the Joneses, the international demonstration effect) and in production (for example, pollution) abound. This in itself is sufficient to preclude economic efficiency as a necessary outcome of a market system. So is the pervasive existence of uncertainty and risk aversion, with the result that on both sides of the market expected profit maximization does not necessarily lead to utility maximization.[3]

Finally, pure competition clearly does not prevail everywhere outside of international commodity markets; therefore, the theory of the "second best"[4] suggests that for purposes of economic efficiency, it may be preferable that it not prevail in international commodity markets either. A simple example may be useful. Assume that there are no externalities, no uncertainty, fixed factor supplies, no intermediate factors of production, and an unalterable monopoly in the rest of the world. Under these very special conditions, attempting to maintain perfect competition in international commodity markets (a "first best" policy in terms of economic efficiency) does not lead to economic efficiency, since the ratios of prices that consumers face are not equal to the ratios of marginal costs from the two sectors. However, the "second best" policy of developing the same degree of monopoly power in international markets as in the rest of the world does lead to equality in these two ratios and to economic efficiency. If specific distortions exist in the real world, second best solutions can be devised to obtain efficient outcomes. The basic problem is that devising such solutions depends upon much more knowledge than policy makers normally have.

Where does this discussion leave us? Unhindered markets may be relatively efficient devices for processing a great deal of information to signal shortages or surpluses through price or inventory changes. However, one has to be quite careful in regard to their normative implications. They lead to maximization of a given social welfare function only with the correct distribution of assets and the satisfaction of all of the other conditions discussed above. The "theory of second best" at worst implies that in the real world, policies directed at economic efficiency should be abandoned. At best, it suggests the advocacy of "third best" very general policies that have a reasonable probability of leading to greater economic efficiency but do not guarantee a step in that direction when applied in a specific case.

Economic theory leads us to this highly qualified view of the normative properties of unhindered market solutions.[5] It is a much weaker view than is suggested by many who oppose international commodity agreements on the basis that they would lessen the gains from free market operations. One can understand why economists from developing countries might question whether the position of the strongest advocate of unhindered free market operations is based on a lack of understanding of underlying economic theory or on a disguised defense of vested interests in the status quo.

2. IMPLICATIONS OF PRICE STABILIZATION ATTEMPTS FOR VARIABILITY AND LEVEL OF REVENUE

The UNCTAD proposal and resolutions recognize that stabilization of export revenues probably is of much more interest to the developing countries than is stabilization of prices. In principle, of course, a buffer stock might buy and sell with the intent to stabilize revenues. It could attempt to act so that the total market demand curve facing producers approached a unit-elastic curve with constant revenue implications.[6] Such an operation would be much more difficult than price stabilization, however, for several reasons. Day-to-day operations would be harder because of the greater lags in the availability of quantity than price data. If such an arrangement were successful, strong inducements would exist for supply reduction, since the same revenues could be earned with lower sales, which would release factors of production for other uses (although such an outcome probably could occur only if supply were organized in other than a purely competitive manner). The concurrence of importing nations with such a revenue-stabilizing scheme seems unlikely.

For such reasons, UNCTAD advocates price stabilization instead of revenue stabilization. But this strategy raises the question, What are the implications of price-stabilization attempts for revenues? Johnson (1977) states that "elementary economic analysis" suggests that the UNCTAD proposal is dubious on these grounds. He claims that under reasonable assumptions concerning the responsiveness of supply and demand to prices, there is a trade-off between the impact of price-stabilization programs on the level and on the stability of revenues. That is, increased revenues result only if there is greater instability in revenues, and vice versa. However, his argument is based on certain critical assumptions—such as high price responsiveness for suppliers and users or shifts in the underlying response relations—that do not seem justified by existing empirical evidence. Careful consideration of the argument reveals that, contrary to Johnson's "elementary economic analysis," conclusive statements may not be made on the basis of theory alone. Empirical estimates are required.

Johnson's argument is illustrated in figure 1,[7] in which the basic average supply and demand curves for a purely competitive international commodity market are given by solid straight lines, equally likely shifts in these curves are indicated by dashed lines, and P_0 is the average price and the price at which a buffer stock scheme would stabilize prices.

Consider first the case of instability due to shifts only in the demand curve (fig. 1A). Without price stabilization, average producers' revenues are $(P_1Q_1 + P_2Q_2)/2$. With stabilization they are P_0Q_0. Therefore, price stabilization in this case reduces the instability of producers' revenues as well as their average level.

Consider next the case of instability due to supply shifts alone. Figures 1B–1E provide illustrations of four cases with, respectively, elastic supply and demand, inelastic supply and demand, inelastic supply and elastic demand, and elastic supply and inelastic demand. Without price stabilization producers' average revenues are $(P_2Q_2 + P_3Q_3)/2$. With stabilization they are $P_0(Q_1 + Q_4)/2$ in fig. 1B, and $P_0(Q_2 + Q_3)/2$ in figures 1C–1E. In each case, price stabilization increases producers' revenues.[8]

In the case of elastic supply and demand, which Johnson considers to be most normal, however, price stabilization increases the instability of revenues (fig. 1B). The instability of revenues also increases in two of the other cases, but not if both curves are sufficiently inelastic (fig. 1C).

Of course, one can also consider mixed cases, in which both demand and

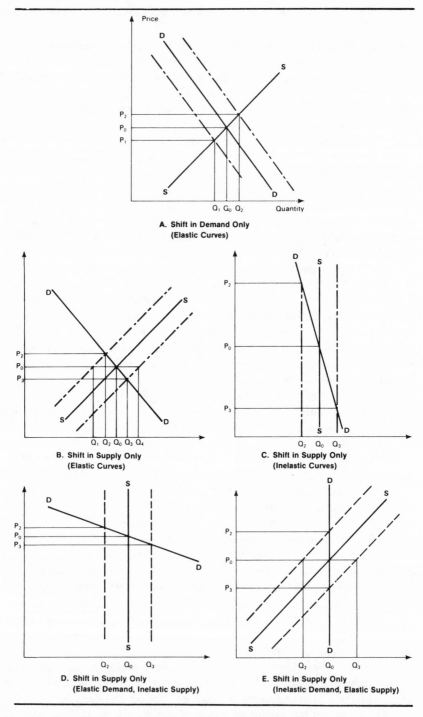

Fig.1. Impact on Revenues of Shifts in Demand and Supply Curves with and without Price Stabilization at P_0 by a Buffer Stock

supply curves shift. The net result depends upon the size of the two shifts and other particulars of the situation. The trade-off between level and instability of revenues nevertheless seems to persist in a number of cases.

Does it follow from Johnson's conclusion that the UNCTAD advocacy of price stabilization lumps together two different economic problems (instability of demand and instability of supply) that require quite different solutions? The answer is that it depends. Suppose, for example, that the producing nations are quite risk averse. Price stabilization reduces revenue instability if demand shifts are dominant relative to supply shifts (fig. 1A). It does the same if supply shifts are dominant and the relevant elasticities are quite low (fig. 1C).[9] If producers are sufficiently risk averse and if the relevant elasticities are sufficiently low, therefore, the producers are better off with price stabilization than without it, no matter which curve shifts most. This example suggests that in important ways the question is an empirical one, depending on the size of the market elasticities.

The UNCTAD resolution also may make sense if shifts in one or the other of the market curves dominate. If shifts in the demand curve dominate and the producers are risk averse, they may be better off even with lower average revenues, as is noted above. If shifts in the supply curve dominate, the curves are sufficiently elastic, and the producers are not very risk averse, they may be better off with the larger average revenues and greater revenue instability. If either of these cases prevails, Johnson is correct that two problems are being lumped together; but the occurrence of large shifts in one curve is sufficiently rare that it can be ignored.

Yet another possibility is that destabilizing speculation[10] causes large price fluctuations that lower the long-run demand curve by inducing substitution by risk-averse manufacturers of synthetics and other goods for the commodities of concern. Producers therefore might rationally prefer price stabilization in order to limit the downward long-run shift in the market demand curve, even if the short-term result is lower immediate revenues or greater instability in revenues.

These possibilities all emphasize that in important respects the matter is an empirical question. "Elementary economic analysis" is not enough. Without empirical knowledge concerning long-run movements, the shapes of the curves, risk aversion, the demand and supply elasticities of price responsiveness, the causes of shifts, whether the movements in supply and demand curves are additive or multiplicative, and so on, we cannot state with assurance what the impacts of stabilization are.

3. WHO GAINS FROM PRICE STABILIZATION?

The question of who gains from price stabilization obviously is related to the previous discussion. Answers to it generally ignore risk, distributional aspects, and general-equilibrium effects and focus on producers' and consumers' surpluses and the financial gains from operating a buffer stock.

The most frequent situation considered is one in which instability is due to

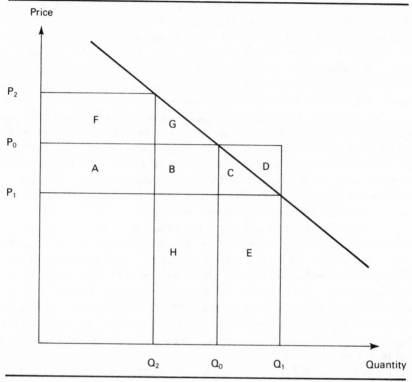

Fig.2. Gains and Losses from Price Stabilization (Shifts in Inelastic Supply Curve Only)

shifts in a completely inelastic supply curve.[11] Figure 2 illustrates the simplest case, in which the price is stabilized by buffer stock operations (with no storage costs) at P_0.

When the supply curve shifts out to Q_1, buffer stocks are accumulated. The consumers' benefit of paying P_0 instead of the price P_1, which would have prevailed without a buffer stock, is negative $(-A, -B, -C)$. The producers' benefit is the gain due to the higher prices $(A + B + C + D)$. The cost to the buffer stock is $-C - D - E$. The total benefit (summing these three components) is $-C - E$.

When the supply curve shifts into Q_2, stocks are sold at P_0. This precludes the price from rising to P_2, as it otherwise would. The benefit to the consumers is $F + G$, due to the lower price and the larger quantity. The benefit to the producers is $-F$, since they receive a lower price than they would without the buffer stock. The financial inflow to the buffer stock is $B + H$. The total benefit is the sum of these three components, $B + G + H$.

If shifts in the supply curve to Q_1 and Q_2 are equally likely and if the sequencing over time is ignored (or a zero discount rate is utilized), the total benefit to each of the three groups is the sum of those obtained from buffer stock operation with supply at Q_1 and at Q_2. For consumers, the sum is $F + G - A - B - C$. For producers, the sum is $A + B + C + D - F$. For the buffer stock, the sum is $B + H - C - D - E$. The total benefit is represented by $B + G + H - C - E$. Under these assumptions, the sum for the buffer stock is zero and the overall sum is positive. However, whether consumers or producers benefit depends on the exact shape of the curves. The issue basically is an empirical one.

The provocative analysis of Newbery (1976) reinforces such an emphasis. He also explores the impact of shifts in the supply curve. He concludes that consumers' surplus, producers' surplus, average price, and average supply each can increase, decrease, or remain the same under stabilization—depending on the form of the demand function (that is, linear versus constant elasticity), the size of the price elasticity of demand, and whether the disturbance term in the supply relation is additive or multiplicative. His analysis sheds considerable doubt on the derivation of strong conclusions from too simple models.

But even Newbery's extended analysis is not the whole story. As he explicitly recognizes, he does not consider the possibilities of risk aversion, distributional aspects, and general equilibrium effects. These considerations may be quite important. But an even more important factor may be in regard to the inflation-output trade-off. What reduction in national output would have to be accepted to offset the inflationary consequences in the consuming nations of upward fluctuations in international commodity market prices together with an oligopolistic "ratchet" effect in the pricing mechanism? Existing estimates suggest that the answer may be sufficiently large to make the changes in consumers' and producers' surpluses second-order considerations.

4. Conditions under Which Collusive Action Can Raise Market Prices

The basic motivation behind the UNCTAD proposal and resolution may have little to do with stabilization per se. Instead, the major concern may be to raise the real resources of the developing countries that export the affected commodities.

Under certain conditions discussed above—for example, dominant supply shifts, subject to the qualifications necessary because of Newbery's (1976) analysis—stabilization itself may lead to increased revenues for the exporters. The content of the UNCTAD documents, however, suggests that the concern goes further than this, to a desire to raise market prices (or prevent real market prices from falling) to levels above those that otherwise would prevail. If market demand curves are inelastic, successful price raising will be rewarded by greater revenues.[12]

The questions then arise, Why try to form commodity agreements that include consumer representation? Why not follow the lead of OPEC? This line of inquiry leads to the further question, Under what conditions are producers likely to be able to collude by themselves to raise prices? This leads us into a much less rigorous area of economic theory: the formation and behavior of oligopolies (that is, markets in which there are sellers with large enough market shares that each one can recognize that his sales affect the market price). Behrman (1978) provides an extensive review of this area and its implications for modeling the structure of international commodity markets. Here I limit the presentation to a checklist of conditions that facilitate oligopolistic coordination of pricing and output decisions: (1) the perception that joint action will lead to greater returns for producers, (2) common output preferences due to similar cost structures and market shares, (3) cheap and rapid communication, (4) high concentration, (5) a small or no competitive fringe, (6) repetitive small transactions, (7) homogeneity and simplicity of products, (8) the willingness to utilize inventory and order backlogs as buffers instead of overly sensitive price adjustments, (9) limited or no substitution for the product, and (10) high barriers to entry (such as restricted technological knowledge, restricted control over exhaustible resources, returns to scale at a high level of production relative to market size). I also note that the dynamics of substitution and of limit pricing (that is, pricing by a collusive group of oligopolists or by a monopolist to discourage too-rapid entry) generally lead to relatively few cases in which prices can be long sustained significantly above marginal costs.

Whatever reorganization of the market structure is proposed by UNCTAD or by anyone else, it is obvious that the starting point is not purely competitive international commodity markets. On the contrary, there are in existence a number of private and national-government organizations that currently can exert market power in international commodity markets.[13] Thus Johnson's (1977) elementary economic analysis may be misleading because it assumes the wrong market structure to start with. Indeed, from this point of view, price stabilization schemes may be a device to increase oligopolistic price discipline (although producer cartels might be as effective as international commodity agreements in attaining this end).

NOTES

1. For details and a range of interpretations concerning the call for a "new international economic order" by developing nations and the reaction of the developed countries to this call see Lewis 1976; Hansen 1976; Smith 1976; Barraclough 1976; Erb and Kallab 1975; Eads 1976; and Michalopoulos 1976. Also relevant are the other studies in the Overseas Development Council's "New International Economic Order" series of publications and the references contained in each of them.

2. Given certain sets of values, of course, this impersonality is a negative, dehumanizing feature.

3. Other questions have been raised about the appropriateness of the assumption of profit and utility maximization. For a good review of the former see Scherer 1970. If such assumptions are not good approximations to real-world behavior, they also point to other than efficient outcomes from a market system.

4. See Lancaster and Lipsey 1956.

5. The above analysis is basically static. Extension to dynamic concerns does not drastically change these conclusions.

6. Elasticity refers to the percentage change in quantity for a given percentage change in price. The higher the elasticity, the greater the responsiveness. If the elasticity of demand is one (or unitary), the change in quantity demanded along the curve just offsets the change in the price, so that revenue (the product of price times quantity) remains constant. If the response is more limited, that is, if the elasticity is less than one, the curve is called inelastic. If the response is greater, that is, if the elasticity is greater than one, the curve is called elastic.

7. For an algebraic demonstration see Johnson 1977.

8. With sufficiently nonlinear curves, even this result does not hold. Consider the case in which the demand curve is very inelastic above P_0 but very elastic below this price, and the inelastic supply curve shifts. In this case, price stabilization may reduce revenues but increase their stability.

9. The available empirical evidence implies that the market price elasticities may be much lower, especially in the short run, than Johnson suggests.

10. The initial speculators may not be driven out of business by such cycles, since they may sell out before the market reverses to those who follow in the bandwagon effect. Therefore, they may continue to speculate again as long as there is some fringe to follow and bear the losses.

11. See, for example, Reutlinger 1976.

12. Lump-sum transfers may be a more efficient means to assure the desired distributional outcomes; but if the developing countries do not see them occurring in sufficient quantities, altering market outcomes may be a second-best solution.

13. For example, Helleiner (1978) reports that almost 70 percent of the imports of the commodities in the UNCTAD resolution into the United States in 1975 were from "related parties" (defined as a firm in which 5 percent or more of the voting stock of one party in the transaction is owned by the other). This high proportion suggests that very considerable trade in these commodities may be undertaken on an intrafirm basis or on some other basis that is not well represented by the purely competitive model.

REFERENCES

Barraclough, Geoffrey. 1976. "The Haves and the Have Nots." *New York Review of Books* 23(4).

Behrman, Jere A. 1978. "International Commodity Market Structures and the Theory Underlying International Commodity Market Models." In *Econometric Modeling of World Commodity Markets,* edited by F. G. Adams and J. R. Behrman, 9–48. Lexington, Mass.: Lexington-Heath Publishing Company.

Eads, George C. 1976. "Address of Executive Director of National Commission on Supplies and Shortages before the Section of Natural Resources Law." American Bar Association Convention, Atlanta, Georgia, 10 August 1976. Mimeo.

Erb, Guy F., and Valeriana Kallab, eds. 1975. *Beyond Dependency: The Developing World Speaks Out.* Washington, D.C.: Overseas Development Council.

Erb, Guy F., and S. Schiavo-Campo. 1969. "Export Instability, Level of Development and Economic Size of Less Developed Countries." *Oxford Bulletin of Economics and Statistics* 31 (May): 263–83.

Hansen, Roger D., and the staff of the Overseas Development Council. 1976. *The U.S. and World Development: Agenda for Action, 1976.* New York: Praeger Publishers for the Overseas Development Council.

Helleiner, G. K. 1978. "Freedom and Management in Primary Commodity Markets: U.S. Imports from Developing Countries." *World Development* 6(1):23–30.

Johnson, Harry G. 1977. "Commodities: Less Developed Countries' Demands and Developed Countries' Responses." In *The New Economic Order and the North-South Debate,* edited by J. N. Bhagwati, 240–49. Cambridge: MIT Press.

Lancaster, Kelvin, and R. G. Lipsey. 1956. "The General Theory of Second Best." *Review of Economic Studies* 24(1):11–32.

Lewis, Paul. 1976. "The Have-Nots are Gaining Ground in Their Drive to Gain Concessions." *National Journal* 8(23):774–82.

Michalopoulos, Constantine. 1976. "U.S. Commodity Trade Policy and the Developing Countries." Washington, D.C.: U.S. Agency for International Development. Mimeo.

Newbery, David M. G. 1976. "Price Stabilization with Risky Production." Palo Alto: Stanford University. Mimeo.

Reutlinger, Shlomo. 1976. "A Simulation Model for Evaluating Worldwide Buffer Stocks of Wheat." *American Journal of Agricultural Economics* 58(1):1–12.

Scherer, F. M. 1970. *Industrial Market Structure and Economic Performance.* Chicago: Rand McNally & Co.

Smith, Gordon W. 1976. "Informational Efficiency in Commodity Markets." Houston: Rice University. Mimeo.

UNCTAD. 1974a. "An Integrated Programme for Commodities." TD/B/C.1/166. Geneva.

————. 1974b. "A Common Fund for the Financing of Commodity Stocks." TD/B/C.1/166, supp. 2. Geneva.

————. 1974c. "The Role of Multilateral Commitments in Commodity Trade." TD/B/C.1/166, supp. 3. Geneva.

————. 1974d. "Compensatory Financing of Export Fluctuations in Commodity Trade." TD/B/C.1/166, supp. 4. Geneva.

————. 1974e. "Trade Measures to Expand Processing of Primary Commodities in Developing Countries." TD/B/C.1/166, supp. 4. Geneva.

————. 1975a. "Progress Report on Storage Costs and Warehouse Facilities." TD/B/C.1/187. Geneva.

————. 1975b. "Recent Development in International Commodity Arrangements Relevant to the Elaboration of an Integrated Programme for Commodities." TD/B/C.1/185. Geneva.

————. 1975c. "The Impact on Imports, Particularly of Developing Countries." TD/B/C.1/169. Geneva.

————. 1975d. "A Common Fund for the Financing of Commodity Stocks: Amounts, Terms and Prospective Sources of Finances." TD/B/C.1/184. Geneva.

————. 1975e. "A Common Fund for the Financing of Commodity Stocks: Amounts, Terms and Prospective Sources of Finances—Addendum." TD/B/C.1/184/add. 1/corr. 1. Geneva.

————. 1975f. "Commodity Trade Indexation." TD/B/563. Geneva.

IV

Agricultural Growth and the Transformation of Rural Institutions

Introduction

In the early 1970s development economists began to deemphasize macroeconomic and agricultural-sector modeling and to concentrate more attention on income distribution, employment, migration, and the microeconomic aspects of agricultural development. Increasing recognition of the complexity of Third World farming systems and rural nonfarm enterprises led agricultural economists to expand their research on farming systems, rural product and input markets, and rural nonfarm employment. The eleven articles in part IV draw heavily on this empirical research to examine the relationships between agricultural growth and changing rural institutions. Several of the articles show that innovations in rural areas have important *inter*sectoral as well as *intra*sectoral effects. As a result, agricultural innovations cannot be analyzed solely in terms of their impact on agriculture. For example, technical change in rice processing may affect the demand for labor in rural areas and the price of rice in the cities. The primary emphasis of the articles in part IV, however, is on understanding the effects of such innovations on farming and related rural industries.

Growth in agricultural output occurs in three ways: through changes in the use of existing factors of production; through the development and use of new factors of production (technological change); and through changes in institutions, which alter the incentives and rights to use existing factors of production and develop new ones.[1] All three sources of agricultural growth are discussed in part IV. Chapters 17–22 examine how the market and other institutional structures influence the use of land, labor, and capital in farming and in rural nonfarm enterprises. Chapters 23–25 focus on technological change in agriculture and on alternative approaches to agricultural research, while chapters 26 and 27 examine the distribution of costs and benefits from technological change in agriculture.

During the 1950s and 1960s many economists investigated land tenure issues in Third World countries. Most concluded that tenure reforms were often necessary in order to give small farmers incentives to invest in land improvements and to adopt new technologies that required more purchased inputs.[2] Although the 1960s and early 1970s witnessed a large number of land reforms in Latin America and Asia, by the late 1970s the pace of land reform in most parts of the world had slowed dramatically.

In chapter 17 de Janvry addresses the question of why land reform, although still important in political rhetoric, is seldom implemented today in the Third World. After reviewing the experience of antifeudal land reforms during the past three decades, de Janvry concludes that most reforms came about because of an alliance between the rural poor and the urban bourgeoisie. The rural poor viewed land reform as a means of achieving a more equitable income distribution, while the urban bourgeoisie believed that land reform would increase agricultural productivity, thereby reducing the price of wage goods and allowing industrial profits to rise. According to de Janvry, the green revolution, with its emphasis on land substitutes like fertilizer and high-yielding grain varieties, shattered the land-reform coalition by permitting large farms to produce cheap wage goods without land redistribution. In fact, because land reform might threaten the short-term productivity of large commercial farms, it came to be opposed by the bourgeoisie. Hence, although land reform remains an important symbol in political rhetoric, de Janvry believes that it lacks the necessary political support among donors and Third World governments to be implemented.[3]

Bromley, in chapter 18, questions whether de Janvry's Marxian framework, with its emphasis on "modes of production writ large," is flexible enough to provide a very insightful analysis of land reform. Lumping all forms of capitalist farming together, Bromley says, obscures many distinguishing characteristics of different tenure arrangements. Bromley argues that it is possible to reach conclusions similar to those of de Janvry from a very different theoretical perspective, particularly since land, compared with other types of assets such as human capital, declines in importance during the course of economic development. Although Bromley agrees that the economic importance of land reform is declining in many countries, he argues that the reasons may be very different from those outlined by de Janvry.

De Janvry and Bromley discuss land reform primarily as it relates to densely populated areas of Asia and Latin America. In more sparsely populated regions of Asia and in much of sub-Saharan Africa, however, where land tenure is moving from communal to freehold systems, land tenure and land settlement will be important issues in the 1980s and 1990s.[4]

During the 1970s, concern for creating productive rural employment stimulated many empirical studies of rural labor markets and the determinants of rural employment. Two areas receiving particular attention were the choice of technique in agriculture and industry and the importance of rural nonfarm enterprises as sources of employment, income, and output.

In chapter 19 Timmer uses the example of rice milling in Java to illustrate the type of economic analysis that is useful in selecting techniques appropriate for agricultural processing in low-wage economies.[5] Timmer undertook this analysis because the government of Indonesia was considering importing several large, capital-intensive milling facilities in order to "modernize" the Javanese rice processing industry. Timmer shows that importing technology appropriate to high-wage countries like the United States would have resulted in much higher milling costs and increased unemployment as expensive imported capital displaced inexpensive domestic labor. Using neoclassical analysis, Timmer demonstrates the superiority, from an efficiency standpoint, of the small-scale rice mills adopted spontaneously in Java during the late 1960s. These mills outperformed both large, capital-intensive facilities and hand pounding of rice.

In their comment on Timmer's paper, Collier et al. point out that although the small mills may have been economically efficient given the existing property rights, the costs and benefits of shifting from hand-pounding to small mechanized mills were borne very unevenly. The major cost was in the form of increased unemployment among low-income women who previously had hand-pounded the crop, while the benefits were in the form of lower consumer rice prices and increased incomes of rice millers.

The Timmer-Collier et al. interchange raises the issue of whether mechanisms are needed to compensate those hurt by technical change (see Schmitz and Seckler 1970). The exchange also demonstrates the large array of detailed engineering, economic, and social data needed to understand the production and income-distribution consequences of technical change.[6]

Development theory and practice during the 1950s and 1960s devoted little attention to rural nonfarm activities such as processing, carpentry, tailoring, and trading. As Chuta and Liedholm point out in chapter 20, however, rural nonfarm enterprises are major sources of output, employment, income, and intermediate demand for agricultural products in many Third World countries. Rural small-scale industry is particularly important in providing agricultural equipment adapted to local conditions and slack-season rural employment for farm laborers. Because of the interdependencies between agriculture and rural nonfarm enterprises, the profitability of nonfarm enterprises is crucially linked to the growth of farm income. Industrial and agricultural policies in many Third World countries, however, either have underestimated the importance of rural nonfarm industries or have assumed that they will (and should) decline relative to urban industry during the course of development. Chuta and Liedholm discuss the importance of rural nonfarm industries in several Third World countries and describe the ways in which government policies can foster the growth of rural enterprises and their linkages with other sectors of the economy.

Development theorists during the 1950s viewed the lack of physical capital as a crucial constraint on both agricultural and industrial growth. Concern about capital shortages in agriculture grew during the 1960s as the green revolution technology, with its heavy reliance on purchased inputs, was introduced into many countries. This concern led to greatly expanded efforts to channel subsidized credit to Third World farmers. In fact, Adams and Graham, in chapter 21, estimate that donors have poured over $5 billion into rural credit projects over the past two decades. A major debate of the 1960s and 1970s centered on whether specialized lending agencies should subsidize interest rates to small farmers in order to offset the administrative costs and risks of loaning to small farmers, the alleged monopoly profits in private capital markets, and the price distortions in other parts of the economy—for example, low farm-gate prices due to heavy agricultural taxation.[7] Many researchers argued that subsidized interest rates were an extremely inefficient means of counteracting price distortions, that they tended to undermine incentives for rural savings mobilization, and that they worsened income distribution. In chapter 21 Adams and Graham review the arguments for and against specialized small-farmer lending institutions and subsidized interest rates, while in chapter 22 Gonzalez-Vega focuses on the income-distribution effects of subsidized credit programs. The work of Adams, Graham, Gonzalez-Vega, and others has led most donors and, to a lesser extent, Third World governments to deemphasize subsidized agricultural credit. But as Adams and Graham point out in the final section of their paper, there are often politically powerful groups that benefit enormously from subsidized credit programs and therefore favor their continuation. Like so many other aspects of agricultural policy, farm credit needs to be analyzed within a broad political economy framework (Von Pischke 1980).

Technical change in agriculture is the subject of chapters 23–27. In chapter 23 T. W. Schultz argues that because agricultural research uses scarce resources to produce valuable outputs, it should be viewed as an economic activity. The determinants of the productivity of agricultural research systems are therefore amenable to economic analysis. Schultz outlines the broad array of factors affecting research productivity, ranging from the degree of decentralization of the research system to the design of incentives that effectively encourage research entrepreneurship.

Schultz maintains that many Third World countries have underinvested in agricultural research because they have erroneously assumed that agricultural technology could be imported from high-income nations and the international agricultural research centers with little or no local adaptive research. Schultz argues that the international agricultural research centers should be seen as complements to, not substitutes for, national agricultural research systems.

Evenson's review of studies on the rates of return on agricultural research in the Third World (chapter 24) supports Schultz's contention that many countries have underinvested in research, particularly on food crops. The figures cited by Evenson have to be interpreted with caution, however, because studies of the rates of return on research typically focus only on successful research programs and may understate the importance of complementary investments in input delivery systems, extension services, and so on. Moreover, the payoff to research sometimes comes only after decades of work. For example, although the internal rate of return on investment in research on hybrid corn in the United States was 35–40 percent (Griliches 1958), plant geneticists created hybrid corn in the 1930s only *after twenty-three years of research* (Schultz 1981, 104). Evenson notes that the *average* lag time between agricultural research spending and the full realization of its benefits is ten years. Therefore, although there may be significant payoffs to agricultural research, it must be viewed in a long-term perspective; research is not a "quick fix" to the food problems of the Third World.

Evenson also points out that because agricultural production is a biological process, local environmental conditions can often hinder the transferability of agricultural technology, particularly of the biogenetic type. This implies that there may be a high payoff to decentralizing a national agricultural research system by developing local research stations in each of a country's major agroecological zones.[8] The contribution of decentralization to the productivity of the U.S. agricultural research system has been documented by Evenson, Waggoner, and Ruttan (1979). These authors stress, however, that research and extension should be viewed as complementary because returns to investment in basic and technological research would be low unless this research were linked to a productive extension system.[9]

The question of how to combine research and extension in a way that overcomes the constraints faced by small farmers is addressed by the CIMMYT Economics Staff in chapter 25. Farming systems research (FSR) is a multidisciplinary approach to problem solving that was developed in the late 1970s in response to the realization that many national agricultural research systems were not producing technologies appropriate to small farmers.[10] In other words, the induced innovation mechanism described by Hayami and Ruttan in chapter 4 was not generating technologies appropriate to small farmers' circumstances. FSR attempts to overcome these barriers by involving farmers directly in the research process, from problem identification to on-farm trials and extension. FSR attempts to develop technologies consistent with farmers' goals and with the constraints imposed by the entire farming system rather than simply to increase production of a few major commodities. Nonetheless, as the CIMMYT researchers point out, FSR is a complement to, not a substitute for, strong national commodity research programs.

FSR provides an example of economists and biological scientists "trespassing" across disciplinary boundaries to work together (Hirschman 1981). Although FSR has promise of strengthening national agricultural research systems, it is still in a pilot stage and in many countries is still funded mainly by external donors. It remains to be seen whether FSR can generate the domestic political support necessary to give voice to the small farmer in influencing the priorities of national agricultural research systems.

The benefits and costs of technological change in agriculture are seldom shared equally among the members of society. During the 1970s many observers expressed the fear that new, high-yielding grain varieties would benefit mainly large farmers living in well-irrigated regions, while smallholders and tenants, particularly in upland areas, would be made relatively, and perhaps absolutely, worse off (see chapter 1). In chapter 26 Scobie and Posada examine the distribution of costs and benefits from rice research in Colombia, while in chapter 27 Hayami reviews evidence on the income-distribution consequences of the green revolution in Asia.

Scobie and Posada isolate, by income strata and by rural/urban residence, the incidence of the costs and benefits of the Colombian research program on new varieties of rice for irrigated land. The rapid and widespread adoption of the new varieties led to a substantial increase in production and a concomitant fall in the price of rice. The authors show that 70 percent of the net benefits of the research program accrued (in the form of lower rice prices) to the over one million low-income Colombian consumers who receive only 15 percent of national income.[11] As a result of the fall in the price of rice, the incomes of small upland rice farmers were substantially reduced from what they would have been without the research program, as were the incomes of some of the large irrigated farmers. The authors stress, however, that the research resulted in a small number of losers relative to winners: over one million low-income consumers benefited from lower rice prices, while fewer than twelve thousand small upland rice farmers had their incomes reduced by the research program.[12] "Hence, under any plausible set of welfare weights the . . . losses would be more than offset by the gain . . . implying an overall gain (albeit uncompensated) in some measure of social welfare." The analysis by Scobie and Posada demonstrates the importance of looking beyond farmers when analyzing the welfare effects of technological change in agriculture.

Although many of the early evaluations of the income-distribution consequences of the green revolution in Asia (such as Frankel 1971; and Griffin 1974) were impressionistic, they did serve as warnings that the high-yielding varieties were not a cure-all for rural poverty. In chapter 27 Hayami reviews more recent evidence on how the modern grain varieties have affected farm output, employment, and income distribution within villages in Asia.[13] Hayami argues that the new seed/fertilizer technology is both scale-neutral and factor-neutral. Therefore, in most instances small farmers have adopted the new varieties as quickly as large farmers, and the new technology has not by itself led to labor-displacing farm mechanization. Where such mechanization has occurred, it has resulted from factor-price distortions (particularly subsidized interest rates and overvalued domestic currencies) unrelated to the new technology.

Hayami does not deny that the income distribution in rural Asia has become more unequal in recent years, but he disputes the conclusion of authors such as Pearse (1980), who attribute the growing inequality to the effects of high-yielding varieties. Rather, Hayami says, the empirical evidence strongly supports the position that growing population pressure on the land is the core problem and that the worsening income distribution has resulted from too little, not too much, technical change in agriculture.

NOTES

1. Although reallocation of existing factors of production in technologically stagnant agriculture may result in small increases in production, reallocation may be a much more important source of growth in technologically dynamic agriculture characterized by frequent disequilibria (Schultz 1964). Increasing farm specialization, often cited as a major source of agricultural growth, frequently involves all three of the mechanisms mentioned above.

2. Raup (1967) presents an excellent summary of these views. Most authors argued that land

ownership per se was not necessary in order for small farmers to be technologically dynamic; many improvements could be achieved through reform of rental agreements.

3. Blackie (1981) reports that several thousand large farmers controlled the most fertile land in Zimbabwe at the time of independence in 1980. Still only one donor—the United Kingdom—has come forth to finance a land transfer program to settle landless families.

4. Neither de Janvry nor Bromley discusses land settlement programs. Lewis (1954) and Nelson (1973) analyze the reasons why land settlement schemes have so frequently failed in the Third World. For a standard reference on land tenure problems in Africa see Cohen 1980.

5. For similar analyses of the choice of technique in farming see Sen 1959, 1960; and Binswanger 1978.

6. Byerlee et al. (1983) extended Timmer's approach to analyze the output, employment, and foreign-trade consequences of alternative rice processing techniques in Sierra Leone. Their analysis showed that if the government continued to subsidize interest rates, there would be a rapid shift from hand pounding to mechanized milling, resulting in the loss of forty thousand person-years of employment—a staggering figure for a country of only three million. The articles by Timmer, Collier et al., and Byerlee et al. are excellent examples of how the debate on the choice of technique can be furthered by empirical research.

7. In theory, real market interest rates have four components: a pure interest rate, an administrative premium, a risk premium, and monopoly profit (Bottomly 1963; Adams 1978).

8. See, however, Schultz's warning in chapter 23 about the dangers of spreading resources so thinly that major economies of scale in research are forgone.

9. The literature on agricultural extension in the Third World is vast, but there has been little consensus about which approaches to extension are most appropriate in different institutional environments. For a sampling of the literature see Jones and Rolls 1982; Roling, Ascroft, and Wa Chege 1976; and Benor and Harrison 1977.

10. For a comprehensive review of farming systems research see Gilbert, Norman, and Winch 1980.

11. This finding is consistent with Mellor's analysis in chapter 10 of the income-distribution consequences of changes in food prices.

12. The new varieties were adopted by farmers only on irrigated land because they were not suited to upland (rainfed) farming systems.

13. Hayami does not discuss how the modern varieties may have affected interregional income distribution.

REFERENCES

Adams, Dale W. 1978. "Mobilizing Household Savings through Rural Financial Markets." *Economic Development and Cultural Change* 26(3):547–60.

Benor, D., and J. Q. Harrison. 1977. *Agricultural Extension: The Training and Visit System.* Washington, D.C.: World Bank.

Binswanger, Hans P. 1978. *The Economics of Tractors in South Asia: An Analytical Review.* New York and Hyderabad: Agricultural Development Council and International Crops Research Institute for the Semi-Arid Tropics.

Blackie, Malcolm J. 1981. "A Time to Listen: A Perspective on Agricultural Policy in Zimbabwe." Working Paper 5/81. Salisbury: University of Zimbabwe, Department of Land Management. Mimeo.

Bottomly, A. 1963. "The Premium for Risk as a Determinant of Interest Rates in Underdeveloped Rural Areas." *Quarterly Journal of Economics* 77(4):634–47.

Byerlee, Derek, Carl K. Eicher, Carl Liedholm, and Dunstan S. C. Spencer. 1983. "Employment-Output Conflicts, Factor Price Distortions and Choice of Technique: Empirical Results from Sierra Leone." *Economic Development and Cultural Change* 31(2):315–36.

Cohen, John. 1980. "Land Tenure and Rural Development in Africa." In *Agricultural Development in Africa: Issues of Public Policy,* edited by R. H. Bates and M. F. Lofchie, 349–400. New York: Praeger.

Evenson, Robert E., Paul E. Waggoner, and Vernon W. Ruttan. 1979. "Economic Benefits from Research: An Example from Agriculture." *Science* 205 (14 September):1101–7.

Frankel, Francine R. 1971. *India's Green Revolution: Economic Gains and Political Costs.* Princeton: Princeton University Press.

Gilbert, E. H., D. W. Norman, and F. E. Winch. 1980. *Farming Systems Research: A Critical Appraisal.* MSU Rural Development Paper, no. 6. East Lansing: Michigan State University, Department of Agricultural Economics.

Griffin, Keith. 1974. *The Political Economy of Agrarian Change: An Essay on the Green Revolution.* Cambridge: Harvard University Press.

Griliches, Zvi. 1958. "Research Costs and Social Returns: Hybrid Corn and Related Innovations." *Journal of Political Economy* 66:419–31. Reprinted in *Agriculture in Economic Development,* edited by Carl K. Eicher and Lawrence Witt, 369–86. New York: McGraw-Hill, 1964.

Hirschman, Albert O. 1981. *Essays in Trespassing: Economics to Politics and Beyond.* New York: Cambridge University Press.

Jones, Gwyn E., and Maurice J. Rolls, eds. 1982. *Progress in Rural Extension and Community Development.* Vol. I. New York: John Wiley and Sons.

Lewis, W. Arthur. 1954. "Thoughts on Land Settlement." *Journal of Agricultural Economics* 11 (June): 3–11. Reprinted in *Agriculture in Economic Development,* edited by Carl K. Eicher and Lawrence W. Witt, 299–310. New York: McGraw-Hill, 1964.

Nelson, Michael. 1973. *The Development of Tropical Lands: Policy Issues in Latin America.* Baltimore: Johns Hopkins University Press.

Pearse, Andrew. 1980. *Seeds of Plenty, Seeds of Want: Social and Economic Implications of the Green Revolution.* New York: Clarendon Press and Oxford University Press.

Raup, Philip M. 1967. "Land Reform and Agricultural Development." In *Agricultural Development and Economic Growth,* edited by Herman M. Southworth and Bruce F. Johnston, 267–314. Ithaca: Cornell University Press.

Roling, N. J., J. Ascroft, and F. Wa Chege. 1976. "The Diffusion of Innovations and the Issue of Equity in Rural Development." *Communication Research* 3(2):155–70.

Schmitz, Andrew, and David Seckler. 1970. "Mechanized Agriculture and Social Welfare: The Case of the Tomato Harvester." *American Journal of Agricultural Economics* 52(4):569–77.

Schultz, Theodore W. 1964. *Transforming Traditional Agriculture.* New Haven: Yale University Press.

———. 1981. *Investing in People: The Economics of Population Quality.* Berkeley: University of California Press.

Sen, Amartya Kumar. 1959. "The Choice of Agricultural Techniques in Underdeveloped Countries." *Economic Development and Cultural Change* 7(3, pt. 1):279–85.

———. 1960. *Choice of Techniques: An Aspect of the Theory of Planned Economic Development.* Oxford: Basil Blackwell.

Von Pischke, J. D. 1980. "The Political Economy of Specialized Farm Credit Institutions." In *Borrowers and Lenders: Rural Financial Markets and Institutions in Developing Countries,* edited by J. Howell, 81–103. London: Overseas Development Institute.

17

The Role of Land Reform in Economic Development: Policies and Politics

As the extensive literature on the subject attests, the issue of land reform has been and remains heatedly debated.[1] The latest massive demonstration of interest came with the 1979 World Conference on Agrarian Reform and Rural Development, where representatives of no less than 145 nations and 3 liberation movements agreed that "equitable distribution and efficient use of land . . . are indispensable for rural development, for the mobilization of human resources, and for increased production for the alleviation of poverty" (FAO 1979). At one time or another, but especially since 1960, virtually every country in the world has passed land reform laws. Yet the record is far more modest than the promise: (*a*) in spite of decades, if not centuries (Tuma 1965), of land reform activities, land ownership remains extremely skewed, concentration of land ownership is almost universally increasing, the mass of landless is growing rapidly, and the extent of rural poverty and malnutrition has reached horrendous proportions; and (*b*) in spite of widespread agreement on the need for land reform, there are virtually no significant ongoing land reform programs except under extreme political pressures (revolutionary in El Salvador and postrevolutionary in Nicaragua, Mozambique, and Angola). In several countries the progressive gains achieved by land reform programs are being either eroded by the forces of economic growth (Mexico, Venezuela, and South Korea) or purposefully canceled by public policies (Chile).

The question I want to explore in this paper is, Why this blatant discrepancy between rhetoric and reality? Or to put it another way, Why is land reform no longer a significant policy issue even though it remains an important political issue in most countries of the world?[2] To answer, we need first to explore, in a

ALAIN DE JANVRY is professor of agricultural and resource economics, University of California at Berkeley.

Reprinted from *American Journal of Agricultural Economics* 63, no. 2 (1981): 384–92, with omissions and minor editorial revisions, by permission of the American Agricultural Economics Association and the author.

positive sense, what has been the nature of different land reforms that occurred in the past; what were their purposes, achievements, and limits. This will be done by developing a typology of land reforms in the Third World, because it is essential to distinguish carefully among a wide variety of reforms. Following the approach of political economy, I will do this in the context of modes of production, social class structure, and types of land tenure.

Using this typology, we can then characterize the nature of past land reforms and identify the needed character of future reforms, given the actual state of the agrarian structure. This gives us a basis from which to seek answers to the question as to why land reform appears to be a dead policy issue in most countries today.

By looking at the wide array of rationales and proposals for land reform, we can identify a number of political stands on the expected role of land reform in the process of economic development. Because land reform is fundamentally a political issue that seeks to achieve or prevent social change, examining explicitly the political expectations from reform for different groups permits us to order and clarify the massive number of arguments advanced on the subject.

TYPES OF LAND REFORM

We need to start first with a few definitions, because part of the difficulty with the debate on land reform comes from using different concepts under the same name.

A reform is an institutional innovation promoted by the ruling order in an attempt to overcome economic or political contradictions without changing the dominant social relations. Thus, reform falls short of revolution (where the dominant social relations are changed) and goes beyond mere disregard of economic problems or repression of political demands. While instituted within the ruling order, the origin of reform can, however, rest just as well in the political pressures of the dominated groups as in the initiative of the dominant classes. Land reform, in particular, aims at transforming the agrarian structure. The agrarian structure is characterized by a system of social relations (modes of production and their corresponding social class composition) and a system of land tenure (ownership and usufruct of land and water by farm sizes). Land reforms consequently can change the modes of production in agriculture (themselves dominated by an eventually different mode in society at large which cannot be altered by reforms), the class structure (and, consequently, control of the state by specific classes and their respective access to public goods and services), and the pattern of land tenure.

Use of the concepts of mode of production, social class, and land tenure thus allows us to characterize a variety of states of the agrarian structure. A land reform is then nothing else than an attempt by the government, through public policies, at either inducing a change among states of the agrarian structure or at preventing such a change. A typology of land reforms, consequently, can be usefully constructed as a matrix of changes/no changes among states of the agrarian structure. This is done in table 1 for the thirty-three most important land reforms in the world over the last seventy years.

For our purpose, we can distinguish between three modes of production in agriculture (semifeudal, capitalist, and socialist) and, under capitalism, three landed social classes (capitalist landed elites, farmers, and peasants).[3] This gives us five relevant initial states of the agrarian structure: semifeudal estates controlled by the traditional landed elite with bonded labor (debt peonage, rent in labor services, and so on) and extraeconomic forms of coercion; capitalist estates controlled by the landed elite-turned-capitalist and using wage labor; capitalist farms and plantations whose owners, the farmers, share in the control of the state with the bourgeoisie-at-large and hire wage labor; peasant farms, ranging from family to subfamily (semiproletarian), where no labor is hired but some may be sold; and socialist farms, either family farms, labor cooperatives, or state farms, imbedded within a socialist mode of production in society at large.

The land reform itself creates a reform sector and transforms or not the agrarian structure into a set of other states. It is important here to distinguish between reform and nonreform sectors, even though most of the literature on land reform considers only the first. The reform sector is the set of farms that is created by expropriation of private lands, distribution of public lands, or colonization of new lands. It is usually organized in the form of family or collective/state farms. The nonreform sector includes the lands retained, subdivided, or sold privately by the former owners. As we will see, most land reforms have sought their economic results in the impact they had on the nonreform sector. After land reform, there are correspondingly five relevant states of the agrarian structure: three where the nonreform sector remains dominant and composed of either semifeudal estates, capitalist estates, or capitalist farms and two where the reform sector is dominant and sometimes exclusive—one under the capitalist mode of production and the other under the socialist mode.

Thirty-three reforms in twenty countries are classified in table 1. A particular country can reenter the matrix more than once if land reform programs are redefined over time, but it must always reenter it in the same state to which it was transformed by the previous land reform.

All the diagonal reforms are essentially redistributive reforms in the sense that they either increase the size of the reform sector without changing the nature of the nonreform sector (reform types 1, 7, and 13 in table 1) or redefine the nature of the reform sector (19 and 25). Reform 1 is redistributive within the semifeudal order. Typical examples are the early reforms in Taiwan (1949–51), Colombia (1961–67), and Chile (1962–67), where the main objective was to distribute to peasants empty public lands or lands abandoned by the landed elite without otherwise questioning the continued domination of semifeudal estates in the agrarian structure.

TABLE 1
A TYPOLOGY OF LAND REFORMS

		Post-Land Reform				
		Semifeudal	**Capitalist**		**Socialist**	
Mode of Production in Whole Society →	Mode of Production in Agriculture →	Semifeudal Estates and Reform Sector	Capitalist Estates and Reform Sector	Capitalist Farms and Reform Sector	Peasant Farms	Socialist Farms
Pre-Land Reform / Land Tenure ↓						
Semifeudal	Semifeudal estates	(1) Mexico, 1917–34; Taiwan, 1949–51; Colombia, 1961–67; Chile, 1962–67	(2) Bolivia, 1952–; Venezuela, 1959–; Philippines, 1963–72; Ecuador, 1964–; Peru, 1964–69; Colombia, 1968–	(3) Mexico, 1934–40; India, 1950–; Guatemala, 1952–54; Egypt, 1952–66; Iran, 1962–67; Chile, 1967–73	(4) South Korea, 1950–; Taiwan, 1951–63; Iraq, 1958–	(5) China, 1949–56
Capitalist	Capitalist estates	(6)	(7) Costa Rica, 1962–76	(8) Peru, 1969–75; Philippines, 1972–79	(9)	(10) Cuba, 1959–63; Algeria, 1961–71
Capitalist	Capitalist farms	(11) Guatemala, 1954–	(12) Chile, 1973–	(13) Mexico, 1940–; Dominican Republic, 1963–; Egypt, 1961–	(14)	(15)
Socialist	Peasant farms	(16)	(17)	(18)	(19)	(20)
Socialist	Socialist farms	(21)	(22)	(23)	(24)	(25) Cuba, 1963–; China, 1952–; Algeria, 1971–77

Reforms 2, 3, and 4 aim at liquidating feudal remnants from agriculture and at inducing a transition to capitalism under domination in agriculture of the landed elite (2), farmers (3), or peasants (4). These reforms, directed against feudalism principally during the 1960s, have been both the most prevalent and the most successful in their intents. Reforms that induce a transformation of semifeudal into capitalist estates (2) seek this result by prohibiting bonded labor and rents in labor services (Bolivia, Philippines 1963–72, Ecuador, and Peru 1964–69) and by imposing minimum productivity levels on land use (Venezuela and Colombia). This is obtained by threats of expropriation if these requirements are not met. The purpose of a reform sector, in this case, is more to demonstrate the seriousness of the threats and to satisfy peasants' clamors for land than to increase production. Land reforms (3) induce the reorganization of the nonreform sector into a system of capitalist farms by imposing ceilings on land ownership in addition to prohibiting semifeudal social relations. Maximum farm size was thus restricted to 200 irrigated hectares in Mexico (1934–40), 50 in India, 42 in Egypt, 150 in Iran, and 80 in Chile. Here, again, expropriations and creation of a reform sector have mainly the political purpose of stabilization, while the economic gains of the reform are sought in the nonreform sector. The case of Mexico, with its remarkable political stability in spite of massive rural poverty and exceptional growth through the 1950s and 1960s obtained mainly in the nonreform sector, is the clearest illustration of the means and ends of this type of reform. Land reforms (4) are integral reforms in the sense that the nonreform sector is secondary or totally eliminated, and peasants acquire control of the land under the form of family farms (South Korea and Taiwan) or collective farms (Iraq). Taiwan and South Korea demonstrated the economic and political gains of integral reforms where family farms are effectively supported by external institutions that allow them to realize their full production potential.

Once the capitalist order has been established throughout agriculture, transition reforms are passé. The only possible reforms under capitalism in society at large are either shifts among types of agrarian structure (8, 9, and 14) or redistributive reforms within a given type of agrarian structure (7, 13, and 19). Historically, the most important type of reform in the first group has been that which transformed the nonreform sector from capitalist estates to capitalist farms (8). Its purpose is evidently to eliminate the landed elite from control of the state and thus achieve a more production-oriented agricultural policy while creating a more competitive environment in agriculture. This was attempted in Peru by imposing a ceiling on land ownership of fifty irrigated hectares on the coast and thirty in the Sierra. In the Philippines, the ceiling was seven hectares on rice and corn land. With the agrarian structure dominated by capitalist farms, subsequent land reforms can only be redistributive (13) or integral (14). Expansion of the *ejido* sector in Mexico from 1940 to 1977 and expropriation of rice farms in the Dominican Republic are instances of redistributive reforms (13).

All reforms can, of course, give way to counterreforms. Chile thus recreated a

landed elite after the military coup of 1973 by abolishing ceilings on land ownership, and Guatemala restored semifeudal social relations in agriculture when the
Arbenz government was overthrown by foreign armed intervention.

Land reforms that are part of a transition to socialism are evidently more than
mere reforms, since the mode of production dominant in society at large is
transformed—a process that is, by definition, nonreformist (5, 10, 15, and 20).
The Chinese land reform was part of a transition from feudal society, while the
Cuban and Algerian land reforms originated in agrarian structures dominated by
capitalist estates and plantations. Subsequent transformations of the agrarian
structure in China, Cuba, and Algeria fall in the category of redistributive reform
(25). For lack of space, we concentrate, in the rest of this paper, on land reforms
achieved under continued domination of the capitalist order in society at large.

LAND REFORM AS A POLICY ISSUE

During the last seventy years most land reforms under the capitalist order in
society at large have been of either one of two broad types: antifeudal reforms
seeking to induce in agriculture a transition to capitalism, with the resulting
agrarian structure dominated by a capitalist landed elite (reform type 2), a farmer
class (3), or a free peasantry (4) and land reforms within capitalist agriculture
seeking to create shifts in the dominant rural class from capitalist landed elite to
farmers and to peasants (8, 9, and 14) or to amplify the reform sector under
domination of either one of these three classes (7, 13, and 19).

ANTIFEUDAL REFORMS

Reforms against feudal bonds and personalized coercion have originated in
peasant rebellions since time immemorial, but they have led to transitions to
capitalist agriculture in the Third World only during the last fifty years, with the
cases of Mexico and Bolivia as outstanding examples. Since the late 1950s,
however, with the rise of surplus labor associated with demographic explosion
and modernization of many landed estates, organization of rural labor markets
with plentiful supplies has created a new rationality to oppose feudalism—this
time on an economic instead of a purely political basis: labor relations in agriculture could be transformed from feudal, where labor is a fixed cost, to capitalist,
where labor becomes a variable cost adjusted to fluctuating weather, market, and
technological opportunities. Antifeudal land reforms thus potentially could
achieve at the same time equity gains in response to peasant pressures and
efficiency gains in response to demands for cheap food originating among urban
industrialists and consumers and for increased export earnings originating in the
urban-industrial sector. This led to formation, in the 1960s, of broad coalitions of
different factions of the bourgeois, peasant, and proletarian classes opposing the
feudal landed elites under the banner of expropriation and redistribution of the
land. In Latin America, this resulted in the 1961 Punta del Este Charter of the

Organization of American States, inducing essentially every country of the continent to pass antifeudal land reform laws.

For the dominant interests in this coalition, the objectives of land reform were to be sought in the combination of nonreform and reform sectors. The economic goal was to be obtained by fomenting the development of capitalism in the nonreform sector through a mix of threats of expropriation and inducements to investment (subsidized credit, infrastructure construction, new technological advances, extension services, and so on). The goal of political stabilization was sought by creating a reform sector of a size commensurate with peasant pressures. Consequently, it was only under the most extreme threats of destabilization, such as in Mexico but even more in South Korea and Taiwan, that the reform sector reached major proportions. The success of antifeudal reforms should, consequently, be assessed not in terms of the extensiveness of expropriations (the publicized explicit goal) but in terms of development of capitalism and political stabilization, with actual expropriations serving only as a means of reaching these goals. In that sense, countries like Colombia, Ecuador, and India had successful antifeudal land reforms without virtually having any in a distributive sense.

In Latin America, the combination of spontaneous development of capitalism and antifeudal land reforms has virtually liquidated feudal remnants and, hence, put an end to this type of reform. The Green Revolution in food production and development of agroexport agriculture, in close connection with international agribusiness, have themselves served as effective surrogates or complements to antifeudal reforms (Arroyo 1979). In most of Asia and the Middle East, similarly, feudal elites have given way to a capitalist elite and a farmer class. In Africa, rapid privatization of the land has also seriously undermined precapitalist forms of land use.

With the end of feudalism, future land reforms must, of necessity, occur within the capitalist order in agriculture. This requires markedly different coalitions and occurs in response to markedly different purposes.

LAND REFORMS WITHIN CAPITALIST AGRICULTURE

Once agriculture is characterized by capitalist social relations, there are four important reasons why land reform becomes an unlikely policy issue. The first is that the political alliance that must be organized to support land reform will need to be capable of opposing the established capitalist interests in agriculture. This is evidently such a formidable undertaking that it usually will require extreme economic or political circumstances. Opposing the landed elite is difficult because it has strong control of the state apparatus; usually has diversified investments in industry, commerce, and finance that give it economic power beyond agriculture; and is closely allied with foreign capital. Opposing the farmer class is equally difficult, since it demands that one fraction of the bourgeoisie oppose another and puts into question the inalienability of property rights. The result is that reforms in capitalist agriculture have only occurred under rather exceptional circumstances, such as government reaction to strong revolutionary pressures (Mexico 1940–77 and the Philippines 1972–75), military interventions in favor of the national bourgeoisie (Peru 1969–75), or external influences (Dominican Republic). Short of this, expropriation of one fraction of the bourgeoisie by another requires full compensation for the former landlords. This, in turn, depends upon availability of a large fiscal surplus or massive foreign aid. This is partly what happened in the Dominican Republic, where the state used the enormous urban and industrial assets confiscated from Trujillo and plentiful foreign aid to compensate fully the expropriated rice farmers. Seen in a historical perspective, however, strong public budgets are unlikely to characterize Third World states.

The second reason why land reforms in capitalist agriculture are an unlikely policy issue is that the former coincidence of efficiency and equity gains that characterize antifeudal reforms not only may have vanished with the development of capitalism in agriculture but may have given way to a trade-off between these two purposes of reform. With agriculture well advanced on the road to modernization in the context of medium- and large-scale capitalist farms and of a close integration with multinational agribusiness, any drastic land redistribution is likely to nullify past technological achievements and imply shortfalls in production, at least in the short run. Where the population is increasingly landless and urbanized, the social cost of higher food prices may be more widespread than the welfare gains of land redistribution. This does not mean, of course, that land reforms cannot be managed to avoid efficiency costs: organization of the reform sector on the basis of state or cooperative farms (Peru's sugar farms and the Dominican Republic's rice sector) and establishment of institutions strongly supportive of family farms (Taiwan and South Korea) have demonstrated feasible solutions. However, it does mean that the margin for successful management of reform in terms of production performance is substantially reduced by the very development of capitalism. Seen in this fashion, the Green Revolution and integration of Third World agriculture with international agribusiness have both been effective surrogates for antifeudal land reforms and obstacles to subsequent progressive reforms.

A third reason why land reforms in capitalist agriculture are unlikely policy issues is that most Third World countries have opted for models of economic development where industrial growth is based on expansion of a market for exports or luxury consumption goods and not principally of a market for wage goods. This implies that progressive income redistribution to create the home market for industry is not essential to avoid underconsumption crises. As in Lewis's (1958) model of the dual economy, where the working class and peasants do not consume the products of the modern sector, income redistribution acts as a brake on the rate of economic growth as it lowers the rate of profit. Income redistribution and, hence, land reforms are not economic needs for the system at large in creating the domestic market for the modern sector but political gains of the working class and peasants. For that reason, again, land reforms will only result from strong pressures that can question the existing social order.

A fourth and last difficulty with capitalist land-reform policies is that the bourgeoisie will respond to social pressures for land reform by conceding it for the sake of legitimation of the dominant social relations. Hence, land-reform programs are expected to be as limited as possible while achieving their political purpose. But any program of land reform tends to unleash redistributive expectations and to stimulate broad mobilization of peasants and workers. Limited land reforms are thus an instrument of both expected stabilization and potential destabilization. As Sorj (1980) observes for the case of Brazil, this implies that "any attempt at limited agrarian reform will be accompanied by the action of a repressive apparatus and by creation of corporativist bodies intended to keep the reform within the limits of the present structure of accumulation and domination" (p. 31). Faced with both the need to create social support in agriculture and the ambiguities of land reform for that purpose, most governments have turned, since the early 1970s, to the alternate strategy of integrated rural development. The political end is here the same: to create a supportive minority of upper peasants. But the instrument for that purpose need not threaten land ownership, because technological change becomes a substitute for land redistribution. The example of Colombia is here clear: in 1973, both liberal and conservative parties agreed to put an end to the destabilizing effects of limited land reform resulting, in particular, in continuing land invasions, while launching an ambitious program of integrated rural development with the support of international lending institutions. At the same time, agricultural growth was stimulated by generous public support to the large-scale capitalist sector.

In a postfeudal capitalist order, consequently, it is no surprise that so few land reform policies are being enacted or implemented. Yet, the symptoms of serious agrarian crises remain unabated: food deficits keep on increasing (IFPRI 1977), and rural poverty keeps on worsening (ILO 1977). For this reason, land reform remains an important political issue even if for a variety of different purposes according to contrasted political programs.

LAND REFORM AS A POLITICAL ISSUE

A wide range of arguments has been advanced justifying the role of land reform in economic development, and it is the prevalence of these arguments that keeps land reform an active political issue. Some of these arguments have been advanced on technocratic grounds, while others have been openly couched in ideological terms. Yet, all land reform programs fit within some global view of the role of agriculture and peasants in economic development. These can be regrouped under four alternative political programs, each of which proposes land reform for sharply contrasted economic and political purposes.

LAND REFORM FOR SOCIAL STATUS QUO: THE CONSERVATIVE MODEL

With capitalism well entrenched in agriculture and the performance of agriculture fundamentally determined by market incentives and public and supporting institutions (infrastructure, technology, credit, and so on), the conservative model allows for land reform only as a minimal concession for political stabilization. Creation of a reform sector is used to defuse social tensions when pressure on the land (land invasions, and so on) or revolutionary threats are excessive. It does this by allowing some vertical mobility to peasants and by creating a buffer minority of privileged peasants in the reform sector whose economic success is tied to the interests of capitalist farming and to continued patronage of the state but whose ideological allegiance is still with peasants. As such, they tend to become the political representatives of the peasantry at large and strong advocates of the social status quo that benefits them so exclusively. Land reform is here purely legitimizing of the dominant social relations. It only becomes an active political issue when this legitimacy is being questioned. It is, consequently, only in the following three positions that land reform is an integral component of political programs.

LAND REFORM FOR NATIONAL BOURGEOIS REVOLUTION: THE LIBERAL MODEL

For as long as feudal elements dominated agriculture, national liberal forces were directed at promoting antifeudal land reforms. With the omnipresence of capitalism, the same forces identify their targets in terms of opposing blatant external dependency relations and in terms of redefining the social and geographical location of the market for the modern sector in wage earnings as opposed to profits, rents, and exports. Both creation of a class of farmers instead of a capitalist landed elite (land reform type 8) and redistribution of land to peasants (7, 9, 13, and 14) are seen as ways of expanding the domestic market for wage goods. Since farmers and peasants are oriented more toward producing basic wage foods than the landed elite, which is principally involved in the production of agroexports, these land reforms also are proposed as mechanisms for reducing the bias against food and alleviating the food crisis. Finally, by reducing inequality, they are seen as essential to creating the economic basis needed for more democratic forms of government. In Brazil today, for example, land reform is an important rallying point for dispersed forces struggling for a return to civilian democracy.

LAND REFORM TO GIVE THE LAND TO THE PEASANTS: THE POPULIST MODEL

This position usually is argued in technocratic terms by pointing at the superior social efficiency of small farms and peasants under conditions of surplus labor:

smaller farms produce more per acre than larger ones and reach a higher total factor productivity when labor and capital are valued at their social opportunity cost, which is virtually zero on small farms. Integral land reforms (types 9 and 14) to create a free peasantry are, thus, advocated since peasants' higher levels of (Chayanovian) self-exploitation lead them to deliver cheaper food on the market than under a system of capitalist farming, at least for as long as the productivity of labor is nearly equal on small and large farms (Dorner and Kanel 1971; Lipton 1977). As Berry and Cline (1979) put it, "Land redistribution should therefore be expected to raise total output by combining underused labor from small farms and the landless work force with underused land on large farms. Nor is there likely to be a sacrifice of potential efficiency from land redistribution, because it is unlikely that there will be significant economies of scale for actual farming operations" (p. 29). Integral land reforms are, in turn, the preconditions for any meaningful effort at rural development (Wortman and Cummings 1978). Efficiency gains are, of course, also accompanied by equity gains, with the result that the populist vision of land reform offers the same logical panacea for capitalism as antifeudal reforms did.

LAND REFORM FOR SOCIAL CHANGE: THE RADICAL MODEL

A radical interpretation of the agrarian crisis in the Third World essentially concludes that (*a*) once feudalism is gone, the problems of agriculture are essentially nonagrarian, as they reflect the contradictions of the global model of economic development pursued—characterized by its export and luxury orientation and by strong relations of external dependency. This, in turn, leads to a bias against food production (cheap food policies) and in favor of agroexports (comparative advantage theory). And (*b*) peasants are rapidly being dispossessed of their status as producers and transformed into a reserve of cheap semiproletarian labor.

Since resolution of the agrarian crisis consequently hinges upon broad social changes in the economy at large, land reform per se is necessary for economic development but certainly is insufficient. Yet, land reform need not wait for broader social changes to occur. The promotion of land reform itself is seen as an important instrument of social change, as a rallying cause to foment the organization of peasants and workers and the emergence of new ideas. Land reform is thus a process that is to be carried on by and not for the peasants and is completed only with the restructuring of the global social system.

CONCLUSION

Transformation of the agrarian structure is today more dependent upon the forces of capitalist development and, in particular, on the industrialization of commercial agriculture and the proletarianization of peasants than upon the application of land-reform policies. Solutions to food and balance-of-payment deficits are sought in medium- and large-scale capitalist farms closely integrated with international agribusiness. Peasants' demands for better living conditions are either neglected and left to the uncertainties of domestic and international migration or severely repressed. Yet, it is abundantly clear that the crises of food production and rural poverty are, if anything, worsening under the current development model. Thus, even if land reform is dead as a policy issue, it remains a key ingredient of any meaningful political program of economic development, be it of liberal, populist, or radical slant.

NOTES

The author is indebted to Karen Lopilato for her assistance in fact finding.

1. For Latin America only see Land Tenure Center 1974.
2. I call "policy" the institutional changes implemented by the government in power, while "politics" are institutional changes proposed by any social group—dominant or not.
3. There are, of course, other modes of production beyond the feudal that are not capitalist or socialist. These include the Asiatic, communal, and lineage modes. Without much loss of generality for the purpose of this paper, I am subsuming all these other modes under the feudal heading.

REFERENCES

Arroyo, Gonzalo. 1979. "The Industrialization of Agriculture." *International Development Review* 21:3–8.
Berry, Albert, and William Cline. 1979. *Agrarian Structure and Productivity in Developing Countries.* Baltimore: Johns Hopkins University Press.
Dorner, Peter, and Don Kanel. 1971. "The Economic Case for Land Reform: Employment, Income Distribution, and Productivity." In *Land Reform in Latin America,* edited by Peter Dorner. Land Economics Monograph, no. 3. Madison: University of Wisconsin.
Food and Agriculture Organization of the United Nations (FAO). 1979. *Report: World Conference on Agrarian Reform and Rural Development.* Rome.
International Food Policy Research Institute. 1977. *Food Needs of Developing Countries: Projections of Production and Consumption to 1990.* IFPRI Research Report, no. 3. Washington, D.C.
International Labour Office. 1977. *Poverty and Landlessness in Rural Asia.* Geneva.
Land Tenure Center. 1974. *Agrarian Reform in Latin America: An Annotated Bibliography.* Madison: Land Tenure Center, University of Wisconsin.
Lewis, W. Arthur. 1958. "Economic Development with Unlimited Supplies of Labour." In *The Economics of Development,* edited by A. Agarwala and S. Singh. New York: Oxford University Press.
Lipton, Michael. 1977. *Rural Poverty and Agribusiness.* Sussex: Institute of Development Studies, University of Sussex.
Sorj, Bernardo. 1980. "Agrarian Structure and Politics in Present Day Brazil." *Latin American Perspectives* 7:23–34.
Tuma, Elias. 1965. *Twenty-six Centuries of Agrarian Reform.* Berkeley and Los Angeles: University of California Press.
Wortman, Sterling, and Ralph Cummings, Jr. 1978. *To Feed This World.* Baltimore: Johns Hopkins University Press.

18

The Role of Land Reform in Economic Development: Comment

DANIEL W. BROMLEY

There are several important points made by de Janvry: one relates to his taxonomy of land reforms by country and over time; another is to be found in his four reasons why land reform will not be an important issue on the economist's research agenda in the coming decade. With respect to his taxonomy, de Janvry claims that it permits the identification of the needed characteristics of future reforms, given the actual state of existing agrarian structures. He then argues that this gives us a basis from which to seek answers to the question about why land reform appears to be a dead policy issue in most countries. I have four concerns with respect to his treatment of the land reform issue and will now turn to a brief discussion of each.

My first concern is that the four reasons he offers for why land reform is a dead policy issue do not follow uniquely from his analysis but could be derived quite independently. De Janvry offers the following reasons: (*a*) the political alliance for land reform must be capable of opposing an established capitalist class in agriculture; (*b*) all efficiency gains from land reform have virtually been exhausted; (*c*) domestic demand is no longer important as a macro consideration, since the industrial sectors are primarily export oriented; and (*d*) the ruling class will only give in grudgingly and in small doses. I see little unique about his paradigm or his taxonomy which produce these findings. To put it differently, it is possible to arrive at these conclusions quite outside of his analysis.

My second concern is that the above list of reasons for the demise of land reform as a policy issue, while rather reasonable in spite of my contention that they do not follow from his analysis, misses one very important consideration. A plethora of changes have occurred in the developing economies which, in combination, drive down the economic role of land to the point that in some instances

DANIEL W. BROMLEY is professor of agricultural economics and chair, Department of Agricultural Economics, University of Wisconsin at Madison.

Reprinted from *American Journal of Agricultural Economics* 63, no. 2 (1981):399–400, with minor editorial revisions, by permission of the American Agricultural Economics Association and the author.

access to—or control over—land may be quite irrelevant. Perhaps this is implicit in his second item on the above list, but its importance warrants special mention.

The advent of high-yielding varieties, increased reliance on chemical pesticides and fertilizers, control over irrigation water, and the multitude of ways in which urban-dominated agricultural policy can countermand any nominal gains from better land-use opportunities all combine to render—in many instances—land ownership quite irrelevant. There are so many factors that impinge upon the economic environment of a newly landed peasant that the legal and/or economic relations of that person to the land resource may be quite beside the point. Land cannot be considered important in either a policy or a political context when it is often so irrelevant in the ultimate reckoning of a peasant's economic position.

My third concern is that his paradigm, which concentrates on modes of production and class relations, is overly rigid and not terribly helpful analytically. For instance, he says that a reform is an institutional innovation promoted [permitted?] by the ruling order to overcome economic and/or political contradictions without changing the dominant social relations. However, if the agricultural population is transformed from one of 10 percent landowners and 90 percent wage laborers/tenants to one of 75 percent landowners and 25 percent wage laborers/tenants, it is hard for me to see how the dominant social relations have not been seriously altered. The fact that the nonagricultural sector remains capitalist, along with the agricultural sector, may make it seem that social relations have not changed much, but this is an artifact of a model that sees social relations as derivative of modes of production writ large. It is the old capitalist/socialist dichotomy, and here it seems to miss very important changes in social relations even in the absence of a transition from capitalist to socialist modes.

To de Janvry, revolution is the only thing that will alter social relations, and here he would mean a transition from feudal to capitalist, or feudal to socialist, or capitalist to socialist. This sort of rigid dichotomy does not seem very helpful analytically, especially when it leads us to conclude that dominant social relations have not been altered by a massive redistribution of land within the capitalist mode. When this new economic opportunity is created for the formerly landless, few could deny that sweeping changes have occurred in the social relations governing rural life.

My fourth and final point concerns de Janvry's use of the term "land tenure." He says that land reform changes the modes of production, the associated class structure, and the pattern of land tenure. I may be missing something here, but I consider it more helpful to say that land reform alters four aspects of the man/land interface: namely, (*a*) who controls land, (*b*) who may use land, (*c*) who reaps the benefits of land use, and (*d*) who bears the costs of land use. Once this is said, we then can explore the nature and extent of changes in these four factors, and we might conclude that the dominant mode of production has been changed and that this will then hold important implications for social relations.

Simply put, I question his causal chain, which runs from modes of production to class structure to land tenure.

In closing, let me emphasize that in spite of my concerns, I found the paper to be useful, informative, and an important new perspective on a topic which—as he indicated—will remain an important political issue. The symbolic value of land will remain prominent even as its economic value may be allowed to decline.

19

Choice of Technique in Rice Milling on Java

C. PETER TIMMER

INTRODUCTION

A technological revolution has swept across Java virtually un-noticed. As recently as 1971 informed estimates assumed that as much as 80 percent of Java's rice crop was hand-pounded, both for subsistence consumption and for local marketings. The figure now (1973) is certainly less than 50 percent and may well be as little as 10 percent—there are no direct statistics from which to judge. The number of small mechanical rice processing facilities has increased dramatically. The economic and social impacts are only beginning to be felt, and any assessments of these changes in the countryside and villages are necessarily for the future.

This article will try to answer three questions about the recent changes in rice processing: (1) What has happened, in rough quantitative terms? (2) What explains the shift from hand-pounding to small rice mills and why not to larger mills? and (3) What has happened to rural employment with the decline of hand-pounding? The secondary data are taken from an anonymous report from the Asian Development Bank (ADB),[1] a Rice Marketing Study (cited in this paper as RMS) prepared by a U.S. consulting firm,[2] and two publications by the author.[3]

TECHNOLOGY CHOICES

Three rice processing techniques are in use in Indonesia: hand-pounding, small rice mills, and large rice mills. The overall patterns of what has happened technologically to rice milling in the past few years are very clear: hand-pounding has declined drastically. But the large-scale mills have not been the beneficiaries of the decline in hand-pounding. Rather, a whole new rice-milling industry has sprung up, composed of an assortment of machinery but with one overriding

C. PETER TIMMER is John D. Black Professor of Agriculture and Business, Harvard University.

This paper and the following comment are reprinted from *Bulletin of Indonesian Economic Studies* 9, no. 2 (1973), with omissions and minor editorial revisions, by permission of the publisher—The Research School of Pacific Studies of the Australian National University—and the authors.

characteristic: all of the facilities are small in scale and labor-intensive. A quote from ADB, *Rice Processing Report,* makes the point explicitly:

> Why pay more than 40,000 Dollars for a large rice mill when the same output can be attained by a small mill with an investment of 7,400 Dollars? Naturally the cheap unit requires far more labour, but that does not count much in a country where a mill labourer is paid Rp 100–150 (U.S. Dollars 0.24 to 0.36) per working day. The differences in rendement (extraction) rate (less than 1 percent) and broken percentage (around 3 percent) are also insignificant compared to the high debt service for the large-scale mill.

The type of machinery used to mill rice—from wooden pounding pole and pestle to large-scale multi-stage mills with bulk storage and drying—is, then, a function of labor and capital costs. The results of a formal economic analysis of the question of choice of technique in rice milling strongly confirm the "common sense" of the quote above.[4] Both the methodology and the results of the analysis are presented below.

The analytical technique is purposefully textbookish: the aim is not to confuse the choice of technique issue with fancy methodology but to be as simple-minded as the problem will allow. Consequently, an isoquant has been constructed that represents several different techniques, in terms of relative capital/labor ratios, of producing a unit amount of value added in rice processing. After going step by step through the data, the assumptions and the manipulations, we end up with a standardized isoquant in the two dimensions of investment cost and labor cost.

Five different processing techniques form the basis of the analysis: hand-pounding (HP); small rice mills (SRM); large rice mills (LRM); small bulk facilities (SBF); and large bulk facilities (LBF). The first three techniques are at present in operation in Indonesia—hand-pounding is millennia old. The small rice mills can be thought of as ranging from the now obsolete double Engelberg-type huller/polisher combinations to the smaller self-contained Japanese rice milling units and rubber roll huskers connected to Engelberg-type or pneumatic polishers. All these together are taken as the class "small rice mills." They do not have mechanical drying equipment but rely solely on sun drying. Consequently, the small rice mills (SRM) are assumed to suffer high physical (and monetary) losses during processing, not so much in the milling per se but from a lack of control in drying.

Fewer "large rice mills" are seen in Indonesia, although all the older multi-stage milling equipment fits this category if mechanical drying facilities are also available. The major feature of this category is the combined use of mechanical and sun drying with modern milling equipment, either Japanese-type or conventional multi-stage.

The fourth and fifth techniques—bulk facilities—represent proposed improvements to the Indonesian rice processing sector. "Bulk facilities" is a shorthand way of describing a rice mill/storage system. The rice mill produces high-quality milled rice which is stored in vertical silos of varying capacity. No example of either the small bulk facility or the large bulk facility exists at present in Indonesia, although even the larger facility is not large by international grain-handling standards. The small bulk facilities have a milling and drying capacity of three tons per hour and forty-five hundred tons of vertical steel storage. The large bulk facilities can process nine tons of rough rice per hour and contain fifteen thousand tons of bulk storage in vertical steel silos for rough rice and conventional multi-stage gradual-reduction milling machinery. A continuous-flow dryer attached has a capacity of twenty tons per hour.

CONSTRUCTING UNIT ISOQUANTS

The data and calculations necessary to construct a unit isoquant in value added by rice processing are presented in table 1. The isoquant is drawn in the two dimensions of total investment cost and number of unskilled workers needed to run each facility. A "budget" or "iso-cost" line can then be drawn tangent to the isoquant in order to determine the optimum capital/labor ratio for rice processing. Since the capital axis is in total investment cost and not in annualized capital charges, it is necessary to convert annual laborers' wages into a lifetime "wage fund." This is done by discounting a laborer's annual earnings for his lifetime (fifty years) by an appropriate discount rate (12 percent, 18 percent, or 24 percent) to calculate the present value of the cost of a laborer.

Three basic steps are necessary to construct the isoquant, and all three are carried out in table 1. The "Data Per Unit" section simply reports capacity, investment cost, and number of manual laborers required per operating shift for each facility. These data are then standardized in terms of one thousand tons of rough rice input per year. The data for the small rice mill do not change, since its initial capacity was one thousand tons per year, but each of the larger facilities is scaled down to a comparable one-thousand-ton capacity. Obviously, it is not possible to install one-twentieth of a large bulk facility, but this technique merely keeps the numbers manageable without losing any of the scale economies of the larger facilities. Conceptually it is equally possible to use the large bulk facility capacity of 21,600 tons as the standard and multiply the number of smaller units to be on a comparable basis. Nothing is lost by keeping the numbers smaller and more manageable.

PHYSICAL AND MONETARY CONVERSIONS

The next step is to convert the one thousand tons of rough rice input per facility into a value of output. There are two parts to the process: a physical conversion and a monetary conversion. Table 1 shows the conversion rates for both processes. The first is merely the rendement or extraction rate. How much milled rice does each facility produce per ton of rough rice? No fixed rates are really applicable, since this extraction rate depends so critically on quality of input, moisture, variety, by-product rates, and so on, but the average figures

TABLE 1

DERIVATION OF A UNIT ISOQUANT IN VALUE ADDED FROM RICE PROCESSING

	Hand-Pounding HP (s)[a]	Small Rice Mill SRM (s)[a]	Large Rice Mill LRM (s/m)[a]	Small Bulk Facility SBF (m)[a]	Large Bulk Facility LBF (m)[a]
	Data per Unit				
Milling capacity (tons per year)[b]	—	1,000[c]	2,500[d]	7,200	21,600
Investment cost (U.S. dollars)[e]	0	$8,049[c]	$90,511[d]	$453,283	$2,605,926
Operative laborers (number per shift)	—	12[c]	16[d]	27	39
	Data per 1,000 Tons of Rough Rice Input per Year				
Investment cost (U.S. dollars)	0	$8,049	$36,204	$62,956	$120,645
Operative laborers (number)	22.00[f]	12.00	6.40	3.75	1.81
Milled rice output (tons)	570	590	630	650	670
Market price (Rp/kg)	40.0	45.0	48.0	49.5	50.0
Value of output (million Rp)	22.8	26.6	30.2	32.2	33.5
Value added[g] (million Rp)	4.8	8.6	12.2	14.2	15.5
	Data per Rp 10 Million in Value Added per Year				
Investment cost (U.S. dollars)	0	$9,359	$29,675	$44,335	$77,835
Operative laborers (number)	45.83	13.95	5.25	2.64	1.17

[a]The *s* in parentheses indicates that the facility uses sun drying; *m* indicates mechanical drying.

[b]Milling capacity is measured in tons of rough rice input per year, assuming the facility can operate 2,400 hours per year.

[c]The technical and cost data for the small rice mill relate specifically to a locally manufactured flash-type husker with an input capacity of three-quarters of a ton per hour linked to an Engelberg-type polisher with an input capacity of half a ton per hour. A thresher shed and sun-drying pad are provided, but no additional storage capacity. This facility is taken as "representative" of the entire range of small rice milling facilities.

[d]The technical and cost data for the large rice mill relate specifically to a Japanese self-contained milling unit integrated with 756 tons of rough rice storage capacity: 300 tons in bagged storage and 456 tons in upright bulk steel bins. A mechanical dryer with three-quarters ton per hour capacity is provided along with a sun-drying pad. The milling capacity is 1.0 to 1.1 tons per hour of rough rice.

[e]Costs of building and machinery (including a diesel-powered thresher) are included, but not land.

[f]This assumes that one worker hand-pounds about 150 kg of *gabah* (unhusked rice) per day, with a yield of approximately 95 kg of rice.

[g]Value added is calculated assuming a *gabah* cost of Rp 18 per kg, which is above the Rp 16 per kg floor price for "village dry" *gabah* at the mill. The cost (and value added) of reducing the moisture to "mill dry" is included as a mill activity.

shown in table 1 capture the general trend: the larger and more capital-intensive facilities have significantly higher extraction rates. Thus the hand-pounding rate is taken as 57 percent, the small mill rate as 59 percent, the large mill rate as 63 percent, and the small and large bulk facility rates as 65 percent and 67 percent, respectively. As was noted earlier, these differences are not to be taken as strictly due to different milling techniques, but capture the overall impact of the significantly different drying and storage facilities connected with each type of mill.

Apart from the sheer differences in physical output per ton of input, each facility is presumed to produce an output with varying consumer value. Thus one kg of hand-pounded rice sells for only Rp 40, while the output of the large bulk facility is valued at Rp 50 per kg; facilities between these two extremes sell at appropriate values in between, as shown in table 1. Although the different values include some allowance for better by-product retention in the larger facilities, these prices still somewhat overstate the actual market differences that consumers are willing to pay for rice quality per se, as opposed to varietal price differentials, which are substantial. Should these consumer preferences ever be developed, however (or if they were ever imposed by Bulog),[5] the smaller facilities and especially hand-pounding would be at a severe disadvantage.

The overall impact of the physical and monetary conversions is that one thousand tons of rough rice input is transformed into Rp 22.8 million in output by hand-pounding, Rp 26.6 million by small rice mills, Rp 30.2 million by large rice mills, Rp 32.2 million by small bulk facilities, and Rp 33.5 million by the large bulk facilities. That is, the large bulk facilities are able to produce 47 percent more value of output from a given input than hand-pounding and 26 percent more output than the small rice mills. The differences in value added, when Rp 18 million is subtracted from value of output as the cost of the rough rice input, are even more dramatic. The large bulk facility produces more than 200 percent more value added than hand-pounding and almost double the value added from small rice mills. At this stage of the analysis the larger facilities seem to hold a very substantial competitive edge.

The last step required in the construction of a unit isoquant is to calculate the investment cost and numbers of laborers needed to produce a given amount of value added—Rp 10 million in table 1. Since the small rice mill produced only Rp 8.6 million in value added from one thousand tons of rough rice input, it is necessary to increase the investment-cost and number-of-laborers data under that heading by 10/8.6, or 1.16. Thus to produce Rp 10 million in value added, the small rice mill needs $8,049 × 1.16 = $9,359 in investment cost and 12 × 1.16 = 13.95 operative laborers. The calculations for the other facilities are similar, and the results are shown in the last two lines of table 1.

These numbers are the "x, y coordinates" of the desired isoquant, which is drawn in figure 1. Because only five different processing techniques are considered in table 1, the isoquant in figure 1 does not look exactly like the smooth, twice-differentiable isoquants shown in neoclassical production economics textbooks. But the family resemblance is very close. In the linear-segmented world shown in figure 1 smooth tangencies of iso-cost lines to the isoquant must give way to corner tangencies. Thus it is that all three iso-cost lines shown are (corner) tangent at small rice mills. This means that small rice mills are the least-cost facilities for producing value added in rice processing, at least at the wage levels shown in figure 1.

The wage level used to calculate the iso-cost lines was $200 per year, a level used by the *RMS Report* to evaluate their recommendations. The present value of a $200 wage payment each year for fifty years was calculated using three different time rates of discount: 12 percent, 18 percent, and 24 percent. Although it is

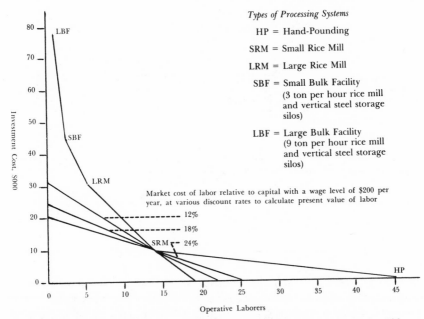

Fig.1. Isoquant Relating Investment Cost and Laborers Needed to Produce Rp 100 Million in Value Added Assuming Rough Rice Costs Rp per kg

always difficult to determine the appropriate social discount rate under any circumstances, investment calculations in rice milling on Java should use something close to the social opportunity cost of capital. Most observers would place this closer to the 24 percent end of the spectrum, with 12 percent clearly reflecting a substantial subsidy. Thus at the opportunity cost of capital—24 percent—a wage of $200 per year has a present value of $833. This means that the slope of the ''24 percent'' iso-cost line in figure 1 is the same as a line that connects $833 on the Investment Cost axis with one Laborer on the horizontal axis.

Table 2 presents the present values of a range of wage levels discounted at the three different discount rates. The three iso-cost lines drawn in figure 1 (all of them with corner tangencies at SRM) represent the highest assumed wage under consideration. Two other wage levels, not shown in figure 1, are also worth consideration. The $80 wage is close to an actual market wage for unskilled labor for large parts of Java. The $40 wage represents a shadow wage to take account of the high degree of surplus labor on Java, although this may actually be close to a market wage in certain heavily populated rural areas.

The striking effect of discounting to obtain the present value of wages stands out in table 2. The $200 wage for fifty years is a total wage bill of $10,000, but its present value is only $1,661 when discounted at 12 percent per year, the subsidized rate of interest for medium-/long-term investment credits in Indonesia. Using a market interest rate of 24 percent reduces the present value wage

TABLE 2
PRESENT VALUE OF ALTERNATIVE WAGE RATES
($)

Assumed Wage Levels per Year	*Discounted Present Value of Wages[a]*		
	12% Discount Rate	*18% Discount Rate*	*24% Discount Rate*
$200 (RMS)	$1,661	$1,111	$833
$80 (Market)	664	444	333
$40 (Shadow price)	332	222	167

[a]The present worth factor for fifty years is as follows:

Discount Rate (%)	Present Worth Factor
12	8.304
18	5.554
24	4.167

to $833. Market wages ($80) discounted at market interest rates have a present value of $333, while a shadow wage rate equal to half the market wage has a present value of less than $200. Whether one is calculating annual capital charges on a fixed investment or determining the present value of a worker's lifetime earnings, the effect of discounting, the only known way of evaluating future economic events at the present time, is dramatic indeed.

What would the optimal rice processing technique be at the lower market and shadow wage rates? And what happens if the selling price of rice is held constant (by competition or imports) while the price of rough rice paid by the milling facilities increases? Both these questions are answered in table 3, where the maximum present value wage at which each rice processing technique remains optimal for various rough rice prices is shown. Thus for a rough rice price of Rp 18 per kg (the value used to construct the isoquant in table 1 and figure 1), hand-pounding is the optimal technique until the present value wage exceeds $294; small rice mills are optimal from a present value wage of $295 to $2,335, beyond which large rice mills, small bulk facilities, and ultimately large bulk facilities become optimal.

Similarly, the maximum present value wage a facility can afford and still remain optimal declines as the rough rice price increases. At a price of Rp 18 per kg for rough rice the small rice mill can pay up to $2,335, but at Rp 20 per kg it can pay only $1,956. At a price of Rp 26 per kg the mill would have to be subsidized to remain optimal; that is, it would have to receive $260 per worker (for his lifetime services). At Rp 28 per kg the small rice mill actually produces a negative value added—its output sells for less than the cost of its rough rice input. No subsidy can make the firm optimal, although obviously a large enough subsidy would permit the small rice mill to survive (but never to produce a positive value added).

TABLE 3

MAXIMUM PRESENT VALUES OF WAGE RATES THAT YIELD CORNER TANGENCIES
TO THE ISOQUANT

	Maximum Present Value Wage at Which the Indicated Facility is Optimal for Various Rough Rice Prices Rp/kg					
Corner Solution at	*18*	*20*	*22*	*24*	*26*	*28*
Hand-pounding (HP)	294	202	70	a	a	a
Small rice mills (SRM)	2,335	1,956	1,457	766	−260	a
Large rice mills (LRM)	5,617	5,034	4,265	3,197	1,699	−728
Small bulk facilities (SBF)	22,789	21,829	20,468	18,879	16,296	12,316
Large bulk facilities (LBF)[b]	greater than the value shown for SBF					

[a]Negatve value added.

[b]Just for comparison, it is worth noting that the discounted present value of $6,000 per year at 8 percent is $73,400. Thus the relative capital intensity of even the large bulk facilities (LBF) is low by U.S. standards.

The interesting comparisons are between the various present value wages of table 2 and the maximum wages that yield optimum solutions in table 3. At either $200 per year or the market wage of $80 per year, for all three discount rates, the small rice mills remain the optimal rice processing facility at normal rough rice prices. When the price reaches Rp 24 per kg, the large rice mills are optimal at the highest wage for all three discount rates. The small bulk facilities only enter the picture when the rough rice price reaches Rp 28 per kg, so high that the small rice mills yield negative value added. This situation could only come about by full "modernization" of Indonesia's entire rice processing system or a rigidly enforced price policy that used imports to control retail price and large-scale facilities (or large subsidies to smaller mills) in connection with a vigorous floor price. The general point ought to be clear, however: in a squeeze between rough rice and milled rice prices only the technically more efficient facilities can survive.

It is not hard to see, then, the reasons why few large rice mills (or bulk facilities) have been installed in recent years in Indonesia. Where market forces—low wages, high interest rates, and cheap rice prices—have been allowed to work, the overwhelming superiority of small rice mills as a generic class has been apparent to investors, and these facilities have indeed mushroomed throughout Java. Some large mills have been built, but they have been either government- or army-sponsored (where market forces are little felt in the decision making) or close to cities, where wage rates are higher than in rural areas, rough rice prices relatively higher, and when investment capital at 12 percent was available. Under these circumstances investment in the larger rice mills is understandable, though it may not be socially desirable.

At the other end of the spectrum, the economies of hand-pounding are fascinating. If market wages of about $80 per year (say Rp 100 to Rp 150 per day) must be paid, hand-pounding is not optimal unless the discount rate is greater than 24 percent or unless the *gabah* price falls below Rp 16 per kg (at the same time that the retail price is Rp 40 for the end product). So it is not hard to see why hand-pounding has virtually disappeared from Java as a cash-hire activity. And yet it is still frequently profitable for the farmer or his family to hand-pound their rice for home consumption if the perceived opportunity cost of the labor spent is not great. For example, hand-pounding is economically optimal (despite the very large losses that are assumed) if the wage rate is $40 per year and is discounted at 18 percent or higher, at least for rough rice prices less than Rp 20 per kg, as they are likely to be in many rural areas. And in fact, of course, some rural families still do hand-pound their rice.

The evidence from the countryside, which showed literally thousands of small rice mills installed on Java in the past three years, is thus strongly corroborated by the economic analysis. Only at the extremes of economic conditions existing in Indonesia could hand-pounding or large rice mills be explained. The small rice mills were shown to be socially and privately optimal over a wide range of circumstances in between. There are exceptions to be sure: exporting rice would require facilities not optimal in this analysis, and Bulog may have some special needs that are not fully met by all small rice mills. But in total this is a good example of where "getting prices right" has helped the development effort at minimal cost and effort to the scarce planning manpower in the government. Not all development problems are so easy, of course, but some are made harder by inappropriate pricing policies. Indonesia's rice processing sector still faces some difficult problems, especially with regard to drying and storage. But milling has been largely removed as a constraint.

EMPLOYMENT IN RICE MILLING

The employment potential of the entire rice economy—production, harvesting, processing, transporting, storing, selling—is enormous. Rice marketing, defined broadly, is only part of this potential, and rice processing but part of this part. Still, the total employment figures must be large indeed. No reliable survey data are available to serve as a reference here, but table 4 provides a rough point of departure for discussion.

The intent of table 4 is to work backward from the input/output micro data used to generate the isoquant to an overall estimate of the number of workers needed to process the entire 1971 rice crop, assuming only a single technique were used. This is obviously an artificial number. But somewhere within the range of the 399,000 full-time workers needed to hand-pound the crop to only 33,000 workers needed if large bulk facilities were used lies the actual full-time employment potential. (If work sharing is pervasive, the numbers actually earning a livelihood in rice processing could be much larger.) It is interesting that each technique employs roughly double the number of workers of the next most

TABLE 4
EMPLOYMENT AND INVESTMENT IN RICE PROCESSING IN 1971 USING
ALTERNATIVE MILLING TECHNIQUES

	Inputs			
	Per 1,000 Tons of Rough Rice		*For 1971 Harvest[a]*	
Milling Technique	*Investment Cost ($)*	*Operatives (Number)*	*Investment Cost ($m)*	*Operatives (Thousands)*
Hand-pounding (HP)	$ 0	22.00	0	399[b]
Small rice mills (SRM)	8,049	12.00	146	217
Large rice mills (LRM)	36,204	6.40	656	116
Small bulk facilities (SBF)	62,956	3.75	1,141	68
Large bulk facilities (LBF)	120,645	1.81	2,186	33

SOURCE: Calculated from table 1.

[a]For each technique, the figure shows the investment and employment required to process the 1971 rice crop of 18.123 million tons of rough rice, assuming that particular technique is the only one utilized. Obviously this shows alternative hypothetical magnitudes only, and none of the figures should be interpreted as representing actual employment levels in rice milling.

[b]This is the full-time equivalent required for hand-pounding. More than 399,000 laborers would be needed if hand-pounding were done only part-time.

capital-intensive technique. This is simply a repeated and forceful demonstration that this particular part of the economic system does not exhibit fixed technical relationships between capital and labor. Indeed, the "wrong" choice of technique, whether by government decision directly or by private investors reacting to inappropriate price signals, can easily cost many thousands, in fact, hundreds of thousands, of jobs. The choice between small and large rice mills depended very much on the government's interest rate policy, its rice price policy, and unskilled wage levels. Given that a dramatic transformation to mechanized rice processing had been occurring at private initiative in Indonesia in the past few years, the country was fortunate indeed to have those critical policies more or less in order.

As recently as 1971 it was assumed that as much as 80 percent of Java's rice crop was hand-pounded. Although hand-pounding has certainly not been eliminated over the past few years, the evidence does suggest that it no longer accounts for more than perhaps 20 percent of the crop, at least in Java. Upwards of 100,000 jobs have probably been lost (at least jobs in the sense of an opportunity cost income—the "return" to milling is no longer captured by the family but by the miller). This loss of jobs should be kept in perspective, however. The farmer having his rice milled is frequently one of the laborers in the small rice mill, thus capturing part of the milling cost for his family. In addition, little cash income has been forgone, and the small rice mills are frequently so much more technically efficient that the farmer appears to go home with more rice than if he had hand-pounded it. (But since the farmer ate the bran if the rice were hand-pounded, whereas the miller keeps it for himself, this is probably not usually the case.) Lastly, hand-pounding is drudgery, and the people obviously welcome its demise.

SUMMARY

Two main points have been presented in this paper. The first is that economic analysis shows labor-intensive (but mechanical) rice processing facilities to dominate both hand-pounding and more capital-intensive techniques. And this is a result amply confirmed by recent evidence on actual investments in rice processing facilities in the countryside. A second, more general point is that the employment effects of the choice of technique decision are not trivial and are critically dependent on parameters directly subject to government control. "Getting prices right" is not the end of economic development. But "getting prices wrong" frequently is.

NOTES

1. Anonymous, "Report from the Agricultural Credit Specialist" (1972); "Rice Processing in Java" (Manila: Asian Development Bank, August 1972), cited hereafter as ADB, *Rice Processing Report.*

2. Weitz-Hettelsater Engineers, "Rice Storage, Handling, and Marketing Study: Economics and Engineering Aspects" (Kansas City, Mo., 1972). The final draft was submitted to the Republic of Indonesia in December 1972. This report is referred to in this article as Rice Marketing Study (RMS).

3. C. Peter Timmer, "Employment Aspects of Investment in Rice Marketing in Indonesia," *Food Research Institute Studies in Agricultural Economics, Trade, and Development* 11, no. 1 (1972); and idem, "Choice of Technique in Indonesia" (Cambridge: Harvard Center for International Affairs, Occasional Papers, 1973).

4. Ibid.

5. Bulog is the Indonesian state rice marketing agency. —ED.

A Comment

WILLIAM L. COLLIER, JUSUF COLTER, SINARHADI,
AND ROBERT D'A. SHAW

Over the past few years, nothing short of a revolution has occurred in the process of rice cultivation in Java. Probably one of the most rapid and widespread changes of all has been the displacement of traditional hand-pounding techniques by rice hullers. Perhaps the most interesting and widely debated aspect of these changes is the distribution of the benefits arising from them. Dr. Timmer implied that villagers in general welcome the demise of hand-pounding, since those engaged in the activity gained little income from it. Recent information suggests that this proposition is not tenable. It appears that Dr. Timmer's estimate of the number of jobs lost is far too low. More important in terms of rural poverty, he has failed to give sufficient weight to the large number of women who, prior to the introduction of hullers, gained a sufficient share of their income from hand-pounding. Moreover, it seems that Dr. Timmer overestimated the amount of employment created by rice hullers.

These differences are very important when they are considered in the aggregate. While Dr. Timmer thinks that "upwards of 100,000 jobs" may have been destroyed, our estimate using the same assumptions is much higher. Of even greater consequence is our estimate that earnings of laborers engaged in hand-pounding of rice on Java may have been reduced by as much as $50 million annually. Most of those losing work and sources of income from hand-pounding are likely to be members of the poorest families in the rural areas, those with little or no access to land. Thus the spread of rice hullers has, in our opinion, probably worsened the distribution of income in a substantial way.

These opinions are based on surveys carried out in twelve villages in Java by the Agro-Economic Survey in August and September 1973. Approximately four hundred farmers, laborers, village leaders, and owners of rice hullers were interviewed in these surveys of rural change in rice production. The interviews included information about hand-pounding and mechanical hulling. The material is supplemented by some additional information from eight other villages in the Agro-Economic Survey sample that were surveyed earlier in 1973.

WILLIAM L. COLLIER is chief of party, Resources Management International, Inc., Bogor, Indonesia. JUSUF COLTER and SINARHADI are economists, Survey Agro Ekonomi, Bogor, Indonesia. ROBERT D'A. SHAW is director of special programs, Aga Khan Foundation.

Even though we believe that more of the total rice crop on Java is still being hand-pounded than is indicated in Dr. Timmer's model, this factor is more than compensated for by his overestimation of the amount of rice that can be hand-pounded per day. The actual amounts hand-pounded per day and per hour by workers in our sample villages vary from 12.5 kg to 50.0 kg of *padi* per day. Much of the variation is accounted for by the number of hours worked by each woman. Usually this is five to six hours per day, but in those villages where women undertake contracts for hand-pounding it stretches up to an amazing thirteen hours per day. Thus the average weight of *padi* pounded per hour per person was 5.0 kg, with the range varying from 3.1 to 7.8 kg. This average translates into 3.9 kg of *gabah* per hour and 2.4 kg of *beras* (milled rice) per hour.

Dr. Timmer, however, used an estimate of 150 kg of *gabah* per day. If we assume that he was using a working day of eight hours, this converts to 18.7 kg of *gabah*, or nearly 12 kg of *beras* per hour. Since this is about five times the figure found in our survey, it seems clear that his estimate has little empirical basis.

However, Dr. Timmer's models have additional problems. In the first place, while the title of his article and his coefficients apply to Java alone, Timmer's employment estimates are made on the basis of the total Indonesian rice crop. Information about the rice processing situation outside Java is, in fact, conspicuously inadequate. At the same time, it is probably fair to assume that the greater availability of employment in the other islands makes this technological change less important in terms of human welfare. We will therefore limit our estimates on the impact of mechanical hulling to Java, where nearly two-thirds of the population of Indonesia live and approximately one-half of the country's rice is produced.

Moreover, within Java, Dr. Timmer's estimates for the employment and income effects of hullers would seem to be misleading in the following important ways:

(1) His estimate of the percentage still hand-pounded is probably too low—20 percent against our estimate of 40 percent.
(2) While the above mitigates the unemployment resulting from the use of hullers, this is more than offset by our findings that he used estimates of the hand-pounding capacity of women that seem to be five times too high.
(3) The importance of wage labor in hand-pounding was almost entirely neglected by Dr. Timmer. In our estimate, this may have involved half the total crop on Java.
(4) Lastly, Dr. Timmer may have exaggerated the employment created by hullers by a factor of two.

Let us now take a schematic look at these changes to give an indication of the orders of magnitude of their importance on Java. For this purpose, we shall use a figure of 12 million tons of milled (*gabah*) rice for the total Indonesian rice crop.[1] Typically, half of the total is produced in Java—a total of 6 million tons in our example. If we further assume that 50 percent of this would have been hand-pounded by wage laborers in the absence of hullers, this amounts to 3 million tons that would have created wage employment. Let us now postulate conservatively that one woman can hand-pound enough *gabah* in an hour to obtain 3 kg of *beras* (cf. our estimate of 2.4 kg). To hand-pound 3 million tons would take 1 billion woman-hours, or 125 million woman-days. At a wage of Rp 180 per day, this amounts to earnings of Rp 22.5 billion in a year, or just under $55 million.

To estimate the earnings of employees in SRM hulling this amount, if we assume that one SRM can hull 1,000 tons of *gabah* per year (or approximately 620 tons of *beras*), then to obtain 3 million tons of *beras* requires nearly 5,000 SRM. At an average wage bill of $80 per month, the annual earnings of laborers in these SRM would be just under $5 million.

Thus the total loss in laborers' earnings attributable to the introduction of hullers seems to be of the order of $50 million annually in Java, where the cash incomes of the rural poor are exceedingly low and where the possibilities of alternative employment opportunities are often slight. This represents a substantial diminution of income for large numbers of households of landless laborers and small farmers. Three million tons of rice could provide wages for one million women every day for four months each year.

The beneficiaries of the new technology are those farmers who would otherwise have hired laborers to pound their rice, the huller operators and the buyers of rice, to whom prices of milled rice may be around Rp 5 per kg lower than those that would have prevailed if hand-pounding had remained in force.[2] The losers, on the other hand, are those wives of small farmers and landless laborers who gain additional income from hand-pounding. These are the people who can least afford such a drop in income, as the number of alternative work opportunities is so limited.

The major conclusion we draw from this rough approximation of the impact of the introduction of hullers is that this redistribution of income in favor or relatively large farmers and SRM operators requires, as an urgent matter of public policy, massive programs to create additional income opportunities for the rural disadvantaged. In contrast to our estimate of 125 million woman-days of wage labor lost in Java, the *kabupaten* public works program in 1972–73 provided 43.6 million man-days of employment throughout Indonesia. This program is an excellent start but clearly inadequate to meet the needs.

NOTES

1. The actual total for 1971–72 was estimated at 12.8 million tons.
2. Whether such potential savings have in fact been passed on to rice buyers is impossible to determine because of the very rapid rise in rice prices over the last two years for other reasons.

A Reply

C. PETER TIMMER

Everyone stands in debt to Messrs. Collier, Colter, Sinarhadi, and Shaw (henceforth acronymized in good Indonesian fashion as CCSS) for providing a substantial body of empirical material dealing with the impact of changing technology in rice milling. I am especially pleased to have some firm evidence on the technical coefficients for hand-pounding. This was a topic of so little concern to the Weitz-Hettelsater Rice Marketing Study[1] that hand-pounding was completely ignored in its report. The coefficient of forty workers per one thousand tons of *gabah* per year used in its final report was picked up from a draft of my first paper on this topic. Since I relied on that document for the great bulk of my technical and cost data, I am not surprised to find that the crude estimates were fairly far off the mark.

Granting this, however, it is important not to lose sight of the meaning and impact of the new numbers provided. The first issue is how they affect the economic analysis of the choice of technique; the second issue is their impact on a social analysis of the economic results. For space reasons, my article does not deal at all with the relationship between an economic analysis and a social analysis, but my earlier article on "Employment Aspects of Investment in Rice Marketing in Indonesia"[2] provided a framework for comparing these two viewpoints (and others). An economic analysis can be done with either market prices or shadow prices. The shadow price analysis results in the largest national income possible when all prices are taken at their social rather than their private values (and this should include distributional and employment considerations). A social analysis carries the concern for unemployment and an unfavorable distribution of income into the political sphere, with the possible (but not inevitable) result that measures taken to relieve unemployment in specific situations would actually have an adverse impact on social equity for the entire economy and society. Only the economic perspective (with both market and shadow prices) is presented in this article, with the exception of the short section on employment impact. It is on this section that CCSS focus their comments, but the data have significant implications for the economic analysis as well.

Picking even the most optimistic figures about hand-pounding productivity and hours worked from the CCSS data, it is evident that the coefficient I used in the choice of technique analysis was off by a minimum of a factor of two and perhaps on average by a factor of four or five. That is, hand-pounding is at most only one-half as productive as I assumed and probably less. The bias I built into the analysis was in favor of hand-pounding entering the tableau of optimal economic activities under a realistic set of circumstances, and, of course, that was one of the striking results of the analysis. Even with the high productivity of hand-pounding that I assumed, it did not become optimal under any set of market price conditions on Java, but it still seemed to make economic sense for farm families (and harvesters) to hand-pound for their own consumption if their opportunity labor costs were below the market wage rate.

The new data provided by CCSS dash this economic rationality completely. Hand-pounding makes sense only when the opportunity wage is virtually zero and not even then if there is any significant squeeze between *padi* prices and milled rice prices. Hand-pounding quickly succumbs to negative value added in such a price squeeze (perhaps caused by a too narrow range between the floor and ceiling prices, which is defended by large-scale imports) because of its technical inefficiency. It is important that we understand this result. From an economic viewpoint, hand-pounding is shown by the CCSS data to be a completely inappropriate technique under present market conditions on Java and would make sense only if labor were shadow priced at a near zero wage (and the government paid the difference between this level and the market wage) and the shadow value of rice were well below present market prices (and the government could afford this subsidy as well).

Having just argued that the only significant effect of the CCSS data is to make hand-pounding nearly impossible to justify on economic grounds, I do agree with the major thrust of the social concern demonstrated by the authors. The distribution impact of the SRM has been higher incomes for large farmers and for operators of the SRM, but at the expense of the poor village women who have lost a source of case income and are now pressing into the harvest labor forces, with serious economic and social consequences. I fully accept the conclusions of CCSS with respect to lost cash income potential of village women. Their numbers without doubt more correctly assess the impact here than my own, based as they were on very rough estimates. But even as early as 1971 hand-pounding as a cash-hire activity was a rapidly vanishing phenomenon. My whole perspective on this particular issue was colored by this *fait accompli*. In historical terms the loss of cash wages by the poor women villagers is important, but it had little immediate bearing on the policy issue directly before the government, namely, Would Indonesia be best served by investment in large bulk facilities for drying, milling, and storing rice? We should all be very thankful to CCSS for redirecting our attention to the neglected social issue now that the immediate policy issue has been resolved. Their pleas for a vastly expanded rural works program will be met with enthusiasm by all whose consciences are moved by the plight of the rural (and urban) poor in Indonesia.

This brings us to the other side of the distributional impact of the new milling technology. Rice prices to consumers are perhaps ''Rp 5 per kg lower than those that would have prevailed if hand-pounding had remained in force'' (CCSS). As CCSS point out, it is hard to tell, with rice prices in such turmoil over the past few years, whether these cost reductions have actually been passed on to the consumer, but let us work with this magnitude and examine its implications. Do lower consumer prices for rice for a given farm price hurt income distribution or help it? I would argue that a Rp 5 per kg reduction in consumer rice price while farm prices are held constant has an enormous welfare impact on the lower half or two-thirds of the income distribution. The magnitudes are just as impressive as the $50 million quoted by CCSS as wage losses to the rural poor. A Rp 5 per kg cost saving to the new technology applied to a 13-million-ton rice crop gives a resulting saving to consumers of $165 million. Even if only half of this benefits the lower half of the population—a very conservative assumption—its gain exceeds $80 million.

We must also put in the balance the greater outturn of rice from a given volume of *gabah* achieved by the machine technology. The new data on extraction rates provided by CCSS raise the savings to truly important proportions. The difference between 57 percent extraction for hand-pounding and 66 percent for the SRM on average is 9 percent. For a *gabah* crop of 18 million tons a milling industry composed entirely of SRM would yield *over one and a half million tons more milled rice* than if the crop were hand-pounded. At a very conservative price of $200 per ton this is an added value to society of more than $300 million. Again, the exact distributional nature of this gain is not clear, but it is hard to see how the unemployed women are made worse off by this aspect of the technological change.

Losing the efficiency gains by banning small rice mills (thus forcing a return to hand-pounding) would be an enormously costly way of helping the displaced women whose primary source of wages during the harvest has been removed. We must set the $50 million loss to these women against a gain to society of more than $450 million, and some of that gain accrues to these very same women, although not a sufficient amount to offset their losses. But surely with the comprehensive and statistically documented picture presented by CCSS to create awareness of the problem, a mechanism can be found that would redistribute 10–15 percent of the social gain to the private losers.[3]

Lastly, I should like to enter a confession. I started this entire area of research with the intention of demonstrating in simple, clear-cut terms that both economic planners and engineering consultants would understand that the large-scale rice mills and bulk silo terminals were inappropriate in the Indonesian countryside. The battle to be fought in the planning agency was not hand-pounding versus small rice mills but large bulk facilities versus small rice mills. I was nearly laughed out of court for defending the small rice mills, but the analysis spoke for itself and still does. In terms of prevailing market prices and any imaginable modification of them (including reasonable shadow prices), facilities in the generic class ''small rice mills'' are the most appropriate for Indonesia.

NOTES

1. Weitz-Hettelsater Engineers, "Rice Storage, Handling, and Marketing Study: Economics and Engineering Aspects" (Kansas City, Mo., 1972).

2. Timmer, "Employment Aspects of Investment in Rice Marketing in Indonesia," *Food Research Institute Studies in Agricultural Economics, Trade, and Development* 11, no. 1 (1972).

3. Another distributional issue that neither CCSS nor I have treated is nutrition. The primary source of the B-vitamin complex for the poorer half of the population is the bran left on hand-pounded rice. A complete shift to the white rice made possible by well-operated small rice mills could easily result in serious vitamin deficiencies that could be offset only by a fairly expensive vitamin fortification program. I have no way of knowing how these costs would compare with the gains already cited nor with the wage losses cited by CCSS.

20

Rural Small-Scale Industry: Empirical Evidence and Policy Issues

ENYINNA CHUTA AND CARL LIEDHOLM

The role of rural small-scale industries in providing productive employment and earning opportunities has recently emerged as an important research concern in development economics and has become a hotly debated topic among policy makers and international donor agencies.[1] This heightened interest has paralleled the increased international concern for equity and employment objectives and the growing realization that the large-scale urban industrialization strategies of the 1960s generally failed to solve the problems of underemployment and poverty. Moreover, empirical research has begun to demonstrate that small rural manufacturing enterprises have been substantially underreported in official publications and that these smaller firms might be more effective vehicles for meeting a country's growth and equity objectives than their larger-scale urban counterparts.

In spite of the importance of this topic, there have been few analytical or empirical studies of small rural industries. The World Bank (1978a), for example, recently concluded that "there is little concrete evidence" on many of the important characteristics of these activities. As a result, policy makers and planners charged with the formulation of policies and programs to assist rural small-scale industry in the Third World are often forced to make decisions that are "unencumbered by evidence."

The purpose of this paper is to fill the gap in the literature on rural industries in the Third World. We shall begin with a descriptive profile of rural industries and then examine the role of rural small-scale industries in development. In the final section we shall discuss major policy and program issues.

ENYINNA CHUTA, formerly an economist with the Technology and Employment Branch, International Labour Office, is senior lecturer, School of Management Technology, Federal University of Technology, Yola, Gongola State, Nigeria. CARL LIEDHOLM is professor of economics and agricultural economics, Michigan State University.

DESCRIPTIVE PROFILE

Rural industries, which include manufacturing, processing, and repair activities, form one component of a larger set of rural nonfarm activities. Commerce, services, construction, and transport are also important activities within the rural nonfarm sector. Although these activities fall outside the purview of this paper, the overall importance of the rural nonfarm sector needs to be examined briefly at the outset, in order to place the discussion of rural industry in a proper perspective.

OVERALL MAGNITUDE

Just how significant are rural industries and other nonfarm activities in developing countries? The evidence available from national censuses and various regional and rural surveys indicates that nonfarm activities provide an important source of *primary* rural employment in developing countries. In the vast majority of the eighteen developing countries where relatively recent data on the subject are available, one-fifth or more of the rural labor force is *primarily* engaged in nonfarm activities (table 1). Although the rural nonfarm percentage ranged from 14 percent to 49 percent, in over three-quarters of the countries the percentage fell between 19 percent and 28 percent.

TABLE 1
PERCENTAGE OF THE RURAL LABOR FORCE WITH PRIMARY EMPLOYMENT IN
RURAL NONFARM ACTIVITIES

Country	Year	Coverage	*Percentage of Rural Labor Force Primarily Employed in Nonfarm Sector (%)*
Guatemala	1964	All rural	14%
Thailand	1970	All rural	18
Sierra Leone	1976	Male–rural	19
South Korea	1970	All rural	19
Pakistan	1970	Punjab only	19
Nigeria	1966	Male—3 districts, Western State	19
India	1966	All rural	20
Uganda	1967	Four rural villages	20
Afghanistan	1971	Male—Paktia Region	22
Mexico	1970	All—Sinaloa State	23
Colombia	1970	All rural	23
Indonesia	1971	All rural	24
Venezuela	1969	All rural	27
Kenya	1970	All rural	28
Philippines	1971	All rural	28
W. Malaysia	1970	All rural	32
Iran	1972	All rural	33
Taiwan	1966	All rural	49

SOURCE: Chuta and Liedholm 1979.

The figures provide a minimal estimate of the magnitude of primary employment in nonfarm activities in rural areas. First, they generally reflect the employment characteristics of the rural villages with populations below five thousand; if the larger rural towns were included, the rural nonfarm percentage would likely be larger.[2] Second, there are certain measurement errors that cause systematic undercounting of nonfarm activities. In some African countries rural respondents will report farming to be their main occupation even if they engage in this activity only part-time. In addition, women's participation in nonfarm activities is often substantial, but frequently it is not measured or included in labor-force figures. In Honduras, for example, a recent study revealed that over 60 percent of the rural industrial entrepreneurs were women, a fact not reflected in the country's official statistics (Stallmann and Pease 1979).

These primary employment statistics also understate the magnitude of rural nonfarm activities, because they fail to reflect those farmers who engage in nonfarm activities on a part-time or seasonal basis. Data on secondary employment are not generally available for most countries. The limited evidence indicates that 10–20 percent of the rural male labor force undertakes nonfarm work as a secondary occupation. In western Nigeria, for example, 20 percent of the rural males engaged in nonfarm work on a part-time basis, while in Sierra Leone, Afghanistan, and Korea the figures were 11 percent, 16 percent, and 20 percent, respectively (Chuta and Liedholm 1979).

There are significant monthly variations in the amounts of rural farm and nonfarm employment over the agricultural cycle. Farm and nonfarm employment move in opposite directions. There is no period when nonfarm employment disappears; thus, nonfarm employment does compete somewhat with farm employment during periods of the peak agricultural demand for labor. Data from Nigeria reveal that the peak in nonfarm labor use is nine times the use in the slack periods (Norman 1973). The fluidity of labor between a number of activities on a seasonal basis is thus a striking feature of rural households.

In summary, rural industries and other rural nonfarm activities appear to provide a source of primary or secondary employment for 30–50 percent of the rural labor force in developing nations.[3] Consequently, in terms of employment, nonfarm activities are quantitatively an important component of the rural economy that should not be overlooked in the design of rural development policies or programs.

In view of the magnitude of rural nonfarm employment, it is not surprising that rural industries and other nonfarm activities also provide an important source of income for rural households. Although data on rural incomes are generally lacking for most countries, evidence from those countries where information is available indicates that nonfarm earnings account for over one-fifth of total rural household income. Indeed, in Sierra Leone, where a detailed rural household survey was recently undertaken, nonfarm income was found to provide 36 percent of rural household income, while in Taiwan the comparable figure was 43 percent (Chuta and Liedholm 1979).

COMPOSITION

What are the most important types of activities within the rural nonfarm sector? In most developing countries, rural industries, rural commerce, and rural services tend to dominate. For the nine countries for which the required data are available, rural industry accounts for 22–46 percent of total nonfarm employment, rural commerce ranges from 11 percent to 35 percent, while rural services range from 10 percent to 50 percent.[4] Other nonfarm activities, such as construction, transport, and utilities, generally account for less than 25 percent of rural nonfarm employment (see Chuta and Liedholm 1979 for more details).

The relative importance of rural industrial activity as a component of the rural economy may appear surprising. Even more surprising, perhaps, is how important rural industries appear when compared with urban industries in many developing countries. Indeed, there is evidence that employment in small rural industrial enterprises often exceeds that in large urban industrial firms. In Sierra Leone 86 percent of the total industrial-sector employment and 95 percent of the industrial establishments were located in rural areas (Liedholm and Chuta 1976). The percentage of rural industrial employment in other countries ranged from 70 percent in Bangladesh (Bangladesh Institute of Development Studies 1979) to 67 percent in Jamaica (Davies et al. 1979) and 63 percent in Malaysia (World Bank 1978b). These figures may actually understate the true magnitude of rural industrial activity, because country censuses often fail to pick up the very small rural enterprises. A recently completed rural industry survey in Honduras (Stallmann and Pease 1979) found that rural industrial employment had been underestimated in Honduras by almost one-half. A similar survey in Bangladesh indicated that in one rural district the number of rural industrial firms was twenty times greater than indicated by the official statistics (Ahmed, Chuta, and Rahman 1978).

Within rural industries there is a surprising diversity of activities. The most important activity in the majority of countries appears to be clothing production, followed by woodworking, metalworking, and food processing. Clothing production, for example, accounted for 53 percent of the rural manufacturing employment in Sierra Leone, 41 percent in Korea, 24 percent in Taiwan, 32 percent in western Nigeria, and 52 percent in rural Bangladesh (Chuta and Liedholm 1979).

SIZE

What is the average size of these rural industries? The available empirical evidence is limited, but it does indicate that the vast majority of rural industrial activities are undertaken by very small-scale artisan and informal enterprises.[5] In Sierra Leone the average rural industrial firm employed 1.6 workers, and 99 percent of the firms employed fewer than 5 individuals (Liedholm and Chuta 1976). In rural Jamaica (Davies et al. 1979) the average rural industrial enterprise engaged 1.8 workers. The results of a similar survey undertaken in rural Honduras (Stallmann and Pease 1979) revealed that 59 percent of the industrial firms were 1-person endeavors and that 95 percent employed fewer than 5 workers. These findings indicate that most rural enterprises are very small and thus may be potentially an important target group for policy makers concerned with the rural poor.

GROWTH POTENTIAL

Finally, in a dynamic sense, do rural industries decrease as rural incomes rise and opportunities for trade increase? On this issue there has been a divergence of views. The issue was sparked by the 1969 paper "A Model of an Agrarian Economy with Nonagricultural Activities," by Stephen Hymer and Stephen Resnick. The authors develop a model of the rural economy in which rural industrial and other nonfarm activities, denoted as Z goods, are hypothesized to decline as rural incomes rise and opportunities for trade increase. Resnick, in a subsequent article (1970), provided empirical evidence for the contention by tracing the decline of rural industry in Burma, the Philippines, and Thailand from 1870 to 1938. Comprehensive times series data were not available, however, and Resnick was forced to rely on fragments of evidence from various sources. Consequently, the results of the study, while interesting, cannot be considered conclusive.

The empirical evidence available for more recent periods indicates that in the aggregate, rural industrial and other nonfarm employment and output have been increasing, rather than decreasing, with development (see Chuta and Liedholm 1979 for details). There do appear to be variations, however, in the growth performance of individual types of rural industries. Although much of the available evidence is anecdotal or episodic, some time series data are available for countries such as the Philippines (Anderson and Khambata 1981), Sierra Leone (Chuta and Liedholm 1982), and Haiti (Haggblade, Defay, and Pitman 1979). The available evidence indicates that tailoring, dressmaking, furniture making, baking, and rice milling have all continued to grow in importance even after large-scale domestic factory production in these subsectors has begun. Shoe production, leather production, and pottery making appear to have generally declined in importance.[6] A mixed record appears for blacksmithing and for spinning and weaving.[7] The kinds of activities undertaken by some of the important artisan groups have also been evolving. In some countries, for example, rural blacksmiths, who previously were primarily engaged in the production or servicing of hand tools, now also produce or service animal-drawn or mechanized farm equipment and irrigation equipment (Liedholm and Chuta 1976; Child and Kaneda 1975). Moreover, several newer types of artisan activities, such as bicycle, auto, and electrical repair, have grown particularly rapidly in recent years. These newer activities reflect the increasing service orientation of many artisan activities as incomes and urban factory production increase. In addition, certain types of craft-oriented artisan activities designed for the international market, such as gara (tie-dye) cloth in Sierra Leone (Liedholm and Chuta

1976) and woodcarving in Haiti (Haggblade, Defay, and Pitman 1979), have also been growing rapidly in certain countries. Finally, a few "modern" factory activities, some of which have emerged from small enterprises such as metal-working factories in India (Berna 1960) and cement block production and essential oils (luxury perfume) production in rural Haiti (Haggblade, Defay, and Pitman 1979), have also begun to increase in importance.[8]

Recognizing these differential growth patterns is important in the design of programs and policies for rural industry. Government policies, particularly with respect to large, modern industries and agriculture, influence growth patterns of individual activities within each country.[9] Although some new rural industrial activities will emerge, the sheer magnitude of existing informal artisan activities in most countries indicates that any major transformation will take many years to complete. Stewart (1972) has estimated that it will take several decades before the "formal" sector will begin to absorb even the additions to the labor force in most developing countries. Consequently, attention must continue to be directed towards enhancing the types of activities represented in the existing structure of rural enterprises, even if, in the longer run, many of them will eventually decline in importance or disappear.[10]

DETERMINANTS OF THE ROLE OF RURAL INDUSTRY

What are the main determinants of both future and existing patterns of rural industrial employment and income? These can be usefully understood by focusing on the set of factors influencing the demand for and supply of these activities.

DEMAND PROSPECTS

Several important issues relate to the nature of the demand for the goods and services produced by rural industries. One crucial issue, on which there has been a divergence of opinion, is whether the demand for these activities increases as rural incomes increase. Hymer and Resnick (1969) have argued that rural industrial activities produce inferior goods, that is, that the demand for them would be expected to decline as rural incomes rise. Mellor (1976), Chuta and Liedholm (1979), and various International Labour Office missions (1972, 1974), on the other hand, have contended that there is a strong, positive relationship between income and the demand for these activities. The few empirical studies of rural demand, particularly that of King and Byerlee (1978), support the latter position.

Another demand-related issue is whether there are strong backward and forward linkages between rural industrial activities and other sectors of the economy, particularly agriculture. Hirschman (1958) has contended that linkages between agriculture and other sectors are quite weak,[11] while others, such as Mellor (1976) and Johnston and Kilby (1975), have argued that the linkages between rural industries and agriculture, in particular, are or could be potentially very strong. The available empirical evidence indicates that these linkages are quite important (see Chuta and Liedholm 1979). Rural industries are influenced by the pattern of agricultural growth and can themselves influence the course and rate of agricultural development. Finally, there is some empirical and analytical evidence that the international market is an important component of demand for certain types of rural industrial products (see Chuta and Liedholm 1979).

SUPPLY FACTORS

With respect to supply, one important issue is whether rural industrial activities are more labor-intensive and thus generate more employment per unit of capital than other components of the economy. The available empirical evidence is generally quite consistent in indicating that small-scale rural enterprises are more labor-intensive than their larger-scale urban counterparts (see Chuta and Liedholm 1979; and table 2).

A key related issue is whether these same labor-intensive rural industries use the scarce factor, capital, more efficiently than do other, larger-scale enterprises. Several international groups and individuals, including Nicholas Kaldor (Robinson 1965), have argued that the capital productivity (that is, the output/capital ratio) of small rural enterprises is lower than that of their larger-scale counterparts. Marsden (1969), Chuta and Liedholm (1979), and others have contended that the reverse situation holds. The available aggregate country data are generally not of high enough quality to provide a conclusive answer to these conflicting views, although there are many instances where the small rural industrial enterprises appear to possess the higher capital productivity.

The results of a comprehensive, micro-level survey of the major economic sectors of Sierra Leone, however, do provide some insights into the factor intensities of various size categories of industrial enterprises within single product groups (Byerlee et al. 1983). In the Sierra Leone survey, five hundred rural households were interviewed twice weekly over a twelve-month period during 1974/75 to obtain daily information on farm and nonfarm production and consumption. In addition, sixty firms in rice milling, the major agricultural processing sector, and one hundred twenty small-scale and large-scale firms in the fishing sector, another important rural sector, were interviewed twice weekly over the year. Finally, surveys were conducted with two hundred fifty small manufacturing firms in both rural and urban areas, and these surveys, together with secondary data on large-scale industries, provided an overview of production in the industrial sector.

Table 2 lists the computed labor and capital intensities for alternative techniques of rice milling, fishing, and baking in Sierra Leone. A wide array of production techniques was found within each of these industrial subsectors, ranging from hand pounding up to large disc sheller mills capable of processing two tons of rice per hour in rice milling; from dugout canoes and paddles valued at less than one hundred dollars to large trawlers costing over ten thousand dollars in fishing; and from simple, mud ovens to modern, electrically operated bakeries with investments exceeding ten thousand dollars in baking. Within each

TABLE 2

OUTPUT/CAPITAL, OUTPUT/LABOR, AND LABOR/CAPITAL RATIOS FOR SELECTED PRODUCTION TECHNIQUES AND INDUSTRIES IN SIERRA LEONE, 1974[a]

Industry and Production Technique	Output/Capital Ratio	Output/Labor Ratio	Labor/Capital Ratio	Economic Rate of Return to Capital (%)
Rice milling				
Hand pounding	57.1	0.06	889	
Small steel cylinder mills	2.98	1.46	3.54	
Small rubber roller mills	1.32	1.34	1.71	
Large rubber roller mills[b]	1.40	9.61	0.34	
Fisheries				
Boat, less than 20 ft, cast net	20.9	0.43	48.12	837%
Boat, 20 ft, ring net	8.1	0.37	21.71	204
Boat, 30 ft, beach seine net	6.1	0.31	19.73	154
Boat, 40 ft, ring net, motor 26 HP	6.7	0.29	22.94	100
Trawler	1.85	0.45	5.10	-3.7
Manufacturing				
Baking				
Rural, small mud oven, traditional	12.4	0.50	24.7	11.5
Urban, small peel oven	9.8	1.00	10.0	109.7
Urban, small multiple-deck oven	2.1	0.60	3.4	3.8
Urban, large tunnel oven	1.7	1.00	1.7	-10.5

SOURCE: Byerlee et al. 1983.

[a]Output is measured in Leones of value added; capital is measured in annual costs at 35 percent opportunity cost; and labor is measured in man-hours.

[b]Potential technique.

of these subsectors, the techniques used by the smaller rural enterprises were not only more labor-intensive (that is, they had higher labor/capital ratios) but also more productive per unit of scarce capital (higher output/capital ratios) than those techniques used by larger-scale enterprises. Consequently, if capital is assumed to be the scarce factor, there would be no conflict between output and employment objectives, at least in a static sense (see Chuta and Liedholm 1979).

These findings are reinforced by evidence on the economic profitability of these small rural industries. The economic rate of return to capital, a measure that reflects profit when all inputs including family labor and capital are valued at their opportunity costs, may be a better measure of economic efficiency than the output/capital ratio, which implicitly assumes that labor and other inputs have a shadow price of zero. Although economic profitability data are not widely available, the results from Sierra Leone (see table 2) are consistent with the previous finding based on the output/capital ratio; in all cases, the smaller-scale, more labor-intensive techniques generated the highest economic rates of return to capital. These findings, while certainly not conclusive, do indicate that in several lines of activity small rural industries are economically viable.[12] Moreover, in a dynamic context, there is no empirical evidence to support Galenson and Leibenstein's (1955) contention that the profit, savings, and reinvestment rates of rural small-scale industries are necessarily lower than those of the larger, capital-intensive enterprises.

MAJOR POLICY AND PROJECT ISSUES

Given the favorable characteristics of rural industries with respect to employment, income generation, and income distribution issues, many governments are showing increasing interest in incorporating these activities into their development strategies. Governments can assist these enterprises by general policy measures that affect the environment in which rural industries operate and by providing direct project assistance.

GENERAL POLICY MEASURES

Several major policy options are available to those governments interested in influencing rural industries. However, great care must be exercised in policy selections, as many government actions seemingly unrelated to rural industrial activities can inadvertently have adverse effects on them.

Factor-Price Distortions

Policies that result in input price distortions, for example, have significant, though often unintended, negative effects on rural industries. Two of the major sources of input price distortions, interest and tariff rates, will be discussed here.

With respect to interest rates, two distinct capital markets—the "formal" and the "informal"—exist in most developing countries.[13] Banks and similar institutions constitute the formal market, while moneylenders, raw material suppliers, and purchasers constitute the bulk of the informal market. Interest rates vary widely between the two. Official interest rates, where government-imposed ceilings frequently exist, generally run from 10 percent to 20 percent, while the nonofficial rates are frequently 100 percent or more (see Chuta and Liedholm 1979). Particularly under inflationary conditions, the formal real rates become very low, sometimes negative. Thus, not surprisingly, banks have tended to lend only to the established, large-scale firms, which may appear to the banks to involve lower risks and lending costs. Most of the recipients are urban-based, and their operations have tended to become more capital-intensive than they would have been had they been forced to borrow at the opportunity cost of capital.

The import duty structure can also be an important source of differential treatment for urban large-scale industries compared with rural small-scale industries. For most developing countries, import duties are lowest for heavy capital goods and become progressively higher through intermediate and consumer durable goods categories. Yet, many items classified as intermediate or consumer goods in tariff schedules are capital goods for rural small-scale firms. In Sierra Leone the sewing machine, an important capital item for tailoring firms, was classified as a luxury consumer good and taxed accordingly (Liedholm and Chuta 1976). Further escalating the distortion in capital cost is the frequent practice of granting concessions or even total waivers of import duties on capital goods or raw materials for specified periods as an inducement for industrial development. In some cases, small firms may technically qualify for similar concessions but may be unaware of this opportunity or, even when they are aware, may find the process so complicated and time-consuming that it is not economic for them to exercise the option. In many other cases small firms do not even qualify.

Other Policy Measures

Government policies with respect to infrastructure, industry, and agriculture also have important indirect effects on the expansion of rural industrial employment and income opportunities. Because of the strong linkages between agricultural and rural industrial activities, agricultural policies and programs, in particular, have a strong influence on rural small-scale enterprises. The analyses in the earlier section of this paper revealed that the primary demand for most rural industrial goods and services stems from the agricultural sector and that this demand is transmitted through both income and production linkages. Since the available evidence indicates that the rural households' income elasticity of demand for rural industrial goods is positive and that agriculture generates the largest share of rural incomes, policies designed to increase agricultural output and/or income have an important indirect effect on the demand for these activities. Consequently, government actions ranging from improvements in the terms of trade between agriculture and the larger-scale urban sector to specific investment programs and policies designed to increase, directly or indirectly, agricultural production and income can generate an increased demand for a wide array of rural industrial goods and services.

The nature and composition of these agricultural policies and programs should also be considered, however, since they can have important, differential effects on the demand for products from rural industries. There is some evidence that higher-income rural residents have a somewhat lower income elasticity of demand for rural industrial products than do lower-income individuals, the majority of whom are small-scale farmers (see King and Byerlee 1978). Moreover, the agricultural inputs, such as tractors and fertilizers, used by large-scale, high income farmers are less likely to be produced in rural localities than are the inputs used by the small-scale farmers (see Johnston and Kilby 1975).[14] Consequently, policies and programs designed to benefit a larger number of small-scale, low-income farmers are likely to generate a larger demand for rural industrial activities and services than those designed to benefit a few larger-scale farmers. These differential effects on rural industrial activities must be recognized when designing agricultural policies or rural development strategies.

DIRECT PROJECT ASSISTANCE

The major types of direct project assistance used to promote rural industries include a broad spectrum of interventions: credit, technical, management, and marketing assistance and the provision of common facilities, usually industrial estates. A crucial element in determining which form of direct intervention is most appropriate is the identification of the key constraints facing the rural enterprise.

Financial Assistance

Credit assistance is one of the most frequently used mechanisms to aid rural small-scale industries. An important issue in the design of such assistance is determining the extent of effective demand for this credit by rural industries. Some evidence appears to indicate that this demand is quite sizable. Rural entrepreneurs, for example, when asked directly to identify their greatest assistance needs and greatest perceived bottleneck, will usually list credit and capital first (see Kilby, Liedholm, and Meyer 1981). There is also evidence that for many types of rural industrial enterprises the rates of return on existing capital are substantial. These high rates of return indicate that the *potential* demand for credit could be quite large.

Yet, other evidence indicates that the rural industrial enterprises' demand for credit may be less extensive than indicated above. Detailed analyses undertaken in Sierra Leone (Liedholm and Chuta 1976) and Kenya (Harper 1978) revealed that although entrepreneurs perceived the lack of credit to be the crucial bottleneck, other problems, such as inadequate management or raw material procurement difficulties, proved to be the crucial basic constraints facing many small enterprises. Unless these other difficulties are recognized and dealt with, the simple provision of credit could, at a minimum, be wasteful and could actually harm rural industrial enterprises by inducing overcapitalization.

Another demand issue relates to the composition of the credit demand from rural industrial enterprises. In particular, is the credit demand primarily for fixed or working capital? The composition of credit demand does appear to vary somewhat depending on the size and type of rural industrial enterprise. For the smallest enterprises, which account for the bulk of the sector, the primary credit demand appears to be for working capital (see Kilby, Liedholm, and Meyer 1981 for more details on working capital). It is important to ascertain how much of the apparent working capital demand is simply a manifestation of some other problem, such as a raw material shortage, inadequate management, or a lack of demand for final products.

An important supply issue centers on determining the appropriate channel for providing financial services to rural industries. In most developing countries, formal credit institutions such as commercial banks, specialized small enterprise banks, specialized divisions of development banks, credit unions, and cooperative and worker banks have typically been used to channel funds to these enterprises. Although such devices as rediscounting facilities, guarantees, and earmarked funds are frequently introduced to entice these ''formal'' institutions into expanding their lending to rural industrial enterprises, it is not yet clear that these inducements have been successful in significantly expanding the amount of formal credit available to these enterprises, particularly the smaller ones.[15] Indeed, the vast majority of these rural industries have never even applied for funds from formal credit institutions.[16] Most of their capital is obtained from family or internal sources. Thus, alternative institutional mechanisms to the formal ones might also be needed. Informal financial institutions such as moneylenders, input suppliers, purchasers, and rotating credit societies provide most of the institutional credit to rural industrial enterprises. Studies of these institutions might reveal how formal and informal financial institutions are linked and how the informal institutions might be better integrated into the formal credit system. Finally, consideration might also be given to establishing new intermediary institutions, possibly linked with private voluntary organizations, which might, in turn, link into the formal financial system. Clearly, no single delivery system has emerged as a solution for providing credit to the wide array of rural industries. Indeed, rural industries may derive greater long-term economic and social benefits from the development of sound rural financial markets than from specialized, rural industry lending schemes.[17]

Nonfinancial Assistance

Technical production assistance has been another popular method of providing direct assistance to small rural industries. One important issue centers on determining the magnitude of the demand for technical assistance. There is some evidence to indicate that entrepreneurs are not aware of their need for technical assistance and the benefits they may derive from it (see Liedholm and Chuta 1976; and Harper 1978). Consequently, case studies may be needed to identify the amounts and required forms of technical assistance, particularly since assistance needs will likely vary by the size and type of enterprise.

A second general issue that deserves careful consideration is determining the most cost-effective institutional mechanisms, if any, for delivering the technical/production assistance to rural nonfarm enterprises. One important channel that has been utilized in developing countries such as Kenya (Harper 1978), Ghana, and India is a rural industry extension service. Unfortunately, there are no systematic analyses of the experiences with or the effectiveness of this approach. Is this active outreach approach as cost-effective as a more centralized assistance approach, in which clients would approach a centrally located institute for assistance? The Rural Industrial Centres in Kenya (see Livingston 1977) and the Industrial Development Centres in Nigeria (see Hawbaker and Turner 1972) are examples of the more centralized approach to the delivery of assistance.

The provision of management assistance is another direct assistance method used to aid small rural enterprises. Once again, a key issue is the nature and extent of the effective demand for this type of assistance. Indeed, most rural industrial enterprises scarcely recognize that lack of management capacity could pose a serious problem. Yet, previous studies of such enterprises have revealed that managerial competence is a key determinant of business success (Harris 1970; Liedholm and Chuta 1976). It is therefore crucial to ascertain the actual as well as the perceived needs in this area.

Another related issue that must be considered is the type of management skills these enterprises really need; these needs will likely vary depending on the size and nature of the enterprises. Most rural industrial entrepreneurs use simple technologies and have little or no formal education. Consequently, it may be important to design management assistance to enable these entrepreneurs to distinguish between personal and business transactions, to evaluate resources (especially their labor input) at their appropriate opportunity costs, to understand effective inventory plans, to adjust their businesses to viable sizes, and to adopt methods of production that utilize local resources.

Marketing assistance projects for rural industries are also found in several developing countries. It is important to ascertain what existing or new sources of domestic demand are available and how these could be further stimulated or developed. In addition, the governments themselves have frequently developed programs to purchase the products of rural enterprises, but sales to governments have frequently been hampered by cumbersome purchase procedures and unrealistic quality standards (Schatz 1977, 199).

It is also important to ascertain the external demand for the products of rural industries. In particular, a key issue is how to develop and deliver information to rural enterprises on the details of both existing and new product demand in foreign markets, as well as information on product handling and financial transactions. In addition, since products of rural industrial enterprises must be competitive in foreign markets, a related issue is how to ensure at least minimum product quality.

Moreover, there is the issue of whether there exists accessible, cost-effective, institutional support that can enable rural industries to purchase raw materials and effectively reach the export markets. In most countries this institutional support is urban-based. Decentralization of such facilities to service the needs of rural industrial enterprises becomes crucial.

Evidence from developing countries such as Haiti and Bangladesh (Haggblade, Defay, and Pitman 1979; Ahmed, Chuta, and Rahman 1978) reveals that the lack of raw materials constitutes a major constraint for rural industries. Therefore, an important issue is determining the forms of delivery channels that will be cost-effective in providing raw materials to these enterprises. Some governments have relied on the formation of rural cooperatives or producer associations for bulk purchasing of inputs in order to lower costs of production. However, evidence from some developing countries reveals that rural cooperatives have often failed due to personal rivalry, lack of effective leadership, and management problems (Shetty 1963, 184–85).

In several developing countries the most popular type of assistance used in providing common facilities for rural industries is industrial estates. In some developing countries industrial estates have been utilized for decentralizing industry towards small rural towns and villages (Dhar and Lydall 1961, 36). An important issue that arises is whether estates located in rural areas where basic infrastructural facilities are lacking can be cost-effective. Experience from India reveals that there are relatively few economic justifications for establishing estates in rural areas (Kochav et al. 1974, 33).

SUMMARY

International donor agencies and the governments of many developing nations have recently begun to devote increased attention to a previously neglected component of the rural economy, rural industries. The accumulating empirical evidence indicates that these activities not only generate a significant amount of rural employment and output but also provide an important source of income for rural households. Moreover, there is mounting evidence that several kinds of rural industries may be more economically efficient than their larger-scale urban counterparts. With judicious governmental policies and carefully formulated direct assistance measures, the contribution of rural industries in providing employment and income to rural households can be significantly enhanced.

NOTES

1. India and China, two countries with relatively long experience in rural industrialization, began to introduce policies and programs to foster village and cottage industries in the 1950s. In India,

cottage or household-type industries, which generally employed traditional technologies, frequently stressed the goal of absorbing surplus labor; these enterprises were frequently protected and subsidized and were not always closely linked with local demands or the agricultural sector. In China, communally organized industries using improved techniques were designed to meet local demands and were generally closely linked to the agricultural sector (see Gupta 1980).

2. See, for example, the evidence cited in World Bank 1978a. The dividing line between "rural" and "urban" is arbitrary, particularly in census data collected in most countries. The boundary lines are often framed in terms of urbanization characteristics rather than minimum size or occupational structure; consequently, settlements of a few thousand are often classified as "urban." The United Nations includes localities with fewer than twenty thousand inhabitants in its definition of "rural areas." This broader definition, which includes small- and medium-sized towns, is used in this paper.

3. There is evidence that the figure may be as high as 50 percent in some countries. Luning (1967), in a survey of villages in northern Nigeria, reports that 48 percent of the employed males engaged either full- or part-time in rural nonfarm activities, while Norman (1973) reports that in the same area 47 percent of male labor *time* was devoted to these activities.

4. Agricultural processing and marketing activities are reflected in these figures; fishing and livestock activities are not. See Chuta and Liedholm 1979 for more details.

5. "Small-scale" is not a precisely defined concept. At least fifty different definitions are used in seventy-five countries (see, for example, Staley and Morse 1965). As a working definition for this paper, "small-scale" includes those establishments employing fewer than fifty persons.

6. Additional evidence on the decline of these particular activities is found for India (Prasad 1963), Ethiopia (Karsten 1972), and Burma (Resnick 1970).

7. Spinning and weaving have declined in the Philippines and Sierra Leone but have increased, since Independence, in India.

8. For an excellent listing of the types of "modern" small enterprises, both urban and rural, likely to increase in importance see Staley and Morse (1965, 97ff.). Locational, process, and market influences are stressed.

9. In India handloom production declined from 1901 to 1947 under colonial rule (Prasad 1963, table 14) but increased after Independence with government encouragement.

10. Investments in most "informal" rural enterprises, for example, would be fully amortized within ten to twenty years.

11. Two reasons why Hirschman perceived few linkages were: (1) he implicitly was using a two-sector model in which *all* rural activities are labeled "agriculture"; and (2) he was writing in the context of a technologically stagnant agriculture.

12. For data on the profits and productivity of small-scale industries in other countries see Chuta and Liedholm 1979.

13. See Adams and Graham, chapter 21, and Gonzalez-Vega, chapter 22, in this volume.

14. Small-scale farmers are also more likely to use primarily smaller rural-based agricultural processing establishments, while large-scale farmers might be expected to make more use of larger-scale urban-based processing plants.

15. For a review of alternative institutional arrangements see Kochav et al. 1974.

16. In Haiti, 94 percent had never applied (Haggblade, Defay, and Pitman 1979), while in Sierra Leone the figure was 96 percent (Liedholm and Chuta 1976).

17. See chapters 21 and 22 in this volume.

REFERENCES

Ahmed, S., E. Chuta, and M. Rahman. 1978. "Rural Industries in Bangladesh: Research and Policy Implications." *Journal of Management* (University of Dacca), April, 25–37.

Anderson, Dennis, and Famda Khambata. 1981. "The Financing of Small and Medium Enterprise in the Philippines: A Case Study." World Bank Working Paper, no. 468. Washington, D.C. Mimeo.

Bangladesh Institute of Development Studies. 1979. "Rural Industries Study Project—Phase I Report." Dacca. Mimeo.

Berna, J. J. 1960. *Industrial Entrepreneurship in Madras State, India.* Bombay: Asia Publishing House.

Byerlee, Derek, Carl K. Eicher, Carl Liedholm, and Dunstan S. C. Spencer. 1983. "Employment-Output Conflicts, Factor Price Distortions, and Choice of Technique: Empirical Results from Sierra Leone." *Economic Development and Cultural Change* 31 (2): 315–36.

Child, Frank C., and Hiromitsu Kaneda. 1975. "Links to the Green Revolution: A Study of Small-scale, Agriculturally Related Industry in the Pakistan Punjab." *Economic Development and Cultural Change* 23:249–75.

Chuta, Enyinna, and Carl Liedholm. 1979. *Rural Non-Farm Employment: A Review of the State of the Art.* Michigan State University Rural Development Paper, no. 4. East Lansing: Michigan State University, Department of Agricultural Economics.

———. 1982. "Employment Growth and Change in Sierra Leone Small Scale Industry, 1974–1980." *International Labour Review* 121 (1): 101–12.

Davies, Omar, Yacob Fisseha, Annette Francis, and C. Kirton. 1979. "A Preliminary Analysis of the Small-Scale, Non-farm Sector in Jamaica." Kingston, Jamaica: Small-Scale Enterprise Survey Unit, Institute of Social and Economic Research, University of the West Indies. Mimeo.

Dhar, P. N., and H. F. Lydall. 1961. *The Role of Small Enterprises in Indian Economic Development.* Bombay: Asia Publishing House.

Galenson, W., and H. Leibenstein. 1955. "Investment Criteria, Productivity and Economic Development." *Quarterly Journal of Economics* 69:343–70.

Gupta, Devendra. 1980. "Government Policies and Programmes of Rural Industrialization with Special Reference to the Punjab Region of Northern India." World Employment Research Paper WEP 2-37/WP5. Geneva: International Labour Organization. Mimeo.

Haggblade, S., Jacques Defay, and Bob Pitman. 1979. "Small Manufacturing and Repair Enterprises in Haiti: Survey Results." Michigan State University Rural Development Series Working Paper, no. 4. East Lansing: Michigan State University, Department of Agricultural Economics. Mimeo.

Harper, Malcolm. 1978. *Consultancy for Small Business.* London: Intermediate Technology Publications.

Harris, John R. 1970. "Nigerian Entrepreneurship in Industry." In *Growth and Development of the Nigerian Economy,* edited by Carl Eicher and Carl Liedholm. East Lansing: Michigan State University Press.

Hawbaker, George, and Howard H. Turner. 1972. *Developing Small Industries: A Case Study of A.I.D. Assistance in Nigeria, 1962–1971.* Washington, D.C.: U.S.A.I.D.

Hirschman, A. O. 1958. *The Strategy of Economic Development.* New Haven: Yale University Press.

Hymer, Stephen, and Stephen Resnick. 1969. "A Model of an Agrarian Economy with Non-agricultural Activities." *American Economic Review* 50:493–506.

International Labour Office. 1972. *Employment, Incomes and Equality: A Strategy for Discovering Productive Employment in Kenya.* Geneva.

———. 1974. *Sharing in Development: A Programme of Employment, Equity and Growth for the Philippines.* Geneva.

Johnston, Bruce F., and Peter Kilby. 1975. *Agriculture and Structural Transformation: Economic Strategies in Late-Developing Countries.* London: Oxford University Press.

Karsten, Detlev. 1972. *The Economics of Handicrafts in Traditional Societies.* Munich: Weltforum Verlag.

Kilby, Peter, Carl Liedholm, and Richard Meyer. 1981. "Working Capital, Rural Nonfarm Firms and Rural Financial Markets." Colloquium on Rural Finance Discussion Paper, no. 5. Washington, D.C.: Economic Development Institute, World Bank. Mimeo.

King, Robert P., and Derek Byerlee. 1978. "Factor Intensities and Locational Linkages of Rural Consumption Patterns in Sierra Leone." *American Journal of Agricultural Economics* 60:197–206.

Kochav, David, Holger Bohlen, Kathleen DiTullio, Ilman Roostal, and Nurit Wahl. 1974. "Financing the Development of Small Scale Industries." Washington, D.C.: World Bank. Mimeo.

Liedholm, Carl, and Enyinna Chuta. 1976. *The Economics of Rural and Urban Small-Scale Industries in Sierra Leone*. African Rural Economy Paper, no. 14. East Lansing: Michigan State University, Department of Agricultural Economics.

Livingston, Ian. 1977. "An Evaluation of Kenya's Rural Industrial Development Programme." *Journal of Modern African Studies* 15:494–504.

Luning, H. A. 1967. *Economic Aspects of Low Labor Income Farming*. Agricultural Research Report, no. 699. Wageningen, Netherlands: Centre for Agricultural Publications and Documentations.

Marsden, Keith. 1969. "Toward a Synthesis of Economic Growth and Social Justice." *International Labour Review* 100:389–418.

Mellor, John W. 1976. *The New Economics of Growth*. Ithaca: Cornell University Press.

Norman, David W. 1973. *Methodology and Problems of Farm Management Investigations: Experiences from Northern Nigeria*. African Rural Employment Paper, no. 8. East Lansing: Michigan State University, Department of Agricultural Economics.

Prasad, Kedamath. 1963. *Technological Choice under Developmental Planning (A Case Study of the Small Industries of India)*. Bombay: Popular Prakashan.

Resnick, Stephen. 1970. "The Decline of Rural Industry under Export Expansion: A Comparison among Burma, Philippines and Thailand, 1870–1938." *Journal of Economic History* 30:51–73.

Robinson, Ronald. 1965. "The Argument of the Conference." In *Industrialization in Developing Countries*, edited by Ronald Robinson. Cambridge: Cambridge University Press.

Schatz, S. P. 1977. *Nigerian Capitalism*. Berkeley: University of California Press.

Shetty, M. C. 1963. *Small-Scale and Household Industries in a Developing Economy: A Study of Their Rationale, Structure and Operative Conditions*. New York: Asia Publishing House.

Staley, Eugene, and Richard Morse. 1965. *Modern Small Scale Industry for Developing Countries*. New York: McGraw-Hill.

Stallmann, Judith, and James Pease. 1979. "Characteristics of Manufacturing Enterprises by Locality Size in Four Regions of Honduras: Implications for Rural Development." East Lansing: Michigan State University, Department of Agricultural Economics. Mimeo.

Stewart, Frances. 1972. "Choice of Technique in Developing Countries." *Journal of Development Studies* 9:99–121.

World Bank. 1978a. *Rural Enterprise and Nonfarm Employment*. Washington, D.C.

———. 1978b. *Employment and Development in Small Enterprises*. Washington, D.C.

21

A Critique of Traditional Agricultural Credit Projects and Policies

DALE W ADAMS AND DOUGLAS H. GRAHAM

INTRODUCTION

In the past several decades aid agencies have spent in excess of $5 billion on rural financial market (RFM) projects. These projects have accompanied substantial increases in the number of institutions providing formal loans in low-income countries (LICs), as well as increases in amounts spent by local governments for agricultural credit. Currently the volume of new agricultural loans in low-income countries is in excess of U.S. $30 billion per year. In several countries, especially Brazil and Thailand, agricultural credit programs currently make up a very large part of the efforts aimed at agricultural development.

In part, this intense interest in agricultural credit projects results from the ease with which they can be carried out and the feeling that loans are a vital part of a package of inputs needed to stimulate change in agriculture. Some policy makers have also felt that cheap credit is an effective way of offsetting policies that penalize agriculture and, at the same time, a convenient way to treat rural poverty. In our opinion, this emphasis on loans to stimulate production and to help the poor has unfortunately diverted attention from the essential properties of finance, the process of financial intermediation, and the basic role that rural financial markets ought to play in development.[1] While some attention has been given to overall resource misallocation caused by RFM policies, little attention is paid to how RFMs intermediate between savers and borrowers (Wai 1972). Likewise, very little attention has been given to how RFM policies affect overall income and wealth distribution and how political forces use financial systems to

DALE W ADAMS and DOUGLAS H. GRAHAM are professors of agricultural economics, Ohio State University.

Reprinted from *Journal of Development Economics* 8(1981):347–66, with omissions and minor editorial revisions, by permission of the publisher, North-Holland Publishing Company, and the authors.

further their own aims. Even less attention has been given to how various policies influence the vitality of RFMs.

TRADITIONAL AGRICULTURAL CREDIT PROJECTS

Many agricultural credit projects carried out in the past twenty years in LICs have been similar,[2] partly due to the replication in these countries of financial institutions that were successful in several developed countries: credit unions, credit cooperatives, private banks, and supervised credit agencies. Additional similarities are due to the common assumptions that underlie most of these projects.[3] These assumptions can be grouped into those relating to saver-borrower behavior, those associated with lender behavior, and those about the performance of rural finance markets. Common assumptions on saver-borrower behavior are that the rural poor cannot save and therefore will not respond to incentives or opportunities to save, that most farmers need cheap loans and supervision before they will adopt new technologies and make major farm investments, and that loans in kind are used in the form granted.

Common assumptions about lender behavior are that most informal lenders are exploitative and charge borrowers rates of interest that result in large monopoly profits, that the rural poor do not receive formal loans because formal lenders are overly risk-averse, that nationalized lenders can be forced to ignore their own profits and losses to service risky customers and the rural poor, and that all formal lenders can be induced to follow government regulations in allocating financial services. At a national level it is commonly assumed that cheap credit is an efficient way of offsetting production disincentives caused by low product prices or high input prices, that loan quotas established in the capital city are efficient ways of allocating loans in the countryside, that loans should be a part of a package of inputs, that only production loans should be made, and that RFM vitality is not related to projects and policies. Recent research is showing that many of these assumptions are either unsubstantiated, weak, or incorrect.

Because so many institutions and assumptions are similar, it should not be surprising that RFM policies and techniques in LICs are also very similar. For example, heavy emphasis has been placed on creating new financial institutions to service particular rural needs or target groups such as the rural poor. Many countries, for example, have created specialized agricultural banks or development banks that lend largely to agriculture. Credit cooperatives, credit unions, and supervised credit programs have also been popular at various times and places. Seldom do these institutions offer financial savings facilities. Instead they depend largely on central banks, government budgets, and foreign aid for funds.

Low interest rates are almost always assigned to formal loans and savings deposits alike, thereby penalizing savers. In nominal terms, rates of interest on agricultural loans may be as low as zero, and seldom do they exceed 12 percent per year in most low-income countries. Typically, the rates of interest paid on rural savings deposits are much less than the concessionary rates charged on loans. Recent inflation has been double-digit in most regions of the world, outside Asia. This has resulted in negative real rates of interest on most formal loans and deposits in rural areas. The real rate of interest is defined as the nominal rate of interest (the contractual rate) adjusted by some overall expected price index change for the economy.[4] Because of the excess demand caused by these negative interest rates, governments have tried to force lenders to allocate loans to priority groups through quota systems, political persuasion, nationalization of banks, or use of other inducements. Lenders quickly find ways to subvert many of these regulations, however. Portfolio quotas result in redefinition of loans by lenders, and loan size limits cause lenders to extend multiple small loans to previous borrowers of large loans, for example.

While some RFMs work better than others, a number of common problems stand out. These include very serious loan repayment problems in all too many countries (Boakye-Dankwa 1979). They also include very little medium- and long-term formal credit and high loan transaction costs for some borrowers and most lenders. These transaction costs discourage some from seeking formal loans and also discourage lenders from serving certain groups. A handful of recent studies show that the lenders' costs of making agricultural loans to medium- and small-sized farmers is 20 percent or more of the value of the loans extended, even in moderately well-run programs (Ahmed 1980; World Bank 1978). If the lender is in a country experiencing substantial inflation, the nominal rates of interest on loans needed to cover these lender costs and also maintain the purchasing power of the loan portfolio can be well in excess of 30 percent of the value of agricultural loans made.

Studies have also shown that the borrowers' costs of acquiring these formal loans can be substantially larger than the nominal interest payments (Adams and Nehman 1979). Total borrowing costs, especially for borrowers of small amounts, may be two or three times as much as the nominal interest payments. These costs include waiting in line, transportation costs, bribes, legal and title fees, paperwork expenses, and time lost from work to deal with these demands.

Even more serious, in all too many countries, policies have been ineffective in allocating a larger share of formal loans to agriculture in general and to the rural poor in particular because the risks, returns, and costs of doing so are unattractive to formal lenders (Vogel and Larson 1980). A less obvious problem relates to the nature of innovations taking place in RFMs. Most of these innovations are increasing rather than decreasing the total cost to society of financial intermediation. Many of these "distorted" innovations are defensive in nature; that is, they emerge in response to various regulations such as loan portfolio quotas and interest rate ceilings (Bhatt 1979). In extreme cases financial markets may overbuild facilities in rural areas to siphon off savings deposits to urban centers. Most serious of all, it appears that operations of RFMs in most countries are resulting in inefficient allocation of resources, causing income and asset ownership concentration, allowing financial resources to flow out of low-income areas, and, in some cases, diverting resources out of agriculture (Araujo and Meyer 1978; Vogel 1977).

Over the past few years an increasing number of observers have criticized RFM performance (Von Pischke 1979; Gonzalez-Vega 1977). They argue that too little attention has been given to the economic and policy environment that influences RFM performance, and they also challenge the validity of many assumptions on which RFM projects are built. In addition, they attack policies commonly used to influence the behavior of lenders, borrowers, and financial markets as a whole. Ubiquitous low interest rate policies have taken the brunt of these attacks (Shaw 1973).

Out of these criticisms, new suggestions have emerged on changes needed in RFM projects so that publicly stated goals and the performance of RFMs can be more closely synchronized. Despite increasing consensus, there have been very few changes in rural financial market projects to date. In the last section of this chapter we speculate on why these changes are so slow in coming.

NEW VIEWS ON RURAL FINANCIAL MARKET PROJECTS

A key element in the new views on RFMs is the identification of the expected real rate of interest as a major determinant of borrower, saver, and lender behavior (Gonzalez-Vega 1976). Real rates are also thought to influence strongly the overall performance of financial markets. Proponents of the new views argue that low real rates of interest seriously disrupt the supply side of the financial system. Because interest rates on savings deposits are low, savers minimize the amount of financial savings they hold. This forces formal lenders to rely on external funds to finance loans. Poor people in rural areas are especially disadvantaged by these low interest rates on savings. In large part the rich evade interest rate restrictions on savings accounts by lending through informal financial markets or by buying nonfinancial assets. The poor, however, find it difficult to assemble sufficient funds to acquire many asset forms, such as large animals, land, gold, buildings, and time certificates of deposit. They are thus forced to hold surpluses in cash, crop inventories, or small animals or to consume what might otherwise be saved. Furthermore, because the funds lent in these programs are not locally mobilized, borrowers feel little obligation to repay funds that are provided by national or foreign governments.

Because the risks and marginal costs of lending to agriculture in general, and to the rural poor in particular, are often higher than for loans to other parts of the economy, formal lenders tend to shy from lending in rural areas, even with government pressure to serve agriculture. Lenders have even less incentive to lend to agriculture and the rural poor when regulations set interest rates lower on agricultural loans than can be charged on other loans (Blitz and Long 1965). The same microeconomic forces cause formal lenders to shorten the loan term structure and shift their funds to a more concentrated and less risky portfolio when expected rates of inflation increase.

Governments have used a number of techniques and policies, up to and including the nationalization of banks, to force formal lenders to ignore their own profit and loss considerations and serve some social objective or target group not reached through market criteria (Shetty 1978). Generally, the results of these efforts have been disappointing. It is virtually impossible for a government to monitor and enforce loan rationing policies when hundreds of thousands of formal loans are made in widely dispersed areas of the country. The essential properties of financial instruments are their fungibility, their divisibility, and their substitutability (Von Pischke and Adams 1980). Lenders, for example, may meet the letter of the law by simply reclassifying loans to meet quota requirements. The lender may also shift small borrowers who are funded with the lender's own resources onto lines of credit provided by the government or by an aid agency. The lender may then lend its own released funds for nonpriority, yet profitable, loans. This may result in little or no additional lending to the priority group or activity specified by the lender. The same problems of fungibility occur among borrowers.

Negative real rates of interest also distort loan demand. If expected interest rates are negative, the borrower may realize an income transfer by taking a loan, investing the money in an asset that increases in value at the same pace as inflation, and later liquidating the asset to repay the loan. With negative real rates of interest, some loan demand may be for acquiring this income transfer rather than for making productive use of loans. These income transfers can be very sizable when real rates of interest are highly negative and sizable formal agricultural credit programs are involved, as in Brazil, for example, where yearly income transfers of U.S. \$3–4 billion may be involved (Sayad 1979). The excess loan demand stemming from the negative interest rates may also cause the lender to create a number of administrative hurdles that raise the loan transaction costs for potential borrowers who are not profitable clients. In this way the lender effectively discourages loan demand from some potential borrowers without violating policy directives. In the end, lenders exclude small borrowers and concentrate their "rationed" loans on large borrowers who have excellent collateral.

The new views also include more positive attitudes about informal financial markets (Barton 1977; Harriss 1979; Bouman 1979). Informal lenders are thought to provide valuable services, and impose lower costs on most borrowers than had been generally thought. The opportunity costs of money lent in the informal market by merchants or farmers are usually ignored by those who criticize informal lenders' charges on loans. Singh (1968), for example, found that the opportunity costs of money informally lent by some Indian farmers amounted, on an annual basis, to 77 percent of the value of the money lent. These opportunity costs made up over half of the interest charges. Harriss's recent work among merchants in southern India, who also extend informal loans, showed that their opportunity costs of lending, instead of using the funds internally to expand their buying and selling operations, amounted to as much as 63 percent of the value of the loans extended. In addition, Nehman (1973) showed that for the rural poor, informal loans may be no more costly than formal loans when total loan transaction costs for the new and small borrower are carefully calculated. In some cases the informal lender is also able to provide more flexible and more desirable financial services than do formal lenders. The fact that borrowers often choose to repay informal loans before they repay formal loans supports this conclusion.

The new views also suggest that the rural poor, when they are given adequate opportunities and incentives to save, may have much larger savings capacities than heretofore recognized. Only a few studies have been done on voluntary rural savings capacities, and some of these have used survey data that may have included underreported income information (Bhalla 1978). Only a handful of studies have used farm record-keeping data or time series survey data that may have given accurate estimates of rural household consumption, savings, and income activities. Studies on time series data from Farm Household Economy Surveys in Japan, for example, showed average propensities to save that grew from 0.10 in 1950 to 0.22 in 1973 (Adams 1978). Marginal propensities to save were significantly higher. Studies using time series data from farm record-keeping families in Taiwan showed even higher average and marginal propensities to save over the period 1960–74 (Ong, Adams, and Singh 1976). Similar studies on time series data from Farm Household Economic Surveys in Korea showed average propensities to save that ranged from 0.15 in 1962 to 0.33 in 1974 (Ahn, Adams, and Ro 1979). Marginal propensities to save were, again, substantially higher. Less comprehensive studies in Kenya, Mexico, Malaysia, the Sudan, the Punjab of India, and Zambia also uncovered substantial voluntary savings capacities (Adams 1978). Several studies of cooperatives and farmers' associations in Korea, Japan, and Taiwan showed that mobilization of voluntary financial savings deposits in these institutions played a major role in their economic strength (Lee, Kim, and Adams 1977; Kato 1972). Borrowers are more likely to repay loans if a substantial part of the money lent is mobilized via savings deposits in the local area.

The new consensus also holds that borrowers' loan transaction costs are more important in determining loan demand among small and new borrowers than are interest rates. In contrast, large and experienced borrowers may be very sensitive to changes in interest rates because interest payments make up a large part of their total borrowing costs and their less obvious loan transaction costs are negligible. In a Bangladesh study, Shahjahan (1968) found that interest payments made up only 17 percent of the total borrowers' transaction costs for those farmers with small loans from the Agricultural Development Bank. At the same time, large borrowers from the same bank incurred interest payments that made up 57 percent of their total borrowing costs. All borrowers from the bank paid the same rate of interest on their loans, 7 percent. Ahmed's (1980) work in the Sudan supports these results.

The new views also posit that overall savings behavior in rural areas is quite sensitive to changes in real rates of interest paid on deposits. The preliminary results from a pilot savings mobilization project in Peru, which involves substantial increases in the interest rates paid on deposits, strongly suggests that people in rural areas will substantially increase their savings deposits if given security, liquidity, and high returns. Earlier rural savings performance in Taiwan, Japan, and Korea reinforce this conclusion.

The new views go on to argue that interest rates and loan supervision have a weak effect on decisions to adopt new technology or make on-farm investments. Loan supervision is often ineffective because the supervisor knows little about the practical problems of farming, has little incentive to provide useful technical assistance, or has few if any profitable new production techniques to extend to the borrowing farmer. Interest payments are only one of several factors that influence loan use decisions. While *everyone* wants to pay the lowest interest rate possible, borrowers may be even more interested in the noninterest borrowing transaction costs, the timeliness of the loan disbursement, the flexibility of loan repayment procedures, and the availability of additional loans from the lender. Many advocates of low interest rates ignore the importance of other borrower loan transaction costs in the borrowing decision, especially for small farmers. For example, if interest payments make up only 25 percent of total borrowing cost, a doubling of interest rates will only increase borrowing costs by one-quarter. It might be argued that if formal lenders were allowed to charge higher rates on agricultural loans, they would eliminate a number of the loan application hurdles and collateral requirements that currently make up a large part of borrowers' loan transaction costs. Higher interest rates may result in lower borrowing costs for some borrowers and tilt the system towards a more equitable inclusion of heretofore excluded smaller farmers.

Low interest rate advocates also ignore that interest payments, especially among small- and medium-sized borrowers, often make up a very small part of total operating expenses. In 1970, small farmers in Taiwan and Korea, for example, spent on the average only 2 percent and 1 percent, respectively, of their total farm and household cash expenditures on interest payments. Among only the borrowing households the percentages were less than 4 percent in both countries. In most cases, product or input prices are much stronger incentives to adopt new technology than are interest rates. Furthermore, these prices have a much wider impact among the farming population than do credit programs.

Interest rates do, however, have a very strong influence on lenders' behavior (formal lenders as well as formal savers). Under normal conditions, receipts from interest payments make up a very large part of a formal agricultural lender's total revenues. Major increases or decreases in interest rates applied on farmer loans, therefore, have dramatic impacts on the marginal as well as total revenues and thus surpluses or deficits of the lender. Even in nationalized banking systems these lender revenues and surpluses or deficits largely determine the overall vitality of RFMs and their ability and willingness to perform financial intermediation in a socially desirable manner (Von Pischke 1979). With long periods

of negative real rates of interest, lenders are forced to rely on permanent subsidies to cover their operating expenses, to cut back on their scale of operations,
or allow the quality and quantity of their financial services to deteriorate.

Critics have also questioned attempts to include loans as part of a package of
inputs. They argue that packaging loans and use of other similar nonmarket
rationing devices diminish the most attractive and useful property of finance,
fungibility. It is the fungibility of money that allows it to be converted into any
good or service available in the market (Von Pischke and Adams 1980). Many
planners try to destroy this essential property of finance by allocating loans in
fixed quotas, making loans in kind, or trying to specify the ultimate use of the
loan. This planning approach to the allocation of loans assumes that a borrower
knows not what is best for him or her, that loans can be allocated like physical
inputs, and that the planner in the capital city can effectively make efficiency and
equity decisions for thousands of heterogeneous borrowers. Pushed to its extreme, the planning approach to loan allocation would result in a return to a
barter economy. Fortunately, the ability of planners to diminish fungibility is
extremely limited; secondary markets for goods lent in kind quickly spring up
(borrowers receiving the rationed input will sell it to others who need it more),
and borrowers can substitute borrowed liquidity for their own liquidity when
planners' priorities do not match those of the borrower. The widespread nonmarket rationing devices used for agricultural loans in many low-income countries are mostly a mirage and have very little impact on the allocation of real
resources (Vogel and Larson 1980). Their main effect is to increase the total
costs to society of financial intermediation and also to undermine the viability of
lenders.

KEY ELEMENTS IN A NEW RFM STRATEGY

The new views on RFM projects challenge many of the assumptions and
policies that have been vital parts of LIC agricultural credit projects in the past.
They also stress that the results from these projects are not consistent with
efficiency or equity goals. While the specific suggestions for improving the
results of RFM projects must be time- and place-specific, a few general suggestions do emerge out of these views.

One of the most prominent suggestions is that more flexible interest rates
could be a key factor in improving the results from most RFM projects
(Gonzalez-Vega 1977; Adams 1978). Nominal rates of interest must be flexible
so that they go up and down with inflation. Interest rate policies on both credit
and deposits should be aimed at maintaining relatively stable and positive *real
rates* of interest. Lenders (banks and individual savers) must expect to receive
positive real returns most of the time from their financial transactions if RFMs
are to function equitably and efficiently.

With more attractive incentives for savers, RFMs could mount major savings mobilization schemes in rural areas. The previously mentioned pilot savings mobilization project in Peru and another pilot project in Bangladesh which is experimenting with more flexible and higher interest rates on both loans and deposits in rural areas should provide insights on how to proceed with larger schemes. As suggested earlier, changing the image of who owns the money lent will improve loan repayment. If formal lenders depended less on central banks, foreign aid, and government budgets for funds, they would experience less political interference (Ladman and Tinnermeier 1981). If lenders, such as cooperatives, were able to provide attractive savings deposit facilities for their members, it would give more cooperative members strong reasons for being active members (Robert 1979). In early stages of development, savings mobilization should receive top priority in RFM activities, and loans should receive secondary attention.

In most cases it also appears that the building of new specialized credit institutions to service fragmented financial needs in rural areas should receive less attention. These institutions usually rely on government subsidies or foreign aid for funds to make up their loan portfolio and to cover operating expenses. The funding source often loses interest in underwriting the costs of the agency after a time and reduces funding. Because the agency does not typically accept deposits, it becomes heavily dependent on the government for continued funding, and political interventions into the operations of the agency become common. Furthermore, the agency is often asked to lend to a relatively narrow target group, such as livestock farmers, organizations undertaking long-term investments, and small farmers. This loan specialization does not allow the agency to diversify its lending risks or to service nonfarm rural enterprises (Meyer 1979). Instead of continuing to emphasize the creation of new lenders, policy makers should direct more attention to diagnosing why existing financial institutions are not providing the types and amounts of services desired. Policy changes should be aimed at providing more incentives to existing lenders to expand their services in the desired directions.

Governments and aid agencies must also use care when they introduce additional loanable funds into RFMs via special rediscount facilities in central banks. For example, why should banks in the Dominican Republic or the Philippines open new savings deposit facilities in their branches and pay 6–8 percent on these deposits when they can get rediscount money from the central bank at lower rates!

Finally, RFM projects would be improved if designers and policy makers stopped viewing loans as inputs similar to fertilizer, labor, seeds, or breeding stock. Rather, loans must be viewed for what they are: claims on resources that allow the borrower command over additional goods and services that may or may not be used for the purposes stated in the loan application. Instead of trying to ration this command over resources in predetermined lumps to thousands of borrowers, policy makers should provide proper incentives for lenders/mobilizers to perform in more socially desirable ways. Stress should be placed on improving the process of financial intermediation and reducing the costs of this process for society. The focus should be on inducing RFMs as a whole to service better the credit and deposit needs of a much broader clientele in rural areas. Along with this, RFMs should also be given strong inducements to adopt innovations that reduce the total costs of financial intermediation. RFMs cannot be used to transfer cheap credit to thousands or millions of small, previously unserviced farmers. If governments attempt to push this strategy, the cheap credit will end up mostly in the hands of the wealthy. Other methods must be used to help the rural poor more directly.

WHY SO LITTLE CHANGE IN RFM PROJECTS AND POLICIES?

There are at least four possible reasons why so little change has occurred in agricultural credit projects during the past several decades, even though a number of people are heavily criticizing the results of traditional projects. The first reason might be that the new views are incorrect or that they are based on faulty research or on research done on cases or areas that make generalization inappropriate. It seems to us that while additional research would be useful, enough information is at hand, and enough knowledgeable people agree on the results of this research, that some experimentation with new policies along the lines presented above is warranted. At the very least, advocates of traditional agricultural credit projects and policies should be required to offer more than received wisdom, horror stories, and seat-of-the-pants empiricism to justify their positions.

A second reason for so little change might be that it takes a good deal of time for policy makers to understand, accept, and adopt the ideas included in these new views. Many of these views challenge dogma about RFMs that have deep historical roots whose "truth" has been reinforced in the minds of policy makers by endless repetition, numerous tales of horror, and religious teachings. Old ideas die very hard! It took Christian societies many centuries to view usury and lending with some logic rather than all passion. Intermediaries, especially lenders, have been viewed with suspicion in almost all societies. Is that because they are often "outsiders" or "foreigners"—Jews in Europe, Chinese in Southeast Asia, people from the Middle East in Latin America? These intermediaries are often targets of criticism stemming from any unexplained economic discomfort experienced by producers and/or consumers. Because most countries are rapidly moving away from subsistence and barter activities into highly monetized economies, we feel that policy makers do not have the luxury of waiting several centuries to understand the importance of finance in development.

A third explanation might be that policy makers understand that RFM projects are not working well and that elimination of some RFM distortions might improve resource allocation and help meet equity goals. The reason these policy changes are not made is that distortions in RFMs are often justified as offsets to other distortions in the economic system that penalize agriculture (Vogel 1983). These other distortions may be overvalued exchange rates, price controls on food, import regulations, taxing policies, or sectoral investment strategies favoring industry. The distortions in RFMs are second-best measures aimed at partially offsetting these other distortions. To the extent that circumstances continually force the adoption of broader macroeconomic policies that penalize agriculture, policy makers may feel compelled to resort to concessionary-priced credit programs to help the sector adapt satisfactorily to these other penalizing measures (Bourne and Graham 1983). Some argue that it would be impossible to substitute appropriate policy adjustment to make RFMs perform more satisfactorily unless these other distortions are also removed. Thus, the prospects for effective reform of RFM policies become inextricably linked to the difficult tasks of reforming the entire structure of the economy.

We agree that adjustments in financial market policies, accompanied by reforms in other economic policies, as in South Korea in the mid-1960s, are the best means of improving the performance of RFMs (Brown 1973). We feel, however, that reforms in RFM policies alone can result in important gains in resource-allocation efficiency and more equitable allocation of income. The complex and often confusing second-best arguments used to justify distortions in financial markets make it difficult for many to understand the vital issues involved. The tax-subsidy framework often used to justify concessionary-priced agricultural loans to offset other adverse policies in agriculture breaks down for at least three reasons: concessionary loans concentrate income, they do not result in more efficient resource allocation, and they discourage savings.

Proponents of this line of argument ignore that low interest rates strongly affect lender behavior, and administrative fiats are largely ineffective in reversing this behavior. With low interest rates the lender often has excess demand for the "sweet money." The lender reacts by transferring part of the loan transaction costs to the borrower, lends to those who present very little default risk, requires substantial collateral, tries to increase the average size of loans made, and excludes new borrowers. The net result is that lenders concentrate cheap loans in the hands of relatively wealthy and experienced borrowers (Vogel 1977). Because the subsidy involved in cheap credit is proportional to the amount of money borrowed, the subsidy also ends up being very concentrated (Gonzalez-Vega 1977). The microeconomic interests of the lender typically overwhelm the effects of policy directives from the capital city aimed at forcing less concentration of loans. It is impossible for policy makers to police administrative fiats in RFMs because of the large number of lenders and borrowers that are usually involved.

It should also be clear that because of fungibility, cheap credit will not help to offset inefficiencies in resource use caused by policies adverse to agriculture. If cheap credit is to offset inefficiencies, the cheap credit must result in additional use of inputs in the production process that is discouraged by the price distortion due to adverse policy. Because loans are claims on real resources and provide additional liquidity, the borrower can choose to use this additional liquidity in *any* economic activity available in the market. If the price of product X is artificially low, why should the borrower choose to buy more inputs to produce more X just because the costs of the additional liquidity provided by a loan are kept low through concessionary interest rates? Economic theory and common sense lead one to expect that the borrower will use the additional liquidity to buy that good or service providing the highest marginal return or utility. The essential property of finance, fungibility, largely dissolves the ability of policy makers to offset inefficiencies in resource allocation in agriculture caused by various policies, with cheap credit.

In our opinion, the strongest case against the second-best argument can be made on what low interest rates on loans, and thus on deposits, do to savers and the overall vitality of rural financial markets. Low interest rates on financial savings seriously weaken the incentive that many people in the society have to postpone consumption. These potential savers are the invisible victims of cheap credit. Low interest rates force many people in the society to use their "surpluses" in economic activities that have low marginal returns. The cheap loans also cause the rich to colonize most formal agricultural credit programs, and the low rates of interest paid on savings reinforce the exclusion of the poor from participating in formal financial intermediation. Economies of scale and widespread popular support for formal financial market activities are impossible to realize under these conditions.

A final reason for the lack of change in RFM policies may be that the political system finds the current performance of RFMs *satisfactory* (Ladman and Tinnermeier 1981). That is, political forces in the country may be more than satisfied with the results of distortions introduced by negative real rates of interest in RFMs because they result in the allocation of political patronage in the form of implied income transfers to those influential people in the economy who end up receiving most of the cheap credit (Robert 1979). Distortions in interest rates as well as other price distortions, caused by fixed exchange rates, import and export regulations, and licenses allow the political system to allocate "administrative profits." If interest rates were raised to equilibrium levels, the political system would have no cheap credit to grant to its favored patrons and strong supporters.

One might ask why individuals in society who are disadvantaged by low interest rate policies do not organize to press for more appropriate policies. An explanation for this is that large numbers of widely dispersed individuals (that is, landless workers and small- to medium-sized farmers) are disadvantaged by current interest rate policies. They are largely excluded from access to formal credit because of the credit rationing process practiced by formal lenders. Others are paid low returns on their small savings or decide not to save at all in financial form because of the low returns. When a large number of people are only hurt a small amount by a policy, it is difficult to mobilize these individuals for political action. The opposite is true for those who benefit from low interest rate policies. Many who receive these benefits are powerful individuals. Any policy change that reduces the benefits they receive through cheap credit draws immediate and strong reactions. This may be one of the reasons why a number of powerful economic interests are so tolerant of inflation. Inflation, along with low and inflexible interest rate policies, allows those with access to concessionary-priced loans to receive large income transfers because of the negative real rates of interest. Inflation also allows the political system to mask the magnitudes and directions of the political patronage transferred through the financial system. In most cases it is not a conspiracy among a few individuals that results in fixed nominal interest rates, inflation pressures, and negative real rates of interest. Rather, it is a *convergence* of interests that results in the popularity of negative real rates of interest once they have become established through rising rates of inflation.

CONCLUSIONS

In concluding this review, it is useful to recognize two broader problems generated by cheap agricultural credit. First, in those countries where the viability of lending institutions may be of secondary importance because they explicitly engage in deficit financing of these programs, extensive subsidized financing of agricultural credit programs can generate significant inflationary pressures. Recent work by the World Bank staff has highlighted the important role that the large volume of rural credit has played in adding to the money supply of Brazil in the mid- to late 1970s and contributing substantially to inflationary pressures.

Second, the degree and magnitude of credit subsidization in most countries has taken its toll on the amount of resources available for other vital programs in such areas as agricultural research, basic infrastructure to lower the costs and risks of marketing, and improved educational services for the rural population, among others. Unfortunately, it is precisely in these areas that major efforts must be undertaken to improve the economic rate of return of farming. It is only when these bottlenecks are reduced that credit can make a substantial difference and in doing so can be priced realistically. Conversely, if these other problem areas are not properly dealt with, credit (subsidized or not) will not make any difference. Credit by itself cannot raise the rate of return to farm investments.

Twenty years ago development experts began to realize that rural people in low-income countries were able to count, even though many were not able to read. Schultz, Hopper, and others did a valuable service by educating the development profession on the rationality of farmers in LICs.[5] Currently, almost all knowledgeable persons working on development respect the ability of farmers in LICs to allocate their resources efficiently and respond to product prices, input prices, and new technology, with all its related risks, in rational ways. It is past time that the development profession recognized that these same individuals make similar rational decisions when they participate in financial markets. Current low interest rate policies are making it virtually impossible to induce formal lenders to provide needed loan and deposit services to the rural poor. Financial systems will not produce the types of services needed to satisfy generally accepted development goals until more enlightened policies along the lines suggested by the new views are adopted.

NOTES

Our colleagues at Ohio State University contributed a number of the ideas summarized here. Also, the Office of Rural Development and Development Administration of the Agency for International Development provided support for the preparation of this chapter.

1. Readers wanting more background on these ignored issues might look at Shaw 1973.
2. Those looking for details on agricultural credit projects might review Donald 1976; and World Bank 1975.
3. For a statement of these assumptions in the 1950s see Technical Cooperation Administration 1952.
4. The real rate of interest is defined as $[(1 + i)/(1 + p)] - 1$, where i is the nominal rate of interest and p is some index of the annual change in prices.
5. See chapter 1 in this volume.—ED.

REFERENCES

Adams, Dale W. 1978. "Mobilizing Household Savings through Rural Financial Markets." *Economic Development and Cultural Change* 26:547–60.

Adams, Dale W, and G. I. Nehman. 1979. "Borrowing Costs and the Demand for Rural Credit." *Journal of Development Studies* 15:165–76.

Ahmed, Ahmed Humeida. 1980. "Lender Behavior and the Recent Performance of Rural Financial Markets in the Sudan." Ph.D. diss., Ohio State University.

Ahn, Choong Yong, Dale W Adams, and Young Key Ro. 1979. "Rural Household Savings in the Republic of Korea, 1962–76." *Journal of Economic Development* 4:53–75.

Araujo, Paulo F. C., and Richard L. Meyer. 1978. "Agricultural Credit Policy in Brazil: Objectives and Results." *Savings and Development* 2:169–94.

Barton, Clifton G. 1977. "Credit and Commercial Control: Strategies and Methods of Chinese Businessmen in South Vietnam." Ph.D. diss., Cornell University.

Bhalla, Surjit S. 1978. "The Role of Sources of Income and Investment Opportunities in Rural Savings." *Journal of Development Economics* 5:259–81.

Bhatt, V. V. 1979. "Interest Rate, Transaction Costs and Financial Innovations." *Savings and Development* 3:95–126.

Blitz, Rudolph, and Millard Long. 1965. "The Economics of Usury Regulation." *Journal of Political Economy* 73:608–19.

Boakye-Dankwa, Kwadwo. 1979. "A Review of the Farm Loan Repayment Problem in Low Income Countries." *Savings and Development* 3:235–52.

Bouman, F. J. A. 1979. "The ROSCA: Financial Technology of an Informal Savings and Credit Institution in Developing Economies." *Savings and Development* 3:253–76.

Bourne, Compton, and Douglas H. Graham. 1983. "Economic Disequilibria and Rural Financial Market Performance in Developing Economies." *Canadian Journal of Agricultural Economics* 31:59–76.

Brown, Gilbert T. 1973. *Korean Pricing Policies and Economic Development in the 1960s*. Baltimore: Johns Hopkins Press.

Donald, Gordon. 1976. *Credit for Small Farmers in Developing Countries.* Boulder: Westview Press.

Gonzalez-Vega, Claudio. 1976. "On the Iron Law of Interest Rate Restrictions: Agricultural Credit Policies in Costa Rica and Other Less Developed Countries." Ph.D. diss. Stanford University.

————. 1977. "Interest Rate Restrictions and Income Distribution." *American Journal of Agricultural Economics* 59:972–76. Reprinted as chapter 22 in this volume.

Harriss, Barbara. 1979. "Money and Commodities, Monopoly and Competition." Paper presented at Workshop on Rural Financial Markets and Institutions, 12–14 June, Wye College, Wye, Wales.

Kane, Edward J. 1975. "Deposit-Interest Ceilings and Sectoral Shortages of Credit: How to Improve Credit Allocation Without Allocating Credit." In Kane, *Government Credit Allocation,* 15–31. Rochester, N.Y.: Center for Research in Government Policy and Business, Graduate School of Management, University of Rochester.

Kato, Yuzuru. 1972. "Sources of Loanable Funds of Agricultural Credit Institutions in Asia: Japan's Experience." *Developing Economies* 10:126–40.

Ladman, Jerry R., and Ronald L. Tinnermeier. 1981. "The Political Economy of Agricultural Credit: The Case of Bolivia." *American Journal of Agricultural Economics* 63:66–72.

Lee, Tae Young, Dong Hi Kim, and Dale W Adams. 1977. "Savings Deposits and Credit Activities in South Korean Agricultural Cooperatives, 1961–1975." *Asian Survey* 17:1182–94.

Long, Millard F. 1968. "Why Peasant Farmers Borrow." *American Journal of Agricultural Economics* 50:991–1008.

Meyer, Richard L. 1979. "Financing Rural Non-farm Enterprises in Low Income Countries." *Financing Agriculture* 11:5–10.

Nehman, Gerald I. 1973. "Small Farmer Credit Use in a Depressed Community of São Paulo, Brazil." Ph.D. diss., Ohio State University.

Ong, Marcia L., Dale W Adams, and I. J. Singh. 1976. "Voluntary Rural Savings Capacities in Taiwan, 1960–70." *American Journal of Agricultural Economics* 58:278–82.

Robert, Bruce L., Jr. 1979. "Agricultural Credit Cooperatives in Madras, 1893–1937: Rural Development and Agrarian Politics in Pre-independence India." *Indian Economic and Social History Review* 16:163–84.

Sayad, João. 1979. "The Impact of Rural Credit on Production and Income Distribution." Paper presented at Conference on Rural Finance Research Issues, 29–31 August, Calgary, Canada.

Shahjahan, Mirza. 1968. *Agricultural Finance in East Pakistan.* Dacca: Asiatic Press.

Shaw, Edward S. 1973. *Financial Deepening in Economic Development.* New York: Oxford University Press.

Shetty, S. L. 1978. "Performance of Commercial Banks since Nationalization of Major Banks: Promise and Reality." *Economic and Political Weekly* 13:1407–51.

Singh, Karam. 1968. "Structural Analysis of Interest Rates on Consumption Loans in an Indian Village." *Asian Economic Review* 10:471–75.

Technical Cooperation Administration. 1952. *Proceedings of the International Conference on Agricultural and Cooperative Credit.* 3 vols. Berkeley: University of California Press.

Vogel, Robert C. 1977. "The Effects of Subsidized Agricultural Credit on the Distribution of Income in Costa Rica." Paper, Department of Economics, Southern Illinois University, Carbondale, Ill.

————. 1983. "Implementing Interest Rate Reform." In *Rural Financial Markets in Low-Income Countries: Their Use and Abuse,* edited by J. D. Von Pischke, Dale W Adams, and Gordon Donald. Baltimore: Johns Hopkins University Press, 1983.

Vogel, Robert C., and Donald W. Larson. 1980. "Limitations of Agricultural Credit Planning: The Case of Colombia." *Savings and Development* 4:52–62.

Von Pischke, J. D. 1979. "The Political Economy of Specialized Farm Credit Institutions in Low-income Countries." Paper presented at Workshop on Rural Financial Markets and Institutions, 12–14 June, Wye College, Wye, Wales.

Von Pischke, J. D., and Dale W Adams. 1980. "Fungibility and the Design and Evaluation of Agricultural Credit Projects." *American Journal of Agricultural Economics* 62:719–26.

Wai, U. Tun. 1972. *Financial Intermediation and National Savings in Developing Countries.* New York: Praeger.

World Bank. 1975. *Agricultural Credit Sector Policy Paper.* Washington, D.C.

———. 1978. "The Financial Cost of Agricultural Credit: A Case Study of Indian Experience." Staff Working Paper, no. 296. Washington, D.C.

22

Interest Rate Restrictions and Income Distribution

CLAUDIO GONZALEZ-VEGA

Credit has been one of the most important components of the strategies for development in the agricultural sector of low-income countries (LICs). Not only is the input provided—money—fungible but an infrastructure, the banking system, is available for its supply, and it can be administered at a significant distance from the farm, unlike other inputs. Because credit programs are easier to implement than other types of programs, there has been a rapid expansion of the volume of agricultural credit granted by formal lenders in LICs.

Two important and universal features of this expansion have been the concentration of the loaned funds to a select clientele and the low rates of interest charged. Only a small fraction of the farmers of LICs receive formal credit—about 5 percent in Africa and 15 percent in Asia and Latin America (Donald 1976; World Bank 1975). This concentration is also reflected in the distribution by size of the loans granted. Consider Honduras, an intermediate case in Latin America, where in rural areas the outstanding credit per capita is thirty-five dollars, but only 10 percent of the rural families have bank loans. Out of every ten families, nine receive no formal loans and one receives a loan of twenty-one hundred dollars. Moreover, among those obtaining credit, 9 percent receive about 81 percent of the total money loaned.

The rates of interest have been low because they have been maintained at about the same rates as those charged in countries rich in capital and have not reflected the opportunity cost of capital or its shadow price in LICs. Also, the supply and demand for formal loans have not been equated, and this has generated excess demand, which has required nonprice rationing. Bank interest rates have been lower than those charged by informal lenders, and they have frequently failed to cover the cost of granting credit. In particular, because they do

CLAUDIO GONZALEZ-VEGA is professor of economics, University of Costa Rica, San José, and professor of agricultural economics and adjunct professor of economics, Ohio State University.

Reprinted from *American Journal of Agricultural Economics* 59, no. 5 (1977): 973–76, with minor editorial revisions, by permission of the American Agricultural Economics Association and the author.

not cover the cost of entering into additional markets, the low rates have prevented formal lenders from serving marginal clients, making it necessary to subsidize financial institutions to ensure their survival. Also, low interest rates have transferred a substantial subsidy to privileged borrowers. Finally, they have implied ceilings on the deposit rates paid on savings, and in inflationary LICs they have been negative and unpredictable in real terms. Interest rates have not been low because of market forces, higher productivity, or financial innovation; rather, they have been imposed by usury laws and central bank regulations. These administered rates have been kept fixed at low levels despite changing circumstances, but their rigidity has not prevented the real rates from affecting allocative efficiency, institutional viability, growth, employment, equity, and income distribution in the rural areas of LICs. This paper examines their impact on the latter.[1]

Any farmer's net income is a function of his internal productive opportunity and other investment options, as well as his command over the resources necessary to take advantage of them. The farmer's control over resources depends in turn on his own endowments and on his access to external finance. In fragmented markets, these opportunities, endowments, and access to finance are poorly correlated. Many farmers, each with his own unique set of endowments and opportunities, are so isolated that they earn a wide range of rates of return (McKinnon 1973). These farmers face very different income growth potentials, which are reflected in the distribution of income.

The impact of access to external finance on income distribution can be illustrated by referring to three financial regimes. First, consider a group of farmers under conditions of self-finance. Each will be able to exploit his opportunities only to the extent allowed by his endowment. Farmers with larger endowments will earn higher incomes. However, the value of the marginal product of the inputs used will be higher, *ceteris paribus,* when the farmer's own endowment is small and his productive opportunity is good. Even though the large farmers may use superior production techniques and have larger endowments, if their endowments are better than their opportunities, given the cumulative biases of income growth processes and of diminishing returns, the value of the marginal product of the inputs employed may be lower for them than for the small farmers.

In the previous case, if direct external finance—the transfer of resources via loans from large to small farmers—is possible, the net incomes of both classes of farmers will increase. Moreover, if the value of the marginal product of the inputs declines faster for the small farmers than for the large farmer, in view of the relatively limited entrepreneurial ability and small endowment of other fixed inputs of the former, the absolute and relative net gain in income as a result of the loan will be greater for the small farmer than for the large. Thus, in addition to its allocative effect, direct external finance improves income distribution (Gonzalez-Vega 1976).

A similarly favorable distributive impact takes place under a regime of indirect external finance, when credit is granted by an institution that also collects savings. Consider two farmers with identical production opportunities. Under the conditions of self-finance, if their endowments are equal, their incomes are equal, but if not, the farmer with the larger endowment earns a higher income. However, if both farmers have access to credit, their incomes are equalized. Therefore, the extent to which differences in income are due to differences in endowments is reduced by access to credit. This distributive property of access to credit is also present in a dynamic situation where opportunities can be improved through technical change. In sum, the more fragmented the capital markets are, the more diverse their individual participants will be; and the more dynamic and risky the available opportunities are, the more crucial access to credit will be in determining income growth potentials and income distribution.

Financial intermediation is not free, however. The costs are social because the resources used for credit could have been devoted to increasing output elsewhere. The costs are private as well because lenders have to compete for the resources in the market. On the other hand, credit is not a homogeneous product. There are many dimensions of credit: short-term and long-term, agricultural and industrial, large and small borrowers. Each one of these dimensions is a separate product with its own peculiar cost function.

Once the various dimensions of credit and their costs are distinguished, it is clear that the socially optimum, competitive rates of interest will not be the same for all farmers. A social optimum requires that each farmer be granted a loan of a size that makes the marginal cost of lending equal to the value of the marginal product of the inputs purchased with it and, in turn, to the rate of interest charged. If a larger loan is granted, more resources are spent in administering it than are generated by the additional production induced, with a net loss to the society. Therefore, the optimum rates of interest will be higher, *ceteris paribus,* the better the productive opportunity, the smaller the endowment, and the higher the marginal cost of lending to a given farmer.

Similarly, a profit-maximizing lender in a monopolistically competitive market will try to segment it on the basis of demand elasticities or of marginal costs in order to charge different rates to different borrowers. If more inelastic demands and higher costs are associated with small farmers, the rates charged to them will be higher.

If restrictions are imposed on rates of interest, lenders may practice credit rationing, that is, granting smaller loans than those demanded at the rate charged. From the point of view of the individual borrower, rationing implies an excess demand for credit at the rate of interest paid (Jaffee 1971; Eckaus 1974). Generally, a lender restricts the size of a loan if the marginal cost of granting it is higher than the imposed rate of interest. If a larger loan is granted, the added costs will be greater than the revenues and his expected profits will decline. Since marginal cost is an increasing function of the size of any individual loan (Gonzalez-Vega 1976), the lender can reduce his marginal cost and equate it to the level of the imposed rate by reducing the size of the loan. If this rate becomes sufficiently low, dropping below the average variable cost, rationing takes the form of exclusion from the lender's portfolio.

The possibility that a borrower will be rationed depends on the relationship between the marginal cost and the demand functions faced by the lender and on the nature of the restrictions. The slower the marginal cost rises as the size of the loan increases, a function of risk, and the faster the quantity of credit demanded increases as the rate of interest declines due to the behavior of the value of the marginal product of the inputs, the less likely is rationing. The low ceilings imposed on rates of interest for loans in LICs have led to widespread rationing, as evidenced by quotas, long delays in disbursement, and other quantitative restrictions. While the loans received by most farmers have been smaller than those demanded at the low rates, a few privileged borrowers have obtained all the credit they have demanded at the restricted rates. Although this preferential treatment reflects the influence of social and political power, the result can also be explained by purely economic considerations related to the profit-maximizing behavior of private lenders and to the survival behavior of public lenders.

Furthermore, given empirically relevant values of certain parameters, when the ceilings imposed on interest rates become more restrictive—due to inflation, increased lending costs, or the frequent practice of granting preferential rates for specific activities—the size of the loans granted to nonrationed borrowers increases and the size of those granted to rationed borrowers decreases. This is the "iron law of interest rate restrictions" (Gonzalez-Vega 1976). As the ceiling becomes lower, nonrationed borrowers move along their demand curves, demanding and receiving larger amounts of credit at the lower rate. On the other hand, rationed borrowers move along the lender's marginal cost curves of granting credit to them, and the lender grants them even smaller loans than before, although they, too, demand larger loans at the low rate.

Small farmers are rationed credit long before the larger farmers because the costs of lending per unit loaned are inversely related to the size of the loan. The iron law of interest rate restrictions implies that the size of loans granted to large farmers increases and the size of loans granted to small farmers decreases as interest ceilings become more restrictive. This, in turn, leads to a reduction in both the absolute and the relative shares of credit small farmers receive from formal lenders; that is, the ceilings redistribute the credit portfolios in favor of large farmers, contributing to the concentration of income in the rural areas of LICs. Moreover, given the high cost of lending to rural clients, the ceilings usually exclude the majority of the small farmers in LICs from access to formal credit. It is these excluded farmers—the smallest and the poorest—who suffer the most significant loss as a consequence of the ceilings.

Subsidized interest rates are frequently recommended as the only mechanism for effecting income transfers to small farmers that is politically feasible or administratively possible. Unfortunately, credit in general and subsidized interest rates in particular are very inefficient for income redistribution. Subsidized rates affect income growth potentials through their effect on the access to resources of different classes of farmers and have a direct effect on income distribution. The grant transferred is directly proportional to the size of the loan. The larger the loan, the greater the unrequited transfer. Since loan size and borrower size are positively correlated, the amount of the grant becomes a direct function of the borrower's wealth. The large borrowers receive large subsidies, while the small borrowers, who consitute the target of the strategy, receive, at best, small loans with their implicit subsidy. Nonborrowers—the smallest and the poorest farmers—receive no subsidy at all. This necessarily makes income distribution worse.

Interest-rate ceilings not only have affected the distribution of agricultural loan portfolios but have influenced the total value of these portfolios as well. Through their impact on the rates of deposit, they have reduced the total volume of savings mobilized by the formal financial markets (Shaw 1973). Also, through their impact on the profitability of lending to agriculture, they have reduced the share of the agricultural sector in credit volume. They have also reduced technical efficiency in lending and have sometimes even induced a reduction in the volume of informal credit.

On the other hand, low interest rates cannot eliminate the monopoly of moneylenders in the rural areas of LICs, since they restrict access to formal credit. Low interest rates cannot create the missing physical inputs, the missing markets, or the missing technologies that keep the productivity of farmers low in many LICs. At a sufficiently subsidized rate, credit could in such cases make an investment opportunity appear profitable to the few privileged borrowers receiving the underpriced resources, but it will not make it more profitable socially. Once the inputs, markets, and technologies are available, subsidized credit cannot stimulate adoption of the innovations unless large enough loans are granted to large numbers of farmers. Subsidized rates, however, lead to rationing and exclusion. Access to formal credit can do a better job even at fairly high rates of interest as long as these rates are lower than those charged by informal lenders.

Therefore, except in a few cases that are empirically unimportant, the arguments for low interest rates are invalid. What is important is access to credit. The policies that have attempted to keep the price of credit low have modified access in undesirable ways. As a result, these policies not only have reduced both the allocative efficiency and the viability of financial institutions but also have contributed to concentration of income within the rural areas of low-income countries.

NOTE

1. For a discussion of the effects of administered interest rates on some of the other dimensions of rural financial market performance see Adams and Graham, chapter 21 in this volume.—ED.

REFERENCES

Donald, Gordon. 1976. *Credit for Small Farmers in Developing Countries*. Boulder: Westview Press.
Eckaus, Richard S. 1974. "Monopoly Power, Credit Rationing and the Variegation of Financial

Structure." In *Documents and Summary Discussions, III Annual Meeting, Capital Markets Development Program, July 31–August 2, 1973. Mexico D.F., Mexico,* 27–50. Washington, D.C.: Organization of American States.

Gonzalez-Vega, Claudio. 1976. "On the Iron Law of Interest Rate Restrictions: Agricultural Credit Policies in Costa Rica and in Other Less Developed Countries." Ph.D. diss., Stanford University.

Jaffee, Dwight M. 1971. *Credit Rationing and the Commercial Loan Market.* New York: John Wiley & Sons.

McKinnon, Ronald I. *Money and Capital in Economic Development.* 1973. Washington, D.C.: Brookings Institution.

Shaw, Edward S. *Financial Deepening in Economic Development.* 1973. New York: Oxford University Press.

World Bank. 1975. *The Assault on World Poverty: Problems of Rural Development, Education, and Health.* Baltimore: Johns Hopkins University Press.

23

The Economics of Agricultural Research

The future food supply depends in large measure on the achievements of agricultural research. Whether or not we will succeed in doing the necessary research is of no concern to the sun, or to the earth and the winds that sweep her face. Our popular doomsday activists know it will be impossible for agricultural research to save us from disaster. For them the history of the struggle of mankind to produce enough food is irrelevant. What is odd, however, is that most scientists have virtually no historical perspective. Why know history?[1] Although the common man cherishes his roots, scientists do not waste their time on history. Nor do most economists as they become more "scientific." This is an age of ever more minute specialization. Avoid history, discover a little new knowledge, and you have the makings of another crisis. Without history, current events produce much pessimism. He who is an optimist is either isolated or committed to the belief that there is a thread of historical continuity in the human condition and that it is not altogether bad.

The case for history and utility of the sciences in agriculture has been stated imaginatively as follows: "Few scientists think of agriculture as the chief, or the model science. Many, indeed, do not consider it a science at all. Yet it was the first science—the mother of all sciences; it remains the science which makes human life possible; and it may well be that, before the century is over, the success or failure of Science as a whole will be judged by the success or failure of agriculture."[2]

I shall treat research as an economic activity because it requires scarce resources and it produces something of value. It is a specialized activity that calls for special skills and facilities that are employed to discover and develop new and

THEODORE W. SCHULTZ is professor emeritus, Department of Economics, University of Chicago, and co-winner of the 1979 Nobel prize for economics.

Originally titled "The Economics of Research and Agricultural Productivity," this paper was presented at the Seminar on Socio-Economic Aspects of Agricultural Research in Developing Countries, 7–11 May 1979, Santiago, Chile, and was subsequently published as an IADS Occasional Paper, 1979. Reprinted with minor editorial revisions by permission of the International Agricultural Development Service and the author.

presumably useful information. The resources that are allocated to various re-search enterprises are readily observed and measured. But the value of the new information that these enterprises produce is always hard to determine. Orga-nized agricultural research has become an important part of the process of mod-ernizing agriculture throughout the world. It has been increasing at a rapid rate, and annual expenditures on research are in the billions of dollars.

The economics of agricultural research has long been high on the Ph.D. research agenda of our graduate students at the University of Chicago. I am much indebted for their contributions and for what others have done.[3] Needless to say, there are unsettled economic questions that await analysis. Much as I am tempted to elaborate on these questions, I have decided to leave this task to others. What I plan to do is of five parts. I begin with some estimates of the growth and magnitude of agricultural research throughout the world and in Latin America and with some observations on the increases in agricultural productivity from this research. I then consider what I deem to be a very important question, namely, Who should pay for agricultural research? I then proceed to the organizational quandary. I close with two very brief parts: one deals with the harm that is done to agricultural research by the distortions in agricultural prices, and the other comments on the function of research entrepreneurship.

EXPENDITURES ON RESEARCH AND AGRICULTURAL PRODUCTIVITY

Annual expenditures on agricultural research throughout the world have in-creased over fivefold since 1951. I view the rapid growth of this sector as a response to its success. When one adds up the annual costs of the national agricultural research systems and of the agricultural research that industrial firms are doing and also that of related university research, as Evenson and Kislev have done,[4] and which has been extended by Boyce and Evenson,[5] the estimates covering the period from 1951 to 1974, in constant 1971 U.S. dollars, show the total world expenditure for this purpose increased from $769 million to $3,841 million. I have extrapolated these estimates under very conservative assump-tions. My estimate for 1979 is $4,130 million in terms of 1971 constant dollars. In 1979 prices it would total over $7,000 million for the current year. By the same reckoning, using the Boyce-Evenson estimates, the expenditures on agri-cultural research throughout Latin America rose from $30 million in 1951 to $183 million in 1979 in constant 1971 dollars. Adjusted for the decline in the dollar, this 1979 figure comes to over $300 million.[6] It should be noted, howev-er, that in terms of the percentage of total research expenditure relative to the value of the agricultural products of a region, Latin America devotes fewer resources to this research than any of the other five major regions of the world.[7] Has this relationship improved in Latin America since Boyce and Evenson made their estimates? Why should this region have lagged as recently as 1974?

Returning to the total world agricultural research expenditures, this research sector is obviously no longer an infant. Many parts of it have all the earmarks of maturity. Are there signs of senility in any of these parts? In the language of economists, are there specific classes of agricultural research where diminishing returns indicate that it is no longer worthwhile? It will not do for us to be silent on this issue. To answer this question requires both the knowledge of scientists pertaining to scientific possibilities and that of economists with respect to the value of the required resources compared with the potential value of such research contributions.

Allocative decisions, however, must be made on the basis of limited information. But in fact we know a good deal. I consider actual allocative behavior as useful information. The observed rapid growth tells me that those who have made and are making the allocative decisions do so because they deem it to be worthwhile. There are also a fairly large number of competent economic studies that show that the rates of return on investment in various specific classes of agricultural research have been much higher than the normal rates of return.[8] It is true that those research endeavors that have not been successful have not been identified in these studies. Nevertheless, the classes of agricultural research that have been analyzed are of major economic importance in agricultural production, especially so in the case of food, feed, and fiber crops.

Recent economic history provides additional useful information on the value of the contributions of agricultural research. I rate this historical information highly for the purpose at hand. The following achievements are pertinent:

1. No doubt agriculture throughout the world will find it increasingly costly to increase the area of cropland. Agricultural research, along with complementary inputs, has been very successful in developing substitutes for cropland (some call this land augmentation). Actual increases in yield per hectare have held and may well continue to hold the key to increases in crop production. For example, during my first year at Iowa State College, 1931, the U.S. yield of maize was 1,500 kg/ha, a normal crop. In 1978 this yield came to 6,300 kg/ha. Although the maize area harvested in 1978 was 16 million hectares less than in 1931, total production was over 175 million metric tons, compared with 65 million tons in 1931. No wonder the estimated rates of return on maize research in the United States are exceedingly high. The achievement with sorghum is even more dramatic. Taking 1929 as a normal year, the yield of sorghum grain rose from 870 kg/ha to 2,800 kg/ha in 1978, despite the fact that the area devoted to this crop increased from 1.8 to 6.7 million hectares. Total production in 1978 was 19 million tons, which is over 15 times as much as that in 1929.

2. It is no longer true that a large per capita supply of beef can be produced only in countries with a sparse population and with a lot of good grazing land. As maize and sorghum have become cheaper than grass in the United States, producers of beef have turned increasingly to feed grains. Here

again, one sees the large economic effects of maize and sorghum research, along with complementary inputs on the reduction of the real costs of production and on increases in supply. The per capita consumption of beef in the United States doubled between 1940 and 1975 (retail weight, civilian population: 19.7 kg in 1940 and 40.3 kg in 1975). The rise in real per capita income tells most of the story of this extraordinary increase in the demand for beef. The decline in the real price of feed grains as a consequence of much higher yields and lower costs in turn tells most of the story that explains the threefold increase in domestic beef slaughter and the approximately constant real producer beef price trend, with fluctuations, to be sure, about that trend.[9]

3. The returns to poultry research are well known.[10] Add to them the effects of research on the supply of poultry feed and the result is a major improvement in the production of poultry products and a large gain for consumers. In the United States per capita consumption in terms of retail weight increased almost threefold between 1940 and 1975 (from 7.9 kg to 22.5 kg).

4. The last item on my partial list of research achievements pertains to wheat. The real costs of producing wheat have declined very much. As yet, the costs of producing rice have not come down as they have for wheat. Back in 1911–15, the world prices of these two primary food grains per ton tended to be equal. During 1947–62 wheat sold at about 60 percent of the price of rice. Between 1965 and 1970 the wheat price had declined to half of that of rice. My most recent date is August 1978, when wheat was quoted at 35 percent relative to rice, namely wheat at $129 and rice at $366 per ton.[11] When will the real costs of producing rice come tumbling down?

WHO SHOULD PAY FOR AGRICULTURAL RESEARCH?

In my view, Who should pay for agricultural research? is a critical question. There is, I regret to say, a good deal of confusion on this issue. The international agricultural research centers have prospered. They are a successful innovation.[12] The donors who have provided the funds have been generous. In view of their success, why not have these centers expand further and do most of the necessary agricultural research? The answer is that these centers, good as they are, will not suffice. They are not a substitute for national agricultural research enterprises. Nor are they capable of doing more than a small part of the required basic research in this area.

Since basic research is very expensive and what may be discovered is subject to much uncertainty, why not let the rich countries do it and pay for it? Presumably they can afford to do it. The implication of this view is that low-income countries can be "free riders" when it comes to basic research related to agriculture. It is, however, a shortsighted view because even to be a free rider requires a high level of scientific competence. To take advantage of such advances in the pertinent sciences, achieved elsewhere throughout the world, calls for a corps of highly skilled scientists. The unique requirements of agriculture by countries is still another important consideration in this context.

The conclusion that I come to at this point is that it would be a serious mistake for Chile or any other major country in Latin America to assume that the international agricultural research centers, along with the ongoing agricultural research in high-income countries, are substitutes for first-rate national agricultural research enterprises.

Who, then, within a country should pay for agricultural research? Economists, of course, will raise their hands wanting to be heard. They will, however, modify the question and proceed to comment on who pays the bill. Accordingly there are two quite different questions.

Consider the activities of experiment stations and those of universities related to agriculture. They do not normally sell their products; they make their findings available to the public. Nor do they provide the funds that cover the costs of doing the research in which they engage. Who benefits and who bears the costs of agricultural research requires some elaboration.

I shall begin with the agricultural research that is being done by private industrial firms, because in terms of economics it is the least difficult to present. Industrial firms, understandably, restrict their agricultural research to projects from which they expect to derive a profit. A good example has been the research of private firms pertaining to the medication of poultry feed and to the optimum mix of poultry feed ingredients at the lowest cost as the relative prices of the various feed ingredients change. Some types of research on insecticides, pesticides, animal antibiotics, drugs, location-specific seeds (such as hybrid maize) and various types of engineering research oriented to the requirements of agriculture are profitable for firms to undertake. In an advanced industrial economy there are many such research opportunities. About 25 percent of all expenditures on agricultural research in North America and Oceania is accounted for by the industrial sector.[13] In Latin America, industrial firms account for about 5 percent of all agricultural research; note, however, that given the state of industrial development, it would not make sense for governments to attempt to mandate that their industrial firms increase their share of expenditures for this purpose. In an open market economy these firms will increase their expenditures on agricultural research when it becomes evident that it is profitable for them to do so.[14]

The same economic logic is sometimes used to argue that farmers should pay for the research from which they profit because private farms are in principle like private industrial firms in undertaking activities that are profitable. It is true that landlords with large land holdings have engaged at various periods in something akin to agricultural research. Landlords in England in the past took much pride in developing various breeds of livestock. What they did, however, required very little scientific knowledge. I recall my impressions while in Uruguay in 1941: the large livestock farms were at best relying on the advice from individuals with a bachelor's degree in livestock with a modicum of knowledge about genetics, whereas at the small agricultural experiment stations in South Uruguay the plant breeding projects were designed and carried out by highly competent geneticists with full knowledge of R. A. Fisher's experimental design.

Individual farms the world over are obviously too small to undertake scientific research on their own. Nor are commodity organizations of farmers capable of doing it. The scale of the research enterprise and the continuity required to recruit and hold competent scientists entails a capability that is beyond that of the individual farmer or that of various farm organizations.

It is necessary at this point to consider who actually benefits from agricultural research. Under the assumption that the contributions of this research reduce the real costs of producing agricultural products, the reduction in cost results in either a producer surplus or a consumer surplus, or some combination of the two. Under market competition over time the benefits derived from this research accrue predominantly to consumers. It enhances their real income and welfare. Some farmers benefit during the early stages, when, for example, a new high-yielding variety is being adopted. Those farmers who are among the first to adopt and who are successful at it benefit, often substantially so. Once, however, all farmers have adopted such a variety, the reduction in costs under competition results in a lower supply price to the benefit of consumers. When such a high-yielding variety is for a time location-specific, as in the case of Mexican wheat in the Punjab of India and Pakistan, farmers, landowners, and farm laborers have profited. In general, if the commodity that is produced is solely for domestic consumption, domestic consumers are the primary beneficiaries; if the commodity is solely for export, consumers abroad benefit, and farmers in other countries who produce the same commodity and sell in the world market will experience a decline in their comparative advantage.[15] Although the consumer surpluses derived from the contributions of agricultural research are real and over time they are large, it is not feasible for consumers here or elsewhere to organize and finance modern agricultural research enterprises. The complexity of university-related basic research raises additional issues on who can and will finance this type of research.

In summing up, the implications of my arguments on paying for agricultural research are as follows:

• International agricultural research centers are not substitutes for national research enterprises.
• Nor do the large agricultural research institutions in high-income countries serve the unique requirements of Latin American countries.
• Industrial firms within any country will undertake only strictly applied research from which they can derive a profit. In countries where the industrial sector is small and not highly developed, their expenditures on agriculturally related research are, for good reasons, a very small part of the total agricultural research that is required.

• It is beyond the capacity of the individual farmer to do the required research on his own; nor are farmers collectively up to organizing and financing national agricultural research.

• Although over time most of the benefits from agricultural research accrue to consumers, it is not feasible for them to organize and finance national agricultural research enterprises.

• The only meaningful approach to modern agricultural research is to conceptualize most of its contributions as *public goods*. As such they must be paid for on public account, which does not exclude private gifts to be used to produce public goods.

THE ORGANIZATION QUANDARY

Organized agricultural research has a long history. Despite the constraints and difficulties that have been encountered and the mistakes that have been made, viewed historically, organized agricultural research has been a remarkable success. I featured some of its achievements at the outset. We may learn from past mistakes, and we should ponder the puzzles. I have a little list.

1. Why did many of the states in the United States establish all too many tiny agricultural substations? Many of them were inefficient. They could not recruit competent scientists. The staff that could be had was in general isolated intellectually, having all too little interaction with the principal experiment station of the state and with scientists at the land-grant college.[16] Whatever the reasons, the same mistake is occurring in many low-income countries.

2. Agricultural laboratories established to do research on agricultural product processing, where the scientists are off by themselves far removed from university scientists, are a mistake. The puzzle is, why did the U.S. government establish various expensive regional laboratories of this type? It behooves other countries not to repeat this costly mistake.

3. There are many examples throughout the world of important crops that are receiving all too little attention when it comes to agricultural research.[17] Who is to blame? For example, in the United States there is, in my view, a gross underinvestment in soybean research, compared with funds devoted to wheat, cotton, and maize research, relative to the value of these various crops. The value of the annual soybean crop is as large as the value of the wheat and cotton crops combined.[18] It is hard for me to believe that geneticists who are also plant breeders have no theories from which to derive research hypotheses to improve the genetic capacity of the soybean. Why this underinvestment in soybean research?

4. All too much of the foreign aid for agricultural development has undervalued agricultural research. When it has supported such research, it has been, in general, short-term aid, notwithstanding the fact that the gestation period in research is a matter of years and it should for that reason be

approached as a long-term investment. The commitment to obtain quick results has been the bane of most foreign aid, not only with respect to research but also in other program areas. It continues to be the better part of wisdom for low-income countries not to rely on foreign aid in financing their agricultural research enterprises.

5. It is difficult to understand why any of the major private foundations should assume that the necessary research results for agricultural modernization are at hand and, on that assumption, proceed to support various communications activities. The dissemination of existing knowledge, better communication, featuring new approaches in extension activities, has been and continues to be the announced policy of the Kellogg Foundation. The prestigious Ford Foundation made the same mistake during its early agricultural programs in India. It is to the credit of the Rockefeller Foundation that it has had the wisdom to see that research must come first, and it has consistently held to that policy.

6. It is still a puzzle for me why so few worthwhile agricultural research results were available throughout Latin America at the time when President Truman's Point Four Program was launched. It was my responsibility during the early 1950s, with ample foundation funds and with five competent colleagues, to make an assessment of the achievements of the Point Four programs. Our studies and publications were labeled TALA, Technical Assistance Latin America. We roamed over all parts of Latin America. We found that in the area of agriculture, Point Four had accomplished very little. Agricultural extension work was prematurely emphasized.[19] In general the extension work supported by Point Four was empty for lack of available agricultural research results. Where some funds were allocated to agricultural research, they were used far less effectively than the funds of the Rockefeller Foundation, jointly with those of the Mexican government in Mexico, in what is now CIMMYT.

7. The last point on my little list reaches far back in time. It serves the purpose of deflating our self-acclaimed importance in augmenting the production of food. It is a disconcerting puzzle. Whereas Neolithic women invented agriculture and developed many of the food crop species that we have today, our highly skilled plant breeders have produced only one new food species, triticale. Norman Borlaug,[20] the agricultural Nobel Laureate, puts it this way, "The first and greatest green revolution occurred when women decided that something had to be done about their dwindling food supply." Neolithic men in hunting for meat were failing to bring home enough to eat. Try to explain and ponder the implications of the achievements then and now.

Important as it is to avoid making the mistakes that other countries have made in organizing and administering agricultural research, there are various other considerations. Robert E. Evenson's recent essay on the organization of research[21] is an important contribution to the allocation of research funds by commodities. By his criteria,[22] the allocation of funds for cotton or rubber research is twice as large as that for wheat or sugar cane, and that allocated for rice, maize, millet, and sorghum or livestock and its products is only half as large as that for wheat or sugar cane. Pulses, groundnuts, oilseeds, and roots and tubers are among the really neglected crops. Evenson also deals competently with the allocation of research resources by environmental regions and with the issues pertaining to single-commodity, multiple-commodity, and discipline-oriented research.

My own experience and observations lead me to stress the following organizing decisions:

1. Once the commodities have been selected, the state of the market for the services of the scientists who are required to do the research becomes an important consideration.
2. It is all too easy to become enamored of the phrase "interdisciplinary research." What matters is the actual value that is to realized from the complementarity between the scientists who have different professional skills. So-called interdisciplinary research is as a rule weak on theory and soft in the quality of research that gets done.
3. The effects of various incentives on research workers' productivity is of major importance. The built-in incentives that characterize organized agricultural research in most low-income countries are bad. All too often agricultural scientists are worse off in this respect than high-class clerks in the bureaucracy of the government.
4. Too little attention is given to the effects of alternative accountability requirements on research efficiency. Unnecessary paper work abounds. Those who provide the funds call the accounting rules; they are rarely aware of the sharply diminishing returns to the burdensome accounting that they call for. On this score, in my view, the international agricultural research centers are no exception.
5. The trade-off, or call it the compromise, between the loss from fragmentation and the gain from location-specific research within a country is a choice that has to be made. The mistakes that various high-income countries have made on this issue in their organizational decisions are instructive on what not to do.
6. Each of the major Latin American countries must have its own corps of competent agricultural scientists. The long-term payoff on this investment is very high. The ever-present strong desire of those who make the organizational decisions for quick results is a serious obstacle in recruiting and maintaining a corps of competent scientists. Where the main agricultural experiment station is an integral part of a university, it is possible to give due attention to basic research oriented to agriculture, provided that the decisions of the government in allocating funds do not thwart the basic research.
7. In economics there is the concept of the optimum scale of an enterprise.

Although it is difficult to apply this concept in determining the optimum scale of an agricultural experiment station, it is nevertheless relevant. There is a strong tendency in agricultural research to violate all scale considerations by ever more centralization of its administration.

THE HARM DONE TO RESEARCH BY PRICE DISTORTIONS

The scarcity of agricultural resources—land relative to labor and both of these relative to the stock of reproducible forms of physical capital—is a major consideration in determining the forms of useful knowledge that are appropriate for an economy. The historical responses of agricultural research in various countries to resource scarcity considerations are presented by Hayami and Ruttan in their well-known book on agricultural development.[23]

The scarcity of the factors of production is not self-evident. It requires proof, and proof calls for measurement. The price of the services of each of the factors of production in an open competitive market is a unit of measurement of scarcity. When the market is rigged, be it by governmental intervention or by means of private monopoly pricing, the resulting prices are distorted. The price signals that are a consequence of such distortions do not reveal the true scarcity of the factors of production, and for that reason they are beset with misinformation. The allocation of funds for agricultural research and the use to which these funds are put are not immune to the adverse effects of such price misinformation.

The overpricing of sugar beets in Western Europe and the United States, and the associated expenditures on sugar-beet research, is a case in point. The expenditures on rice research in Japan have not been immune to the enormous overpricing of rice in that country.[24] In India on rice it is the other way around. There is no end to examples. The harm that is being done throughout the world to agricultural research as a consequence of the distortions in prices and in agricultural incentives is very substantial.[25]

THE FUNCTION OF RESEARCH ENTREPRENEURSHIP

The dynamic attributes of research are pervasive both in the domain of economic growth and in the conduct of actual research. Advances in useful knowledge are compelling dynamic forces. Such new knowledge is the mainspring of economic growth. Were it not for advances in knowledge, the economy would arrive at a stationary state and all economic activities would become essentially routine in nature. Over time, new knowledge has augmented the productive capacity of land, and it has led to the development of new forms of physical capital and of new human skills. The fundamental dynamic agent of long-term economic growth is the research sector of the economy.

The concept of meaningful research conducted to enhance the stock of knowledge that is useful in production and consumption is inconsistent with static, unchanging, routine work on the part of scientists. The very essence of research is in the fact that it is a dynamic venture into the unknown or into what is only partially known. Research, in this context, is inescapably subject to risk and uncertainty. Whereas funds, organization, and competent scientists are necessary, they are not sufficient. An important factor in producing knowledge is the human ability that I shall define as *research entrepreneurship*. It is an ability that is scarce; it is hard to identify this talent; it is rewarded haphazardly in the not-for-profit research sector; and it is increasingly misused and impaired by the over-organization of our research enterprises. What is happening in agricultural research is on this score no exception.

Who are these research entrepreneurs? In business enterprises that are profit-oriented, the chief executive officers perform the entrepreneurial function. The skilled factory worker is not an entrepreneur in doing his job. In research it is otherwise. Whereas administrators who are in charge of a research organization may be entrepreneurs, much of the actual entrepreneurship is a function of the assessment by scientists of the scientific frontiers of knowledge. Their professional competence is required to determine the research hypotheses that may be worthwhile pursuing.

Briefly and much simplified, my argument is that in the quest for appropriations and research grants, all too little attention is given to that scarce talent which is the source of research entrepreneurship.[26] The convenient assumption is that a highly organized research institution firmly controlled by an administrator will perform this important function. But in fact a large organization that is tightly controlled is the death of creative research, regardless of whether it be the National Science Foundation, a government agency, a large private foundation, or a large research-oriented university. No research director in Washington or Santiago can know the array of research options that the state of scientific knowledge and its frontier afford. Nor can the managers of foundation funds know what needs to be known to perform this function. Having served as a member of a research advisory committee to a highly competent experiment station director for some years and having observed the vast array of research talent supported by funds that we as a committee had a hand in allocating, I am convinced that most working scientists are research entrepreneurs. But it is exceedingly difficult to devise institutions to utilize this special talent efficiently. Organization is necessary. It too requires entrepreneurs. Agricultural research has benefitted from its experiment stations, specialized university laboratories, and from the recently developed international agricultural research centers. But there is the ever-present danger of over-organization, of directing research from the top, of requiring working scientists to devote ever more time to preparing reports to "justify" the work they are doing and to treat research as if it were some routine activity.

NOTES

1. Here I draw briefly on my "What Are We Doing to Research Entrepreneurship?" in *Transforming Knowledge into Food in a Worldwide Context*, ed. William F. Hueg, Jr., and Craig A. Gannon (Minneapolis: Miller Publishing Co., 1978), 96–105.

2. André Mayer and Jean Mayer, "Agriculture: The Island's Empire," in *Science and Its Public: The Changing Relationship, Daedalus* 103, no. 3 (1974): 83–95. See also the excellent essay by Edward Shils, "Faith, Utility and the Legitimacy of Science," in the same issue of *Daedalus*.

3. My debt to scientists concerned about agricultural research is large. I learned much from R. E. Buchanan, who was director of the Agricultural Experiment Station during my years at Iowa State College. Albert H. Moseman has also contributed a good deal to my understanding of agricultural research in low-income countries, beginning with the symposium that he organized for the American Association for the Advancement of Science, "Agricultural Sciences for the Developing Nations," 1964, and his comprehensive analysis, *Building Agricultural Systems in Developing Nations* (New York: Agricultural Development Council, 1970).

4. Robert E. Evenson and Yoav Kislev, "Investment in Agricultural Research and Extension: An International Survey," *Economic Development and Cultural Change* 23, no. 3 (1975): 507–21. [See also Evenson, chapter 24 in this volume—ED.]

5. James K. Boyce and Robert E. Evenson. *National and International Agricultural Research and Extension Programs* (New York: Agricultural Development Council, 1975).

6. This estimate for expenditures in Latin America does not account fully for the recent expansion of the agricultural research activities in Brazil, and to this extent it is too low for Latin America.

7. See table 1.5, for the year 1974, in Boyce and Evenson, *National and International Agricultural Research.*

8. See Evenson, chapter 24 in this volume, table 3.—ED.

9. T. W. Schultz, "The Politics and Economics of Beef" (Paper presented at the Conference on Livestock Production in the Tropics, 8–12 March 1976, Acapulco, Mexico, published by the Banco de Mexico in their *Proceedings,* 1976).

10. See Willis Peterson, "Returns to Poultry Research in the United States" (Ph.D. diss., University of Chicago, 1966).

11. Why there has not been more substitution of wheat for rice, inasmuch as the nutritive value of rice and wheat is virtually the same, presents a puzzle (see my "Reckoning the Economic Achievements and Prospects of Low Income Countries" [James C. Snyder Memorial Lecture, Purdue University, 22 February 1979]).

12. It should be noted that the Rockefeller Foundation, in cooperation with the government of Mexico, was the first to launch this type of venture. Not to be overlooked is the importance of research entrepreneurship in this connection. I shall have more to say on research entrepreneurship later on in this paper.

13. This estimate and the one that follows for Latin America are for 1974, based on table 1.1 in Boyce and Evenson, *National and International Agricultural Research.*

14. Experience in the United States indicates that there is a tendency for some agricultural experiment stations to hold on to research work that has been successful and has reached the point where private firms would continue it, as occurred in the case of the southern experiment stations' holding on to the development and production of hybrid maize for seed.

15. The economics of the preceding arguments is greatly simplified. Much depends on the elasticity of the demands and on the shifts in the supply curves as a consequence of the production effects of the contributions derived from agricultural research.

16. These interactions among agricultural scientists contribute to their research productivity, but they do not solve the institutional requirements for useful working interactions between farmers and scientists. The more centralized the agricultural research establishment of a country becomes, the fewer the actual working contacts that scientists have with on-farm production problems. Robert E. Evenson, in his comments on this paper, made the point that one of the overlooked contributions of substations is that they provide agricultural scientists with some of these much needed working contacts with actual farm conditions (see also Robert E. Evenson, Paul E. Waggoner, and Vernon W. Ruttan, "Economic Benefits from Research: An Example from Agriculture," *Science,* 14 September 1979, 1101–7).

17. See Evenson, chapter 24 in this volume.—ED.

18. The size of the area devoted to soybeans in the United States in 1979 exceeded that devoted to

maize. The anticipated soybean crop at current future harvest prices implies a value of $14,000 million.

19. See chapter 1 in this volume.—ED.

20. Norman E. Borlaug, "The Green Revolution: Can We Make It Meet Expectations?" *Proceedings of the American Phytopathological Society* 3 (1976).

21. Robert E. Evenson, "The Organization of Research to Improve Crops and Animals in Low Income Countries," in *Distortions of Agricultural Incentives,* ed. Theodore W. Schultz (Bloomington: Indiana University Press, 1978).

22. See table 3, p. 230, in Evenson, "The Organization of Research." [See also Evenson, chapter 24 in this volume, table 2.—ED.]

23. Yujiro Hayami and Vernon W. Ruttan, *Agricultural Development: An International Perspective* (Baltimore: Johns Hopkins Press, 1971). See also Hans P. Binswanger et al., *Induced Innovation: Technology, Institutions and Development* (Baltimore: Johns Hopkins University Press, 1978). [See also chapter 4, by Ruttan and Hayami.—ED.]

24. Keijiro Otsuka, "Public Research and Rice Production—Rice Sector in Japan, 1953–76" (Ph.D. diss., University of Chicago, 1979).

25. For an extended treatment of the adverse effects of the distortions in agricultural incentives, including their effects on the adoption of research results and on providing the wrong price signals to guide agricultural research, see Schultz, *Distortions of Agricultural Incentives.*

26. This closing paragraph is a slightly revised part of my "What Are We Doing to Research Entrepreneurship?" cited in n. 1 above.

24

Benefits and Obstacles in Developing Appropriate Agricultural Technology

Agricultural production is based on biological processes. Both plant and animal commodities require the growth and reproduction of living organisms. These organisms are subject to disease and insect problems. Their growth processes are affected by differences in soil qualities, temperatures, water availability, and a number of other environmental factors. Appropriate agricultural technology thus encompasses a biological dimension as well as the mechanical and chemical dimensions dominating many other classes of production technology.

In this discussion of the benefits and obstacles to appropriate agricultural technology, I will be specifically concerned with the developing regions of the world and with the possibilities for transfer of technology produced in other regions to these regions. The relevance of the biological process dimension to the issue is so dominant that it requires attention first. After considering both the premodern and modern approaches to plant and animal improvement through selective breeding, I will discuss the chemical and mechanical dimensions. Finally, I will review the evidence regarding the effectiveness of research by agricultural scientists to produce more appropriate agricultural technology for different parts of the world.

BIOLOGICAL PROCESS COMPONENT OF AGRICULTURAL TECHNOLOGY

Charles Darwin, in his famous *The Origin of Species,* established a number of principles guiding the relationships between types of biological technology and

ROBERT E. EVENSON is professor of economics, Yale University.

Reprinted from *Annals of the American Academy of Political and Social Science* 458 (1981): 54–67. Originally titled "Benefits and Obstacles to Appropriate Agricultural Technology." Copyright © 1981 by The American Academy of Political and Social Science, with permission of Sage Publications, Inc. Published with minor editorial revisions by permission of the author.

the environments in which they have a comparative advantage. The fact that the world's surface provides a rich variety of environments—environmental niches—in which plants and animals grow has been well documented for centuries. Similarly it has long been known that an incredibly rich array of differentiated species of plants, animals, insects, parasites, and pathogens has been associated with these variegated environments. Darwin clarified this relationship between species and environments with an argument that has economic characteristics. The survival of species is a matter of appropriateness of biological technology to environmental conditions. Given that natural mutation processes were continually producing new genetic variation in living matter—that is, new technology—there was a natural sorting out of the "fittest" for each environmental niche.

Human populations have produced an economic system of plant and animal husbandry designed to alter the natural system in order to produce more valuable products. Important and economically valuable plants and animals are domesticated.[1]

This altered the Darwinian equilibrium in several ways. First, the valued species evolved through time under husbandry selection pressures. Second, in response to this selection, the equilibrium was altered regarding nondomesticated species. Pests, parasites, and pathogens associated with cultivated crops found improved environments in which to survive as new types of crops were developed.

The development of crops and animals over the period of husbandry selection, which lasted until the nineteenth century, when modern scientific breeding methods were utilized, shows a pattern of changing comparative advantage of regions for crop production. Husbandry-selected technology tended to be adapted to environments other than those best suited to its origination. Virtually all modern crop and animal species and types eventually had better economic performance outside their "centers of origin" than in the regions where they first emerged. This appears to be the result of selection and reproduction, which allowed crops to "escape" from their most serious pests and pathogens. Since most centers of origin were in tropical and subtropical climate regions, the temperate-zone regions of the world were favored during the premodern period of plant improvement.

The modern period in agricultural science dates from the early 1800s, when the field of agricultural chemistry was established in Germany and the Rothamstead Experiment Station in England began the systematic application of scientific methods to agricultural production technology. This early work moved agricultural improvement efforts out of the "country gentlemen" societies and the botanical gardens and into the laboratories. By the early 1900s hundreds of agricultural experiment stations had been established in many parts of the world. The United States had by then a well-established State Agricultural Experiment Station system. The European countries and Japan also had well-developed research programs. In today's developing world, with few exceptions, the only significant research programs prior to 1950 were directed toward the improvement of the "colonial crops": sugar, tea, coffee, cocoa, and cotton.[2]

In general the developed countries of the world entered the modern period with a large comparative advantage in most crop and animal production. Except for a few crops specific to the tropics, yields per acre were higher in temperate-zone countries. This advantage was greatly increased with the advent of modern agricultural science, for two reasons. First, the temperate-zone developed countries invested in agricultural science and built effective research systems, while the developing countries did not—except for the colonial crops. Second, modern agricultural science only partly overcame the fundamental linkage of technology to the environment that characterized premodern development. As a consequence, very little of the technology produced in the developed countries was actually transferred to the developing world.

The strength of the technology-environmental linkages has been persistently underestimated by policy makers. In the early years of the modern period of emphasis on economic development, policy makers emphasized agricultural extension and community development programs as the means to rapid development. Agricultural extension advisers swarmed over the developing world in the 1950s, bringing U.S. and European "know-how" to the farmers of the tropics. By the early 1960s, it was clear that U.S. know-how, including virtually all aspects of agricultural technology—varieties, machines, and even chemicals— was simply not transferable to environments that differed greatly from those for which it was developed.[3]

This led to a modification of the earlier development strategy. International agencies have, for the most part, pursued a bifurcated strategy over the past two decades. The emphasis on technology transfer has continued in the form of "rural development" projects, which continue to receive the bulk of development aid. These projects are greatly varied in nature but often have a know-how transfer component. The second component of the strategy has been the support of agricultural research systems in the developing countries.

This research system development strategy has taken primary form in the building of the system of International Research Centers. The International Center for Wheat and Maize Improvement (CIMMYT) in Mexico and the International Rice Research Institute (IRRI) in the Philippines are the oldest and best known of these centers.[4] In recent years a number of national research programs have also attained significant research capacity, and current international policy is slowly shifting toward further strengthening of these programs.

Tables 1 and 2 provide a summary of comparative research expenditures. Table 1 reports research spending as a percent of the value of agricultural product for five categories of countries, grouped according to per capita income in 1971. The disparity in spending patterns between rich and poor nations is readily apparent. It is also clear that this disparity is less severe for national public expenditures than for the total of national public, international, and industrial expenditures. Table 2 shows the commodity orientation of research spending in

TABLE 1
RELATION OF WORLD EXPENDITURES ON AGRICULTURAL RESEARCH TO THE
VALUE OF AGRICULTURAL PRODUCT, BY INCOME GROUP

Group	Per capita income (U.S. dollars)	Percentage of total and public research expenditures to value of agricultural product[a]			
		1971 total	(Public)[b]	1974 total	(Public)[b]
I	1750	2.48	(1.44)	2.55	(1.48)
II	1001–1750	2.34	(1.76)	2.34	(1.83)
III	401–1000	1.13	(0.86)	1.16	(0.92)
IV	150–400	0.84	(0.71)	1.01	(0.84)
V	150	0.70	(0.65)	0.67	(0.62)

SOURCE: Adapted from table 1.7 in James K. Boyce and Robert E. Evenson, *Agricultural Research and Extension Programs* (New York: Agricultural Development Council, 1975), 11.

[a]Total expenditures include: (1) national public agriculture, (2) national public agriculture related, (3) industry, and (4) international. Excludes People's Republic of China.

[b]National public agriculture (national public agriculture related is not included).

all developing countries; the data on proportion of product value are not fully comparable with the data in table 1 because general research that cannot be associated with commodities is included in table 1. The research emphasis of the international centers is shown in the table. It is clear that while some major commodities have reasonable research programs in place, others do not. The negligible research on important commodities such as cassava, coconuts, sweet potatoes, groundnuts, and chick peas is especially noteworthy. Cotton appears to be the only commodity in the developing world with research emphasis comparable to the emphasis placed on it by developed countries.

It is, of course, natural to ask whether, in the absence of an elaborate system of agricultural research centers, each serving a particular environmental region, the limited research capacity in the developing world might produce more widely adaptable technology. In other words, could a research center produce technology that would be worth more in environments other than those of its immediate location? The answer to this question is that in plant breeding there is scope for a trade-off between adaptability and local effectiveness. The degree to which this trade-off can be made varies according to the crop. In some crops, such as maize, there is little scope for wide adaptability; in others, the scope is considerable.[5]

The basis for this trade-off exists because plants do differ in their tolerance of environmental factors such as temperature, humidity, and soil salinity. Breeders can select varieties that are more tolerant, or, to use a modern term, have lower interaction. Selecting for low interactions usually means giving up some other desirable trait. Most advanced-country research systems have had little incentive to seek aggressively widely adaptable materials because they have elaborate regional research systems designed to produce varieties for relatively small regions.

TABLE 2
ESTIMATES OF INTERNATIONAL AND NATIONAL RESEARCH INVESTMENT BY
MAJOR COMMODITIES, 1971 CONSTANT DOLLARS

Commodity, in Order of Value of Production	Value of Commodity in All Developing Nations ($ billions)	Estimated Research Investment		
		International Centers (1976)[a] ($ millions)	National Centers (1976)[b] ($ millions)	National Investment as Proportion of Product Value (percentage)
1. Rice	Over 13	7.9	34.7	0.26[d]
2. Wheat	5–6	3.8	35.9	0.65
3. Sugar cane	5–6	0	30.2	0.50
4. Cassava	5–6	1.9	4.0	0.07
5. Cattle	5–6	7.9	54.8	0.88
6. Maize	3–4	4.1	29.6	0.75
7. Coconuts	3–4	0	2.0	0.06
8. Sweet potatoes	3–4	0.6[c]	3.4	0.09
9. Coffee	2	0	8.5	0.40
10. Grapes	2	0	6.9	0.35
11. Sorghum	1–1½	1.2	12.2	0.77
12. Barley	1–1½	0.5	9.4	0.62
13. Groundnuts	1–1½	0.5	4.0	0.13
14. Cotton	1–1½	0	60.1	3.50
15. Dry beans	1–1½	1.5	4.0	0.25
16. Chick peas	1–1½	1.2	3.0	0.18
17. Chilies and spices	1–1½	0	4.0	0.25
18. Olives	1–1½	0	5.0	0.33
19. Grain legumes	1	1.6	(25.3)	(2.00)
20. Potatoes (white)	1	2.0[c]	8.2	0.68

SOURCE: Reproduced from *Supporting Papers, World Food and Nutrition Study, Vol. 5* (1977), 51, by permission of the National Academy Press, Washington, D.C.

[a]Centers and programs sponsored by the Consultative Group on International Agricultural Research.

[b]Rough estimate derived by allocating total research expenditures by country according to the proportion of standardized publications. Standardized publications are converted into constant scientist-years.

[c]Additional funds also were spent on these crops at the Asian Vegetable and Research Development Center.

[d]The proportion varied sharply by type of rice: shallow water, 0.40; upland rainfed, 0.16; intermediate, 0.16; and deep water, 0.05. The international center investment was principally in the first two types.

There are at least three important cases, each associated with a "green revolution," in which wide adaptability of plant material played a major role. In each case the supporters of the research system had motives that gave much weight to wide adaptability. Also, each case demonstrated another principle or facet of wide adaptability: when superior technology is made available to a region by transfer from another region, the receiving region has good potential to add to the value of the transferred material through local adaptive research.

The first of these cases is the development of improved sugar-cane varieties in the 1920s. After the discovery of methods to induce the cane plant to flower and reproduce sexually in 1887—prior to this, all cane was propagated asexually, except for rare natural sexual reproduction—a number of improved varieties were developed in the experiment stations of the major producing countries. These modern varieties were susceptible to many local diseases and were not widely adapted. In the early 1920s, the experiment stations in Java (P.O.J.) and India (Coimbatore) developed "interspecific hybrids," which incorporated genetic material from hardy, noncommercial canes. This was almost an accidental production of widely adapted varieties, although the colonial interests of the period that supported this research were generally interested in improving technology in a number of countries.[6]

The early interspecific hybrids, especially P.O.J. 2878, were planted in a number of countries and quickly replaced local canes in much the same fashion as the Mexican semidwarf wheats and the semidwarf rices did some decades later. At one point in the 1930s, the P.O.J. 2878 variety probably accounted for 40 percent of the world's sugar-cane production.

The widely adapted international varieties were initially superior to local varieties over a wide range of environments, probably in as much as 90 percent of the world's sugar-cane area. Virtually every country or region that developed a local program to improve further on these varieties, and to "target" them to local regions, however, was successful in doing so. By the 1960s, the world's sugar-cane acreage was almost entirely planted with varieties that utilized the 1920s material in breeding programs but were highly targeted to local environments. No single variety accounted for a large part of the world's production.

The development of the semidwarf wheats in the Mexican Rockefeller Foundation program—later to become CIMMYT—has clear parallels. The funders of this research gave high priority to producing new technology that would be widely available to countries that generally had not improved their own varieties and did not have a strong capacity to do so. Wide adaptability was emphasized, and screening procedures were devised to obtain it. The introduction of the first Mexican wheats in Pakistan and India in 1965 or so was followed by their rapid adoption. Within a short period those countries with a strong research capacity produced local adaptations to these international varieties. Varieties produced at CIMMYT probably accounted for 25 percent of the wheat acreage in the developing regions by the mid-1970s but were being rapidly replaced by local adaptations. Dalrymple's most recent data suggest that the combined total of CIMMYT varieties and CIMMYT-induced adaptations accounts for more than 70 percent of the acreage for Asian countries.[7]

The third case, that of semidwarf rices, is similar. By the early 1960s the semidwarf technology was ripe for development. In the Philippines the University of the Philippines College of Agriculture had already produced several important high-yielding varieties, particularly C4-63, which was bred at the same time as IR-8. The establishment of IRRI provided major new impetus to the development and spread of the high-yielding rice varieties. As with wheat, national research programs quickly incorporated the IRRI and other materials into local breeding programs and have now produced hundreds of locally bred semidwarf varieties. Since the semidwarf rices are suited only to environments with a high degree of water control, their maximum transfer potential across environments, even when locally modified, is to only approximately 50 percent of the rice-producing regions of the tropics and subtropics.

These three are not the only successful cases of development of appropriate technology in the developing world. Many more cases, including localized success in corn, have been documented. Corn, however, is subject to strong genotype-environment interactions, and this has blocked the development of anything parallel to wheat, even though CIMMYT has been pursuing a corn research program for many years. The cases discussed above illustrate the natural evolution of at least some of the international centers. CIMMYT and IRRI have evolved into "wholesalers" of specialized technology and technology components.

CHEMICAL TECHNOLOGY IN AGRICULTURE

The major types of chemical technology of relevance to agriculture are fertilizer, herbicides, insecticides, and animal pharmaceuticals.[8] Fertilizing materials have long been used in agriculture. Most inorganic fertilizers have been in production for a century or more. Major advances in efficiency of production and in handling technology have taken place, however. The real prices of most fertilizers have fallen over most of the twentieth century. During the 1950s and 1960s, price declines were quite dramatic for nitrogenous materials. Since most nitrogen-producing processes rely on oil or gas raw materials, prices have risen in the 1970s.

With declining fertilizer prices, consumption increased and agronomists and soil scientists developed more effective systems for applying fertilizers and minerals—chiefly lime and trace minerals—to compensate for soil deficiencies. The implementation of these systems was assisted by extension and education programs. Plant breeders also responded to declining fertilizer prices by placing more weight on fertilizer responsiveness in breeding programs. Both the high-yielding green revolution rices and wheats were selected and designed for high fertilizer responsiveness. Most of the yield increases in crops throughout the world in the past century have been associated with fertilizer. Fertilizer itself is not new technology. The complementary varietal improvements and managerial improvements are the sources of the yield improvements.

Insecticides, herbicides, and related chemicals have been important in the past three decades or so. Most of these chemicals have been developed by private industry rather than by the public sector, which has dominated biological technology development. Herbicides have, for the most part, been economically important only where labor costs are high. In general, hand labor can achieve the same or better control of noxious weeds in most situations; thus we find that they have little impact on most of the tropical developing countries.

Insecticides, on the other hand, can achieve pest control not possible by hand labor. They are important in the developing countries, particularly when new crop varieties are introduced. The experience with many improved varieties is that they are often susceptible to pests and disease attacks—often unimportant and unnoticed pests and disease—within a year or two after introduction. The brown plant hopper, which became a serious problem in rice production after the introduction of IR-8, had not previously been a significant pest because older rice varieties had cell characteristics that gave them resistance. Chemical methods of control were utilized but in this case were not highly effective.

In a remarkably short time breeders at IRRI were able to develop a cross between IR-8–type material and genetic material from its genetic collection resistant to brown plant hopper attacks. Many thousands of varieties were screened before a resistant noncommercial source was found. Within a year or so after first indications of the severity of the problem, IRRI released new resistant varieties. Today IRRI screens its varietal materials for resistance to several diseases and several insects, including four "bio-types" of the brown plant hopper. As a consequence, the importance of chemical control has been lessened.

In the animal health field, however, genetic improvement has not been as effective a substitute for chemical or pharmaceutical control. The sensitivity of animals to environments is not always fully appreciated by students of technology. One has only to travel short distances in the tropics to note marked differences in the size and strength of work animals living in uncontrolled environments. The work horse, for example, does not thrive in the tropics. Work cattle vary greatly in size, and in some regions native cattle are too small to be useful work animals. Yet they are the only types that can survive in particular insect and disease environments.

MECHANICAL TECHNOLOGY IN AGRICULTURE

Machines for land preparation, weed control, and harvesting and processing of agricultural crops have long been the objective of inventive activity by farmers, blacksmiths, industrial firms, and public-sector agencies. Prior to the nineteenth century, considerable development of plows, animal harnesses, and the like had taken place. The nineteenth century, however, was the age of invention for agricultural implements. The first patent granted by the U.S. Patent Office was for an improved plow in 1796. By the end of the century some sixty thousand to seventy thousand patents had been granted for hundreds of types of plows, cultivators, specialized planting machines, and weed control devices and for numerous types of harvesting and threshing machines. Not only were thousands of patents granted for inventions of new agricultural machines, but many were produced and sold. Paul David, in his analysis of the adoption of the reaper, notes that farm machinery was the largest industry in the United States by the end of the nineteenth century.[9] In the twentieth century a new series of inventions was induced by the development of the tractor as a power source.

It is relevant to ask why, throughout much of today's developing world, land is still prepared with a bullock pair pulling a simple wooden plow, grain is harvested with a simple scythe or hand-held knife, threshing is done by hand, and rice is hand pounded. Machines for all of these tasks have been under a continuous state of improvement for more than a century in other parts of the world. Why are they not being used?

The answer appears to be that in spite of the extensive improvements in all of these forms of mechanical technology, none are economically relevant in settings where the real value of human labor is extremely low. The difference between the real value of human time in economies like Indonesia or Bangladesh and the more advanced developing countries or the modern developed country is huge. In the lowest-wage economies, hand processes for almost all activities are the lowest-cost technologies. A century of intensive mechanical improvement activity has not yielded anything to change this.[10]

We do observe that when wage rates rise, labor-saving machines not only are adopted in low-income countries but are improved as well. It appears that for much of the mechanical technology in agriculture, there is scope for adaptive invention. This invention is important because, even though it does not produce major changes in machines, it produces improvements in "appropriateness."

Very few developing countries have as yet derived effective policies to encourage this adaptive invention. Some countries, such as India, rely heavily on public research investments. Others, such as the Philippines, encourage it by operating a "petty" patent or utility model patent system. The latter approach appears to be quite effective in stimulating this type of invention.

EVIDENCE ON RETURNS TO INVESTMENT IN PRODUCTION OF APPROPRIATE AGRICULTURAL TECHNOLOGY

This article has stressed the extent to which genotype-environment interactions and differences in geoclimate environments limit the diffusion and spread of biological technology. A similar interaction effect with the economic environment, chiefly the abundance or scarcity of labor and mechanical technology, has also been noted. It was further noted that there appeared to be scope in both biological and mechanical technology for technology developers to increase the adaptability of technology, and for recipients of this technology to adapt and modify it further to local conditions.

A summary of investment in agricultural research by developing countries showed that prior to 1950 the tropical developing countries of the world had significant research programs in place only for those commodities important in colonial trade, along with a few small programs on rice and wheat. No significant work on root crops, oilseeds, pulses, sorghum, millets, and other feed grains was being undertaken. During the 1950s and 1960s a number of research institutions were built in the developing world, usually with international support.

National governments in the 1950s were not according high priority to the agricultural sector—this was the period of import substitution policies to stimulate industrial growth—and certainly not to agricultural research.[11] Even when international aid financed the training of agricultural scientists, many national governments failed to provide research facilities and other support. In response to this situation, the International Centers Research System was developed during the 1960s and 1970s. The centers were interdisciplinary and directed attention to a limited number of commodities. They took an international perspective and thus placed emphasis on wide adaptability.

We now have sufficient experience with the agricultural research systems in both the developed and developing countries to evaluate at least partly whether they have produced research products of value. It would be useful to know how the national programs in the developing countries have performed relative to the international centers and to systems in developed countries.

Table 3 provides a summary of a number of studies of agricultural research programs, including a number of specialized commodity programs. These studies used two basic methods for evaluation: imputation and statistical. Approximately half were based on developing country experiences. All reported an estimated "internal rate of return" on investment. This computation treats research expenditures as an investment. The flow of increased commodity production, holding all inputs constant, is treated as the benefits stream. The internal rate of return is the rate realized over the entire period during which costs are incurred and benefits received. Some studies estimated the time lag between the time costs are incurred and benefits realized. The average estimated time lag between research spending and the full realization of its effect is roughly ten years.

The imputation studies each attempted to measure the costs and benefits to a particular program of research conducted over the time periods indicated. Different methods and data were utilized to measure the benefits. Sometimes statistical procedures were used; in other cases data comparing production using old and new technology were used. These studies reported what might be termed average rates of return—that is, rates of return that hold for the entire research investment.

The statistical studies, on the other hand, estimated a rate of return to an additional or marginal dollar of research spending. They generally employed an aggregate production function that was estimated utilizing data on production, inputs, and public-sector programs such as research and extension. In such studies, the research variables have to be specified carefully as to their timing and spatial dimensions. They are subject to the normal statistical bias for drawing inferences. Most of the studies reported "statistically significant" estimates of research effects.

TABLE 3
SUMMARY STUDIES OF AGRICULTURAL RESEARCH PRODUCTIVITY

Study	Country	Commodity	Time Period	Annual Internal Rate of Return (%)
Index Number				
Griliches 1958	USA	Hybrid corn	1940–55	35–40%
Griliches 1958	USA	Hybrid sorghum	1940–57	20
Peterson 1967	USA	Poultry	1915–60	21–25
Evenson 1969	South Africa	Sugar cane	1945–62	40
Ardito Barletta 1970	Mexico	Wheat	1943–63	90
Ardito Barletta 1970	Mexico	Maize	1943–63	35
Ayer 1970	Brazil	Cotton	1924–67	77+
Schmitz & Seckler 1970	USA	Tomato harvester —With no compensation to displaced workers	1958–69	37–46
		—Assuming compensation of displaced workers for 50 percent of earnings loss		16–28
Ayer & Schuh 1972	Brazil	Cotton	1924–67	77–110
Hines 1972	Peru	Maize	1954–67	35–40[a] 50–55[b]
Hayami & Akino 1977	Japan	Rice	1915–50	25–27
Hayami & Akino 1977	Japan	Rice	1930–61	73–75
Hertford, Ardila, Rocha, & Trujillo 1977	Colombia	Rice	1957–72	60–82
	Colombia	Soybeans	1960–71	79–96
	Colombia	Wheat	1953–73	11–12
	Colombia	Cotton	1953–72	0
Pee 1977	Malaysia	Rubber	1932–73	24
Peterson & Fitzharris 1977	USA	Aggregate	1937–42	50
			1947–52	51
			1957–62	49
			1957–72	34
Wennergren & Whitaker 1977	Bolivia	Sheep	1966–75	44.1
		Wheat	1966–75	−47.5
Pray 1978	Punjab (British India)	Agricultural research and extension	1906–56	34–44
	Punjab (Pakistan)	Agricultural research and extension	1948–63	23–37
Scobie & Posada 1978	Colombia	Rice	1957–74	79–96
Production function				
Tang 1963	Japan	Aggregate	1880–1938	35
Griliches 1964	USA	Aggregate	1949–59	35–40
Latimer 1964	USA	Aggregate	1949–59	Not significant
Peterson 1967	USA	Poultry	1915–60	21
Evenson 1968	USA	Aggregate	1949–59	47
Evenson 1969	South Africa	Sugar cane	1945–58	40
Ardito Barletta 1970	Mexico	Crops	1943–63	45–93

(continued)

TABLE 3—*Continued*

Study	Country	Commodity	Time Period	Annual Internal Rate of Return (%)
Duncan 1972	Australia	Pasture improvement	1948–69	58–68
Evenson & Jha 1973	India	Aggregate	1953–71	40
Kahlon, Bal, Saxena, & Jha 1977	India	Aggregate	1960–61	63
Lu & Cline 1977	USA	Aggregate	1938–48	30.5
			1949–59	27.5
			1959–69	25.5
			1969–72	23.5
Bredahl & Peterson 1976	USA	Cash grains	1969	36[c]
		Poultry	1969	37[c]
		Dairy	1969	43[c]
		Livestock	1969	47[c]
Evenson & Flores 1978	Asia—national	Rice	1950–65	32–39
	Asia—		1966–75	73–78
	international	Rice	1966–75	74–102
Flores, Evenson, & Hayami 1978	Tropics	Rice	1966–75	46–71
	Philippines	Rice	1966–75	75
Nagy & Furtan 1978	Canada	Rapeseed	1960–75	95–110
Davis 1979	USA	Aggregate	1949–59	66–100
			1964–74	37
Evenson 1979	USA	Aggregate	1868–1926	65
	USA	Technology-oriented	1927–50	95
	USA—South	Technology-oriented	1948–71	93
	USA—North	Technology-oriented	1948–71	95
	USA—West	Technology-oriented	1948–71	45
	USA	Science-oriented	1927–50	110
			1948–71	45
	USA	Farm management research & agricultural extension	1948–71	110

SOURCES: The results of many of the studies reported in this table have previously been summarized in the following: Thomas M. Arndt, Dana G. Dalrymple, and Vernon W. Ruttan, eds., *Resource Allocation and Productivity in National and International Agricultural Research* (Minneapolis: University of Minnesota Press, 1977), 6, 7; James K. Boyce and Robert E. Evenson, *Agricultural Research and Extension Systems* (New York: Agricultural Development Council, 1975), 104; Robert E. Evenson, Paul E. Waggoner, and Vernon W. Ruttan, "Economic Benefits from Research: An Example from Agriculture," *Science* 205 (14 September 1979): 1101–7; Robert J. R. Sim and Richard Gardner, *A Review of Research and Extension Evaluation in Agriculture* (Moscow: University of Idaho, Department of Agricultural Economics Research Series 214, May 1978), 41, 42; R. Hertford, J. Ardila, A. Rocha, and G. Trujillo, "Productivity of Agricultural Research in Colombia," in Arndt, Dalrymple, and Ruttan, *Resource Allocation and Productivity*, 86–123; J. Hines, "The Utilization of Research for Development: Two Case Studies in Rural Modernization and Agriculture in Peru" (Ph.D. diss., Princeton University, 1972); A. S. Kahlon, H. K. Bal, P. N. Saxena, and D. Jha, "Returns to Investment in Research in India," in Arndt, Dalrymple, and Ruttan, *Resource Allocation and Productivity*, 124–47; R. Latimer, "Some Economic Aspects of Agricultural Research and Extension in the U.S." (Ph.D. diss., Purdue University, 1964); Y. Lu and P. L. Cline, "The Contribution of Research and Extension to Agricultural Productivity Growth" (Paper presented at summer meetings of the American Agricultural Economics Association, San Diego, 1977); J. G. Nagy and W. H. Furtan, "Economic Costs and Returns from Crop Development Research: The Case of Rapeseed Breeding in Canada," *Canadian Journal of Agricultural Economics* 26 (February 1978): 1–14; T. Y. Pee, "Social Returns from Rubber Research on Peninsular Malaysia" (Ph.D. diss., Michigan State University, 1977); W. L. Peterson, "Returns to Poultry Research

(*continued*)

TABLE 3—*Continued*

in the United States,'' *Journal of Farm Economics* 49 (August 1967): 656–69; W. L. Peterson and J. C. Fitzharris, ''The Organisation and Productivity of the Federal–State Research System in the United States,'' in Arndt, Dalrymple, and Ruttan, *Resource Allocation and Productivity*, 60–85; C. E. Pray, ''The Economics of Agricultural Research in British Punjab and Pakistani Punjab, 1905–1975'' (Ph.D. diss., University of Pennsylvania, 1978); A. Schmitz and D. Seckler, ''Mechanized Agriculture and Social Welfare: The Case of the Tomato Harvester,'' *American Journal of Agricultural Economics* 52 (November 1970): 569–77; G. M. Scobie and R. Posada T., ''The Impact of Technical Change on Income Distribution: The Case of Rice in Colombia,'' ibid. 60 (February 1978): 85–92, reprinted as chapter 26 in this volume; A. Tang, ''Research and Education in Japanese Agricultural Development,'' *Economic Studies Quarterly* 13 (February–May 1963): 27–41, 91–99; and E. B. Wennergren and M. D. Whitaker, ''Social Return to U.S. Technical Assistance in Bolivian Agriculture: The Case of Sheep and Wheat,'' *American Journal of Agricultural Economics* 59 (August 1977): 565–69.

The sources for the individual studies are as follows: H. W. Ayer, ''The Costs, Returns and Effects of Agricultural Research in São Paulo, Brazil'' (Ph.D. diss., Purdue University, 1970); H. W. Ayer and G. E. Schuh, ''Social Rates of Return and Other Aspects of Agricultural Research: The Case of Cotton Research in São Paulo, Brazil,'' *American Journal of Agricultural Economics* 54 (November 1972): 557–69; N. Ardito Barletta, ''Costs and Social Benefits of Agricultural Research in Mexico'' (Ph.D. diss., University of Chicago, 1970); M. Bredahl and W. L. Peterson, ''The Productivity and Allocation of Research: U.S. Agricultural Experiment Stations,'' *American Journal of Agricultural Economics* 58 (November 1976): 684–92; R. C. Duncan, ''Evaluating Returns to Research in Pasture Improvement,'' *Australian Journal of Agricultural Economics* 16 (December 1972): 153–68; Robert E. Evenson, ''The Contribution of Agricultural Research and Extension to Agricultural Production'' (Ph.D. diss., University of Chicago, 1968); idem, ''International Transmission of Technology in Sugarcane Production'' (New Haven: Yale University, 1969, mimeo); Robert E. Evenson and D. Jha, ''The Contribution of Agricultural Research Systems to Agricultural Production in India,'' *Indian Journal of Agricultural Economics* 28 (1973): 212–30; Z. Griliches, ''Research Costs and Social Returns: Hybrid Corn and Related Innovations,'' *Journal of Political Economy* 66 (1958): 419–31; idem, ''Research Expenditures, Education and the Aggregate Agricultural Production Function,'' *American Economic Review* 54 (December 1964): 961–74; and Y. Hayami and M. Akino, ''Organisation and Productivity of Agricultural Research Systems in Japan,'' in Arndt, Dalrymple, and Ruttan, *Resource Allocation and Productivity*, 29–59.

[a]Returns to maize research only.

[b]Returns to maize research, plus cultivation ''package.''

[c]Lagged marginal product of 1969 research on output discounted for an estimated mean lag of five years for cash grains, six years for poultry and dairy, and seven years for livestock.

Without discussing each study in detail, the following characterizations may be made.

—Only four studies reported low rates of return. All others were in excess of 20 percent, in real terms.

—The imputation and statistical studies yielded similar estimates.

—Estimates for research programs in developing countries were of roughly the same order of magnitude as those for more advanced countries.

—The estimate for international rice research was one of the highest reported. A similar estimate for wheat research at CIMMYT, while not made, would be similar. However, other centers have not produced results of this type.

These studies are open to criticism on a number of points, but even taking criticism into account, the general results hold. We are left, then, with the conclusion that policies toward investment in the production of appropriate agricultural technology are far from optimal. The fact that most such investment probably has to take place in the public sector is important to understanding why this is so. Public-sector policy makers, often taking cues from international advisers, have persistently overestimated the extent to which appropriate technology will "spill in" to the sector and can be had at low cost. The agricultural sector in many developing countries, until recently, has not had high priority in national plans. The most successful agricultural research systems have a local clientele of farmers who support them and alter their programs. The political mechanisms for such support systems are not established in much of the world.

Nonetheless, the picture is not wholly pessimistic. It seems clear that with a few exceptions, developing countries are moving in the direction of more optimal investment. The long and difficult process of building research institutions is proceeding in many countries. The international centers—at least some of them—are filling in many gaps in the system and are channeling valuable genetic and other raw materials to these programs. In some countries invention is being encouraged. The result of this progress toward a more optimal investment program is revealing itself in improved productivity growth in many agricultural sectors.

NOTES

1. It is of some interest to note that with only one exception, all modern crops and animals were domesticated from preexisting material centuries ago. The one exception is triticale, a crop originating in the early twentieth century from the work of modern plant breeders.

2. See James K. Boyce and Robert E. Evenson, *Agricultural Research and Extension Programs* (New York: Agricultural Development Council, 1975), 78–100.

3. See chapter 1 in this volume.—ED.

4. International Center for Tropical Agriculture (CIAT), Palmira, Colombia; International Institute of Tropical Agriculture (IITA), Ibadan, Nigeria; International Potato Center (CIP), Lima, Peru; International Crops Research Institute for the Semi-Arid Tropics (ICRISAT), Hyderabad, India; International Center for Agricultural Research in the Dry Areas (ICARDA), Beirut, Lebanon; International Laboratory for Research on Animal Diseases (ILRAD), Nairobi, Kenya; and International Livestock Center for Africa (ILCA), Addis Ababa, Ethiopia.

5. See Robert E. Evenson, J. C. O'Toole, R. W. Herdt, W. R. Coffman, and H. E. Kauffman, "Risk and Uncertainty of Factors in Crop Improvement Research," in *Risk, Uncertainty and Agricultural Development* (College Laguna, Philippines: Southeast Asian Regional Center for Graduate Study and Research in Agriculture [SEARCA], 1979).

6. For a discussion of this see Robert E. Evenson and Yoav Kislev, *Agricultural Research and Productivity* (New Haven: Yale University Press, 1975), 34–57.

7. See Dana Dalrymple, *Development and Spread of High Yielding Varieties of Wheat and Rice*, Foreign Agricultural Economic Report, no. 94 (Washington, D.C.: U.S. Department of Agriculture, 1978), 125.

8. The biological technology referred to earlier is also biochemical in nature. The term "chemical technology" as used here refers to industrial chemical products.

9. Paul A. David, *Technical Choice Innovation and Economic Growth* (Cambridge: Cambridge University Press, 1975), 197–200.

10. See Timmer, chapter 19 in this volume.—ED.

11. See chapter 1 in this volume.—ED.

25

The Farming Systems Perspective and Farmer Participation in the Development of Appropriate Technology

CIMMYT ECONOMICS STAFF

It is now widely accepted that technological change is the basis for increasing agricultural productivity and promoting agricultural development. Improved agricultural technologies are, for the most part, the product of formal agricultural research systems. In recent years, significant advances have been made in developing the capacity of agricultural research systems to deliver technologies appropriate to the needs of the major client of agricultural research systems, the farmer.

In this paper we describe research methods conducted with a farming systems perspective that emphasize farmer participation in the research process. First we treat conceptual and definitional issues related to this approach. Then we present research methods that are consistent with the resources of national agricultural research programs and that, we argue, should receive high priority in efforts to improve the effectiveness of these programs.

CONCEPTS AND DEFINITIONS

ON-FARM RESEARCH WITH A FARMING SYSTEMS PERSPECTIVE

Since 1978, when several publications (for example, CGIAR; Norman) drew attention to Farming Systems Research (FSR), interest in FSR has increased

CIMMYT is the International Maize and Wheat Improvement Center in El Batan, Mexico.

This is an abridged version of "Farming Systems Research: Issues in Research Strategy and Technology Design," by Derek Byerlee, Larry Harrington, and Donald Winkelmann, *American Journal of Agricultural Economics* 64, no. 5 (1982), and *Planning Technologies Appropriate to Farmers: Concepts and Procedures,* by Derek Byerlee, M. Collinson, et al. (Mexico: CIMMYT, 1981). Reprinted with revisions by permission of CIMMYT, the American Agricultural Economics Association, and the authors.

dramatically. This is reflected in a growing number of publications and workshops on the theme (such as Byerlee, Collinson, et al. 1981; Zandstra et al. 1981; and Shaner, Philipp, and Schmehl 1982) and a sharp increase in the commitment of resources to its implementation in developing countries.

The term "farming systems research" has been applied to a wide variety of activities, leading to confusion about its objectives and methodology. In its broadest sense, FSR is any research that views the farm in a holistic manner and considers interactions in the system (CGIAR 1978). We will define this explicit recognition of the importance of interactions in the farming system as the *farming systems perspective* (FSP).

Research with a farming systems perspective can have various objectives. The major objective, however, is to increase the productivity of farming systems by generating appropriate new technologies. This, in turn, is often further divided into location-specific research, having a short-term objective of developing improved technologies for a target group of farmers, and research conducted with a longer time perspective to overcome major, widespread limitations in farming systems.[1] Location-specific research is best implemented through *on-farm research methods* (OFR), where farmers are involved in identifying potential technological improvements, which are then tested under their conditions.

In this paper, we discuss a subset of FSR, referred to as *on-farm research with a farming systems perspective* (OFR/FSP), which has the following characteristics: (1) it aims to generate technology to increase resource productivity for an identified group of farmers, especially in the short term; (2) it is conceptually based on a farming systems perspective; and (3) it uses on-farm research methods.

THE NEED FOR A FARMING SYSTEMS PERSPECTIVE

The farming systems perspective is especially important when conducting research for small farmers in subtropical and tropical environments of developing countries. Several characteristics of the small farmers' environment lead to complex farming systems and add to the importance of interactions in farmer decision making.[2] Some of the most important elements leading to this complexity are: (1) a long growing season, which increases the range of potential crops and the possibilities of multiple cropping, including intercropping; (2) unreliable input and output markets, uncertain climate, and low farm incomes, which increase the importance of risk in farmer decisions; (3) high marketing margins and price variability, which encourage farm households to produce what they consume, contributing several additional elements to the farm household's objective function, such as production of preferred foods and a balanced seasonal distribution of food supplies; (4) low average productivity of family labor, the major factor of production, often combined with seasonal labor shortages, which encourage such practices as intercropping; and (5) heterogeneity of resources employed by the farm household (for example, land may be differentiated by quality, or labor may be provided by men, women, and children).

These considerations make for complex farming systems with a wide range of enterprises and even a range of production practices for a given enterprise, such as the use of more than one variety or planting date for a given crop. Complexity in most cases results from (1) direct physical interactions between production activities, generated by intercropping and crop rotation practices; (2) interactions due to competition and complementarity in resource use between different production activities; and (3) interactions due to trade-offs among the multiple objectives of the farm household. These interactions, from both biological and socioeconomic sources, underlie the need for a farming systems perspective and a multidisciplinary approach to research on improved technology.

THE NEED FOR EFFICIENT ON-FARM RESEARCH METHODS

On-farm research methods provide a means for introducing a farming systems perspective to research. Agricultural research has traditionally been organized along disciplinary or commodity lines without involvement of social scientists. It has typically been conducted on research stations under conditions not representative of farmers' fields and has had little or no farmer involvement. In on-farm research, direct communication of a multidisciplinary research team with farmers increases understanding of the farmers' decision-making environment and enables identification of technological alternatives that are more consistent with that environment. Experiments under farmers' conditions lead to estimates of yield and cost changes that better reflect what farmers can expect from using these alternatives.

The reorientation of an agricultural research system so that it is firmly based on on-farm research methods requires changes in research structures, organization, and incentives (see, for example, Moscardi et al. 1983). Moreover, methods used in OFR/FSP must be efficient in terms of resources, especially human resources but also financial resources and data processing facilities.[3] Because these programs initially tend to have only the partial support of research administrators, convincing results are needed early in the research program to ensure continuation and full integration into the research system.

There is an apparent anomaly between our advocacy of a farming systems perspective as a holistic view of an often complex farming system and the use of research methods that are cost-effective and emphasize rapid results. However, small farmers with scarce capital and with risk-avoidance objectives tend to favor a cautious learning process and, as a consequence, rarely make drastic changes in their farming system. Rather they proceed in a stepwise manner to adopt one and sometimes two new inputs or practices at a time.[4] Hence, a research strategy should focus on a very few—perhaps two to four—research opportunities that offer potential to increase resource productivity in a way acceptable to farmers. This narrow focus on a few priority research themes also enables national programs to use OFR/FSP despite shortages of skilled manpower and other resources. The identification of research opportunities and their development into technologies acceptable to farmers can and should be done using a farming systems perspective. However, since farmers rarely adopt farming systems as such (Collinson 1981), an OFR/FSP program should not seek as an immediate objective the development of completely new farming systems.[5] Rather, in the long run, a new farming system may evolve as the result of a series of discrete changes to the existing system.

RESEARCH PROCEDURES FOR NATIONAL AGRICULTURAL RESEARCH PROGRAMS

In this section we present the elements of an integrated agricultural research system based on on-farm research methods applied with a farming systems perspective. We assume that researchers are interested in developing a technology for a target crop in the system. In many cases the need to focus research on high-priority research opportunities will result in research on one crop, usually a major resource user. Many programs also have mandates for research on a specific crop, so researchers can select regions in which it is highly probable that research on that crop will increase system productivity. In any event, even research on a target crop requires a farming systems perspective that considers interactions with other crops in the system.

OVERVIEW OF PROCEDURES TO DEVELOP TECHNOLOGIES FOR FARMERS

A research process involving collaboration among scientists as well as between scientists and farmers is essential for rapid development of technologies that are appropriate to farmers' circumstances and that help to meet national goals.

A *technology* is a combination of all the management *practices* for producing or storing a crop or crop mixture. Each practice is defined by the timing, amount, and type of various *technological components,* such as varieties, land preparation, fertilizer, or weeding. A subsistence farmer who purchases no inputs is nevertheless using a technology—sometimes quite a complex one. Agricultural researchers should be particularly concerned that technologies developed are appropriate to the circumstances of target groups of farmers. *Farmers' circumstances* are all the factors that influence farmers' decisions about a crop technology—the natural environment (such as rainfall), the economic environment (such as product markets), and the farmers' goals, preferences, and resource constraints. If technologies are *appropriate* to farmers' circumstances, they will, by definition, be rapidly adopted by farmers.

Agricultural researchers should also seek a technology that helps meet *national policy goals.* Most governments want increases in food production; therefore any technology that increases production and is rapidly adopted by farmers will help meet this goal. Most governments also hope to reduce income inequalities among their citizens. This might require technologies that are adapted to small farmers or poorer regions or that provide cheap food to low-income urban consumers.[6]

The main participants in on-farm research are applied scientists—scientists of different disciplines who work together to solve immediate, high-priority problems—*and* farmers. Typically, the scientific team will include a biological scientist, usually an agronomist, to assemble information and analyze the physical and biological aspects of crop production, and a social scientist, usually an agricultural economist, to assemble information and analyze farmers' resource endowments, economic goals, and market environment. For specific problems, the scientific team may call on other specialists, such as entomologists or anthropologists, to supplement its skills. Fundamentally, however, it is essential that the agronomist and the agricultural economist *collaborate* in all phases of the research and jointly make decisions on such major topics as the content of on-farm experiments.

An overview of an integrated research program is shown in figure 1. At its heart is on-farm research, which is linked to two other important factors in

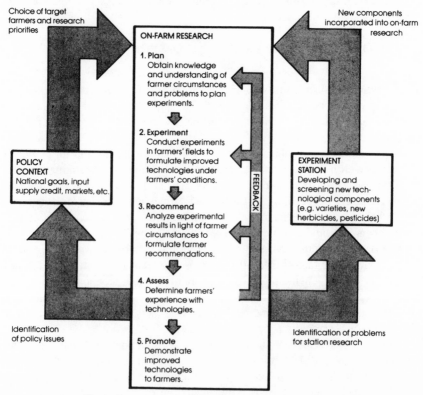

Fig.1. Overview of an Integrated Research Program

developing technologies. On one side is experiment-station research, which emphasizes the development of new technological components, such as new varieties. On the other side is agricultural policy, which influences much of the economic environment—such as national goals, input prices and supply, product markets and infrastructure—in which researchers and farmers make decisions.

On-Farm Research

On-farm research is research conducted in farmers' fields with the participation of farmers. Frequent contact between researchers and farmers makes it more likely that the constraints and problems of farmers will be considered in the design of technologies. Experimentation in farmers' fields ensures that technologies are formulated under farmers' conditions. It also provides a check on experiment-station results; otherwise, the results may be uncritically used to make recommendations to farmers even when the stations are in sites unrepresentative of the surrounding area or are operated with highly intensive management practices.

Because of its farmer orientation, on-farm research must carefully identify the farmers for whom the research is intended. It is most efficient when focused on a specific group of farmers who have similar problems and potentials.

Various activities or stages of on-farm research are indicated in figure 1. In the planning stage, the research team tries to describe and understand farmer circumstances. This information is used to identify priority technological components that increase productivity and that are consistent with the circumstances of target groups of farmers.

The priority technological components are then further investigated in the experimental stage in order to formulate improved technologies, that is, to construct, from known technological components and known biological relationships, technologies that improve upon farmers' existing practices. These experiments are conducted in farmers' fields so that new technologies are formulated under conditions similar to those in which farmers will use them.

Technologies are then recommended to farmers after careful testing against farmers' technologies in several locations and after economic analysis of the results (procedures are described in Perrin et al. 1976).

The final phases of the on-farm research are assessing farmers' experiences with the recommendations and promoting the recommendations to farmers. Farmers' reactions to the recommended technologies when they themselves pay the cost of inputs and bear the risks constitute important information that should be fed into the research process. If farmers are accepting the recommendations, researchers can turn to other problems while extension workers focus on promoting the technologies. If farmers are rejecting or substantially modifying the recommendations, then learning why may suggest appropriate changes in the recommendations or even in the experiments.

The on-farm research focus is continually changing as information is accumulated about farmers' circumstances, the performance of various technologies in experiments, and farmers' experiences with the technologies. Over time some problems may be solved (or set aside because of a lack of solution) and new problems added. The system provides for continual improvement in technologies as researchers apply results gained from past research to the planning of future research.

Experiment-Station Research

If a strong on-farm research program exists, research on experiment stations is primarily aimed at developing new technological components that require more closely controlled conditions, such as the development of new varieties. Also, experiment-station research can be used to screen technological components that might have undesirable effects on farmers' fields, such as herbicides that might leave residues. Promising technological components arising from experiment-station research are further refined and evaluated in on-farm experiments for their appropriateness for farmers.

A two-way flow of information should exist between on-farm research and experiment-station research. Information generated by on-farm research is important for guiding experiment-station research. For example, information about farmers' circumstances from on-farm experiments may indicate the type of variety that performs well under farmers' conditions and that meets farmers' preferences for varietal maturity, yield, storage quality, and cooking quality.

Information summarized from on-farm research in several regions can help in setting broad priorities for experiment-station work. It can provide a valuable base for assessing the impact of alternative breeding decisions—for example, the relative emphasis that should be placed on early maturity versus disease resistance. The information from assessments of farmers' circumstances and from experiments will help establish the production benefit of each technological component, the associated risks, and the types of farmers likely to realize the benefit.

Information fed back to the experiment station is often as important as the technologies recommended to farmers. Many experiment-station research programs lack mechanisms for relating research decisions to farmers' needs. In this situation, the on-farm research program should initially focus on screening the technologies developed at the station for relevance to farmers. The results can be extremely useful for evaluating the appropriateness of existing priorities in experiment-station research.

Policy Context of Agricultural Research

Government policies that shape the economic environment in which researchers and farmers make decisions are another important influence on agricultural research (fig. 1). Some policies directly influence the production decisions of farmers, such as a policy to make available compound fertilizers but not single-nutrient fertilizers. Most policies, however, influence farmers' behavior indirectly through their effects on input prices (for example, through subsidies) or product prices (for example, through marketing boards). The effects of policy on farmers' decision making in turn have implications for agricultural research. In countries where herbicides are expensive or difficult to obtain, researchers might orient research on weed-control problems differently than they would in a country where herbicides are cheap.

Policies may also directly influence research decisions. For example, many governments express the desire to make the distribution of real income more equal. This might influence the orientation of research programs toward poorer rural areas if most of the poor are in agriculture, or toward regions with high production potential if most of the poor are in urban areas. In fact, in most countries many geographical regions need assistance, but research resources are insufficient to initiate research programs in all regions. Measuring the characteristics of regions against national priorities such as increased food production and equalizing income distribution is one way to narrow the choice of target farmers for a research program.

Agricultural research, particularly on-farm research programs, can also provide valuable information to the policy maker that might encourage a change in policies to facilitate the adoption of improved technologies by farmers. For example, on-farm experiments may demonstrate the superiority of an input that is not available to farmers because of import restrictions. Or information on farmer circumstances might reveal important discrepancies between stated policy goals and policy implementation—for example, the late arrival of credit leading to untimely use of inputs.

Agricultural researchers must subjectively decide which elements of the policy environment to consider fixed and which to consider variable for the planning horizon of the research program. Researchers might justifiably experiment with inputs that are not currently available, under the assumption that if they can demonstrate a high payoff technology, they will be able to convince policy makers to make the inputs available. Other policies, such as price policies, might also vary as governments try to adjust to changing supply-and-demand conditions. However, there will be many other elements of the policy environment that reflect basic government strategy or that can only change slowly as agricultural development expenditures increase (for example, for infrastructure), and these policies must generally be taken as fixed when researchers are making decisions.

FARMERS' CIRCUMSTANCES AS A BASIS FOR PLANNING RESEARCH

Farmers' circumstances are those factors that affect farmers' decisions about what technologies to use in growing a crop. Expressed this way, farmers' circumstances explain both a farmer's current technology and his decisions about changes in that technology. Various farmers' circumstances are shown in figure 2. They include natural and economic circumstances. Economic circumstances

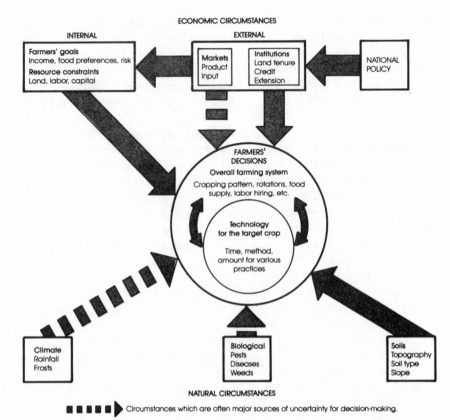

Fig.2. Circumstances Affecting Farmers' Choice of a Crop Technology

can be divided into those that are internal to the farmer, and over which he has some control (such as his goals and resources), and those that are external (such as markets).

Almost all farmers wish to increase their income—broadly defined to include production for home consumption. Generally, too, small farmers want security in meeting subsistence requirements of their preferred foods. Most also want to avoid taking risks that might endanger their subsistence supplies or sources of cash income.

Farmers have relatively fixed quantities of resources—land, family labor, and capital—that they can allocate to meet these goals. (Capital resources here include both durable equipment and cash.) Farmers may allocate these resources to different uses. Within limits, they may also adjust the amount of a resource available—for example, they may use some of their cash to rent more land or hire more labor.

Many circumstances also shape the economic environment in which farmers make decisions. These include the prices and price variability for inputs and products, access to inputs and product markets, land tenure systems, credit facilities, physical infrastructure (roads, dams, irrigation channels), and so on. This economic environment is largely outside of the farmer's control; it is influenced by policy decisions about distribution of inputs, price supports, infrastructure development, and so on. A large number of natural circumstances also condition the farmer's decision making, such as soil slope and depth, climate, weeds, and pests.

In making decisions, the farmer generally accepts as fixed external natural factors, such as rainfall, and economic factors, such as prices, though he may be able to modify their effects. A farmer may know, for example, that he has soils of different fertility, so he may decide to plant crops that meet his subsistence food preferences on the best soils to meet his goal of food security. Many external factors, particularly rainfall and prices, are variable and unknown to the farmer when he makes decisions. (In figure 2, factors that are major sources of uncertainty are marked with a broken line.) They provide an element of *risk*, which may have important effects on farmers' decision making. For example, although a farmer may not be able to predict rainfall, he is aware of its likely variability and therefore may plant a crop on several different dates to spread the risk of a dry period's striking at a critical stage in the crop cycle.

Most of these factors have *direct* effects on farmers' decisions about a technology for an individual crop. Late-season frosts might cause farmers to seek an early-maturing variety to reduce risks. Expensive labor encourages farmers to use a less labor-intensive weeding method, such as herbicides.

Many factors affect the choice of a technology for the target crop *indirectly* because of interactions in the farming system (fig. 2). The farming system is the totality of production and consumption activities of the farm household, including the choice of crop, livestock, and off-farm enterprises, as well as food consumed by the household. For example, a farmer may choose to plant maize late because he is planting beans early to avoid late-season disease problems in beans. Or he may plant an early variety of maize in order to have food early in the season before other crops mature. The point is that crop technologies often result from decisions made for the farming system as a whole. Consequently, planning technologies for one crop requires knowledge of important interactions in the farming system that potentially influence that crop.

The environment in which farmers make decisions changes over time. In particular, the external economic environment is characterized by changes in the ratios of input prices to product prices, which affect farmers' decisions. Changes in the external economic environment may also directly affect farmers' goals and resources. For example, as the market for food staples expands, farmers usually become more willing to depend on it for food supplies, and hence the influence of farmers' food preferences on their production decisions declines.

In the same way that farmers' circumstances determine a current crop technology, they are also important in a farmer's decision to change his technology. If a change in a technology conflicts with any of the circumstances of farmers, that technology may be rejected. For example, new varieties may be rejected because they are not suited to the soil conditions or because they ripen too late for the planting of the next crop. Fertilizer recommendations that aim for maximum yields are usually rejected because they are not consistent with either the farmer's income-increasing objectives or his risk-avoiding objectives.

When farmers reject technologies, it is not because they are conservative or ignorant. Rather, they rationally weigh the likely changes in incomes and risks associated with the technologies under their natural and economic circumstances and decide that for them the technology does not pay. The researchers' task is to incorporate knowledge of these circumstances into the design of technologies.

Decisions in Planning an On-Farm Experimental Program

Researchers must make a series of decisions in planning an on-farm experimental program. First, researchers must determine whether farmers in the region are sufficiently alike to allow a common set of experiments and a common recommendation. If there are significant differences among farmers, researchers must somehow divide farmers into more homogeneous groups and design experiments for each group. Second, they must decide which problems are going to be investigated and which technological components will be included in experiments for each group of farmers. For each technological component chosen, the levels, timing, and type of input or practice must be decided. Third, for each set of experiments, researchers must determine the level of nonexperimental variables, or those variables that are fixed for all treatments in the experiments. Finally, the researchers must choose farmers and sites on which to locate the experiments. The circumstances of the farmers for whom the technology is intended will be a key factor in all of these decisions.

Grouping Farmers into Recommendation Domains

Obviously, no two farmers have *identical* circumstances and therefore identical needs for technology. On the other hand, a research program cannot be established to provide recommendations for each farmer. It is therefore necessary to classify farmers with *similar* circumstances into *recommendation domains,* groups of farmers for whom more or less the same recommendations can be made. At least a tentative delineation of these recommendation domains is necessary for planning on-farm experiments, since the research priorities and consequent experiments might be different in each domain.

The proper number of recommendation domains depends on the amount of variation in farmers' circumstances (the more variation, the more domains needed) and on the amount of research resources (the more resources, the more domains that can be afforded). The final decision on the number of domains will be a trade-off between these two factors. However, it should be remembered that the researcher does not need to seek precise recommendations; general guidelines that the farmer can adjust to his own circumstances will be sufficient.

Recommendation domains can be defined on the basis of the various farmers' circumstances. They may be determined by variations in the natural circumstances of the farmer, such as rainfall, soils, or diseases. A region may contain many *agroclimatic environments*. In an agroclimatic environment a crop exhibits rather uniform biological expression, so that varietal or fertilizer responses would be similar, *everything else being equal*. Within an agroclimatic environment, however, there may be groups of farmers with differing socioeconomic circumstances that require different recommendation domains. For example, close to a large town, maize may be grown for sale as fresh ears, while further away it is grown as a subsistence grain. Such differences may impose modifications on varietal selection and planting date. More commonly, even among farmers who are in the same agroclimatic environment, differences in resource endowments may lead to different technological needs. For example, small farmers who have scarce capital relative to labor and who place more emphasis on food security may follow cropping patterns and practices quite different from those of large farmers in the same agroclimatic environment.

At times, a recommendation domain may result from a complex interaction of agroclimatic and socioeconomic factors. For example, within an agroclimatic environment for maize there may be different disease incidences for beans which cause farmers in one part of the agroclimatic environment to plant beans early, therefore delaying maize plantings. In this case recommendation domains result from natural circumstances, diseases, which affect bean production, and an economic circumstance, labor scarcity, which conveys this effect to maize practices.

Recommendation domains are not necessarily continuous geographical areas. For example, two neighboring farmers may be in different recommendation domains because of large differences in available resources. Even within a farm there may be different recommendation domains due to variation in soil type or topography.

It is clear, then, that knowledge of farmers' circumstances and how they affect crop technologies will be necessary in defining recommendation domains.

Identifying Farmers' Problems and Prescreening Technological Components

Many things directly limit farmers' production and incomes, such as weeds, pests, diseases, inferior varieties, and drought. Few research programs can investigate all of these constraints simultaneously. Priorities must be set by selecting those few most important problems limiting farmers' production and incomes for which there are technological components that promise speedy solutions.

For each important problem, there may be several technological components that could contribute to its solution. A weed problem, for instance, might be alleviated by instituting a crop rotation, by altering the time and method of land preparation, by raising the crop seeding rate, by improving manual weeding techniques, or by using a herbicide. In planning experiments, it is necessary to *prescreen* the various components to select those few *"best-bet"* components that have a high probability of success. Since the components finally picked for on-farm experiments must be compatible with farmers' circumstances, knowledge of those circumstances is essential not only to identify problems but also to prescreen technological components. Information on farmers' circumstances also helps the researcher define the range over which to test the technological component. When fertilizer is expensive, rainfall is variable, and farmers have little cash, the relevant range of levels for on-farm fertilizer trials will be lower than when conditions are more favorable for fertilizer use.

Establishing Representative Practices and Sites

An important reason for conducting experiments in farmers' fields is to be able to formulate technologies under farmers' conditions. Information on farmers' practices helps researchers design experiments in which nonexperimental variables reflect farmers' conditions. For example, in a research program emphasizing variety, fertilizer, and weed control, nonexperimental variables such as time and method of land preparation, planting method, and pest control should be maintained at farmers' levels. If farmers interplant maize and beans and researchers do not, weed control recommendations arising from research may be inappropriate for farmers, and without effective weed control, the profitability of fertilizer recommendations can be markedly altered.

It is likewise important that sites selected for on-farm experiments be representative of the recommendation domain with respect to soils, crop rotations, topography, location, and farm size. If maize is grown on a particular soil type, then fertilizer experiments on maize should be conducted on fields of this soil type. While it is convenient to choose sites that are easy to reach or are identified by cooperating extension personnel, these sites may not be representative of farmers' fields in the area.

Identifying Problems for Station Research and Policy Making

Because on-farm research is closely linked to experiment-station research and to policy decisions, knowledge of farmers' circumstances obtained in on-farm research plays another role in guiding these two activities. A major activity of experiment stations is the development of new varieties. Knowledge of farmers' circumstances is important for setting priorities among various breeding objectives. Do farmers need earlier varieties to increase cropping intensity or to reduce late-season weather risks? Do they need varieties with resistance to an insect or lodging resistance? Or do they need varieties with improved storage characteristics because of difficulties in the marketing system? The answers to these questions depend on the circumstances of farmers for whom the varieties are intended.

Sometimes information on farmers' circumstances will have to be quite detailed. In Egypt farmers regularly strip the lower leaves from their growing maize to feed animals. Experiments by researchers had demonstrated that leaf stripping sharply reduces yields, and they recommended against the practice. The researchers had been working on new, high-yielding, short varieties of maize. When experiments were conducted using the *farmers'* time and method of leaf stripping, it was found that farmers' varieties, which tend to be taller and more leafy than the new varieties, permit leaf stripping with little effect on yields. With information on the value of leaves and the real yield loss when the leaves of existing varieties are stripped, researchers now have a measure of the amount by which yields of grain must be increased if farmers are to adopt new varieties that do not tolerate stripping.

Information on farmers' circumstances also helps researchers identify policy problems that may impede successful introduction of new technologies. In one country, decision makers believed that insecticides were easily available to farmers. However, information obtained from farmers demonstrated that this was not true. Some insecticides were available in one place, some in another, and the distribution of insecticides did not at all coincide with the distribution of insects. This information convinced administrators to reexamine the input distribution system. Often information from research will show policy makers the potential benefits from changing policies. For example, if fertilizer is in short supply, researchers may want to conduct some experiments to provide information to policy makers on fertilizer response. In effect, these become experiments for recommendations to policy makers, rather than for recommendations to farmers, because farmers do not yet have access to fertilizer.

OVERVIEW OF THE PROCEDURES FOR OBTAINING AN UNDERSTANDING OF FARMER CIRCUMSTANCES

The planning of experiments as described above clearly requires an in-depth understanding of farmer circumstances. With experience in many countries, we have found that an efficient approach to obtaining this knowledge begins with collection of secondary data (such as rainfall statistics), followed by an exploratory survey and a verification survey. This is a sequential process in which information becomes more detailed and focused at each subsequent step in the process.

The exploratory survey is a useful technique for rapidly understanding the farming system and indentifying key research priorities. The essential characteristics of this technique are its relatively unstructured approach and the high degree of researcher participation in field interviews and observations (see Collinson 1981; and Hildebrand 1976). A multidisciplinary team of researchers interviews farmers in an informal and iterative manner, guided by a systems perspective of farmer decision making and oriented by a list of topics. Meanwhile, the biological dimensions of crop or livestock production are observed in farmers' fields. The whole survey is usually completed in two to three weeks in a given recommendation domain.

The exploratory survey is a sequential data-collection technique. The research team analyzes and evaluates information on a *daily* basis in order to make a decision on what further data need to be collected. Initially, researchers try to obtain a broad description and understanding of the farming system. They then focus on research opportunities for increasing productivity and, finally, the assessment of possible technological alternatives to be included in on-farm experiments.

The exploratory survey is followed by a well-focused "verification" survey using a short, structured questionnaire (sometimes only one to two pages long) and a random sample. This information allows formal testing of hypotheses and provides greater confidence in the conclusions reached in the exploratory survey. Note that even in the verification survey the emphasis is on a low-cost method that provides information in a few weeks.

SUMMARY

On-farm research with a farming systems perspective is a subset of FSR that can be used by national agricultural research programs to generate new technology appropriate for representative farmers. OFR/FSP is characterized by two-way linkages with the national policy framework and with experiment-station research. It also emphasizes the need to understand farmers' circumstances when planning the various stages of research, including forming recommendation domains, setting research priorities, and selecting representative sites and farmer-collaborators for on-farm experiments.

Although international agricultural research centers such as CIMMYT can make important contributions to the development of new agricultural technologies, the location-specific nature of most technologies means that much of the research and development work in agriculture must take place in national research systems.[7] The procedures for on-farm research discussed above were designed for use by national agricultural research programs that have limited resources because these programs must take the lead in developing technologies for farmers around the world.

NOTES

1. These are referred to as "downstream" and "upstream" FSR by some authors (e.g., CGIAR 1978). We disagree with this terminology—"downstream" is hardly consistent with the "bottom-up" philosophy of FSR. We are also confused by the definition of "upstream" research. Gilbert, Norman, and Winch (1980) limit it to research to overcome major *resource* constraints, such as soil moisture conservation or fertility maintenance.

2. Although we emphasize small farmers in this paper, we feel that OFR also has substantial value in commercial agriculture. Payoffs may, however, be less because of less complex farming systems and because commercial farmers may already have considerable influence on research decisions.

3. To some extent the increasing availability of microcomputers will help overcome the constraint on data processing.

4. It is sometimes assumed that OFR can make significant gains by a reallocation of existing resources, such as changing plant spacing or extra weeding, without introducing new inputs to the system. We believe that this is an exceptional case and in fact is contrary to the systems perspective of a rational farmer. For more on the stepwise adoption behavior of farmers see Mann 1978; and Byerlee and Hesse de Planco 1982.

5. Development of new farming systems may be appropriate where there is a drastic change in the farmers' external environment, such as the introduction of irrigation or a colonization program.

6. See Timmer, chapter 8 in this volume, for a discussion of other common goals governments establish for their food systems.—ED.

7. See Schultz, chapter 23, and Evenson, chapter 24, in this volume.—ED.

REFERENCES

Byerlee, Derek, M. Collinson, et al. 1981. *Planning Technologies Appropriate to Farmers: Concepts and Procedures.* El Batan, Mexico: CIMMYT.

Byerlee, Derek, and E. Hesse de Planco. 1982. *The Rate and Sequence of Adoption of Improved Cereal Technologies: The Case of Rainfed Barley in the Mexican Altiplano.* CIMMYT Economics Working Paper. El Batan, Mexico.

CGIAR, Technical Advisory Committee. 1978. *Farming Systems Research at the International Agricultural Research Centers.* Rome: TAC Secretariat.

Collinson, M. P. 1981. "A Low Cost Approach to Understanding Small Farmers." *Agricultural Administration* 8:433–50.

————. 1982. *Farming Systems Research in Eastern Africa: The Experience of CIMMYT and Some National Agricultural Research Services, 1976–81.* MSU International Development Paper, no. 3. East Lansing: Michigan State University, Department of Agricultural Economics.

Gilbert, Elon H., David W. Norman, and Fred E. Winch. 1980. *Farming Systems Research: A Critical Appraisal.* MSU Rural Development Paper, no. 6. East Lansing: Michigan State University, Department of Agricultural Economics.

Hildebrand, P. E. 1976. *Generating Technology for Traditional Farmers: A Multidisciplinary Methodology.* Guatemala City: ICTA.

Mann, C. K. 1978. "Packages of Practices: A Step at a Time with Clusters." *Gelisme Dergisi Studies in Development* (Middle East Technical University, Ankara) 21:73–80.

Moscardi, E., et al. 1983. *The Establishment of a National On-Farm Research Entity in Ecuador.* CIMMYT Economics Working Paper 83/1. El Batan, Mexico.

Norman, D. W. 1978. "Farming Systems Research to Improve the Livelihood of Small Farmers." *American Journal of Agricultural Economics* 60:813–18.

Perrin, R. K.; D. L. Winkelmann; E. R. Moscardi; and J. R. Anderson. 1976. *From Agronomic Data to Farmer Recommendations: An Economics Training Manual.* CIMMYT Information Bulletin 27. El Batan, Mexico.

Shaner, W. W., P. F. Philipp, and W. R. Schmehl. 1982. *Farming Systems Research in Development: Guidelines for Developing Countries.* Boulder: Westview Press.

Zandstra, H. G.; E. C. Price; J. A. Litsinger; and R. A. Morris. 1981. *A Methodology for On-Farm Cropping Systems Research.* Manila, Philippines: International Rice Research Institute.

26

The Impact of Technical Change on Income Distribution: The Case of Rice in Colombia

GRANT M. SCOBIE AND RAFAEL POSADA T.

The contribution of technical change to agricultural productivity in developing countries has been widely recognized and increasingly documented (Arndt, Dalrymple, and Ruttan 1977). The generation of that technical change through agricultural research is now viewed as an economic activity to which scarce resources can be devoted and measurable output defined (Schultz 1970).[1]

In the appraisal of potential or past research strategies, two central economic issues arise: efficiency and equity. While earlier studies were concerned primarily with the efficiency goal, increasing attention has been given to the distribution of social benefits stemming from programs of agricultural research. Akino and Hayami (1975) and Ramalho de Castro and Schuh (1977) examine the distribution of the gross social benefits between consumers and producers, while others have considered the impact on the functional distribution of income (for example, Ayer and Schuh [1972]; Schmitz and Seckler [1970]; Wallace and Hoover [1966]).

In this chapter we analyze the impact of technological change in the Colombian rice industry, giving particular attention to the consequences for the household income distribution. "It appears that relatively little theoretical and empirical work has been done on the welfare or income distributional effects of technical change. This is unfortunate, for a considerable amount of research funds is spent each year by both private and public institutions to develop new technologies for agriculture" (Bieri, de Janvry, and Schmitz 1972, 801). After sketching the background of the research program, we present some estimates of

GRANT M. SCOBIE is senior agricultural economist, Ruakura Agricultural Research Centre, Ministry of Agriculture and Fisheries, Hamilton, New Zealand. RAFAEL POSADA T. is economist, Centro Internacional de Agricultura Tropical (CIAT), Cali, Colombia.

Reprinted from *American Journal of Agricultural Economics* 60, no. 1 (1978): 85–92, with omissions and minor editorial revisions, by permission of the American Agricultural Economics Association and the authors.

the social benefits: the rate of return (efficiency) and the impact on household income distribution (equity). Both costs and benefits of the research program are used in deriving the distributional consequences.

BACKGROUND

In 1957 a national rice research program was formed within the Ministry of Agriculture with the cooperation of the Rockefeller Foundation (Rosero 1974). At that time, the tall U.S. variety, Bluebonnet-50, was extensively grown; but in 1957 it was attacked by a virus disease, causing extensive losses. Imports of rice rose substantially, and the real domestic retail price was higher in 1957 than in any year since 1950 (and in fact up to 1974). These events stimulated the formation and funding of a national rice research program whose primary objective was the selection of varieties resistant to the virus. It was hoped that by this mechanism domestic output could be increased, partly eliminating the need for rice imports and lowering domestic prices. This strongly consumer-oriented policy contrasts with the vacillation of public policy between a consumer and a producer focus which had been characteristic of Colombian rice policy since the thirties (Leurquin 1967).

This research effort did produce new varieties, but their impact was limited (Hertford et al. 1977). In 1967 the newly formed rice program of the Centro Internacional de Agricultura Tropical (CIAT) joined in a collaborative effort with the Colombian program, and dwarf lines from the International Rice Research Institute in the Philippines were introduced. This was followed by the local development and release of four disease-resistant dwarf rices. The rate of adoption of these modern varieties has been spectacular. In 1966, 90 percent of the irrigated sector was sown to the traditional variety (Bluebonnet-50); by 1974 virtually all the irrigated rice production came from dwarf varieties.

The research program, especially since 1967, has been oriented to the irrigated sector. The modern varieties have been best suited to areas with good water control and high levels of other inputs. Thus, while widely adopted throughout the irrigated sector, they had little impact in the upland or rainfed sector. Given the possibility of rapid increases in national output through the introduction of new varieties suited to irrigated culture, this was undoubtedly a rational choice. The rate of technological progress that could have been achieved with the same research resources surely would have been less had attention been directed to the upland sector. A further explanation of the particular ecological orientation adopted lay in the close collaboration between FEDEARROZ, the National Rice Growers' Federation (founded and supported principally by the large rice growers), and the research program.

As a consequence of the emphasis on the irrigated sector, together with the rapid adoption of the modern varieties, yields and production in that sector rose dramatically. In contrast, the relatively disadvantaged upland sector, which experienced little or no technical change, declined in importance from 50 percent of the national output in 1966 to 10 percent in 1974.

THE MODEL

The approach taken closely follows the formulation of Ayer and Schuh (1972).[2] A more detailed statement of the model and the estimation of the parameters is given in Scobie and Posada 1977. The model estimates the total gross social benefits and their division between Colombian rice producers and consumers, but extends existing formulations (following Bell 1972) by distinguishing between upland and irrigated producers. This distinction was made as a consequence of the differential impact of the research program on the two sectors. It is suggested that the proposed formulation would have general applicability in analyzing the differential impact of a new technology whose relevance is restricted for whatever reason to a subset of the producing firms.

RESULTS

GROSS BENEFITS

The changes in consumer and producer surpluses resulting from the introduction of modern varieties were estimated for each year from 1964 to 1974 and are summarized in table 1. Consumer benefits are positive because in the absence of modern varieties, the volume of rice entering the domestic market would have been much lower, with a concomitant higher internal price. However, for the same reason, both upland and irrigated producers have forgone rents to factors of production whose mobility is limited in the short run. Changes in producer surplus follow as a consequence of some inelasticity in the supply of rice, imparted by rising short-run marginal cost curves. Such changes would be transitory if, in the long run, the supply elasticities of all factors to rice production approached infinity. Despite the overall reduction in short-run producer sur-

TABLE 1
GROSS BENEFITS OF NEW RICE VARIETIES IN COLOMBIA TO
CONSUMERS AND PRODUCERS, IN $(COL.) MILLION

Gross Benefits Accruing to:	1964–69	1970–74
Consumers	1,404	17,542
Producers		
Irrigated	−368	−6,468
Upland	−517	−3,878
Total	519	7,196

NOTE: Each entry is the sum of the annual deflated values (1964 = 100).

pluses, some gains undoubtedly accrued to "early adopters" in the irrigated sector.

Had Colombia not mounted a successful research program, upward pressure on domestic prices may well have been contained by allowing rice imports. Even in the absence of such approval, illegal imports from neighboring Venezuela and Ecuador may have had a price-depressing effect. Higher imports of rice would have reduced the amount of foreign exchange available for other imports and put upward pressure on the exchange rate. However, estimating the distributional consequences of this scenario would lead us far beyond the more modest scope of this investigation.

NET BENEFITS

The distribution of gross benefits between producer and consumer groups is a relatively blunt tool for analyzing the distributional impact of technological change. We attempt two extensions: first we will consider the incidence of the research costs, and so derive net benefits to producers and consumers; subsequently we will examine the distribution of the gross benefits and research costs by income level within groups.

The costs of the research program were borne by three entities: (a) the national rice program of the Instituto Colombiano Agropecuario (ICA); (b) the contribution of the growers through FEDEARROZ under Law 101 of 1963, which created the *Cuota de Fomento Arrocera*. This law requires the collection of $(Col.) 0.01/kg from all growers and authorizes FEDEARROZ to administer the funds for support of research, regional testing, publishing technical bulletins, presenting training courses to field agronomists, and financing the Technical Division of FEDEARROZ; and (c) international cooperation, originally through the Rockefeller Foundation and subsequently through the rice program of the Centro Internacional de Agricultura Tropical (CIAT).

No attempt is made to include any costs incurred by the International Rice Research Institute (IRRI) in the development of IR-8 and IR-22, which occupied up to almost 60 percent of the area sown in Colombia. Hence, for these varieties we will overstate the net global benefits by allowing their contribution to production without discounting their full costs. However, if the measurement of net benefits is viewed from Colombia's standpoint, then it is valid to include only those costs incurred by Colombia in testing, multiplying, and releasing the IRRI materials.

The distribution of gross social benefits, research costs, and net benefits for producers and consumers is shown in table 2. The gross social benefits were totaled for the period 1964–74 and expressed in $(Col.)million 1970, compounding forward the years 1964–69 and discounting 1971–74, both using an estimate of 10 percent for the social opportunity cost of capital in Colombia (Harberger 1972, 155).

In a similar manner the costs of the research from the three sources were

TABLE 2
SIZE AND DISTRIBUTION OF BENEFITS AND COSTS OF MODERN RICE VARIETIES IN
COLOMBIA, 1957–74, IN $(COL.) MILLION

	Producers				Total	International
Item	*Upland*	*Irrigated*	*Total*	*Consumers*	*Colombia*	*Cooperation*[a]
Gross benefits	−3,542	−5,293	−8,835	14,939	6,104	—
Costs of research						
FEDEARROZ	8	30	38	—	38	—
ICA[b]	1	2	2	22	25	—
Total	9	32	40	22	63	19
Net benefits	−3,551	−5,325	−8,875	14,917	6,042	—

NOTE: All data expressed in 1970 pesos; minor discrepancies are due to rounding.
[a]From Ardila 1973, and personal communication from the Centro Internacional de Agricultura Tropical (CIAT).
[b]From Ardila 1973, and personal communication from the Instituto Colombiano Agropecuario (ICA).

summed and are shown in table 2. The costs of the ICA program were assumed to come from general tax revenue and were divided between consumers and producers on the basis of urban and rural proportions of total tax revenues in 1970 (Jallade 1970). The producer contribution was further broken down between upland and irrigated producers on the basis of the production coming from each sector in 1970. The contributions from FEDEARROZ were distributed between the upland and irrigated sectors assuming a 45 percent collection rate (FEDEARROZ 1975), except that no contributions were assumed for upland producers with less than ten hectares. Expressed in 1970 pesos, $(Col.)82 million were devoted to rice research between 1957 and 1974.

In order to assess the sensitivity of the net benefits to varying assumptions about the supply-demand elasticities, the results in table 3 were calculated. The demand elasticity was varied from −0.3 (a typical lower bound found in a review

TABLE 3
NET BENEFITS IN 1974, IN $(COL.) MILLION, AND INTERNAL RATES OF RETURN
FOR DIFFERING ELASTICITIES OF SUPPLY AND DEMAND

	Elasticity of Demand (η)		
Elasticity of Supply (ϵ)	*−0.300*	*−0.449*	*−0.754*
	9,052	3,981	2,174
0.235	89%	94%	89%
	8,627	3,556	1,749
1.500	96%	87%	79%

NOTE: In each cell, the upper figure is the net benefits to Colombia of the rice research program in 1974, and the lower figure is the internal rate of return based on the period 1957–74, with the last year's costs and returns assumed to continue until 1986. The combination of $\eta = -0.449$ and $\epsilon = 0.235$ was used in calculating the results presented in tables 1 and 2.

of numerous studies in developing countries) to −0.449 (based on Pinstrup-Andersen, de Londoño, and Hoover 1976, 137) to −0.754 (Cruz de Schlesinger and Ruiz 1967). The supply elasticity of 0.235 is from Gutiérrez and Hertford 1974, with an arbitrarily chosen upper value of 1.5.

The internal rate of return on the investment is consistently high and relatively insensitive to varying elasticities. These high returns are not uncommon in agricultural research. Ayer and Schuh (1972, 581) report an internal rate of return of 89 percent for cotton in São Paulo, Brazil; Akino and Hayami (1975, 8) report values up to 75 percent for rice in Japan; Peterson (1967, 669) reports 20–30 percent for poultry in the United States; Barletta (1971) reports 75 percent for wheat in Mexico; Griliches (1958) reports 35 percent for corn in the United States; Ardila (1973) reports 58–82 percent for rice in Colombia up until 1971; and Montes (1973) reports 76–96 percent for soybeans in Colombia.[3] One should resist the conclusion that all agricultural research would show such payoffs—the literature is not replete with evaluations of failures.

While the net benefits in 1974 vary little with the supply elasticity, they fall markedly with higher absolute demand elasticities (table 3). However, our concern here is more with the relative distribution of the net benefits within groups than with establishing their absolute magnitude.

DISTRIBUTION OF NET BENEFITS BY INCOME LEVEL

To evaluate the distributional impacts of the technological change, the gross benefits, the costs of the research program, and the consequent net benefits were distributed across income groups for consumers and upland and irrigated producers. In each case the annual average impact for 1970 was estimated by summing the gross benefits and costs (expressed in 1970 pesos) and dividing by the appropriate number of years.

Gross benefits to consumers were assumed to be directly proportional to the quantity of rice consumed, while their contributions to the research costs were distributed in proportion to tax receipts from each income stratum. The resulting net benefits to consumers by income level are shown in table 4.

Rice is now virtually the most important foodstuff in Colombia; between 1969 and 1974 total domestic consumption doubled (U.S. Department of Agriculture 1976, 11), and rice is the major source of calories and the second major source of protein (after beef) in the Colombian diet (Departamento Nacional de Planeación 1974). As rice is disproportionately consumed by the lower-income groups, who make limited tax contributions, the net benefits of the research program were strongly biased toward them in both absolute and relative terms. While the lower 50 percent of Colombian households received about 15 percent of household income, they captured nearly 70 percent of the net benefits of the research program.

In the case of producers, the annual average change in producer surplus was distributed across farm sizes in proportion to estimates of the production based

TABLE 4
DISTRIBUTION OF NET BENEFITS, HOUSEHOLDS, AND HOUSEHOLD INCOME

1970 Income Level ($[Col.]000)	Annual Average Net Benefits ($[Col.])	Net Benefits as a Percentage of Income[a]	Cumulative Percentage of:		
			Net Benefits	Households[b]	Household[b] Income
0–6	385	12.8%	18%	19%	2%
6–12	642	7.1	50	39	8
12–18	530	3.5	67	52	15
18–24	333	1.6	77	64	23
24–30	348	1.3	83	71	29
30–36	353	1.2	88	76	35
36–48	342	0.8	93	82	43
48–60	200	0.4	95	86	51
60–72	128	0.2	96	89	57
72+	138	0.2	100	100	100

[a]Relative to the midpoint of the interval.
[b]From Jallade, 1974, 22.

on census data. The research costs were also distributed by farm size, assuming that tax payments were proportional to production (in the case of the ICA costs), and by the method already discussed for the research levy. The sum of the forgone income and the research costs were then expressed as a percentage of the estimated 1970 average net income by farm size for the entire rural sector (table 5). This last step is clearly less than satisfactory.

Ideally, income distribution data are required for upland and irrigated rice producers by size of farm. As no such data are known to exist, resort was made to a distribution of rural income by farm size for 1960 (Berry 1974, 610), inflated to 1970 values. We have no basis for knowing whether rice producers would have higher or lower incomes than the rural average for each farm size group. However, again, our principal interest is in the relative rather than absolute distribution of benefits by income level.

The group most severely affected was the small (that is, low-income) upland producers. For these producers, the annual average income forgone through lower rice prices (and no compensating technological change) represented a high proportion of their assumed 1970 income. To the extent that their incomes were below the rural sector average, this impact would have been even more pronounced. On the other hand, the forgone income to the irrigated producers varied more erratically depending on the size group, with the heaviest relative burden falling on the 200–500- and 500–1,000-hectare groups. However, the absolute impact may well be overstated if irrigated producers had incomes above the national average for rural income earners.

In summary, the net benefits of the technological change accrued to consumers, with the lowest-income households capturing a disproportionate share. As Hayami and Herdt (1977) note, "The decline in the price of a food staple due

TABLE 5
ANNUAL AVERAGE DISTRIBUTIONAL IMPACT OF RICE RESEARCH PROGRAM ON
UPLAND AND IRRIGATED PRODUCERS

Farm Size (hectares)	Average Income[a] ($[Col.])	Change in Producer Surplus Plus Research Costs as a Percentage of 1970 Income	
		Upland Sector	Irrigated Sector
0–1	1,500[b]	−58%	−56%
1–2	3,647	−53	−39
2–3	5,330	−60	−25
3–4	6,508	−71	−38
4–5	7,406	−75	−53
5–6	10,295	−60	−43
10–20	15,652	−48	−47
20–30	18,934	−41	−48
30–40	23,394	−35	−47
40–50	28,620	−30	−45
50–100	35,904	−29	−48
100–200	66,759	−26	−53
200–500	155,398	−18	−79
500–1,000	287,513	−21	−69
1,000–2,000	532,389	−19	−49
2,000+	1,480,199	−11	−36

[a]From Berry 1974, 610, adjusted to 1970.
[b]Assumed value.

to technical progress in its production has the effect of equalizing income among urban consumers'' (249).[4] The forgone income to producers appeared to fall most heavily on the small upland producers. Even if the average annual consumer benefits are included as benefits to upland producers, the small upland producers still appear as the most severely affected, a not surprising result, given the orientation of the research program toward the irrigated sector. However, some notion of the relative magnitudes of the different groups should be borne in mind. In 1970 (prior to the major impact of the modern varieties) there were only an estimated twelve thousand upland producers with less than five hectares. Hence, under any plausible set of welfare weights, their losses would be more than offset by the gain to more than one million low-income consuming households, implying an overall gain (albeit uncompensated) in some measure of social welfare.

Caution should be exercised in generalizing from this conclusion. The urban and nonlandowning rural poor of Colombia are very much more numerous than the small farmers. In less urbanized countries with a large semisubsistence rural population, the lowest-income households may benefit from technological advances specifically designed for the small-farm sector (Valdés, Scobie, and Dillon 1979).

CONCLUDING COMMENTS

Concern is periodically voiced for the distributional implications of technological change in developing agriculture. One is often led to feel that the introduction of new technology has been only a qualified success because of its apparent failure to solve a broad spectrum of social ills. But frequently it is the well-being of only the rural poor (both the small farmer and the landless worker) that is the focus of attention. The presence of large concentrations of urban poor who are potential beneficiaries of expanded production of basic foodstuffs is sometimes neglected when castigating the "green revolution."

Throughout much of Latin America, the rural poor tend to be concentrated (for historical reasons) in the less favored ecological zones. The development of technology suited to such areas is presumably a more difficult process, which, *ceteris paribus,* would divert research resources from the discovery of technologies that can result in rapid increases in total output from the more favored commercial agricultural sector.

The results presented for the case of Colombian rice exemplify this trade-off. By focusing on the distribution between consumers and producers and, more important, by isolating both the costs and benefits by income strata, we have endeavored to quantify some of the dimensions of this trade-off. Concentrating the research on the upland producers would presumably have entailed forgone benefits to the numerous urban poor (without guaranteeing that small upland producers would have benefitted in the long run).

In this chapter we have attempted some preliminary extensions of the commonly used approaches to analyzing the distributional impact of technological change: (*a*) a model that allows for differential impact of technological change on two classes of producers is introduced; (*b*) the incidence of research costs is considered in the distribution of the social benefits to different groups; and (*c*) the distributional impact (at the national level) on consumer and producers households by income strata is analyzed. These extensions have come only at a price. We have ignored the consequences for the employment of resources released from the rice sector due to the differential impact of the new technology; and the lack of data to analyze the distributional consequences for household income led us to a formidable number of assumptions, we hope not excessively cavalier.

NOTES

The research on which this article is based was conducted while the authors were economists in the rice program of the Centro Internacional de Agricultura Tropical (CIAT), Cali, Colombia. The support and interest of John L. Nickel, director-general of CIAT, and Peter R. Jennings, of the Rockefeller Foundation, are gratefully acknowledged. G. Edward Schuh, Paul R. Johnson, Alberto Valdés, Per Pinstrup-Andersen, Reed Hertford, and two reviewers for the *American Journal of Agricultural Economics* all offered insightful comments.

1. See also Schultz, chapter 23, and Evenson, chapter 24, in this volume.—ED.
2. Hertford and Schmitz (1977) provide a review of the procedures involved in estimating changes

in consumer and producer surplus. A valuable survey with discussion of some of the contentious issues is given by Currie, Murphy, and Schmitz (1971), while some apparent inconsistencies between alternative formulations are noted by Scobie (1976).

3. See Evenson, chapter 24, table 3.—ED.
4. See also Mellor, chapter 10 in this volume.—ED.

REFERENCES

Akino, M., and Y. Hayami. 1975. "Efficiency and Equity in Public Research: Rice Breeding in Japan's Economic Development." *American Journal of Agricultural Economics* 57:1–10.

Ardila V., J. 1973. "Rentabilidad social de las inversiones en investigación de arroz en Colombia." Master's thesis, ICA-Universidad Nacional, Bogotá.

Arndt, T. M., D. G. Dalrymple, and V. W. Ruttan, eds. 1977. *Resource Allocation in National and International Agricultural Research*. Minneapolis: University of Minnesota Press.

Ayer, H. W., and G. E. Schuh. 1972. "Social Rates of Return and Other Aspects of Agricultural Research: The Case of Cotton Research in São Paulo, Brazil." *American Journal of Agricultural Economics* 54:557–69.

Barletta, A. N. 1971. "Costs and Social Benefits of Agricultural Research in Mexico." Ph.D. diss., University of Chicago.

Bell, C. 1972. "The Acquisition of Agricultural Technology: Its Determinants and Effects." *Journal of Development Studies* 9:123–59.

Berry, R. A. 1974. "Distribución de fincas por tamaño, distribución del ingreso y eficiencia de la producción agrícola en Colombia." In *Lecturas sobre desarrollo económico colombiano*, edited by H. Gomez O. and W. Weisner D. Bogotá: Fundación para la Educación Superior y el Desarrollo.

Bieri, J., A. de Janvry, and A. Schmitz. 1972. "Agricultural Technology and the Distribution of Welfare Gains." *American Journal of Agricultural Economics* 54:801–8.

Cruz de Schlesinger, L., and L. J. Ruiz. 1967. *Mercadeo de arroz en Colombia*. Bogotá: Centro de Estudios sobre Desarrollo Económico.

Currie, J. M., J. A. Murphy, and A. Schmitz. 1971. "The Concept of Economic Surplus and Its Use in Economic Analysis." *Economic Journal* 81:741–800.

Departamento Nacional de Planeación. 1974. *Plan nacional de alimentación y nutrición*. Doc. DNP-UDS-DPN-011. Bogotá.

FEDEARROZ (Federación Nacional de Arroceros). 1973. *Informe de Gerencia*. Bogotá.

———. 1975. *Informe de Gerencia*. Bogotá.

Griliches, Z. 1958. "Research Costs and Social Returns: Hybrid Corn and Related Innovations." *Journal of Political Economy* 66:419–31.

Gutiérrez A., N., and R. Hertford. 1974. *Una evaluación de la intervención del gobierno en el mercadeo de arroz en Colombia*. Technical pamphlet no. 4. Cali, Colombia: Centro Internacional de Agricultura Tropical.

Harberger, A. C. 1972. *Project Evaluation*. London: Macmillan and Co.

Hayami, Y., and R. W. Herdt. 1977. "Market Price Effects of Technological Change on Income Distribution in Semisubsistence Agriculture." *American Journal of Agricultural Economics* 59:245–56.

Hertford, R., and A. Schmitz. 1977. "Measuring Economic Returns to Agricultural Research in Colombia." In Arndt, Dalrymple, and Ruttan 1977.

Hertford, R., J. Ardila V., A. Rocha, and C. Trujillo. 1977. "Productivity of Agricultural Research in Colombia." In Arndt, Dalrymple, and Ruttan 1977.

Jallade, J. P. 1974. *Public Expenditures on Education and Income Distribution in Colombia*. World Bank Staff Occasional Paper 18. Washington, D.C.

Leurquin, P. P. 1967. "Rice in Colombia: A Case Study in Agricultural Development." *Food Research Institute Studies in Agricultural Economics, Trade and Development* 7:217–303.

Montes, G. 1973. "Evaluación de un programa de investigación agrícola: el caso de la soya." Master's thesis, Universidad de Los Andes, Bogotá.

Peterson, W. L. 1967. "Return to Poultry Research in the United States." *Journal of Farm Economics* 49:656–69.

Pinstrup-Andersen, P., N. R. de Londoño, and E. Hoover. 1976. "The Impact of Increasing Food Supply on Human Nutrition: Implications for Commodity Priorities in Agricultural Research and Policy." *American Journal of Agricultural Economics* 58:131–42.

Ramalho de Castro, J. P., and G. E. Schuh. 1977. "An Empirical Test of an Economic Model for Establishing Research Priorities: A Brazil Case Study." In Arndt, Dalrymple, and Ruttan 1977.

Rosero M., M. J. 1974. *El Cultivo del arroz.* ICA-FEDEARROZ, Assistance Manual, no. 9. Palmira, Colombia: FEDEARROZ.

Schmitz, A., and D. Seckler. 1970. "Mechanized Agriculture and Social Welfare: The Case of the Tomato Harvester." *American Journal of Agricultural Economics* 52:569–77.

Schultz, T. W. 1970. *Investment in Human Capital: Role of Education and Research.* New York: Free Press.

Scobie, G. M. 1976. "Who Benefits from Agricultural Research?" *Review of Marketing and Agricultural Economics* 44:197–202.

Scobie, G. M., and R. Posada T. 1977. "The Impact of High-Yielding Rice Varieties in Latin America with Special Emphasis on Colombia." Centro de Documentación Económica para la Agricultura Latinoamericana (CEDEAL), Series JE-01. Cali, Colombia: Centro Internacional de Agricultura Tropical.

U.S. Department of Agriculture. 1976. *Foreign Agricultural Circular* FR-1. Washington, D.C.

Valdés, A., G. M. Scobie, and J. L. Dillon, eds. 1979. *Economic Analysis and the Design of Small-Farmer Technology.* Ames: Iowa State University Press.

Wallace, T. D., and D. M. Hoover. 1966. "Income Effects of Innovation: The Case of Labor in Agriculture." *Journal of Farm Economics* 48:325–36.

27

Assessment of the Green Revolution

YUJIRO HAYAMI

In a recent article, Richard Grabowski has applied the Hayami-Ruttan model of induced innovation to the analysis of dualistic rural economies.[1] He argues that in the economies characterized by bimodal distribution of assets and income, the induced innovation mechanism dictates technological change in a socially inefficient and inequitable direction.

According to Grabowski, in the situations of dualistic agrarian structure, large landholders, because of their monopsonistic position, face factor price relations different from those faced by smallholders and the landless. Landed elites usually have better access to institutional credit and subsidized modern inputs, with the result that the cost of nonlabor inputs for them are lower than social opportunity costs. The demand of the powerful elites based on such biased price signals will induce public agencies to generate technological changes that have labor-saving effects and are more profitable for large-scale operations. Thus the inequitable distribution of assets and income in dualistic rural communities will be aggravated further.

I share Grabowski's perspective that in situations where the rural sector consists of more than one group with sharply conflicting interests, the induced innovation mechanism may fail to generate socially desirable innovations or may produce innovations that are socially undesirable for both efficiency and equity reasons—a perspective entirely consistent with Hayami and Ruttan's.[2] Indeed, we often observe cases in which factor market distortions and concentration of land assets induced inefficient and inequitable technological changes, such as labor-displacing mechanization in the labor-abundant economies where the social opportunity cost of labor was low relative to that of capital.

YUJIRO HAYAMI is professor of economics, Tokyo Metropolitan University.

Originally titled "Induced Innovation, Green Revolution, and Income Distribution: Comment," this paper was a comment on the article by Richard Grabowski cited in n. 1, below. Reprinted from *Economic Development and Cultural Change* 30, no. 1 (1981): 169–76, by permission of the University of Chicago Press. Copyright © 1981 by The University of Chicago. All rights reserved. Published with minor editorial revisions by permission of the author.

My major disagreement is with Grabowski's view on the "green revolution"—the development and diffusion of modern semidwarf varieties of rice and wheat in tropical countries with heavier application of fertilizers and chemicals. He considers that the green revolution represents a typical case in that the induced innovation mechanism worked in dualistic situations to generate technological changes biased toward the benefit of landed elites at the expense of smallholders and landless laborers.

THE GREEN REVOLUTION AND INCOME DISTRIBUTION

In this perspective Grabowski shares the criticism of the green revolution popular among radical political economists and sociologists.[3] Their arguments run as follows: Green revolution technology tends to be monopolized by large commercial farmers, who have better access to new information and better financial capacity. Modern varieties (MV) can profitably use higher applications of modern inputs such as fertilizers and chemicals. Adoption of MV is said to be difficult for small subsistence farmers, who have little financial capacity to purchase these inputs. A large profit resulting from the exclusive adoption of MV technology by large farmers stimulates them to enlarge their operational holdings by consolidating the farms of small nonadopters through purchase or tenant eviction. As a result, polarization of rural communities into large commercial farmers and landless proletariat is promoted. Furthermore, the large commercial farms have an intrinsic tendency to introduce large machinery for ease of labor management, which reduces employment opportunities and wage rates for the landless population, resulting in more inequitable income distribution.

Indeed, such arguments are not groundless. It is not difficult to find cases in which significant trends toward polarization and more inequitable income distribution have developed side by side with the diffusion of MV. The point of major controversy is the causal relationship between new technology and polarization phenomena. Does the adoption of new technology in fact tend to be monopolized by largeholders because of constraints in information and credit to smallholders? Does the technology make large-scale operations relatively more efficient and profitable? Is there a technical complementarity between MV and labor-displacing machinery? The issue is an empirical question and should be resolved as such.

EMPIRICAL EVIDENCE

In fact, there is little empirical evidence that the use of MV has been monopolized by large farmers. Of thirty-six villages selected throughout Asia for the international cooperative study on the process of adoption of new rice technology, in only one village was there a significant lag of small farmers behind large farmers in the adoption of MV.[4] This exceptional village was the one studied by

Cumulative adoption (%)

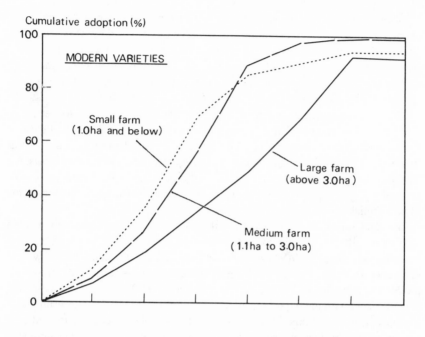

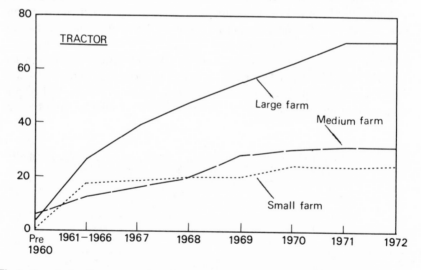

Fig.1. Cumulative Percentage of Farms in Three Size Classes Adopting Modern Varieties (*top*) and Tractors (*bottom*) in Thirty Selected Villages in Asia (IRRI, *Interpretive Analysis of Selected Papers from Changes in Rice Farming in Selected Areas of Asia* [Los Baños, Philippines, 1978], 91)

Parathasarathy and Prasad, which Grabowski refers to as evidence for the differential adoption of MV technology among large and small farms.[5] On the average, small farmers were even faster in the MV adoption than were large farmers (fig. 1, top). The pattern of MV diffusion paths among different farm-size classes contrasts sharply with the diffusion pattern of tractors, in which large farmers achieved distinctly faster and higher rates of adoption (fig. 1, bottom).

The contrasting diffusion patterns suggest that the MV and tractor technologies differ. The latter is characterized by indivisibility or lumpiness in the use of capital and therefore requires a large farm size for efficient operation, whereas the former is divisible into any small units at equal efficiency. In other words, the MV technology is neutral with respect to scale. The scale neutrality of MV technology was supported by the econometric test on wheat production in the Indian Punjab by Sidhu.[6]

The data collected from selected villages in Asia show that there was no significant difference in the average paddy yield per hectare between large and small farmers adopting MV, except for Indonesia (table 1), and that there was no difference in the percentage of farmers reporting increases in profit from rice production and level of living after the introduction of MV (table 2). These data are consistent with the hypothesis of scale neutrality of MV. Similar results were reported in a large body of micro studies for various parts of Asia.[7] After a careful review of those micro studies, Ruttan concluded that "neither farm size nor tenure has been a serious constraint to the adoption of new high yielding grain varieties" and "neither farm size nor tenure has been an important source of differential growth in productivity."[8]

Another major issue revolves around the direction of factor-saving bias in new technology. If MV has a labor-saving bias, the income position of landless laborers and tenants will deteriorate even if the technology is neutral with respect to the scale of farm operation. The econometric test by Sidhu for Punjab wheat

TABLE 1

AVERAGE PADDY YIELDS ON LARGE VERSUS SMALL FARMS ADOPTING MODERN VARIETIES IN 32 ASIAN VILLAGES, 1971/72 WET SEASON

		Paddy Yield (tons per hectare)		
	Villages (N)	Small Farmer (S)	Large Farmer (L)	Difference (L–S)
India	12	3.8	4.6	0.8
Pakistan	2	3.3	3.4	0.1
Indonesia	5	3.7	5.3	1.6
Malaysia	2	2.4	2.2	−0.1
Philippines	9	2.7	2.5	−0.2
Thailand	2	3.4	3.4	0
All villages	32	3.3	3.8	0.5

SOURCE: IRRI, *Interpretive Analysis*, 96.

TABLE 2
AVERAGE PERCENTAGE OF FARMERS REPORTING INCREASES IN PROFIT FROM
RICE AND LEVEL OF LIVING AFTER THE INTRODUCTION OF MODERN VARIETIES IN
SELECTED ASIAN VILLAGES, 1971/72

| | | Farmers Reporting Increases (%) | | | |
| | | Profit from Rice | | Level of Living | |
	Villages *(N)*	*Small* *Farmer*	*Large* *Farmer*	*Small* *Farmer*	*Large* *Farmer*
India	8	72%	80%	59%	73%
Indonesia	5	80	79	65	67
Malaysia	2	67	53	60	45
Philippines	12	62	64	55	62
Thailand	3	41	55	41	48
All villages	30	66	69	57	63

SOURCE: IRRI, *Interpretive Analysis*, 108.

production shows that the MV wheat represented a neutral technological change not only with respect to scale but also with respect to the factor use.[9] A similar study by Ranade and Herdt on rice in the Philippines suggests that the MV technology is biased in the direction of land saving, though the evidence is not conclusive.[10]

Although more econometric investigations are required to confirm the direction of factor-saving bias in the MV technology, available evidence shows that the MV technology resulted in a significant increase in labor demand. Data assembled for various parts of Asia by Barker and Cordova show that labor inputs per hectare for rice production were higher for MV than for traditional varieties (TV) in the order of 10–50 percent.[11] Similar results were drawn from a review of other studies.[12] Typically, labor application for land preparation was reduced by use of tractors, but the reduction was more than compensated for by increases in labor use for weeding, other crop-tending requirements, and harvesting. Grabowski cites Bardhan's study for North India as evidence for the insignificant effect of the green revolution on the increase in rural labor demand.[13] However, Lal shows clearly that Bardhan's study refers to an early period of MV diffusion and that as MV diffused more widely, the net effect of the resultant increase in labor demand was a significant rise in real wages in the Punjab and Haryana at a time when real wages were constant or declining in parts of India where the diffusion of MV technology was limited.[14]

A popular argument is that even if the MV technology may itself be neutral or labor using, it stimulates the introduction of labor-displacing machinery through either technical complementarity or increased income for large farmers to enable the purchase of machinery. However, such a conjecture is not supported by empirical observations. As shown in figure 1, increases in the adoption of tractors by large farmers began earlier than the introduction of MV, and there was no sign that the tractor adoption was accelerated by the dramatic diffusion of MV from the late 1960s to the early 1970s. Much of the growth in the use of tractors in South and Southeast Asia can be attributed to distortions in the price of capital by such means as overvalued exchange rates and concessional credits from national governments and international lending agencies.[15]

CONDITIONS OF GREATER INEQUALITY

The empirical evidence fails to identify the MV technology (in combination with irrigation and fertilizer) as a factor promoting more unequal income distribution. In general, both small and large farmers adopted MV at more or less equal rates and achieved efficiency gains of the same order. It is likely that the MV technology was neutral or biased in the land-saving and labor-using direction and resulted in increases in labor demand, despite the concurrent progress of mechanization.

Contrary to the popular belief, we see a real danger of growing inequality in rural communities not because of the MV technology but because of insufficient progress in the technology under the strong population pressure on land. Rural economies in developing countries have been experiencing a rapid deterioration in the man/land ratio, with the average number of workers per hectare of arable land doubling within a fifty-year period.

If technological progress of the land-saving type is not sufficiently rapid, the increase in labor demand will fail to keep up with the increase in labor supply arising from rapid population growth. The wage rate is bound to decline and the return to land to rise, and the income position of laborers and tenants will deteriorate relative to that of landowners. The process approximates the world predicted by classical economists such as Ricardo.[16] As the growth of population presses hard on limited land resources under stagnant technology, cultivation frontiers are expanded to more marginal land and greater amounts of labor are applied per unit of cultivated land. The cost of food production increases, and food prices rise. In the long run, laborers' income will be lowered to a subsistence minimum barely sufficient to maintain stationary population, and all the surplus will be captured by landlords in the form of increased land rent.

The increased rate of return to land provides a strong incentive to accumulate more land, especially under the condition of an underdeveloped capital market, because alternative investment opportunities, such as stocks and securities, are not easily available. The concentration of landholdings induced by the higher rate of land rental makes the income distribution more skewed, which promotes further concentration of land—a vicious circle producing polarization. This trend is accelerated further by the development of labor-displacing mechanical technology which was induced by market distortions through a process correctly perceived by Grabowski.

It is not easy to stop this process simply by instituting land-reform laws and regulations or trying to protect smallholders by such means as subsidizing credit and input prices, as Grabowski suggests. In many cases such government interventions have the effect of promoting polarization, especially under the power structure of dualistic rural communities. It is a common observation that rich farmers, through their pull with government agencies, manage to receive a disproportionate share of subsidized input and credit.[17] Land-reform regulations, such as rent control or the prospective plan of land confiscation and redistribution, often have resulted in the large-scale eviction of tenants in order to establish landlords' direct cultivation by use of agricultural laborers.[18]

If the real economic force underlying polarization and greater misery of the poor is the decline in the return to labor relative to the return to land, agrarian reform programs—such as land reform and institutional credit programs—will have little chance of success to achieve more equitable distributions of income and assets unless they are supported by the efforts to counteract the decreasing return to additional labor applied per unit of land area. One critical effort, among others, should aim at the development of labor-using and land-saving technology or at least neutral technology in order to raise the marginal productivity of labor on limited land resources.

In this perspective, the green revolution can be considered an induced innovation in response to growing population pressure on land, without which both the wage rate and the income level of the rural working population would have degenerated further. As Grabowski rightly points out, in dualistic situations the induced innovation mechanism can dictate technological change toward a socially inefficient and inequitable direction, as evidenced by labor-displacing mechanization in many parts of labor-abundant economies. However, the green revolution does not seem to represent such a case. Empirical evidence is more consistent with the hypothesis that the growing inequality in the rural sector of developing countries has been a result not of green revolution technology but of insufficient progress of the green revolution technology in overcoming growing population pressure on land.

NOTES

1. Richard Grabowski, "The Implications of an Induced Innovation Model," *Economic Development and Cultural Change* 27 (July 1979): 723–34. [For details on the induced innovation model see chapter 4 in this volume.—ED.]

2. "Usually the gains and losses from technical and institutional change are not distributed neutrally. There are, typically, vested interests which stand to lose and which oppose change. There are limits on the extent to which group behavior can be mobilized to achieve common or group interests. The process of transforming institutions in response to technical and economic opportunities generally involves time lags, social and political stress, and, in some cases, disruption of social and political order" (Yujiro Hayami and Vernon W. Ruttan, *Agricultural Development: An International Perspective* [Baltimore: Johns Hopkins Press, 1971], 61).

3. The radical criticism is typically expressed in Harry M. Cleaver, "The Contradictions of the Green Revolution," *American Economic Review* 72 (May 1972): 177–88; Ali M. S. Fatami, "The

Green Revolution: An Appraisal,'' *Monthly Review* 2 (June 1972): 112–20; Francine R. Frankel, *India's Green Revolution: Economic Gains and Political Costs* (Princeton: Princeton University Press, 1971); and Keith Griffin, *The Political Economy of Agrarian Change: An Essay on the Green Revolution* (Cambridge: Harvard University Press, 1974).

4. International Rice Research Institute (IRRI), *Changes in Rice Farming in Selected Areas of Asia* (Los Baños, Philippines, 1978).

5. G. Parathasarathy and D. S. Prasad, "Response to, and Impact of, HYV Rice according to Land Size and Tenure in a Delta Village, Andhra Pradesh, India," *Developing Economies* 12 (June 1974): 182–98; see also G. Parathasarathy, "West Godavari, Andhra Pradesh," in IRRI, *Changes in Rice Farming,* 43–70.

6. Surjit S. Sidhu, "Economics of Technological Change in Wheat Production in the Indian Punjab," *American Journal of Agricultural Economics* 56 (May 1974): 217–26.

7. For India, Bandhudas Sen, *The Green Revolution in India: A Perspective* (New Delhi: Wiley Eastern, 1974); for Indonesia, Irlan Soejono, "Growth and Distributional Changes in Paddy Farm Income in Central Java," *Prisma,* May 1976, 26–32; for Pakistan, K. M. Azam, "The Future of the Green Revolution in West Pakistan: A Choice of Strategy," *International Journal of Agrarian Affairs* 5 (March 1973): 404–29; for the Philippines, Mahar Mangahas, Virginia A. Miralaro, and Romana P. de los Reyes, *Tenants, Lessees, Owners: Welfare Implications of Tenure Change* (Quezon City, Philippines: Ateneo de Manila University Press, 1974).

8. Vernon W. Ruttan, "The Green Revolution: Seven Generalizations," *International Development Review* 19 (August 1977): 16–23.

9. Sidhu, "Economics of Technological Change."

10. Chandra G. Ranade and Robert W. Herdt, "Shares of Farm Earnings From Rice Production," in IRRI, *Economic Consequences of the New Rice Technology* (Los Baños, Philippines, 1978), 87–104.

11. Randolph Barker and Violeta Cordova, "Labor Utilization in Rice Production," in ibid., 113–36, esp. 117.

12. Ifmkhar Ahmed, "Employment Effects of the Green Revolution," *Bangladesh Development Studies* 4 (January 1976): 115–28; William G. Bartsch, *Employment Effects of Alternative Technologies and Techniques in Asian Crop Production: A Survey of Evidence* (Geneva: International Labour Office, 1973); and Edward J. Clay, "Institutional Change and Agricultural Wages in Bangladesh," *Bangladesh Development Studies* 4 (October 1976): 423–40.

13. Pranab Bardhan, "Green Revolution and Agricultural Laborers," *Economic and Political Weekly* 5 (July 1970): 1239–46.

14. Deepak Lal, "Agricultural Growth, Real Wages, and the Rural Poor in India," ibid. 11 (June 1976): A47–A61.

15. Randolph Barker et al., "Employment and Technological Change in Philippine Agriculture," *International Labour Review* 106 (August–September 1972): 111–39; and Bart Duff, "Mechanization and Use of Modern Rice Varieties," in IRRI, *Economic Consequences,* 145–64.

16. David Ricardo, *On the Principle of Political Economy and Taxation,* ed. Piero Sraffa, 3d ed. (Cambridge: Cambridge University Press, 1951).

17. See Adams and Graham, chapter 21, and Gonzalez-Vega, chapter 22, in this volume.—ED.

18. Kalyan Dutt, "Changes in Land Relations in West Bengal," *Economic and Political Weekly* 12 (December 1977): A106–A110; P. C. Joshi, "Land Reform in India and Pakistan," ibid. 5 (December 1970): A145–A152; Dharm Narain and P. C. Joshi, "Magnitude of Agricultural Tenancy," ibid. 4 (September 1960): A139–A142; Doreen Warriner, *Land Reform in Principle and Practice* (Oxford: Clarendon Press, 1969).

V

Case Studies:
China and Africa

China and Africa

Introduction

Because China and sub-Saharan Africa will be of special interest to development economists and international donors for the rest of this century, the experiences of the People's Republic of China and Africa are highlighted in this section. China is important to agricultural development specialists for at least three reasons. First, China is the largest country in the world,[1] and it desires to become the leading spokesman for the Third World. Second, China's rural development experience has aroused keen interest in other Third World nations and has been held up as a model for emulation in many countries. Because of a lack of reliable data, however, there has been a great deal of speculation about the sources of agricultural growth, the role of agriculture in Chinese development over the past three decades, and the transferability of the Chinese model. Third, since the death of Chairman Mao in 1976, Chinese economic policy has been in a state of flux. An important question is whether China has altered its long-run development strategy since the death of Mao.

We believe that the chapters by Lardy and Tang (chapters 28 and 29, respectively) provide an objective assessment of China's agricultural development experience over the past thirty years. They also offer some valuable insights into the role that the state will likely assign to the agricultural sector in China's development strategy of the 1980s.[2]

As Tang and Lardy point out, the performance of China's agriculture over the past three decades has been impressive, but not as spectacular as commonly believed by many Western scholars during the mid and late 1970s. The major lessons from the literature on agricultural development in China can be summarized as follows:

1. Despite the common view that China is giving priority to meeting basic needs and helping the rural poor, the state has consistently directed the agricultural sector to support the priority sector—heavy industry—in China's drive to become a major industrial power by the year 2000.
2. The performance of China's economic system since the Communist party took power in 1949 has been impressive. Although food production has grown only slightly faster than population (about 2 percent per year), per capita GNP tripled (in constant prices) between 1952 and the late 1970s.
3. Thirty years ago China correctly perceived that for one to two generations rural surplus labor would have to be absorbed in rural areas rather than in urban industry.

399

4. China has pragmatically assembled a food-security system that stresses the interdependence of food production, access to food, and international trade. China has stressed local self-sufficiency in food production, food rationing in urban areas, and the sale of food at subsidized prices in urban areas to allow the urban poor to purchase a bland but adequate diet. Despite comments by food self-sufficiency advocates who hold up China as a model of food self-sufficiency, China never totally abandoned the concept of comparative advantage and has given it greater emphasis in recent years. For example, China regularly sells rice in international markets and buys wheat.

5. The food/population race and the level of food subsidies will be key issues in the debates on China's agriculture in the 1980s and 1990s. The Chinese have stepped up their family planning efforts because of land constraints and the rising share of the national budget being channeled into food subsidies.

6. Despite major social and economic achievements, China is still a low-income country faced with the classic struggles of a food/population race, providing employment in rural areas for its surplus population and trying to transfer resources to the industrial sector. Moreover, because conditions in China are unique, there may be severe limitations to the transferability of the Chinese model of development to other Third World countries. The sobering assessments by Lardy and Tang are the starting points for a more informed debate on China's agricultural development experience and its implications for the Third World.

Case studies of Africa are included in this section because Western development economics textbooks generally give little attention to Africa, because an increasing proportion of foreign development assistance is being channeled to Africa in the 1980s, and because problems unique to Africa will dominate Third World development debates in the 1980s and 1990s (Leys 1982). Although Africa is a tremendously diverse continent, most sub-Saharan countries face several common problems, which are discussed by Lele and Eicher (chapters 30 and 31, respectively).

Twenty-five years ago, when African states started to reclaim their political independence from the Europeans, there was a great deal of optimism among African political leaders and Western donors about skipping stages of growth and "catching up" with the industrial nations. Today, however, the forty-five countries of sub-Saharan Africa collectively comprise the poorest part of the world's economy. Moreover, despite vast amounts of idle land and record levels of foreign aid, most countries in Africa are unable to feed themselves. In fact, Africa is the only continent where per capita food production actually declined during the 1960s and 1970s (U.S. Department of Agriculture 1981).

The study of the economic "backwardness" of Africa must begin with an analysis of Africa's eighty years under colonial rule (1880–1960) and an examination of the post-independence record of Africa's economies. The historical record clearly shows that over the past one hundred years both colonial powers and leaders of Africa's independent states have viewed agriculture as a backward sector that could be exploited for the benefit of industrial/urban development. Although African political leaders widely proclaim agriculture to be the priority sector, most states allocate only 5–10 percent of their government investment to agriculture—the sector that employs 60–80 percent of the population.

The performance of Africa's agriculture over the past two decades has been poor. In addition to stagnant per capita food production and rising food imports, the average rate of growth of agricultural output declined from 2.7 percent in the 1960s to 1.3 percent in the 1970s (World Bank 1982). The prognosis for agricultural development in Africa in the 1980s is also poor, for numerous reasons outlined by Lele and Eicher. Whether agriculture can be harnessed as an engine of growth for development and whether the agricultural sector can provide more productive employment depends on whether African governments face up to a number of economic, technical, social, and political issues and address them in a consistent manner over the next ten to fifteen years.

Donors have responded to Africa's development crisis by greatly accelerating per capita aid flows, to the point where Africa is the most aid-dependent continent in the Third World. But as Lele points out, there is a severe shortage of human resources to absorb present aid flows. The limited pools of trained African scientists and managers partially are a legacy of colonialism and partially reflect the failure of independent African states to give priority to training professional agriculturalists. Although the World Bank has requested donors to double aid to Africa in real terms over the 1980–90 period, Eicher contends that increased aid flows will be ineffective unless there is a fundamental reordering of development priorities.[3]

NOTES

1. One billion of the 4.6 billion people in the world in 1983 were Chinese.

2. The literature on China's development experience is vast. The reader is directed to the classics by Ho Ping-ti (1959), Dwight Perkins (1968), and Alexander Eckstein (1975). For optimistic assessments of China's development experience during the mid-1970s see Gurley 1976; and Timmer 1976.

3. For a recent review of agricultural development literature in sub-Saharan Africa see Eicher and Baker 1982.

REFERENCES

Eckstein, Alexander. 1975. *China's Economic Development*. Ann Arbor: University of Michigan Press.

Eicher, Carl, and Doyle C. Baker. 1982. *Research on Agricultural Development in Sub-Saharan Africa: A Critical Survey*. MSU International Development Paper, no. 1. East Lansing: Michigan State University, Department of Agricultural Economics.

Gray, Jack, and Gordon White, eds. 1982. *China's New Development Strategy*. London: Academic Press.

Gurley, John G. 1976. *China's Economy and the Maoist Strategy*. New York: Monthly Review Press.

Ho Ping-ti. 1959. *Studies on the Population of China, 1368–1953*. Cambridge: Harvard University Press.

Leys, Colin. 1982. "African Economic Development in Theory and Practice." *Daedulus* 111(2):99–124.

Perkins, Dwight H. 1968. *Agricultural Development in China, 1368–1968*. Chicago: Aldine Publishing Company.

Timmer, C. Peter. 1976. "Food Policy in China." *Food Research Institute Studies* 15(1):53–69.

U.S. Department of Agriculture. 1981. *Food Problems and Prospects in Sub-Saharan Africa: The Decade of the 1980s*. Washington, D.C.

World Bank. 1981. *Accelerated Development in Sub-Saharan Africa: An Agenda for Action*. Washington, D.C.

———. 1982. *World Development Report 1982*. New York: Oxford University Press for the World Bank.

28

A Critical Appraisal of the Chinese Model of Development

ANTHONY M. TANG

DEVELOPMENT STRATEGY

The values and goals of the leadership of the Chinese Communist Party (CCP) are aptly summarized in the aphorism attributed to A. Bergson: "To the Communist Party leadership, bread is only an intermediate product; steel is the final good."[1] Such values are apparent in the aspiration of party leaders, present and past, to turn China into a modern, front-ranking power by the end of the century.

This goal dictates a "maximum-speed selective growth" development strategy and requires a command economic system for implementation. Main emphasis is placed on growth of the modern heavy industrial sector, whose absolute size is the preponderant determinant of power. The command system in important ways allocates key economic resources administratively rather than by markets and prices. A liberal democracy also is apt to embrace a command system in times of crisis, as was done by the United States and the United Kingdom in World War II.

The role of agriculture in such a system is to support priority (selective) industrialization with resource transfers in the form of food (the principal wage good), labor, raw materials, and exportable farm products. When agriculture fails to meet the full resource transfers required by the priority industrial sector for maximum growth, the agricultural sector becomes a drag on industrialization. Such was the case in the People's Republic even during the relative economic heyday of 1952–58.[2]

With agriculture limiting industrial growth, the two sectors were no longer separable for operational planning. To develop industry, it was necessary to

ANTHONY M. TANG is professor of economics, Vanderbilt University.

Excerpts, with revisions and updating, from Anthony M. Tang, *Food and Agriculture in China: Trends and Projections, 1952–77 and 2000* (1980), in *Food Production in the People's Republic of China,* a two-monograph volume by Tang and Bruce Stone, International Food Policy Research Institute Research Report 15 (Washington, D.C., May 1980), 11–81. Published by permission of the International Food Policy Research Institute and the author.

403

develop agriculture. Fostering development in agriculture while extracting the maximum surplus from it proved a difficult policy problem. In passing, we note that Russia, during its comparable development phase under Stalin, faced a much simpler task, thanks to its large initial agricultural surplus.[3] The Chinese problem was dramatized by the slogan "agriculture is the foundation" in the early 1960s following the disastrous Great Leap Forward (1958–60). This was not a complete policy reversal as is often argued by China watchers. The Chinese policy had never been "industry only," and the renewed emphasis on agriculture was not a reordering of priorities in a value-goal context; rather, it was concerned with ways to develop the sector. The strenuous development efforts for agriculture in the 1950s were limited primarily to bootstrap operations utilizing the sector's own indigenous resources of low or no opportunity cost. The 1960s saw the beginnings by the state of large-scale commitments of scarce resources to the production and import of modern inputs for agriculture. The current Chinese official view is that both the First Five-Year Plan period (1953–57) and the readjustment period of 1962–65 were comparable periods, noted for their "fairly well-coordinated relations between agriculture, light industry, and heavy industry."[4]

Much has happened in China since then. Books will be written on the Cultural Revolution and the massive leadership shakeup that followed. The radical phase which came to a head with the deaths of Zhou Enlai and Mao Zedong and the Gang of Four episode was followed by the counterrevolution. Deng Xiaoping, who once was sacked for holding the view that "it matters not whether the cat is white or black as long as it catches mice," is now being praised for it in and outside the country. China analysts generally have been inclined to view the struggle over economic growth between Mao and his associates and those of moderate, pragmatical persuasion as involving value-and-goal disputes. A common notion is that with the start of the Cultural Revolution in 1966 Mao and followers tempered the growth objective with liberal doses of nonmaterial, ideologically appealing values, for example, distributional equity, self-reliance, anti-elitism, and work ethic for service as opposed to material incentive.[5] The present leadership is seen as returning (selective) economic growth to its earlier position of primacy and as freely adopting pragmatic methods to achieve it.

It seems more plausible that the contending leadership factions have shared common values and goals but have differed over methods. Clearly Deng was condemned by his detractors for his "cat and mouse" statement not because of the end he advocated but because of the means. Moreover, the quarrel over the means was concerned not so much with ideological or nationalistic considerations as with their economic efficacy. If this were not so, how can one explain Mao's unfailing willingness to tolerate (indeed, to insist on) the continuance of Hong Kong and Macao as Western colonies—an anomaly not many other independent nations would be likely to accept without duress. We know that Portugal tried to return Macao to China but was refused. The Chinese attitude can only be explained by the fact that the two colonies earn for Peking some U.S. $6 billion a year in (hard currency) foreign exchange (revised 1982 estimate), or 30 percent of China's total export earnings.[6] This speaks to the centrality of the economic growth goal and to Peking's inclination to wink at what must be a most unpleasant means to an end.

Consumption was seen by Peking's leaders as an intermediate process rather than as the final act of an economic system. Consequently they tended to think of efficiency in consumption more in relation to workers' productivity than in relation to utility. Here is the clue to much of the policy dispute. It turns on what constitutes the right mix between material incentives and nonmaterial (normative) substitutes. Mao and associates inclined toward the nonmaterial substitutes. Their periodic attempts to impose their view were checked by real-world intractability, giving rise to the well-known policy cycles in China.[7] Each time Mao retreated from his "high tide," material incentive would regain lost ground in the wage-consumption policy debate.

Even so, Mao's dominant power and personality, together with China's thin cushion of agricultural surplus, largely determined wage-consumption policy during 1953–77. The level of consumption for the masses was adequate but strictly without frills. Virtually no improvement occurred.[8] Differences among income classes were kept small. To this end, Peking rationed the basic necessities—grains and cooking oil, cotton cloth, and housing, the latter through government allocation. Convenience and luxury goods were priced extremely high to absorb income not spent on necessities. This policy background provides a perspective for viewing the salient features of economic development. As a rough, perhaps generous statement of magnitude, China's real domestic product may be said to have increased at an average rate of perhaps 5 percent a year between 1952 and 1977, whereas per capita product rose 3 percent. If one takes as plausible an income elasticity of demand for all food of 0.8 for countries like China and a Chinese population growth rate of 2 percent a year, the expected annual increase in demand for food in China would have been on the order of 4.4 percent (0.8×3 percent + 2 percent) under the further plausible assumption of a relatively stable population age structure during the period. Yet the People's Republic managed with a food output growth that did little more than keep pace with population growth, and without massive imports or food crises. It is clear that much of the rise in per capita output was kept by policy from reaching the households, farm and nonfarm, through higher personal incomes.[9] Where some trickling down materialized, continuing rationing helped maintain the basic food balance.

CHINA'S RECORD IN AGRICULTURE, 1952–77

Starting with an output of 154.4 million metric tons in 1952, total grain output in the People's Republic of China reached 270 million metric tons in 1977, an increase of 75 percent (table 1). The average annual rate of increase of 2.3 percent exceeded the average population growth rate of 2.1 percent.[10]

TABLE 1
QUANTITY AND VALUE OF OUTPUT OF GRAINS, SOYBEANS, AND COTTON, 1952–77

Year	Rice (million metric tons)	Wheat (million metric tons)	Coarse Grains[a] (million metric tons)	Potatoes (Grain Equivalent)[b] (million metric tons)	Total Grain Quantity[c] (million metric tons)	Total Grain Value[d] (billion yuan)	Soybeans Quantity (million metric tons)	Soybeans Value[d] (billion yuan)	Cotton Quantity (million metric tons)	Cotton Value[d] (billion yuan)
1952	68.45	18.10	51.50	16.35	154.40	18.95	9.52	1.33	1.30	2.21
1953	71.25	18.30	50.70	16.65	156.90	19.23	9.93	1.39	1.17	1.99
1954	70.85	23.35	49.25	17.00	160.45	19.88	9.08	1.27	1.06	1.81
1955	78.00	22.95	54.95	18.90	174.80	21.56	9.12	1.28	1.52	2.58
1956	82.45	24.80	53.40	21.85	182.50	22.61	10.23	1.43	1.44	2.45
1957	86.80	23.65	52.65	21.90	185.00	22.81	10.04	1.41	1.64	2.78
1958	93.00	25.00	52.00	30.00	200.00	24.70	10.50	1.47	1.60	2.72
1959	79.00	24.00	41.00	21.00	165.00	20.34	11.50	1.61	1.35	2.29
1960	73.00	21.00	36.00	20.00	150.00	18.50	8.20	1.15	0.90	1.53
1961	78.00	16.00	44.00	24.00	162.00	19.97	7.90	1.11	0.89	1.51
1962	78.00	20.00	53.00	23.00	174.00	21.45	7.70	1.08	1.00	1.70
1963	80.00	22.00	56.00	25.00	183.00	22.56	7.04	0.99	1.10	1.87
1964	90.00	25.00	59.00	26.00	200.00	24.66	6.94	0.97	1.50	2.55
1965	90.00	25.00	60.00	25.00	200.00	24.66	6.84	0.96	1.65	2.80
1966	96.00	28.00	66.00	25.00	215.00	26.51	6.80	0.95	1.81	3.07
1967	100.00	28.00	76.00	26.00	230.00	28.36	6.95	0.97	1.94	3.29

1968	95.00	25.00	70.00	25.00	215.00	26.51	6.48	0.91	1.81	3.07
1969	99.00	27.00	69.00	25.00	220.00	27.13	6.20	0.87	1.77	3.01
1970	110.00	31.00	75.00	24.00	240.00	29.59	6.90	0.97	2.00	3.40
1971	117.00	31.00	75.00	23.00	246.00	30.33	7.90	1.11	2.32	3.77
1972	112.00	36.00	69.00	23.00	240.00	29.59	8.70	1.21	2.13	3.62
1973	118.00	35.00	73.00	24.00	250.00	30.82	10.00	1.40	2.55	4.33
1974	127.50	38.00	74.50	25.00	265.00	32.67	9.50	1.33	2.50	4.24
1975	126.50	41.00	77.50	25.00	270.00	33.29	10.00	1.40	2.40	4.08
1976	125.50	45.00	76.50	25.00	272.00	33.54	9.00	1.26	2.35	3.99
1977	126.50	40.50	76.50	26.50	270.00	33.29	9.50	1.33	2.20	3.74

SOURCES: The 1952–57 figures for all categories except soybeans were taken from the People's Republic of China, State Statistical Bureau, *Ten Great Years* (Peking: Foreign Languages Press, 1960); the 1958–77 figures are from U.S. Department of Agriculture, Economics, Statistics, and Cooperatives Service, *People's Republic of China Agricultural Situation: Review of 1977 and Outlook for 1978* (Washington, D. C.: USDA, 1978). The 1952–77 figures for soybeans are from People's Republic of China, Ministry of Agriculture, Bureau of Planning, *A Collection of Statistical Data on Agricultural Production of China and Other Major Countries* (Peking: Agricultural Publishing House, 1958); the 1958 figures are from State Statistical Bureau, *Ten Great Years*; the 1959 figures are from *Jen-min Jih-pao*, 22 January 1960.

NOTES: Values are in constant 1952 prices. Additional notes on this table appear in Anthony M. Tang, *Food and Agriculture in China: Trends and Projections, 1952–77 and 2000*, in *Food Production in the People's Republic of China*, by Tang and Bruce Stone, IFPRI Research Report 15 (Washington, D.C., May 1980), 73, hereafter cited as Tang 1980.

^{a}Coarse grains include millet, corn, kiaoliang, barley, buckwheat, oats, proso-millet, small beans, green beans, broad beans, and peas.

^{b}Potatoes were converted at a ratio of four metric tons of potatoes to one metric ton of grain.

^{c}Total grain does not include soybeans.

^{d}The 1952–57 figures were calculated with 1952 prices developed in Ta-Chung Liu and Kung-Chia Yeh, *The Economy of the Chinese Mainland: National Income and Economic Development, 1933–59* (Princeton: Princeton University Press, 1965).

TABLE 2
VALUE OF AGRICULTURAL OUTPUT AND VALUE ADDED BY AGRICULTURE, 1952–77
(IN BILLIONS OF 1952 YUAN)

Year	Value of Livestock Produced, Grain, Soybeans, and Cotton	Gross Value of Agricultural Output[a]			Insecticides and Miscellaneous Inputs	Chemical Fertilizers	Cake Fertilizers	Value Added by Agriculture[b]
		Pre-1957 Coverage	Post-1957 Coverage	Adjusted Post-1957 Coverage				
1952	27.47	48.39	40.05	37.76	0.070	0.169	0.690	36.83
1953	27.78	49.91	41.31	38.93	0.088	0.253	0.719	37.87
1954	27.86	51.57	42.69	40.26	0.191	0.354	0.659	39.06
1955	29.82	55.54	45.97	43.54	0.313	0.519	0.662	42.05
1956	31.37	58.29	48.25	45.71	0.742	0.659	0.741	43.57
1957	33.06	60.35	49.95	47.21	0.695	0.892	0.728	45.07
1958	36.32	67.27	55.68	52.86	1.148	1.041	0.761	49.91
1959	29.82	54.29	44.94	42.32	1.521	1.219	0.833	38.75
1960	25.72	46.11	38.17	35.76	1.890	1.197	0.595	32.08
1961	26.92	48.50	40.14	37.83	1.195	1.288	0.573	34.77
1962	29.62	53.89	44.61	42.21	1.321	1.650	0.558	38.68

Year								
1963	32.15	58.94	48.79	46.22	1.456	2.211	0.511	42.04
1964	34.73	64.09	53.05	50.34	1.610	3.043	0.503	45.18
1965	36.72	68.06	56.34	53.42	1.773	3.854	0.496	47.30
1966	38.88	72.38	59.91	56.92	1.960	4.616	0.493	49.85
1967	41.07	76.75	63.53	60.49	2.165	5.206	0.505	52.61
1968	38.83	72.28	59.83	56.82	2.389	5.773	0.471	48.19
1969	39.52	73.65	60.96	57.91	2.637	6.698	0.450	48.17
1970	43.39	81.47	67.44	54.32	2.912	7.742	0.500	53.17
1971	45.55	82.81	68.54	65.27	3.215	8.928	0.573	52.55
1972	46.02	86.63	71.71	68.26	3.551	10.248	0.631	53.83
1973	46.59	91.25	75.53	72.17	4.009	10.942	0.725	56.49
1974	49.55	93.68	77.54	74.04	4.527	11.646	0.688	57.18
1975	50.61	95.79	79.29	75.67	5.110	12.208	0.725	57.63
1976	50.71	95.99	79.45	75.77	5.768	11.895	0.653	57.45
1977	50.33	95.23	78.82	75.09	6.510	15.555	0.690	52.33

NOTE: For detailed explanation of data sources and estimation procedures see Tang 1980, 68–73.

[a] According to the State Statistical Bureau (*Ten Great Years*, 16), pre-1957 coverage includes "handicrafts consumed at the rural source of production and preliminary processing of agricultural products." This coverage is estimated by us to be 20.8 percent broader than the post-1957 coverage. The "adjusted" coverage removes some of the double counting implicit in Chinese gross output value aggregates by deducting the value of feed and seed from the total.

[b] Value added by agriculture (or income originating in agriculture) is calculated by deducting the value of insecticides and other inputs, chemical fertilizers, and cake fertilizers from "adjusted" gross value of output.

Gross value of agricultural output (GVAO) rose from 40.05 billion yuan (in constant 1952 prices) to 78.82 during 1952–77, a jump of 97 percent, or 2.9 percent per year (table 2). Unlike other agricultures characterized by dynamic technological and productivity advances, growth in inputs outpaced growth in output, rising 131 percent. As a consequence, total factor productivity dropped 15 percent, or 0.6 percent per year. The increase in resource use was particularly noteworthy for inputs purchased from outside agriculture: power equipment and machinery and modern current inputs such as fuel and electric power, chemical fertilizers, and chemicals for disease and insect control. Modernization in this context refers not only to the provision of inputs from the outside that are qualitatively new and different from the traditional varieties but also to the transfer of traditional agricultural activities and processes of an intermediate character from agriculture to the outside. Thus not only did mechanical power replace draft animals but the feed (or equivalent farm output) saved became an end product of the sector, either directly or indirectly when used to produce livestock products. Under the Chinese-Soviet definition of gross output value, grain imports when embodied in livestock products also add in a round-about way to gross value of agricultural output. Because of such double counting, the Chinese-Soviet–type GVAO becomes increasingly misleading if taken as a measure of the sector's contribution to the national economy. Part of the difference between the higher growth rates for GVAO and for grain output is attributable to double counting.

This is not to minimize the critical importance of modern nonfarm-supplied inputs in agricultural development. Two aspects may be noted. First, modern inputs underpin the generation of new, low-cost income streams in agriculture. Second, they ease the two principal constraints operating in agriculture—time and space as imposed by nature—by permitting transfers of agricultural activities to an industry not so constrained.[11] In other words, the final output of the sector is the joint product of internal and external inputs. As such, the problem does not lend itself to neat accounting solutions.

With this caveat in mind, the value added by agriculture (VABA) as an indicator of the sector's contribution to the gross domestic product of the People's Republic is used. As defined here, VABA is gross of depreciation charges on farm machinery, equipment, and service buildings. VABA increased by only 42 percent, as compared with 97 percent for GVAO, or by an annual rate of 1.7 percent (table 2). In net terms (not attempted in this study), the pace of increase in sector contribution to the Chinese national economy would be slower still. The modest rate of increase in agriculture's contribution has meant a stagnant or declining value added per worker. Value added is traceable to the employment of so-called primary factors of production: labor, land, and capital. These factors as an aggregate rose by 62 percent during 1952–77, as compared with a 42 percent increase in VABA, leading to a primary factor productivity decline of 12 percent.

The Chinese productivity under the Communist Party was none too impressive. Nor was the sector's capacity to generate value added. Agriculture managed to deliver end products in quantities large enough and rising fast enough to meet the elemental needs of China's expanding population, not through finesse, but by a massive injection of resources, particularly those of external origin. This record contrasts sharply with the agricultural productivity gains of Japan, South Korea, and Taiwan.[12]

It is our belief that the command economic system and the Chinese development strategy tend to undermine producer-worker incentive, discourage cost-reducing innovations, distort economic signals, and exaggerate aversions toward risk. The bureaucratic controls needed to ensure compliance with the central design mean irresponsiveness, rigidity, and delay. In the larger (power-oriented) Communist context, decentralization cannot be total. When attempted as a localized partial scheme giving local responsibility without commensurate power to command resources or to alter mixes for greater efficiency, it often destabilizes the system. Decentralization and recentralization and other forms of experimentation come and go in Soviet-type economies without notable success. This is not to say that such a system is about to break down. In fact, the central virtue of the system from the standpoint of Communist leadership values and goals is its ability to bring about larger savings ratios than are generally attainable in market economies and to concentrate resources thus mobilized for the central purposes of the state. Given high savings ratios, growth is going to be vigorous despite difficulty in resource allocation and management. It will be especially so in high-priority heavy industries, where output is more standardized and workers and management are more easily monitored. Difficulties are much greater in consumer goods industries, where the problems of a bureaucracy mediating between millions of consumers and thousands of state enterprises become intractable. And the more affluent the consumers, the greater are the difficulties for the planners, as quality, diversity, assortment, and after-sales service become increasingly more critical.[13]

The planner's problems are compounded in agriculture. The producers are more numerous, the spatial dimension looms large, and parameters are location-specific. The human factor takes on singular importance at the local level. Organizational and incentive issues require a delicate balancing between private and collective activities, compulsory and free marketing, and agricultural and nonagricultural production.

In farming, the basic clash between the household and the planner tends to be at its sharpest. The fact that the Chinese household has seen little increase in income and consumption since the 1950s does not make the planner's task any easier. The Chinese worker, in or out of agriculture, rates low in effort compared with his East Asian counterparts. The Chinese enterprise manager does not perform any better. These problems have been generally observed in all the Marxist-socialist economies modeled after the Soviet command system.[14] Against this background, caution may be well advised in projecting the efficacy of China's post-Mao policies under Deng.

AGRICULTURAL POLICY UNDER DENG XIAOPING

The new leaders in Peking have set out in earnest to "de-Maoify" the Chinese economy. Higher worker incomes are displacing normative appeals and coercive pressures as ways of inducing greater effort and productivity. The first significant wage increase since the 1950s was granted in the fall of 1977 to a broad segment of nonsupervisory, noncadre workers.[15] Deng and his followers are certain that this is the surer and quicker way to the power goal set for the year 2000. Note that the worth of the new policy turns on its ability to raise manager and worker productivity enough to more than offset the increases in wages and salaries. How this evolves depends critically on the new regime's ability to expand consumption opportunities.

While Peking will no doubt continue to rely on high pricing (to soak up purchasing power) of sewing machines, bicycles, watches, cameras, and, most recently, television sets, it is difficult to see how it can avoid providing more agricultural products. Moreover, since direct grain consumption is already adequate, especially in cities and farm areas with relatively high incomes, increased food consumption will consist primarily of meats, fruits, and what the Chinese call subsidiary foods—all of which use much more land than grains for equivalent caloric production. Shigeru Ishikawa estimates that for China the resource cost in terms of scarce land required for equivalent production is 3.3 times higher for meats than for foodgrains and 10 times higher for vegetable oils.[16] Thus the prospects of solving the supply side of the equation under the new economic policy are not reassuring because of the scarcity of land. Furthermore, the collective form of organization and the Chinese system of farming are better suited for grain production. The problem is further aggravated by rural-urban differences in consumption. Urban areas, according to Ishikawa's estimates, consume 20 percent less grain per capita but two to three times more meats and oils. If Deng's liberalization means an end to the coercive "send down" program and a return to city-oriented industrialization and migration, as seems to be the case, further demand pressure will fall on the grain-centered Chinese agriculture.[17]

Peking's new production goals in the Ten-Year Plan (1976–85) unveiled before the Fifth National People's Congress on 26 February 1978 are ambitious in light of China's production record since 1952, when recovery from 1937–49 wartime dislocations was largely completed.[18] Grain output is to reach 400 million tons by 1985 (43 percent above 1977, or a rise of 4.5 percent per year). Gross value of agricultural output (GVAO) is to increase 4–5 percent per year between 1978 and 1985. The grain output target would require growth twice the 1952–77 rate, and the value output target a rate of growth 45–80 percent higher than the historical average. The targeted increases for grain output apparently reflect a large increase in use of grain for livestock feed. Although China will likely continue to export rice, probably in increasing quantities, in exchange for wheat and coarse grain imports, it is inconceivable that the country will import feedgrain on a scale proportional to that of Japan or Taiwan during the Plan period. This need not rule out (in absolute terms) large feedgrain imports from the United States in the 1980s.

It is reasonable to suppose that the People's Republic, because of its size and development strategy, must depend on domestic production for its basic food balance and can turn to foreign trade for marginal adjustments only. Therefore the production targets for 1985 probably take into account the demand implications of the unfolding wage and consumption policy.

It has been widely reported that several major readjustments in prices received and paid by farmers have been put into effect by the government since the 1950s. These are said to correct the longstanding imbalance against agriculture. As such, it was taken to be an incomes policy. These assumptions were checked by calculating the value-added ratio and per-worker figure in current prices, deflating them by a general consumer price index and comparing them with the constant-price magnitudes. This indicated that the changes of terms of trade made by Peking since the 1950s apparently did little more than maintain the meager incomes of the farm people in China. These alternative price-value calculations, which are being made for a study of the People's Republic agricultural historical record, are still tentative. However, the falling value added per farm worker in constant prices is clear from table 2. The decline would be sharper if noncurrent nonfarm-supplied cost items, such as depreciation of tractors, were deducted in arriving at value-added totals.

The pragmatic appraisal by the present leadership of the role of science, technology, and laws of economics in nation building has been received with widespread approval in and outside China. The new course is seen as foretelling the emergence of a progressive, responsible, and stable China with a clear understanding of the advantages of peaceful, mutually beneficial economic, scientific, and technological interchange with the world community. The road ahead, however, may be neither straight nor smooth. The cost of a crash, outward-oriented modernization program may turn out to be higher than Deng has allowed. And the results may not meet the high expectations. Such a combination of outcomes could again set in motion the familiar pattern of policy reversal and leadership change. Derek Davies, of the Hong Kong–based *Far Eastern Economic Review,* characterized Deng's "New Deal" as "a gamble of enormous proportions."[19]

There are many reasons for uncertainty about Deng's approach, particularly in agriculture. The targets of the Ten-Year Plan for 1985 (until recently revised downward) are 400 million metric tons of grain and 60 million metric tons of steel. Although the agricultural target calls for a smaller percentage growth, official Peking considers it the more difficult.[20] Under Mao's policy, China was effectively shielded from comparison with the accomplishments of other countries or from competing ideas and thoughts that might have cast doubt on the design of the Chinese Communist Party (CCP) for nation building. Considering these circumstances, it is not surprising that the past accomplishments, now seen as slow and disappointing, generated a strong sense of pride and dignity among the Chinese people. When the author visited China in January 1973, after an absence of thirty years, he found an insular, self-satisfied, proud, and confident people. The new open policy will call these attitudes into question. It will bring hordes of foreign visitors and residents and send thousands of Chinese abroad on missions and for training. These changes cannot but benefit China and the world in the long term, but in the shorter term the story may be quite different, particularly in relation to the specific goals of Peking.

Modernization, the centerpiece of the new policy, is in itself destabilizing. If recklessly carried out at the expense of the old policy of "walking on two legs," it may slow rather than hasten growth. Thoughtlessly implemented, it may become equated with change for the sake of change. The elementary fact that the technology that makes economic sense in other countries may not be suitable for China could be overlooked. This is all the more likely since decisions to adopt technology in China are not based on local profitability tests but tend to be made administratively from the center.

The material-incentive component of the new economic policy also will add to Deng's burden. The raises in income beginning in late 1977 without commensurate increases in consumer goods supply have increased inflationary pressure. Use of the material incentive to induce greater productivity inevitably means increased economic disparity in income and consumption because of the unequal capacity of people to respond. The presence of an affluent class and the desire to join it will weaken the inhibitions against corruption and crimes fostered by the egalitarianism and pervasive social control of the recent past.

To the student of economic development, the opening up of an insular society, the demonstration effects, modernization, and the play of new economic incentives are all positive and necessary ingredients for change and progress. Although the process is not costless, no one has seriously argued against it. But the Chinese leaders are primarily interested in making China a superpower no later than the year 2000. As Derek Davies's assessment clearly implies, the verdict as to the efficacy of the new policy in this regard is far from being in. It is also useful to remember that in a command economy the question of corruption and ossification of a massive bureaucracy looms much larger than in more decentralized systems. And an unresponsive, self-centered bureaucracy can make the best-laid plans go astray.

The people's perception of the system will be a key to the success or failure of the new policy. Their perception in turn will depend on the relationship of performance to expectations. The newly announced goals for 1985 and for the end of the century, the new openness, the "four modernization" campaign, and the reintroduction of material incentive all tend to raise the expectations of the people and to lower their estimate of China's achievements. This may be compounded by cynicism resulting from past reversals in government policy of which Deng Xiaoping, the twice-resurrected new architect, is a living reminder. In time, prudence, skepticism, and formal compliance (known as simulation in the Soviet Union and Eastern Europe) could replace enthusiasm and commitment as dominant attitudes.

Modernization and liberalization for agriculture need not mean an end to scarcity, if the Soviet experience is any guide. Despite local reorganization, liberalization, and decentralization, the major decisions affecting agriculture are centrally conceived and directed by a vast bureaucracy. Thus the efficacy of price will depend on decisions of central planners.

The values and goals that have served the CCP since the framing of its First Five-Year Plan are not likely to undergo major changes in the years ahead. Consequently agriculture will remain in difficult circumstances. Its role will resemble that exhibited by the familiar two-sector model of the Lewis-Fei-Ranis type, but exacerbated by the CCP's commitment to the "Principle of Maximum-Speed Selective Growth under Austerity."[21]

If the material incentive provided by the new economic policy is to be effective, supplies of consumer goods must be increased. Otherwise higher wages and salaries can only lead to inflation and the frustration of queues and depleted ration coupon books. Since much of the increase in supplies must be filled by agriculture, it is reassuring that the new agricultural growth targets were both prominent and explicit in the Party's economic message to the Fifth National People's Congress in February 1978. As was suggested earlier, the 1985 targets may be plausibly taken to indicate the food demand implications of the new economic policy that the new Ten-Year Plan (1976–85) is to satisfy.

POSTSCRIPT

The study from which this overview is drawn was completed in 1979, with minor updating and revisions undertaken in the spring of 1980. Since that time many changes have taken place in China. This postscript reviews the salient changes and relates them to the earlier analysis and conclusions. There are two main concerns. First, do the statistics subsequently released by Peking alter China's agricultural performance record as we have earlier described it? Second, both the historical record and the future outlook of China's agriculture depend critically on the economic development strategy fashioned by the political leaders and on the requirements and constraints this strategy places on agriculture. Does the repudiation of Mao's economic policy suggest ascendancy of a different development strategy or merely a change in approach?

EVALUATION OF THE NEW STATISTICS

The evaluation of China's historical record in agriculture presented above is based on the author's empirical estimates, built up from available official data for 1952–77 (mostly fragments since 1958). These are time-series estimates of input-output data, organized in a manner suitable for growth accounting. Analysis of these data leads to the conclusion that while China's agricultural output growth, estimated at about 3 percent per year, was creditable, its cost has been inordinately high compared with other East Asian experiences.

The estimated time series have been updated and revised through 1980 in light of new releases by the State Statistical Bureau.[22] In 1981–82 there appeared several official Chinese publications in the style of yearbooks. Of special note is the complete index of "total value of agricultural output" for 1949–79. On the input side, there are complete time series on livestock numbers and certain types of farm machinery, as well as scattered new information on the agricultural labor force, sown and cultivated areas, and some modern inputs. The official output value series appears to be readily reconcilable with the author's estimates used in the analysis above after noting that (1) the official series includes (since 1977) the output of "brigade-run industries," whose growth tends to outstrip agricultural output growth; and (2) the author's estimated series is in constant 1952 prices, while the official series is a chain index across three subperiods, each with its own constant price base (1952, 1957, and 1970, respectively). The (incomplete) Chinese input series for the most part show larger increases in input use during the period than do the estimated series. Thus, the cost implications of the growth in agricultural output, the falling total factor productivity, the sharp decline in value-added ratios, and the noteworthy drop in value added per worker (in constant 1952 prices), as well as the unfavorable farm income picture over time (despite several relative price adjustments by Peking in favor of agriculture), would all seem to stand as earlier sketched.

CHANGES IN ECONOMIC POLICY

The "democracy walls" and other forms of liberalization in speech, art, and literature that flourished in 1978–79 have been replaced by a return to repression. Deng's economic liberalization and reform, however, continue unchecked, tending to place unprecedented emphasis on material incentives and household consumption, decentralization in economic decision making, and external contacts. Fundamental readjustment has been pushed to reduce accumulated price distortions and, more important, to restructure an economy beset with mounting bottlenecks. The restructuring seemingly reversed the established order of priority for agriculture, light industry, and heavy industry. Special economic zones have been set up which economically and administratively amount to states within the state and go far beyond the exceptions that other countries grant to their special zones, usually export-processing zones. The purpose of the Chinese zones, generally located adjacent or close to Hong Kong and Macao, is to encourage direct foreign investment by promulgating special regulations and legal codes so far-reaching as to make the zones virtually autonomous entities. For agriculture, Deng has taken contracting (a euphemism for quota-fixing) and accounting down to the household level, although collective ownership of land is retained subject to "private right of use." The relationship between the prices paid and the prices received by farmers has been restructured in favor of agriculture. Rigid restrictions on output mix have been eased, and the scope of free markets for agriculture has been enlarged and their legal standing affirmed.[23]

Compared with the insular, austere, and rigidly controlled system under Mao, the new Chinese economic setting invites the question whether the post-Mao leadership has embraced a basically new development strategy. If so, a favorable break in the agricultural trends as sketched in this chapter may emerge, and the chapter's conclusions would be subject to modifications accordingly. As argued at length in the author's forthcoming work cited earlier,[24] a judicious view is that, the weight of the changes notwithstanding, they do not seem to suggest an end to the old value and goal imperatives. The power goal for the year 2000 still remains. Readjustment and easing of bottlenecks need not imply a reordering of priorities in the context of the development strategy.

In fact, these acts are best seen as setting the stage for a renewed push for heavy industrialization. Already the Baoshan steel complex, well-known because of its size and sudden construction halt, is back on track. As bottlenecks stemming from agriculture, transport, and power (especially electric power) are eased, the priority sector, heavy industry, will once again assert itself. Nor does Peking's recent declaration that China intends to regain sovereignty over Hong Kong suggest diminished importance of the territories' foreign-exchange contributions to Peking. The declaration was accompanied by a notable statement that Peking will take the necessary steps to ensure Hong Kong's "continued stability and prosperity," presumably by conferring some kind of special autonomous status to take effect upon the termination of the lease in 1997, which affects 90 percent of the territories. It seems clear to this writer that economic growth considerations continue to enjoy primacy in Peking's nation-building design, much as they did under Mao, attesting to the continued validity of the development strategy as sketched above.

NOTES

1. Bergson referred to the Stalinist strategy of economic development that was embraced by Peking's leaders. For a fuller development of the model see my earlier writings: "Policy and Performance in Agriculture," in *Economic Trends in Communist China,* ed. Walter Galenson, Ta-Chung Liu, and Alexander Eckstein (Chicago: Aldine, 1968), 459–509; "Agriculture in the Industrialization of Communist China and the Soviet Union," *Journal of Farm Economics* 49 (5):1118–34; and "Organization and Performance in Chinese Agriculture" (Paper presented at the joint session of the AEA, AAEA, and ACES on the Economic Organization of Socialist Agriculture at the annual meetings of Allied Social Science Associations, 29 December 1975, Dallas, Texas), 1–21.

2. See Tang, "Policy and Performance in Agriculture." For a Chinese statement on the agricultural bottleneck see "Second Shanghai Talk on Mao's Major Relationships," 25 February 1977, Foreign Broadcast Information Service, *People's Republic of China: Daily Report,* E6. Recent statements from Peking remind us of how dependent China's industrialization is on agriculture. In a special feature article on disproportions in the national economy in *Beijing Review,* 29 June 1979, 15, Shi Zhengwen pointed out that agriculture contributes about 85 percent of the people's means of subsistence, 70 percent of the raw materials needed by light industry, 40 percent of raw materials needed by all industry, and "a considerably large part of our financial revenue."

3. Tang, "Policy and Performance in Agriculture," 466–80.

4. *Beijing Review,* 3 August 1979, 10.

5. A more radical but perhaps the best-articulated version is by John G. Gurley: see his *China's Economy and the Maoist Strategy* (New York: Monthly Review Press, 1976).

6. This is a judgmental estimate. The popularly cited media figure of U.S. $8 billion in the fall of

1982 and early 1983 turned out to be a high "estimate" without foundation. Hong Kong is useful to the PRC in other ways: it is the third most important financial center in the world (after New York and London), where Communist-owned banks operating freely account for more than 20 percent of total bank deposits; it is a pool of Western expertise; and it is a near-Friedmanian free market system allowing the PRC to reap considerable benefit offered by an unfettered capitalist market system at its doorstep without having to suffer the system's "undesirable fallout."

7. For seminal writings on the Chinese policy cycle see G. William Skinner and Edwin A. Winckler, "Compliance Succession in Rural Communist China: A Cyclical Theory,"in *A Sociological Reader on Complex Organizations,* ed. Amitai Etzioni, 2d ed. (New York: Holt, Rinehart and Winston, 1969), 410–38; and Alexander Eckstein, *China's Economic Development* (Ann Arbor: University of Michigan Press, 1975), 311–22 and 332–38. Eckstein's excellent work on China is particularly sensitive to the clashes between planner and household, between resource mobilization and economic efficiency (incentive), and, in his language, between the Communist Man and the Economic Man. These clashes create policy dilemmas and policy cycles.

8. See Lardy, chapter 29 in this volume.—ED.

9. See below for Peking's 1979 release of per capita income data.

10. Average growth rates presented in this study are all compounded rates estimated from the standard exponential function fitted to the time series shown in the tables. The population series by Aird appears in Anthony M. Tang, *Food and Agriculture in China: Trends and Projections, 1952–77 and 2000* (1980), in *Food Production in the People's Republic of China,* by Tang and Bruce Stone, IFPRI Research Report 15 (Washington, D.C., May 1980), 60.

11. For a fascinating discussion of the differences between agriculture and industry see Nicholas Georgescu-Roegen, "Process in Farming versus Process in Manufacturing," in *Economic Problems of Agriculture in Industrial Societies: Proceedings of the International Economic Association, Rome, 1965,* ed. Ugo Papi and Charles Nunn (London: Macmillan, 1969), 497–528.

12. For empirical evidence on productivity growth in these countries see Vernon W. Ruttan and Yujiro Hayami, eds., *Agricultural Growth in Japan, Taiwan, Korea and the Philippines* (Honolulu: University of Hawaii Press, 1979), and Anthony M. Tang, "Research and Education in Japanese Agricultural Development," *Riron Keizai Gaku* (Economic Studies Quarterly), February 1963, 27–41; May 1963, 91–99.

13. The literature on the strong and weak points of the Soviet-type command economies is vast. A good, concise introduction is Robert W. Campbell's *The Soviet-type Economies: Performance and Evolution,* 3d ed. (Boston: Houghton-Mifflin, 1974). For an authoritative summary on conditions tending to reduce efficiency of centralized planning see Bela Balassa, "Proposals for Economic Planning in Portugal," *Economica* 43 (May 1976): 117–24. Balassa discusses the need for simple, overriding goals for effective centralized planning and the difficulties created by rising affluence, by increasing sophistication of the economy, and by the mixture of ownership and control of the means of production as is found in Chinese agriculture (119–23).

14. Balassa's recent work ("Proposals for Economic Planning") sheds considerable light, consistent with this generalization, on comparative growth performance between socialist and nonsocialist economies classified by level of development.

15. This is being followed by other income adjustments, planned or already implemented, through further wage-price reform and reinstatement of bonus payment. Vice-Premier Yu Qiuli revealed (*Beijing Review,* 29 June 1979, 11) that the average annual wages of workers and staff in state enterprises rose from 602 yuan in 1977 to 644 yuan in 1978, while peasant collective income rose 13.7 percent. Planned increases in wages and incomes for 1979 are 7 billion yuan for urban families and 13 billion for rural families (the latter amount reflecting not only an improved price relationship but anticipated increases in farm output).

16. See Shigeru Ishikawa, "China's Food and Agriculture: A Turning Point," *Food Policy* 2 (May 1977): 90–102.

17. Ibid.

18. Pre-1937 norms suggest that in 1952, output and the stocks of the means of production were still some 6 percent lower than they would have been had the war not hampered development. [For

more details on the growth record of Chinese agriculture from 1952 to 1979 see Lardy, chapter 29 in this volume.—ED.]

19. See Derek Davies, "Putting People in the Picture," *Far Eastern Economic Review*, 6 July 1979, 13. See also his fuller analysis in "China: Rebirth of the Individual," ibid., 17 August 1979, 51–56.

20. *Beijing Review*, 20 October 1978, 8.

21. For a theoretical exposition of the general principle of maximum-speed development (without selectivity, which was our adaptation to fit the Chinese strategy) see John C. H. Fei and Alpha C. Chiang, "Maximum-Speed Development through Austerity," and Anthony M. Tang, "Comment," in *The Theory and Design of Economic Development*, ed. Irma Adelman and Erik Thorbecke (Baltimore: Johns Hopkins Press, 1966), 67–99.

22. See Anthony M. Tang, "Chinese Agriculture: Its Problems and Prospects," in *China: The 1980's Era*, ed. Norton Ginsberg and Bernard A. Lalor (Boulder: Westview Press, in press).

23. See Lardy, chapter 29 in this volume.—ED.

24. Tang, "Chinese Agriculture."

29

Prices, Markets, and the Chinese Peasant

Since the Communist party rose to power in 1949 per capita economic output in China has accelerated rapidly. This is in marked
contrast to the first half of the twentieth century, when there was no
evidence of sustained per capita economic growth. The period of
rapid recovery from war and civil war was largely complete by the end of 1952,
and per capita output (measured in constant prices) tripled between 1952 and the
late 1970s. Although aggregate output accelerated, agricultural output per capita
rose by less than one-half of 1 percent per year between 1952 and the late 1970s.
Industrial output, by contrast, grew in excess of 11 percent annually, 9 percent in
per capita terms.[1] This chapter attempts to explain the unusual disparity between
the rates of agricultural and industrial growth up to the late 1970s, and ventures a
preliminary assessment of the changes in economic policy since 1978.

AGRICULTURE IN HISTORICAL PERSPECTIVE

In his remarkable study *Agricultural Development in China, 1368–1968*,
Dwight Perkins explains how Chinese agriculture supported, at what appears to
be a constant level of per capita output, a population that grew from 65–80
million at the beginning of the Ming dynasty in 1368 to 540 million by the
middle of the twentieth century.[2] His research suggests that only about half of
that population increase was supported by simply expanding the crop area.
Although agriculture was still traditional by the usual criteria—that farmers'
knowledge about the factors they used had not changed in generations and net per
capita savings was low[3]—Chinese agriculture had become increasingly complex
and sophisticated.

Increased sophistication of Chinese agriculture was evident primarily in the
evolution of cropping and marketing patterns rather than in changes in mechanical technology.[4] Improved cropping patterns were stimulated by the introduc-

NICHOLAS R. LARDY is associate professor in the Henry M. Jackson School of International Studies,
University of Washington.

tion of early-maturing rice varieties from central Indochina during the Sung dynasty (960–1126).[5] Initially planted in limited areas in the southeast, within two centuries these varieties were grown widely in Chekiang, southern Kiangsu, and Kiangsi (see map), and by the Ming dynasty (1368–1643) they had diffused to large parts of the southwest and to central China, especially the present-day provinces of Hupeh and Hunan.[6] The introduction of new crops such as corn, peanuts, Irish potatoes, tobacco, and cotton also contributed to agricultural growth and changes in patterns of regional settlement. The cultivation of new crops was not distributed randomly across China. Rather, innovative Chinese peasants tended to adopt the cultivation of these crops in regions of comparative advantage.

The process of increasing specialization, which contributed to growing productivity per unit of land, was dependent on increased marketing, not only locally but interregionally as well. Beginning in the ninth century the subsistence basis of the rural economy was replaced by an increasingly complex marketing system. Periodic market sites developed into market towns, linking intrarural marketing with interregional markets.[7] The development of interregional markets in rice and other staple commodities began as early as the T'ang dynasty (618–906) and accelerated in the twelfth and thirteenth centuries as increased cultivation of sugar cane and other cash crops on the southeast coast was facilitated by shipments of rice and other staples from surplus regions in the Yangtse Delta, the Canton Delta, and the central Kan River Valley.[8] It appears that even prior to the Ming dynasty the great southward migration of population had led to population densities on the southeast coast far in excess of those that could have been supported on the basis of rice cultivation alone. The population densities were sustained on the basis of cash crops that were traded for rice from surplus grain regions.

Although the regions of surplus cereal production shifted over time, interregional trade increased. By the first half of the seventeenth century the lower Yangtse region itself had become grain-deficit because of increased allocation of land to the production of noncereal crops and a rising rate of urbanization. The lower Yangtse had come to depend on rice supplied from Kiangsi, Anhwei, and especially present-day Hunan and Hupeh, where rice production had increased substantially because of the introduction of early-maturing varieties.[9]

In the late seventeenth century Hunan and Hupeh had become the rice bowl of the Chinese empire. Their rice surpluses flowed down the Yangtse River to supply cereal-deficit regions in the lower Yangtse basin. Part of that supply moved north to the Ch'ing capital at Peking via the Grand Canal. Hunan also shipped somewhat smaller quantities to the southeast coastal provinces of Fukien and Kwangtung and to western regions such as Shensi and possibly to Kweichow in the southwest. Chuan Han-sheng and Richard Kraus estimate that rice exports from Hunan and Hupeh ranged from 500,000 to 750,000 metric tons during the first half of the eighteenth century.[10]

Another major example of interregional trade was the grain-cotton trade be-

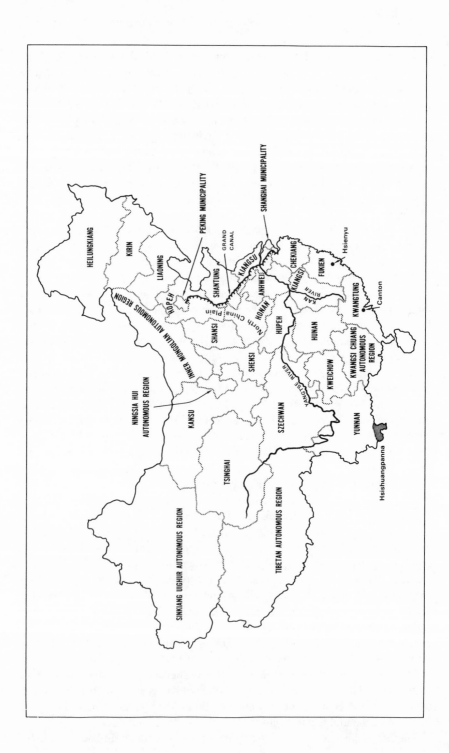

tween north and central China. By the seventeenth and eighteenth centuries textile production in the lower Yangtse Valley was dependent on raw cotton produced in Shantung, Honan, and present-day Hopeh.[11] These regions in north China, in turn, depended for part of their food on cereals that originated in the middle Yangtse region, flowed downriver, and then were shipped north via the Grand Canal or coastal routes from the lower Yangtse. The lower Yangtse, in turn, supplied processed goods, handicrafts, and some luxury items to the middle Yangtse region. Thus a triangular trade based on relatively inexpensive water transport emerged, facilitating productivity growth. The expansion of specialized production based on comparative advantage and interregional trade in agricultural products was stimulated further by railroad development after 1895, particularly in north and northeast China, where water transport was limited.

Thus, between the fourteenth and early twentieth centuries, Chinese peasants responded positively to the availability of improved seed varieties, new crops, improved transport systems, and rising urban demand. In most respects peasant behavior appears consistent with the view advanced by T. W. Schultz and others that peasants respond efficiently to new production and marketing opportunities.[12] For reasons that are not well understood, however, the increased physical productivity of Chinese farming and absorption of new crops did not lead to a growth of per capita incomes; "on the contrary, growth was all demographic."[13] Thus, by the mid-twentieth century, arable land per agricultural worker was only six-tenths of a hectare, substantially less than the levels in other developing countries in East Asia, Southeast Asia, and South Asia and of course much less than at the onset of modern economic growth in Japan or Western Europe.[14]

DEVELOPMENT POLICY SINCE 1949

The thesis of this chapter is that persistent undervaluation of agriculture by the Chinese Communist party, not increased population growth, was the major reason for the relatively slow growth of agriculture after 1949. Undervaluation of agriculture is reflected in policies that on balance have reduced peasant income-earning opportunities, inhibited efficient resource allocation within agriculture, and depressed agricultural investment.

PRICE AND MARKETING ARRANGEMENTS

With a few exceptions, agricultural price policies and marketing arrangements in China have undervalued agriculture since 1953. Between 1949 and 1953 the state met the growing demand for agricultural raw materials to be used in state manufacturing establishments by purchasing commodities on rural periodic markets. State influence on the production decisions of peasants during this time was indirect and depended on peasant responsiveness to changes in relative prices. Throughout this period, however, meeting demands for agricultural products for industry and, increasingly, for urban consumption became more expensive because the terms of trade between the state and peasantry were turning in favor of the peasantry. As a result, the state introduced the producer levy system, a system of compulsory deliveries at fixed prices, in the autumn of 1953.

In theory, the average price received by peasants in a system combining forced deliveries to the government at low prices and an unrestricted market for the remainder of peasant surpluses does not need to be less than the price that would be received in a single free market with no government intervention.[15] The average price received under a system with delivery quotas is simply the weighted average of the price received for sales to the state and the price received in the open market for the residual of peasant marketings. In the short run, if the government separates the market for grain into high- and low-income consumers and supplies the grain purchased at below-market prices to low-income consumers, while the rich compete for the reduced supply on an open market, the average price received by producers will be higher than that under a free market with no government intervention. This result depends only on the (absolute) price elasticity of demand of the rich being less than that of the poor.[16] This rationale has been advanced in favor of the Indian system of producer levies. Its attraction is the ability to use the price mechanism to transfer income from the urban rich to the urban poor, and perhaps even to the peasantry, without imposing a negative incentive effect on producers.

In practice, the conditions necessary for this outcome have never been fulfilled in China.[17] Within a year of introducing the producer levy system, the government initiated a system of coupon rationing to distribute state-procured cereals and edible vegetable oils to the urban population. By 1955 that system was extended to the entire urban population. Since the ration quantities were relatively generous and market prices were higher than ration prices, the volume of market sales by peasants directly to urban consumers was reduced substantially. Moreover, prior to the imposition of the producer levy system, the state converted traditional grain markets in cities and in market towns to state grain markets, where prices were subject to state control. The only remaining outlets for cereals were informal markets in the countryside. But even here sales were limited, and private millers could no longer purchase grain in these markets for ultimate resale to market-town or city dwellers. Private traders and processors came under the administrative jurisdiction of the Ministry of Food and its local agencies. All of these restrictions were necessary to ensure that grain taxes, paid in kind, and delivery quotas, at fixed prices, were fulfilled. Consequently, the volume of grain transactions in rural markets fell substantially, from about 7–8 million metric tons in the early 1950s to 2–3 million metric tons by 1954/55. Restrictions on cotton, which was also subject to delivery quotas, were more severe. Cotton growers were prohibited from privately selling raw cotton, handicraft yarn, or homemade cloth.

Since the imposition of the producer levy system in 1953, rural markets have been liberalized during three periods: from the middle of 1956 through the fall of

1957, from 1961 or 1962 through 1965, and since 1979. In the first two periods the emergence of the opportunity to sell products privately raised the government's procurement costs under the producer levy system, and the scope of the markets subsequently was curtailed. Thus for most of the years since 1953, government procurement policy has tended to lower the prices received by farmers, reducing production incentives. Moreover, the curtailment of local marketing reduced not only the prices peasants received but also their opportunities to earn income from the sale of commodities not subject to state producer levies.

COMPARATIVE ADVANTAGE, INTERREGIONAL TRADE, AND PRODUCTIVITY GROWTH

While constraints on local marketing imposed efficiency losses because of the inhibition of specialization within local marketing regions, curtailment of interregional trade was far more serious, since it interrupted the patterns of regional specialization that were such an important source of agricultural growth during the Ming and Ch'ing dynasties. The suppression of interregional trade was not much in evidence prior to the Great Leap Forward (1958–60). Indeed, during the early 1950s and continuing through the First Plan (1952–57), state policy facilitated trade and specialization. The interprovincial flow of cereals recovered rapidly in the early 1950s, reaching a peak of 7.85 million metric tons in 1953, about 5.5 percent of cereal production.[18] The average annual interprovincial flow for the period 1953–56 was only slightly less than this.

While some of these flows actually went to feed the two largest cities, Peking and Shanghai, a significant share supported patterns of comparative advantage in farming that had evolved over previous centuries.[19] For example, production of cotton, China's most important nonfood crop, was concentrated in traditional cotton-growing regions of the North China Plain. Among all Chinese provinces, Hopeh had the strongest comparative advantage in cotton production (as measured by cotton yields relative to cereal yields) and the highest share of land sown to cotton (8.5 percent). Moreover, 80 percent of provincial cotton production was concentrated in the southern and central portions of the province. In the eight major producing counties, the area sown to cotton absorbed, on average, one-third of the total crop area, and in one county, Ch'engan, it accounted for 51 percent. During the mid 1950s, 40 percent of the cotton produced in the region was shipped to other provinces or to Shanghai, China's traditional textile center. Because of its concentration in crops such as cotton, and to a lesser extent peanuts and sesame, the province was the single largest importer of cereals during the First Plan. The state supplied on average over a million tons of foodgrains per year, slightly over 10 percent of provincial production.

Cotton was not unique; interregional trade facilitated specialized production of soybeans, other oilseed crops, sugar cane, sugar beets, and so on. Nor was specialization limited to the Yangtse River region or the cotton-producing regions of the North China Plain, where water or rail transport facilitated low-cost movement of bulky crops. Even remote regions sometimes specialized in high-value crops. In mountainous southern Yunnan Province, on the border with Laos and Burma, grain production in Hsishuangpanna, an administrative unit comprising three counties, historically never met consumption needs. In the 1950s, prior to the emergence of the policy of local cereal self-sufficiency, this area continued its historic specialization in the production of high-value noncereal crops such as tea, shellac, tropical fruits, and medicinal herbs. These were traded for 70 percent of the region's food requirements. Similarly, prior to liberation, a single prefecture in remote, mountainous northwest Hunan produced one-fifth of China's tung oil. In the 1950s, tung oil, tea, oil, timber, medicinal herbs, and forestry products were the basis of this region's local economy and were traded for rice produced in the alluvial regions of Hunan.

Although its origins are not well understood, local self-sufficiency in cereals became an important policy objective during the Great Leap Forward (1958–60). The communes formed at that time were to become largely self-contained local communities, each with its own small-scale industries, food processing plants, and social services. Provincial party secretaries on the North China Plain, such as Wu Chih-p'u in Honan, sought to demonstrate the success of the Great Leap locally by proclaiming that their provinces had become self-sufficient in cereals. Policies of self-sufficiency were abandoned in the first half of the 1960s as economic policy formulation was increasingly influenced by Ch'en Yün, the highest-ranking party economic specialist.[20] In the 1950s Ch'en had promoted more favorable price and marketing policies, but he was pushed aside by Mao Tse-tung when the Great Leap Forward got under way.[21]

Interregional trade never recovered to the level of the 1950s, however, and after the onset of the Cultural Revolution in 1966 trade declined further. In 1965 interprovincial cereal exports were 4.7 million metric tons, or 2.8 percent of production. Trade presumably was up from a trough in 1960 and 1961, but data for those years are not available. By 1978, after a decade of emphasis on cereal production and self-sufficiency, interprovincial exports had shrunk to 2.05 million tons, less than 1 percent of output. And of this flow, 1.68 million tons were destined for international markets.[22] Thus, internal trade of domestically produced cereals had shrunk to a few hundred thousand tons, less than one-tenth of 1 percent of domestic production. Even in Hunan Province (population 50 million), where 1978 per capita grain production (80 percent of it rice) was more than 25 percent above the national average, rice exports were only 500,000 tons, less than 3 percent of production. These exports were probably less, in absolute terms, than they had been two and one-half centuries earlier. Rice exports from other provinces, where the widespread adoption of high-yielding short-stemmed rice varieties had contributed to rapid growth of output and high levels of per capita production, were also low.[23]

Suppression of interregional marketing by the government appears to have substantially reduced the efficiency of farm production. The reduction or loss of external sources of cereals led to less efficient production in regions that historically had tended to specialize in nongrain crops. The Traditional cereal producers, in response to reduced supplies of nongrain crops from external sources, sought to increase their own production of sugar and other crops, although the opportunity cost in terms of lost cereal production was considerable. Increased inefficiency is reflected in Western estimates of declining total factor productivity between 1965 and the late 1970s and in a sharp rise in the cost of purchased inputs as a share of the gross value of output.[24]

Although the effects of reduced interregional trade cannot be separated from those of egalitarian distribution policies within individual producing units that may have diminished X-efficiency, the significance of reduced interregional trade is suggested by the experiences of a few localities. In 1979 in northwest Hunan, average per capita incomes in 915 production teams (with a population of about 170,000) were below the nationally established poverty line for rural areas (forty yuan per annum).[25] As external supplies of grain had been curtailed, the region had been forced to pursue cereal self-sufficiency, largely through slash-and-burn methods that destroyed the basis of the region's traditional economy, tung oil and other forestry products.

In the 1950s, Fukien Province, on the southeast coast of China, was a major sugar producer and exporter and grain importer, continuing a pattern that had been evident since the twelfth century.[26] Sugar production was concentrated in a few densely populated counties, where as much as 20 percent of the land was devoted to sugar-cane cultivation. Sugar production fell in the late 1950s because of the crisis brought on by the Great Leap Forward. As a result of the more favorable price and market policies adopted in the early 1960s, however, production had fully recovered by 1965. Just before the onset of the Cultural Revolution in 1966 Fukien exported 100,000 tons of refined sugar, over half of total provincial production. By 1976, sugar yields, production, and exports had all fallen sharply. Exports, for example, fell to 22,000 tons, one-fifth the 1966 level. Naturally, as trade opportunities were reduced, peasant incomes and consumption fell sharply.

In Hsienyu, the most important sugar-cane producing county in Fukien, per capita cereal consumption averaged only 145 kilograms (unprocessed) in 1976, well below the 200-kilogram level that constitutes the official poverty standard in south China. The depression of farm income and consumption through the curtailment of long-established trade patterns is not surprising. The dense population on the southeast coast could only be sustained on the basis of specialized production of high-value crops such as sugar cane and fruit. Arable land in Hsienyu County in the late 1970s was extremely limited—0.03 hectares per capita, about one-third the national average.[27] Although rice yields in the region were among the highest in all of China, trade curtailment reduced the region to abject poverty.

The high cost of trade restriction is confirmed by recent developments. In 1981, four years after the central government guaranteed the availability of cereals for sale to cane producers, cane area in Fukien Province increased by two-thirds, yields more than doubled, refined sugar output more than tripled, and shipments of refined sugar out of the province rose over sevenfold. The average income of peasants in sugar-cane producing regions roughly doubled. Moreover, this transformation was achieved solely through restoration of marketing opportunities. In peasant transactions with the state there was no change in the relative sugar-cane/grain price. As discussed later in this chapter there is some question whether the government will continue to encourage the regional specialization that led to these efficiency gains.

State Investment and Credit Policy

Undervaluation of agriculture by the state is also reflected in state investment and credit policy.[28] From 1953 through 1978 the state allocated 12 percent of its investment to agriculture and 60 percent to industry, even through in 1952 agriculture's share of the economy was over three times that of industry.

Naturally, a complete analysis of the sectoral allocation of investment requires an analysis of the flows of credit from the state banking system and internal reinvestment by collective farms and state industrial enterprises. The main points that emerge from that analysis are as follows. First, on balance the Chinese have pursued a classic Stalinist industrialization strategy. Credit supplied by the state banking system to collective agriculture has been quite modest. Moreover, in the aggregate, internal reinvestment in agriculture by collective units has been low, averaging, for example, 6–7 percent of value added in agriculture in 1977 and 1978. On the other hand, internal reinvestment of retained earnings by state industrial enterprises has been quite high, especially after 1966. For complex bureaucratic reasons, however, these data are not reflected in state investment allocation data. Thus the 1:5 ratio of the share of state investment going to agriculture (12 percent) and industry (60 percent) actually understates the priority given to industry. However, policy has not been uniformly unbalanced in favor of industry. From 1952 to 1957 and again since 1978, state credit flows to agriculture have been significant.

The niggardly flow of resources to agriculture from the state and the high degree of taxation embedded in the price structure governing peasant-state transactions is reflected in the very modest stock of assets in collective agriculture. Fixed assets (excluding land) per agricultural laborer in 1977/78 averaged only 275 yuan ($160), less than one-tenth the level in industry.[29] The model communes to which Western visitors are taken invariably have innumerable tractors, pumps, and a vast array of other types of farm machinery and equipment. But the total fixed assets of the average production team, which in 1977/78 had sixty workers and twenty hectares of cultivated land, were less than 17,000 yuan ($9,800). At the farm-level prices prevailing in China in the late 1970s, these assets would typically be a single large tractor or three hand tractors per team.

Synthesis, 1949–77

The inability of agriculture after 1949 to improve upon its historic pattern of simply keeping up with the rate of population growth seems paradoxical. During the first three decades of Communist rule there were important improvements in irrigation systems, the early development and wide diffusion of high-yielding rice varieties, the development of a large modern chemical fertilizer industry, and major improvements in transportation systems, particularly roads and railroads. These developments should have facilitated agricultural output growth in excess of population growth, but that did not, in fact, occur.

Instead, what emerged was a classic pattern of unbalanced growth, whether measured in output trends or in investment shares. Not only did industrial growth accelerate while agriculture stagnated but it was concentrated in heavy industry. Coal, steel, and electric power output soared, while by 1978 per capita cotton cloth consumption was significantly lower than two decades earlier. In short, Chinese agriculture proved far too complex to thrive in an economic system in which the state systematically interfered with local markets, reduced interregional flows of agricultural commodities, curtailed agricultural investment, and suppressed farm prices.

DEVELOPMENT POLICY SINCE THE END OF THE CULTURAL REVOLUTION IN 1977

In many respects 1977/78 appears to mark a major watershed in Chinese agricultural development policy. The major factors retarding growth in the past now have been modified. State purchase prices for farm products rose, on average, more than 40 percent between 1977 and 1981, and the state bank substantially increased the flow of credit to agriculture. Rural markets were reopened by 1979, and producers may sell grain and edible vegetable oils in private markets after government delivery quotas are filled. The state also has sought to guarantee the sale of adequate supplies of cereals to producers of cotton, sugar cane, and some other crops, when necessary, through interregional grain transfers.

These policy changes have led to a remarkable acceleration of agricultural growth. Between 1977 and 1981 cotton output rose by 45 percent; sugar cane, 70 percent; sugar beets, 160 percent; oilseed crops, 155 percent; and meat, 62 percent.[30] While part of this growth in output of noncereal crops reflects an increased allocation of area to these crops, much of the growth is accounted for by increased yields. Cereal area fell by more than 5 million hectares (4.5 percent) between 1977 and 1981 because of both increased area devoted to other crops and a large reduction in multiple cropping of grain. (In an attempt to increase grain output in the first half of the 1970s the government had strongly encouraged multiple cropping of grain without adequate consideration of its costs.) Despite these losses in grain area, output rose by 15 percent between 1977 and 1981 because yields rose by more than 20 percent.

A large portion of the growth in yields and output can be explained by the reemergence of more rational cropping patterns. State guarantees of adequate cereal supplies to sugar producers, undertaken as early as 1977 in Fukien, were extended to the other major producing regions, Kwangtung and Kwangsi, in 1980 and 1981, respectively. The increases in yields, output, and shipments of refined sugar from regions with a comparative advantage in sugar production have so reduced production of cane in noncomparative advantage regions that output has risen by 70 percent, while the total national cane acreage has remained stable.

Reemergence of specialized cotton production in northwest Shantung Province is another success story.[31] In the 1950s four prefectures in the northwest portion of the province continued their historic specialization in cotton production. In 1956 and 1957 their cotton area was 590,000 and 530,000 hectares, respectively, 80 percent of the provincial total. At the trough of production in the 1970s, in response to the attempt to become self-sufficient in grain, cotton area dropped to 250,000 hectares, and most of the counties in the area fell into poverty. In 1977–79 the greatest concentration of rural poverty in terms both of numbers of counties and of people was in the North China Plain, and a large number of these were in areas of historic cotton specialization. Twenty-two of the twenty-four counties classified as chronically poor in Shantung Province were in the cotton-growing regions of the northwest.

The success of the reemergence of cotton specialization is reflected in a quadrupling of provincial cotton output between 1978 and 1981 and the re-emergence of Shantung as a major supplier of cotton to textile centers in other provinces. By 1981 exports were almost 60 percent of output—ahead of the 40 percent share exported during the 1950s.[32] More significantly, by 1980 cotton acreage in the northwest part of the province had recovered to the level of the 1950s. As a result, twenty of the twenty-two counties, with a total population in excess of twenty million, escaped the poverty level; average incomes rose from two-thirds of the national average in 1978 to 10 percent above the national average in 1980.

Much of the growth in crop yields and output since 1978 can be accounted for by the reemergence of specialized production, as in central coastal Fukien and northwest Shantung; the renewal of marketing opportunities; and the sale of adequate food supplies to peasants producing nonfood crops. Although the new policies seem initially successful in terms of their positive stimulus to output, efficiency, and farm income, it is not clear that they will be sustained. Already there is evidence that the party is retreating from the policies embraced in 1978 and 1979. Promised reductions in the prices of major agricultural inputs have not been forthcoming. On the contrary, prices of some inputs, notably diesel fuel, have been raised sharply; promised increases in state funds for agricultural investment have fallen victim to the general budgetary retrenchment of the early 1980s; rural markets, which have contributed so much to the revival of the farm economy, have come under increased state regulation and price controls; and even the higher prices the state has paid since 1979 for farm products are under pressure. Perhaps most surprising, in the spring of 1982 the government froze the existing levels of interprovincial cereal transfers for a three-year period.[33]

The reasons for these signs of policy reversal are the party's commitment to insulate urban consumers from higher farm-level prices and its unwillingness to reduce further the state's rate of investment. The policy of providing staple foods at fixed prices to the urban population is not new. It began in 1952, and because the markup between state purchase prices for unprocessed grain in the countryside and the ration prices for wheat flour and polished rice in the cities was 100–150 percent, these transactions were highly profitable to the state treasury.[34] Even after absorbing processing losses, transport costs, and marketing costs, profits, which were channeled to the state treasury in the form of budgetary revenue, were about one-half billion yuan ($190 million) per year.[35] By the mid-1970s, however, losses to the treasury had become significant because farm-level prices had been raised considerably in the first half of the 1960s and retail price adjustments had been quite modest—3–7 percent for major cereals. Between 1974 and 1978 losses on the purchase and resale of cereals and vegetable oils cumulated to 20.8 billion yuan ($10.7 billion).[36]

Treasury losses on staple food transactions increased rapidly after 1978 because retail prices remained fixed while procurement prices rose sharply. Losses on cereals and edible vegetable oils were 0.2 yuan and 0.8 yuan, respectively, per kilogram in 1982, meaning that the ration price covered only 60 percent of the cost of procurement and marketing of cereals and an even lower percentage for edible vegetable oils.[37] By 1981, total subsidies, almost all of which accrued to the urban population and the small number of state employees in rural areas, were about 25 billion yuan. That is the equivalent of fully one-quarter of state budgetary revenues (combining all levels of government), more than 6 percent of national income (based on the Chinese national income concept), or 30 percent of the wage bill of state workers and employees.

These subsidies dwarf those for the Food Corporation of India, which manages India's Fair Price Shop System with budgetary losses well under 1 percent of national income. More significantly, Chinese subsidies far surpass those prevailing in other centrally planned systems. In the most notorious case, Poland, food subsidies soared in the 1970s and by 1980 were 18 percent of the wage fund, only about two-thirds the Chinese level.

By 1981 the financial burden on the state of food subsidies had grown so large that it became clear that the state could no longer simultaneously maintain an investment rate of 30 percent or more of national income, provide basic food commodities to urban consumers that in nominal prices were only about 5 percent higher than in 1952 and in real terms had fallen for three decades, and sustain the higher prices offered to peasants.

Most of the measures taken since 1980 suggest that the adjustment to this financial burden is being borne by the peasantry. Ch'en Yün, the chief architect of the reforms instituted since the end of the Cultural Revolution and an ardent supporter of more favorable price and marketing incentives for peasants, has ruled out upward adjustment of urban retail prices for staple foods, recognizing that this could cause urban chaos.[38] Although state investment declined from its peak of 37 percent of national income in 1978 to 28 percent in 1981, it appears to have remained stable since then. Moreover, the newly announced plan to quadruple agricultural and industrial output by the year 2000 has created a new impetus for high investment rates. On the other hand, no less an authority than Hu Yao-pang, the chairman of the party, has stated that future growth of farm income must be derived from increased output and productivity growth, not higher prices.[39] Since 1981, in particular, the state has made a concentrated effort to reduce average farm prices, not by reducing posted prices but by reducing the share of output purchased at higher above-quota and negotiated prices, which have been of increased importance since 1979. Ironically, the state's curtailment of some rural marketing opportunities, primarily in order to reduce the main competitor to deliveries to the state, and its curtailment of long-distance trade in cereal crops substantially reduce the prospects for peasants' achieving income gains through productivity growth.

One is left with the impression that the policy changes since 1978 have only modestly affected the historic emphasis on urban and industrial development. Preservation of retail price stability for staple foods consumed by the urban population has been revealed as being more important to party leaders than providing incentive prices to agricultural producers. In some societies such a choice might be justified on equity grounds, since the absolute level of income of the urban poor is low and they frequently suffer sharp declines in their real income and consumption when food prices increase.[40] Such an argument does not appear persuasive for China. In the first place, rationing of subsidized foods is not targeted to the urban poor but is universal for nonagricultural households. Second, according to official data, only about 2 percent of the urban population is poor.[41] That is hardly surprising, because for two or more decades rural-urban migration has been strictly controlled by the Ministry of Public Security, and secure urban employment has been the most important prerequisite for approval to change place of residence. Third, not only is rural poverty more pervasive than urban poverty, but the living standards of the rural poor are probably lower than those of their urban counterparts. Thus, it is likely that China's system of producer levies and urban rationing worsens rather than improves the distribution of income. Production incentives and equity would be jointly improved by imposing fewer constraints on farm-level prices and providing subsidized food only to the small share of the urban population that falls below the existing urban poverty standard.

NOTES

1. Between 1952 and 1978 value added in agriculture rose by 2.3 percent per year, in industry by 11.3 percent (Yang Ch'ien-pai and Li Hsüeh-tseng, "The Relations between Agriculture, Light Industry and Heavy Industry in China," *Social Sciences in China*, no. 2 [1980]: 183). This measure of agricultural growth is somewhat lower than that usually cited, since it is based on a net value measure and thus takes into account the trend of rising production costs between 1952 and 1978. The sources of increased costs and declining efficiency of production are explained below. Over the same period the population increased from 575 million to 958 million, or 2.0 percent per annum.

2. Dwight H. Perkins (*Agricultural Development in China, 1368–1968* [Chicago: Aldine Publish-

ing Co., 1968]) estimates that per capita output in agriculture was unchanged between the beginning of the Ming Dynasty in 1368 and the middle of the twentieth century.

3. T. W. Schultz, *Transforming Traditional Agriculture* (New Haven: Yale University Press, 1964). See also the discussion of Schultz's work on traditional agriculture in chapter 1 of this volume.

4. The best single study of these changes is the classic by Ho Ping-ti, *Studies on the Population of China, 1368–1953* (Cambridge: Harvard University Press, 1959).

5. Ho Ping-ti, *Studies on the Population of China, 1368–1953;* and idem, "Early Ripening Rice in Chinese History," *Economic History Review* 9, no. 2 (1965): 200–218.

6. Diffusion of early-maturing varieties contributed to increased gain output for two reasons. First, their shorter growing season allowed the spread of multiple cropping from south to central China. Second, because their water requirements were less than those of the traditional rice varieties, early-maturing varieties allowed the spread of rice cultivation to higher lands where water supplies were previously insufficient to support rice cultivation.

7. Yoshinobu Shiba, "Urbanization and the Development of Markets in the Lower Yangtze Valley," in *Crisis and Prosperity in Sung China,* ed. John Winthrop Haeger (Tucson: University of Arizona Press, 1975), 13–48.

8. Mark Elvin, *The Pattern of the Chinese Past* (Stanford: Stanford University Press, 1973), 129, 166–72; Evelyn S. Rawski, *Agricultural Change and the Peasant Economy of South China* (Cambridge: Harvard University Press, 1972).

9. Perkins, *Agricultural Development in China, 1368–1968,* 144; Gilbert Rozman, *Urban Networks in Ch'ing China and Tokugawa Japan* (Princeton: Princeton University Press, 1973), 130; Chuan Han-sheng and Richard Kraus, *Mid-Ch'ing Rice Markets: An Essay in Price History* (Cambridge: Harvard East Asian Research Center, 1975), 59.

10. Chuan and Kraus, *Mid-Ch'ing Rice Markets,* 70–71.

11. Elvin, *The Pattern of the Chinese Past,* 213; Ho Ping-ti, *Studies on the Population of China, 1368–1953,* 201–2.

12. Schultz, *Transforming Traditional Agriculture;* Samuel Popkin, *The Rational Peasant: The Political Economy of Rural Society in Vietnam* (Berkeley and Los Angeles: University of California Press, 1979).

13. Eric Jones, *The European Miracle: Environments, Economies, and Geopolitics* (Cambridge: Cambridge University Press, 1981), 217.

14. In 1955 the arable land per worker was 1.8 hectares in Pakistan and 1.2 hectares in India. Among developing countries in Asia, only in Bangladesh (then East Pakistan), with 0.54 hectares per worker, was the man-land ratio less favorable than in China (Yujiro Hayami and Masao Kikuchi, *Asian Village Economy at the Crossroads* [Baltimore and Tokyo: Johns Hopkins University Press and University of Tokyo Press, 1981], 40). Data for China are in Nicholas R. Lardy, *Agriculture in China's Modern Economic Development* (Cambridge: Cambridge University Press, 1983), table 1–3.

15. Yujiro Hayami, K. Subbarao, and Keijiro Otsuka, "Efficiency and Equity in the Producer Levy of India," *American Journal of Agricultural Economics* 64, no. 4 (1982): 655–63.

16. In the longer run the stability of the system and the persistence of a weighted price above a single free market price also requires that the (absolute) demand elasticity of the rich exceed the elasticity of supply of producers.

17. For documentation of the developments summarized below see Lardy, *Agriculture in China's Modern Economic Development,* chap. 2.

18. Ibid., table 2-2.

19. Ibid., chaps. 2 and 4, includes a detailed discussion of the evolution of cropping patterns and interregional trade and complete citations to the relevant primary source materials.

20. Wu's claims and Ch'en's response are analyzed in ibid., chap. 4.

21. This period is analyzed in Nicholas R. Lardy and Kenneth Lieberthal, *Chen Yün's Strategy for China's Development: A Non-Maoist Alternative* (Armonk, N.Y.: M. E. Sharpe, 1983).

22. Interregional commodity flow data are from Lardy, *Agriculture in China's Modern Economic*

Development, chap. 2; Chinese cereal export data are from Hsüeh Mu-ch'iao, ed., *Annual Economic Report of China, 1982* (Hong Kong: Hong Kong Modern Culture Co., 1982), VIII–47.

23. Provincial cereal output data are from provincial newspapers and broadcasts. Rice export data are from Kwangtung Economics Society, *An Economic Investigation of Kwangtung* (Canton: Kwangtung People's Publishing House, 1981), 195. Other rice exports (that is, shipments out of the province) were: Kwangsi, 25,000 tons; Hupeh, 300,000 tons; Kiangsu, 350,000 tons; Chekiang, 100,000 tons. Per capita output in these four provinces ranged from 340 to 408 kilograms, compared with the national average of 318. Exports as a share of grain production ranged from two-tenths of 1 percent in Kwangsi to 1.7 percent in Hupeh.

24. See Anthony M. Tang, chapter 28 in this volume; idem, *Food and Agriculture in China: Trends and Projections, 1952–1977 and 2000*, in *Food Production in the People's Republic of China*, by Anthony Tang and Bruce Stone, International Food Policy Research Institute Report 15 (Washington, D.C., May 1980), 28; and Lardy, *Agriculture in China's Modern Economic Development*, table 2-8. The latter shows costs rising from 27.3 percent in 1965 to a peak of 35.7 percent in 1977.

25. Wu Yün-ch'ang, "Speed Up Agricultural Development in the Hsianghsi Minority Autonomous Region," *People's Daily*, 15 September 1980, 4.

26. This summary analysis of sugar production is drawn from Lardy, *Agriculture in China's Modern Economic Development*, chap. 2.

27. Hsiao Hui-chia and Chang Jui-san, "Link Up Sugar and Grain, Develop Comparative Advantage," *People's Daily*, 18 August 1980, 2, gives Hsienyu County population and cultivated area data as 720,000 and 24,000 hectares, respectively. For the national data see Lardy, *Agriculture in China's Modern Economic Development*, chap. 1.

28. Investment and credit policy are discussed more fully and data sources provided in Lardy, *Agriculture in China's Modern Economic Development*, chap. 3.

29. Dollar values are calculated on the basis of the average official exchange rate prevailing in 1977 and 1978, 1.74 yuan per U.S. dollar. Because the Chinese currency is not fully convertible and the exchange rate is managed by the Chinese government and because the relative price structure in China is so different from that of the United States, the analytical value of U.S. currency equivalents is limited, and therefore they should be interpreted cautiously.

30. State Statistical Bureau, *Chinese Statistical Yearbook* (Peking: Statistical Publishing House, 1982), 144–45, 163; Chinese Agriculture Yearbook Compilation Commission, *Chinese 1980 Agriculture Yearbook* (Peking: Agricultural Publishing House, 1981), 34–35, 359; idem, *Chinese 1981 Agriculture Yearbook* (Peking: Agricultural Publishing House, 1982), 49.

31. This summarizes discussion in Lardy, *Agriculture in China's Modern Economic Development*, chaps. 2 and 4.

32. Pai Ju-ping, "Adjust the Structure of Agriculture, Raise Economic Efficiency," *Red Flag* (1982): 15–19.

33. Ministry of Food, Rural Procurement and Sales Office, "Purchased Grain May Not be Used to Fulfill Procurement Tasks," *Agricultural Work Bulletin* 3 (1982): 45.

34. This analysis of prices and consumption subsidies draws on Nicholas R. Lardy, *Agricultural Prices in China*, Staff Working Paper no. 606 (Washington, D.C.: World Bank, 1983).

35. Dollar value calculated on the basis of the official exchange rate at the time, 2.62 yuan per U.S. dollar.

36. Dollar value calculated on the basis of the average official exchange rate prevailing in 1974–78, 1.94 yuan per U.S. dollar.

37. Loss data per unit of ration sales are provided in Ministry of Food Research Office, *A Discussion of Policy on Procurement and Sales of Cereals and Oils in Rural Areas* (Peking: Finance and Economics Publishing House, 1981), 4.

38. Quoted in Teng Li-ch'ün, *Study How to Do Economic Work from Ch'en Yün* (Peking: Chinese Communist Party Central Party School Publishing House, 1981).

39. Hu Yao-pang, "Create a New Situation in All Fields of Socialist Modernization," *Beijing Review*, 13 September 1982, 18.

40. See Mellor, chapter 10 in this volume.

41. The urban poverty standard is generally 240 yuan per family member per year. In 1981 a survey of 8,700 urban households showed that only 2.1 percent of the families fell below this level (Li Ch'eng-jui and Chang Chung-li, "Remarkable Improvement in Living Standards," *Beijing Review*, 26 April 1982, 15–18, 28, a summary of a longer text with more details in *Red Flag* 8 [1982]: 25–28).

30

Rural Africa: Modernization, Equity, and Long-Term Development

UMA LELE

INTRODUCTION

Within less than a decade Africa is facing a second severe food crisis. The poor crop can yet again be explained as a result of drought. But the continent's growing vulnerability to crop failures is by no means unexpected. In most African countries it appears to be part of a long-term trend. Data on African countries, especially for subsistence production, are too poor to permit precise estimates,[1] but annual rates of increase of major staple food crops in sub-Saharan African seem to have been about 2 percent during the 1960s and early 1970s, compared with almost 3 percent in Asia and over 3 percent in Latin America.[2] Productivity increases in hybrid maize in some selected areas, such as the highlands of Kenya, have been impressive. However, on the whole, increases in the production of major cereals and root crops—maize, sorghum, millets, and cassava—have come about through increases in the area under cultivation rather than through gains in productivity per unit of input. This is in sharp contrast to even South Asia, which is generally perceived as laggard in development but where substantial productivity gains were experienced in food crop production in the 1970s. Per acre yields of many subsistence food crops appear to have stagnated or even declined in many African countries, as, for instance, in Ghana, Mali, Nigeria, and Sudan.

Because of higher population growth, the annual rates of increase in production required to meet consumption needs by 1990 are also estimated to be higher for sub-Saharan Africa—about 4.4 percent, compared with 4 percent for Asia.[3]

UMA LELE is division chief, Development Strategy Division, Economics and Research Staff, The World Bank, Washington, D.C.

Reprinted from *Science,* 6 February 1981, pp. 547–53, with omissions and revisions, by permission of the American Association for the Advancement of Science and the author. Copyright © 1981 by the American Association for the Advancement of Science.

If present trends continue, Africa will increase its dependence on food imports both over time and relative to other developing continents. Undernourishment is expected to become far more widespread, even though alternatives to cereals and staples, such as bananas and other fruit, fish, and animal products, have been far more important sources of calories in many parts of Africa than in South Asia, which has a similar per capita income. Indices of ill health and infant mortality in Africa are already among the highest in the developing world and are not expected to decline significantly in the next decade.

Compared with the poor performance of food production, export crop production has been more varied among African countries since independence. Production of cotton, tobacco, cocoa, and coffee rose significantly in some countries until the 1960s, but during the 1970s and early 1980s production of major export crops has either stagnated or declined in many countries.[4] Nigeria, for instance, became a substantial net importer of edible oils, of which it was previously a net exporter. Groundnuts in Mali, cocoa in Ghana, cotton in Sudan, and cotton, sisal, coffee, and cashews in Tanzania all provide examples of stagnancy or decline in production.

Rural-urban income disparities are already high in Africa, the ratios typically ranging between 1:4 and 1:9, compared with many countries in Asia, with ratios between 1:2 and 1:2.5. But because agricultural sectors have been stagnant or slow-growing even relative to the poorly performing industry and services sectors, these disparities are worsening in many cases. Kenya, Malawi, and Ivory Coast are the few exceptions where until recently economic growth has been impressive, but the distribution of benefits between agriculture and industry and within agriculture in these countries has been particularly unequal. The World Bank's *World Development Report 1981* estimates that per capita incomes in low-income African countries—countries where the annual per capita income is less than three hundred dollars—decreased 0.4 percent per year on the average during the 1970s, compared with a 1.1 percent annual increase in low-income Asia. Even the middle-income African countries experienced per capita income growth rates of only 0.4 percent per annum, compared with 5.7 percent in the corresponding countries in East Asia and the Pacific.[5]

Worse yet, prospects for overall economic growth in low-income Africa appear much poorer than in the rest of the developing world. The *World Development Report* projects the likely average annual growth rates of per capita income for the high case in low-income Africa during the 1980s to be only 0.1 percent, compared with 2.1 percent for low-income Asia. To reverse these long-term trends requires a clear understanding of the causes of poor past performance.

This chapter argues that most African countries are not giving priority to the development of peasant agriculture. There is not even much understanding of what is required to develop it. As a result, the domestic resources that are spent on agriculture go largely to pay for the growing wage bill of an inadequately equipped and inadequately operating public sector or to ineffective subsidies.

The fragmented donor community has focused largely on project financing, mainly of capital expenditure and technical assistance. Project financing has been rapidly increasing over time, directed mainly toward the rural poor. Current and past donor investments are having little impact, however, not only in the short run but in laying the foundations for long-term development as well. The project approach often results in poor policies, a shortage of maintenance and operating funds, and a shortage of qualified staff, hence often a major depletion of capital.

The Asian experience suggests that agricultural development requires large amounts of resources. Donors should give special attention to broadening their support of education substantially and supporting not just primary but middle- and high-level of training of nationals in technical fields to develop a science-based peasant agriculture. This not only would help to create national policy, planning, and implementing capacity but would support a diverse network of institutions required for development, in addition to those operated by governments. Major investments are also needed in transport and communications, many of which will have to be highly capital-intensive. With such a reoriented emphasis, and guaranteed long-term assistance tied to concrete indications of national commitment, at least long-term prospects could improve significantly.

THE CRUCIAL ROLE OF PEASANT AGRICULTURE

As in many parts of low-income Asia, such as Nepal, Sri Lanka, India, Bangladesh, and Thailand, in Africa concern for economic development is primarily a concern about agricultural and rural development. Between 80 and 90 percent of the nearly 400 million people in sub-Saharan Africa live in rural areas. Most derive their subsistence from meager crop and livestock production and survive on annual per capita incomes of less than U.S. $150. Although production is geared largely to subsistence, the rural sector is also the major source of food for urban consumption and of raw materials for exports and for domestic manufacturing. Except in a few mineral-producing countries such as Zaire, Zambia, and Nigeria, agriculture constitutes the largest income-generating sector, contributing up to 40 percent of the gross national product of many African countries. Between 70 and 80 percent of the annual export earnings of many countries is derived from three to six agricultural commodities. Direct and indirect taxes on agriculture are the most important source of government revenues. Although the estate sector is an important producer of marketed surpluses of certain crops in certain countries, a major share of the total production and marketed surplus nevertheless comes from the smallholder sector. Not only is broad-based agricultural development thus crucial for increasing incomes, employment, and export earnings, but raising the incomes of the rural poor is essential for raising government revenues and creating a domestic market for the goods and services produced in a growing urban manufacturing sector.

POLICIES AND STRATEGIES SINCE INDEPENDENCE

THE "MODERNIZATION NOW" APPROACH

Rhetoric and plan documents in almost all African countries make reference to the key role of the agricultural and rural sector in Africa's modernization. Since the disastrous drought of 1973–74 self-sufficiency in food has become a major objective, often supported by donor-financed projects. The need for increasing export earnings is also being recognized more urgently, the balance of payments difficulties having grown with the rising cost of imported energy and manufactured goods. Despite the growing awareness and the increased number of projects, however, unlike in Asia, there is not yet the basic conviction among many African policy makers that the smallholder agricultural sector can and will have to be the engine of broad-based economic development and eventual modernization.

Modernization is taken to mean mainly industrialization and the commercialization of agriculture, largely through mechanized, large-scale farming. The fluctuating prices of primary exports explain the desire to industrialize, as does the relative ease of setting up factories and state farms compared with the organizationally far more demanding development of peasant agriculture. In its broadest sense the objective of modernization is, of course, shared extensively throughout the developing world. It is the short time perspective of the African expectations that poses a problem, especially given the much poorer institutional and trained manpower base that Africa inherited at independence. Goren Hyden aptly contrasts the eloquent Tanzanian President Nyerere's slogan "We must run while others walk" with China's strategy of modernization by the year 2000.[6] The frequently noted perception of peasant agriculture as a "holding sector" is, however, by no means unique to Africa. At an earlier stage, India's first five-year plan (1951–56) incorporated community development and promotion of cottage and small-scale industry essentially as stopgap arrangements to ensure rural welfare and employment until industrialization could absorb the growing pool of surplus agricultural labor.[7] The more dynamic development strategy, oriented toward small-farmer productivity, which is now being implemented successfully in many parts of India came into ascendancy only in the mid-1960s, with technological change made possible by the new high-yield cereals. As is argued below, in Africa the view of agriculture as a holding sector and the "Modernization Now" strategy have had many of the same consequences for the development of peasant agriculture in more free-enterprise, growth-oriented Nigeria and Zambia as in Ethiopia and Tanzania, which show greater concern about income distribution and class formation.

GOVERNMENT INVESTMENT IN AGRICULTURE

Planning the use of government finances for agricultural development is, of course, not easy for most African countries because of great fluctuations in their export earnings. Their bureaucracies are less experienced than those of their Asian counterparts, which experience similar fluctuations in earnings. Lately their ability to plan has been further eroded, as has that of other developing countries, by the declining purchasing power of their export earnings, as import prices of oil and industrial goods have soared. In constant dollars, the purchasing power of exports from fourteen principal countries in Africa fell by about 40 percent from 1973 to 1980.[8]

Even within these all too obvious constraints, however, far fewer resources are plowed back into agriculture by most African countries than would seem justified. Intercountry comparisons are exceedingly difficult, owing to definitional, data, and other measurement problems, but in the 1970s around 10 percent or less of the planned development expenditure was allocated to the agricultural sectors in Kenya and Mali, compared with 31 percent in India during its first five-year plan in 1951 and 20 percent of the much larger absolute investment in the subsequent three plans. In Zambia the total agricultural budget may have decreased in real terms by an annual average of slightly over 9 percent in the late 1970s, reflecting general budgetary cuts. Malawi is one of the few exceptions in Africa; it appears to have allocated close to 30 percent of the known planned public expenditures to agriculture. However, even there, because of the more favorable tax, wage, and pricing policies toward the estate sector, large-scale production has grown at an annual rate of close to 17 percent since 1968, with 70 percent of the share in exports. The corresponding production increase in the smallholder sector has been only 3 percent a year, even though services to peasant agriculture generally operate far more effectively in Malawi than in several neighboring African countries.

Large-scale farming per se is far less important a portion of total production or exports in Tanzania than in Malawi. However, government policies of "villagization" of peasant producers, combined with pronouncements of the need for cooperative cultivation and actual haphazard attempts to introduce it, have had an adverse effect on smallholder incentives and production. Several other seemingly well-motivated government initiatives to raise peasant productivity have ended up being poorly implemented. These have led, for instance, to unrealistically high production and input-use targets, the consequent indiscriminate promotion of fertilizer use, and discouragement of interplanting of crops (which is traditionally done by peasants to reduce risks of crop failure) as not being "modern." These government initiatives, combined with unreliable provision of agricultural extension, credit, and output marketing, rather than enabling producers to raise overall agricultural productivity, have resulted in producers' responding mainly to changing relative prices of food and export crops. The failed government initiatives have in turn led to an increased official tendency to look toward large-scale mechanized and irrigated production to guarantee food and export surpluses. Like Tanzania, many other countries have already invested or have plans to invest substantial resources in large-scale state farms, but the record of public-sector farming is very poor throughout Africa, and large subsidies are required for these operations.

Irrigation will have to become important ultimately, as the vast, less costly possibilities of increasing production under rain-fed conditions begin to be exhausted. For the short run, however, in most of Africa there is not the complex institutional and managerial capacity to operate irrigation systems indigenously. The frequently costly rehabilitation (at five thousand to fifteen thousand dollars per hectare) being undertaken in many of the existing schemes illustrates the problem.

INCENTIVES TO PEASANT PRODUCERS

Peasant agriculture is highly taxed by fixing low prices for its products and by overvaluing the national currencies vis-à-vis those of importing countries. Agricultural taxation helps keep urban food prices low and finances modernization through many capital-intensive investments, such as construction of new capital cities, stadiums, manufacturing and processing plants, and airports. Agriculture is, of course, the most important sector and hence has to be the major source of revenue. However, traditionally it was taxed because peasants were perceived as irrational, lazy, and unresponsive to price incentives. The resulting tax practices were inherited by independent governments from colonial administrations. Evidence of producer response has mounted, however. In turn, relative official producer prices of food and export crops have been changed in many countries in the last decade, first in order to achieve food self-sufficiency and more recently to promote exports. Relative prices have in fact been easier for governments to influence than technology or quality of services. Thus, while the composition of food and export crops has changed, overall productivity has stagnated. The producer's share in the total net market value of the output has frequently remained very low. In Sudan, the rate of taxation on cotton farmers reached 35 percent in the late 1970s; in Mali it ranged from 36 to 69 percent on cotton, 52 to 65 percent on groundnuts, and 23 to 63 percent on sorghum and millets. Even after allowance is made for the subsidies received by farmers on fertilizer and credit, the effective rate of taxation amounted to 24–61 percent for cotton and 48–65 percent for groundnuts in Mali.

Again, the inadequate recognition of producer incentives is by no means confined to Africa. Theodore W. Schultz's *Transforming Traditional Agriculture,* which included examination of the peasant irrationality hypothesis, was prompted by similar observations in developing Asia in the early post-independence period.[9] In Asia these attitudes, trends, and perceptions have been muted, however. In fact, an articulate pro-agriculture lobby has been created within most governments in Asia. What accounts for these differences? In comparison with Africa (with a few exceptions, such as Kenya), in most of Asia there has been greater overt discussion of policy issues, both domestically and between domestic and outside scholars. More widespread formal education and training of policy makers and administrators in Asia has been helpful, as has their greater exposure to the farming communities through longer practical work experience.[10] New technological possibilities and increased use of purchased inputs have also changed the perspective on price incentives. Now several rural development projects in Africa have gradually begun to produce a similar cadre of knowledgeable Africans in several countries, but their numbers are small because of government and donor policies to be described later.

A large part of the agricultural budget in many countries is spent on subsidies—over 70 percent in Zambia. But contrary to general opinion, many of the subsidies provided in the agricultural sectors in the hope of increasing overall peasant production do not compensate effectively for high rates of taxation. For instance, fertilizer subsidies frequently only help alleviate the high cost of production of inefficient domestic fertilizer plants or the high cost of their local distribution. General subsidies on interest rates and inputs largely benefit the already better-off commercial farmers.[11] A policy followed in many African countries of uniform pricing of output, involving complex cross-subsidies of transport and other handling costs across regions, has achieved regional equity, especially where few attractive enterprises exist, but has discouraged crop specialization to exploit different natural resources among regions.

INPUT AND OUTPUT MARKETING

Input and output marketing and processing facilities are almost always operated by semiautonomous government or parastatal agencies or by largely government-initiated cooperatives on a monopoly basis. Public marketing agencies tend to be high-cost operations because of overstaffing, poor financial control and accountability, and inexperienced management. If an informal traditional market operates, it is only tolerated rather than helped to improve.[12] Frequently it is actively discouraged. The eviction of largely Asian-dominated trade through Operation Maduka in Tanzania and the massive expulsion of Asians in Uganda illustrate the point. A strong desire to abolish exploitation of nationals by other races is understandable, even if such exploitation is imputed rather than real. But even Nigeria, which has a buoyant, largely indigenous small-scale traditional trading sector, adopted a policy of public-sector monopoly of the distribution of fertilizer. Tanzania has similarly discouraged its own enterprising tribes from trading, among other things by instituting some four hundred parastatals and over eight thousand village cooperatives, which are expected to provide most of the public services.

Some of these same policies are followed for almost the same political and bureaucratic reasons in most Asian countries, but the consequences there are far less severe. The degree of government control is more limited, there is greater administrative capacity to exercise it, and there has been more development of private institutions and transport and communication networks. In Africa, inputs are more frequently late, inadequately labeled and packaged, and in wrong combinations. Marketed surpluses are often not picked up on time, first payments to farmers are inordinately late, promised second payments rarely materialize, and damages to crops in storage and handling are extensive. Discouragement of private retail trade has affected rural supply of even the most basic day-to-day necessities in some countries, thus further reducing incentives for producers to consume, save, or invest. Institutional pluralism needs to be given major consideration as an element of development strategy in Africa.

AGRICULTURAL RESEARCH, EXTENSION, TRAINING, AND SOCIAL SERVICES

Whereas there is indiscriminate government intervention in some areas of policy, there is neglect of others, for instance, agricultural research, extension, and development of trained manpower. This neglect is due partly to inadequate recognition of the importance of these services and of the time required to establish effective institutions and delivery systems and partly to preoccupation with politically more expeditious short-run objectives. The role of donors in this regard should not be underrated and is discussed later. The diversion of scarce financial and manpower resources to purposes that the private sector could well be allowed to serve is also a handicap.

Because of the inadequate provision of recurrent resources, the research, extension, and training facilities that do exist are frequently underfinanced and poorly maintained. As President Nyerere observed in his famous speech "The Arusha Declaration: Ten Years After," the pressure to maintain and even expand public-sector employment is so high that the wage bill is difficult to control.[13] Consequently, there are not enough public funds for transport allowances for field staff to carry out research trials and extension demonstrations nor for spare parts, maintenance and operation of stores, processing facilities, research stations, vehicles, and roads. The general situation is one of ill-trained, unmotivated, unsupervised, and demoralized field staff in many sectors. Of course there are notable exceptions, such as the Kenya Tea Development Authority and the Agricultural Marketing Corporation in Malawi. Inadequacy and depletion of capital and government services over time are far more severe in areas where donor projects do not exist, inasmuch as these areas do not benefit from priority budgetary allocations. But the implementation of budgets also needs to be improved, as frequently even the resources allocated are not spent.

Social services suffer from many of the same problems. For example, lack or poor quality of water supply in many rural areas of Africa leads to ill health. Time spent in fetching water reduces time available for agricultural activities. Lack of health facilities similarly reduces labor productivity in agriculture. Absence of primary education results in limited access to services and employment opportunities in towns. Demand for social services is therefore widespread throughout Africa. On the other hand, public resources to pay the recurrent costs of providing social services are generally too limited to permit blanket coverage. Either a high degree of selectivity or greater direct cost recovery is therefore required in the provision of such services. As many *harambee* ("self-help") schemes in Kenya illustrate, rural people are glad to contribute their own resources, provided the services are responsive to precise local demands and reliable, low-cost delivery is assured. Tanzania's example indicates, however, that for a combination of welfare and political reasons, governments refrain from cost recovery and genuine local involvement in planning and implementation. Tanzania's policy of universal provision of services through central financing has undoubtedly achieved results in some areas. According to official data, the proportion of the eligible population enrolled in primary schools went up from 28 percent in 1960 to over 90 percent in the late 1970s. The ratio of population with access to safe water has gone up from 13 percent in 1960 to about 40 percent in the 1970s. To a lesser extent, most African countries have expanded coverage of social services in a similar way, but the overall result is still inadequately financed services, with substantial demands on government resources.

Government objectives of modernization also exacerbate manpower shortages in the traditional sector. The low status of the traditional rural sector and the unattractive living conditions and facilities, in contrast to the urban or the large-scale agricultural sector, often deter qualified nationals from serving the needs of peasant agriculture. On the other hand, demand for education in Africa is one of the strongest in the developing world. The governments have allocated substantial portions of their own resources to education, with different emphases on primary and higher education, depending on their ideology. Because Tanzania has largely emphasized primary education, the enrollment ratio in secondary schools in Tanzania only went up from 2 percent at independence to 4 percent by the late 1970s, and from nearly zero to 0.3 percent in higher education. The shortage of middle- and higher-level technical and administrative manpower is consequently extremely severe. In Kenya, budgetary allocations to secondary and higher education have been expanding more rapidly, and private-sector expansion is permitted more liberally. As a result, 18 percent of the eligible population is enrolled in secondary schools and 1 percent is in higher education. Even then, middle- and higher-level manpower shortages are considerable, especially in technical fields such as accountancy, financial aid and physical resources management, agronomy, plant breeding, and mechanical and civil engineering. On a unit basis, skilled labor in African countries typically costs between three and ten times as much as in many Asian countries. The average annual salary of a research scientist in the 1970s was below ten thousand dollars in Asia, compared with thirty-four thousand dollars in East Africa.[14] And, of course, not nearly enough scientists are available even to rehabilitate, let alone to expand, the national research systems in Africa.

To summarize, the "Modernization Now" objective and the consequent national policies, investment priorities, and attitudes toward smallholder agriculture explain the poor performance of the agricultural and rural sectors in many African countries. In contrast, the Asian and, to a very limited extent, the African experience indicate that greater trained manpower, combined with longer developmental experience by nationals, leads to a better time perspective on modernization and more support of peasant agriculture.

AFRICA'S SPECIAL CHALLENGES

The frequent comparisons with low-income Asia in the previous discussion should not lead one to overlook the problems peculiar to Africa. Low rainfall, poor soils, and the highly diverse ecological conditions within individual countries make raising agricultural productivity much more difficult in many parts of sub-Saharan Africa than in Asia, with its extensive scope for small- and medium-scale irrigation and its more fertile soils.

Several seemingly favorable natural features of Africa, such as the low density of population, pose difficult rural development problems in the short run. In the late 1970s, population densities ranged from 6 persons per square kilometer in Somalia and Sudan to 90 in Nigeria. This is in contrast to a density of 155 in the Philippines, 200 in India, and 620 in Bangladesh. Farms are considerably larger and landlessness is less prevalent in Africa than in most Asian countries. However, extensive land use is itself a result of the unreliable and low rainfall and poor soils referred to above, which lead to shifting cultivation and widespread nomadism in many parts of Africa. Low density also makes for much higher per capita costs of providing roads, schools, and agricultural services in Africa than in Asia.

There are also apparent contradictions. In the African farming system seasonal labor shortages are a far more limiting factor in increasing productivity than in Asia, especially in view of the low level of African agricultural technology. Thus, selective use of mechanization in the private sector may be economically justifiable. And yet unemployment and underemployment of rural labor are also increasing, particularly where population pressure on land is rising rapidly. With rising fuel costs, mechanization—now often operated through the public sector—is frequently highly uneconomical. The more intermediate forms of technology that are used extensively in Asia, such as the ox plow, would be far more efficient where the tsetse fly has been controlled.[15]

Cattle are an important element of Africa's agriculture. The tradition of individual ownership of cattle, combined with communal grazing rights, has resulted in overgrazing and declining productivity. For decades technicians have stressed the need for destocking and pasture improvement, but these have proved elusive because of the complex sociocultural and environmental factors that operate in nomadic social systems and the absence of more profitable and less risky ways of investing the surplus resources of cattle owners.

Low population density also explains the extreme inadequacy of roads, railways, and waterways, although even in this respect there is considerable diversity. Small countries with greater population density, such as Kenya and Malawi, are less hampered by inadequate transport than are large countries such as Sudan, Somalia, Ethiopia, and Tanzania. And yet investments in the road system have also been greater in Kenya and Malawi than in many other African countries. Road mileage per square mile of land area is only 0.02 in Sudan, 0.1 in Zambia, and 0.15 in Zaire, compared with 0.23 in Kenya and 0.31 in Malawi.

Limited growth of sedentary cultivation has also meant more limited evolution of indigenous technology and skills in blacksmithing, carpentry, crafts, man-ufacturing, and trading than is typical of most Asian countries, though there are distinct differences between the more developed West African societies and those of East Africa. The range of farm implements, ox plows, and animal-driven modes of transport used extensively in other parts of the developing world are not prevalent even today in much of traditional rural Africa. On the contrary, with the advent of colonialism there was a "technological leap" toward tractors, combine harvesters, and modern means of transport, so that at independence Africa was left with greater technological dualism than was prevalent in most of colonial Asia.

For these various reasons, the challenges to agricultural research systems in Africa are by far the greatest in the world, combining constraints posed by ecological, demographic, technical, and institutional factors.[16] International ag-ricultural research institutes such as the International Institute of Tropical Agri-culture in Nigeria and several others, financed by the Consultative Group on International Agricultural Research, have already begun to address some of these problems. However, substantial additional investment is required in scientific research at the national and regional levels to develop profitable technological packages to suit the highly diverse conditions and reduce the risks now encoun-tered in their adoption by low-income farmers. In some extremely marginal areas, such as parts of the Sahel in the north and Lesotho in the south, it may not be possible to increase productivity in present subsistence crops enough to make them a primary source of livelihood. Alternatives, including migration to more productive areas where labor-intensive, high-value horticultural crops can be produced, may have to be examined. These are costly options demanding consid-erable organization.

The situation with respect to trained manpower can be best appreciated by some comparisons with Asia at the time of independence. In 1960 even the educationally most advanced African countries, Ghana and Nigeria, had only 4 percent of the population of secondary-school age enrolled in school, compared with 8 percent in Bangladesh, 10 percent in Burma, 20 percent in India, and 26 percent in the Philippines. By the late 1970s the percentage in Nigeria had gone up to 13; by then it was 23 for Bangladesh, 22 for Burma, 28 for India, and 56 for the Philippines.

However, as may be seen in Ghana, Uganda, and Ethiopia, which have been better endowed with trained manpower than other African countries, without a conducive political environment little development is possible even with trained manpower. Many African countries have not yet fully achieved national unity or gained domestic political stability, the colonial powers having established na-tional borders without regard to traditional land rights and tribal cohesion. Re-sources and attention sorely needed for rural development have often been diver-ted to internal conflicts, border wars, and maintenance of domestic political control.

Development of administrative capability will also take a long time. At independence, often there was a virtual absence of strong national, regional, and local government administrations of the types that existed in South Asia. Colonial agricultural development policies were geared almost exclusively to the expansion of export crop production for the metropolitan countries. Research was largely concentrated on export crops.[17] Agricultural extension, input supply, credit, and marketing and processing facilities were also highly fragmented. Recent efforts—for example, in Tanzania and Kenya—to decentralize administrative systems to make them more responsive to rural people's needs, while justified in the long run, have only exacerbated administrative weaknesses in the short run because the existing administrative manpower has had to be spread thinly between the central ministries of agriculture and transport and the provincial administration.

Africa thus starts with considerable odds against development. And yet there is immense potential for productivity increases, not simply in Sudan and the highlands of eastern and southern Africa, where it is commonly recognized, but in much of the rest of Africa, in the humid and semihumid tropics and the parts of the savanna that receive adequate rainfall.

THE DONOR'S ROLE

The experience of Asian countries indicates that in addition to providing direct financial support, international assistance can play an important role in the long run by increasing national consciousness about peasant agricultural development, by improving the rationale for policies, by making the effect of alternative policy options on different sectors or income groups more explicit, and by gradually strengthening those national forces that can lobby for policy changes. Changing the distribution of basic assets or political power so that, for instance, cooperatives will effectively include the poor and subsidies will not go to the rich is far more difficult to achieve from outside. National will and capacity are needed to this end.

Concern and debate about the equity issue in the international donor community have been extensive since the "green revolution" and the perceived failure of the trickle-down approach to reach the poor.[18] Since the world food crisis of 1973–74 the objective of national self-sufficiency in food, and subsequently a broader set of issues such as assurance of basic needs, environmental protection, and women's rights, have begun to receive international attention. The seemingly long time required to achieve the green revolution in Asia has created impatience in the donor community to achieve results, and with the widening scope of the development debate, the areas for achieving results have broadened.

Aid in the form of grants and low-interest loans has increased substantially over time in Africa. For the late 1970s, aid ranged between $20 and $40 a year per capita in Sudan, Kenya, Tanzania, Burundi, Ivory Coast, Mali, Cameroon, Zambia, and Malawi to as high as $50 to $120 in the smaller countries of

Botswana, Lesotho, and Swaziland.[19] In many countries it constitutes a quarter or more of the total annual investment and over half the investment in agriculture and rural development. Even Bangladesh, which is one of the largest recipients of aid in Asia, received only about $10 of concessional aid a year per capita in the late 1970s.

Large numbers of aid agencies are involved in assistance to Africa, with relatively little coordination as to objectives, strategy, degree of continuity, and areas of assistance. Coping with the complex and differing procedures and large flows of aid is exceedingly difficult for the inadequately staffed bureaucracies of most African countries.[20]

Apart from targeting more donor-financed projects toward the rural poor, there has been much evolution in the concept of project assistance in recent years.[21] Projects no longer focus solely on export crops, but are increasingly concerned with development of food crops for domestic consumption. They are more strongly geared to institution building, such as strengthening the project-planning and implementing capacity of the national ministries of agriculture, of the provincial, regional, district, and local administrations, and of the financing and marketing entities that provide field services. This is in contrast to the earlier approach of "enclave" projects, which were implemented mainly through separate autonomous entities created for the purpose. The new projects also show greater concern for employment, training of local staff, and the use of local materials and techniques. They also anticipate more explicitly the need for recurrent financing and for financing of several time phases. They are also more likely now to include support for policy units and monitoring and evaluating to ensure greater flexibility and learning by doing.

Despite these major improvements, donor-financed projects are having a very limited impact, especially in light of the resources expended. This holds irrespective of whether their achievements are judged by inputs, such as numbers of local and expatriate staff recruited, research trials carried out, amounts of fertilizer and other inputs distributed, vehicles purchased, buildings and roads constructed or maintained, or amount of data collected or analyzed by evaluation units; or by the end results, such as increases in yields, numbers of staff trained, or administrative and financial procedures instituted.

What explains the limited impact? The gulf between the donors' largely equity-oriented objectives and the national government's goal of modernization has remained wide in Africa. Instead of examining the actual policies, strategies, and institutional frameworks of national governments and assessing the extent to which they are conducive to rural development, donors have largely taken government rhetoric and plan documents as indications of national commitment and priorities. Donors have concentrated on project aid as a way of influencing these priorities; in so doing, they frequently have exacerbated the problems of Africa's rural development in a variety of ways.

First, the simultaneous shift by much of the international community to the alleviation of rural poverty, in the face of obvious shortages of national manpower, resources, and institutional capacity, has led to underutilization and poor maintenance of donor investments. For a variety of reasons, donors have generally preferred to finance mainly capital expenditures, that is, equipment and civil works, rather than the recurrent expenditures required to maintain or operate these and other related investments.

Second, despite much evolution in the right direction, the need for assistance in increasing national capacity for policy development has been underrated. In addition, a number of questionable showpiece investments by governments have been made possible by generous financial support from the donor community. There are a number of reasons for such assistance: a wish to respond to national desires; an expectation of quick, visible results; the promotion of exports from donor countries; the vying among donor agencies to finance projects likely to appeal to their own domestic constituencies; the donors' need to meet their own quotas of assistance; and some understandable errors in judgment. However, there are other factors: the first relates to the provision of technical assistance in the short run, the second to the expansion of secondary and higher-level education to help broaden the capacities of nationals over the long run.

According to some estimates, as much as 75 percent of the technical assistance used in the developing world is used in Africa. In the short run, technical assistance has permitted the planning and implementation of development projects on a scale that would not have been possible otherwise. However, expatriates are becoming less acceptable in sensitive managerial or policy-making positions in most African countries. Their numbers have been growing for more than a decade since independence, mainly in technical and advisory positions. Their high salaries and benefits create resentment among nationals. Even when highly qualified in their specialties, they are not generally effective in working in an alien environment.

Increasing high-level education and training of nationals is therefore critical for augmenting Africa's managerial and policy-making capacity, even though the results will take a long time to achieve. Expansion of basic, primary, vocational, and adult education has been supported strongly by donors as a way of increasing the supply of field staff, meeting the basic-needs objectives, and increasing the receptivity of rural populations to agricultural and other innovations. Some high-level technical training of Africans is also being undertaken by several bilateral donors, such as the U.S. Agency for International Development and the British Overseas Development Ministry, which have traditionally supported this activity. But on the whole, expansion of secondary and higher education has not received the priority it requires from donors. Frequently the shortage of people with the necessary educational qualifications is so great that even those funds that are provided by donors for higher-level on-the-job training remain unused.

The gains to be had from basic, adult, and primary education are undoubtedly considerable, as evidence from Asia indicates. It is also clear, however, that in Africa at present the shortage of educated and technically trained cadres of nationals who can devise effective national strategies and policies is a far greater constraint to the alleviation of rural poverty than is the illiteracy or lack of receptivity of the rural population. Once again, the question is one of balance and priorities at a given stage of development. Evidence, mainly from Asia and Latin America, has also led to anxiety about increasing the ranks of the educated unemployed in developing countries. The perceived indifference of some of the educated urbanites to the largely rural needs of their own countries has led the international community to a general disenchantment with higher education. Perhaps implicit in this is the feeling that in comparison with the need to train lower-level staff, expanding the supply of high-level educated personnel is unnecessary or antithetical to the egalitarian objectives of rural poverty alleviation.

Contrary to these perceptions, an increase in the supply of educated personnel would not only improve national systems but also reduce salaries of the educated, including those of teachers, thus reducing income inequalities as well as the cost of further investment in education and a range of other development activities. By far the most unquestionable though unquantifiable benefit of higher education to Africa would be that of learning by doing, which is now lost to the ever-growing and changing expatriate technical community. It is ironic that most African countries do not have the capacity to propose alternative plans to those presented by donors for using aid funds—plans that would reflect the countries' own long-term needs for higher education.

The need for substantial investment in physical infrastructure in larger countries such as Sudan and Tanzania and in landlocked countries such as Zambia also requires critical examination by donors. Maintenance of past infrastructure has frequently been neglected, and not enough resources have been devoted to development of trunk roads, railways, and waterways by national governments and donors. Feeder road development has received considerably more support, but the lack of an effective national transport network makes investment in feeder roads ineffective. Again, some of the same reasons that apply to education and training explain this neglect, in particular the perception that capital-intensive infrastructure is not so necessary for reaching the poor, especially in the short run. A more appropriate balance between the objectives of immediate alleviation of poverty and the long-term development needs of more resource-intensive investments is required.

IMPLICATIONS FOR LONG-TERM DEVELOPMENT

The problem of Africa's rural development is not one of not knowing in broad terms what needs to be done to support peasant agriculture. The prospects for turning the present gloomy trends around are considerable. At the national level, the most fundamental problems are attitudes and vested interests. The subsistence rural sector must be seen as critical for economic development and must be given the priority that it urgently requires. At the international level, it is evident that current donor approaches of project aid, although perhaps far more essential in Africa than in many countries in Asia, are by themselves not enough to deal with Africa's complex developmental needs. A major reconsideration of the balance of assistance, including the donors' role in education, infrastructure, and long-term policy planning and implementation, is required. Only then can there be a useful discussion of development priorities with nationals. The question of reordering priorities will require a major review by the donor community as a whole, and even if the question is resolved adequately, the reordering of priorities will take at least a decade to show major results. But the prospects for the 1990s will then be considerably better than those for the 1980s. It is also the only way to reduce Africa's growing dependence on outside aid.

NOTES

I thank H. S. Bienen, W. V. Candler, J. M. Cohen, S. D. Eccles, A. O. Falusi, L. S. Hardin, G. Hyden, N. Islam, W. A. Lewis, J. W. Mellor, M. Mensa, T. W. Schultz, and two unknown reviewers for *Science* for comments on earlier drafts. The views expressed in the article are my own and do not necessarily represent those of The World Bank.

1. Uma Lele and Wilfred Candler, "Food Security in Developing Countries: National Issues," chapter 14 in this book; International Food Policy Research Institute, *A Comparative Study of FAO and USDA Data on Production, Area, and Trade of Major Food Staples,* Research Report no. 19 (Washington, D.C., October 1980).

2. International Food Policy Research Institute, *Food Needs of Developing Countries: Projections of Production and Consumption to 1990,* Research Report no. 3 (Washington, D.C., December 1977), table 3, p. 33.

3. Ibid., table 1, p. 22.

4. K. Anthony; B. F. Johnston; W. O. Jones; and V. Uchendu, *Agricultural Change in Tropical Africa* (Ithaca: Cornell University Press, 1979).

5. World Bank, *World Development Report 1981* (New York: Oxford University Press for the World Bank, 1981), table 1.1, p. 3.

6. Goren Hyden, in *Papers on the Political Economy of Tanzania,* ed. K. S. Kim, M. Mabele, and M. J. Schultheis (Nairobi: Heinemann, 1979), 5–14.

7. See Holdcroft, chapter 3 in this volume.—ED.

8. The countries are World Bank members with populations over two million, plus Mauritius.

9. T. W. Schultz, *Transforming Traditional Agriculture* (New Haven: Yale University Press, 1964). [See also the discussion of Schultz's work in chapter 1 of this book.—ED.]

10. Journals such as the *Economic and Political Weekly* have provided an important forum for a vigorous domestic discussion of planned priorities in India in which innumerable external analysts have participated.

11. Uma Lele, "Cooperatives and the Poor: A Comparative Perspective," *World Development* 9 (1981): 55–72. [See also Adams and Graham, chapter 21, and Gonzalez-Vega, chapter 22, in this volume.—ED.]

12. "There is no official policy towards the unofficial market," a comment of a senior official of one of the African ministries of agriculture, states the problem well.

13. Goren Hyden, *Beyond Ujamaa in Tanzania: Underdevelopment and an Uncaptured Peasantry* (Berkeley and Los Angeles: University of California Press, 1980), 133; Uma Lele, *The Design of Rural Development: Lessons from Africa,* 2d ed. (Baltimore: Johns Hopkins University Press, 1979).

14. P. Oram, *Current and Projected Agricultural Research Expenditures and Staff in Developing*

452   *Uma Lele*

Countries, International Food Policy Research Institute Working Paper, no. 30 (Washington, D.C., 1978), 6.

15. The tsetse fly is the vector of trypanosomiasis, or sleeping sickness, which is endemic in many parts of sub-Saharan Africa.—ED.

16. H. Ruthenberg, *Farming Systems in the Tropics* (Oxford: Clarendon, 1976). See also Schultz, chapter 23; Evenson, chapter 24; and CIMMYT Economics Staff, chapter 25, in this volume.

17. See Evenson, chapter 24 in this volume.—ED.

18. See chapter 1 in this volume.—ED.

19. World Bank, *Accelerated Development in Sub-Saharan Africa: An Agenda for Action* (Washington, D.C., 1981).

20. See Eicher, chapter 31 in this volume.—ED.

21. Lele, *The Design of Rural Development.*

31

Facing Up to Africa's Food Crisis

The most intractable food problem facing the world in the 1980s is the food and hunger crisis in the forty-five states in sub-Saharan Africa—the poorest part of the world.[1] Although the crisis follows by less than a decade the prolonged drought of the early 1970s in the Sahelian states of West Africa, weather is not the main cause of the current dilemma.[2] Nor is the chief problem imminent famine, mass starvation, or the feeding and resettling of refugees. Improved international disaster assistance programs can avert mass starvation and famine and assist with refugee resettlement. Rather, Africa's current food crisis is long-term in nature, and it has been building up for two decades; blanketing the entire subcontinent are its two interrelated components—a food production gap and hunger. The food production gap results from an alarming deterioration in food production in the face of a steady increase in the rate of growth of population over the past two decades. The hunger and malnutrition problem is caused by poverty: even in areas where per capita food production is not declining, the poor do not have the income or resources to cope with hunger and malnutrition.

Twenty of the thirty-three poorest countries in the world are African (World Bank 1982).[3] After more than two decades of rising commercial food imports and food aid, the region is now experiencing a deep economic malaise, with growing balance-of-payment deficits and external public debts. The world economic recession has imposed a severe constraint on Africa's export-oriented economies. Prospects for meeting Africa's food production deficit through expanded commercial food imports thus appear dismal. Donors have responded to these difficult problems by increasing aid flows to the point where African countries now lead the list of the world's aid recipients in per capita terms.[4] Furthermore, the World Bank report *Accelerated Development in Sub-Saharan Africa* (1981)[5] advocates a doubling of aid to Africa in real terms by the end of the 1980s. But the crisis cannot be solved through crash food production projects

CARL K. EICHER is professor of agricultural economics, Michigan State University.

Reprinted by permission from *Foreign Affairs* 61, no. 1 (Fall 1982): 154–74. Copyright by the Council on Foreign Relations, Inc., 1982. Published with revisions.

or a doubling of aid. Since the food and hunger crisis has been in the making for ten to twenty years, solutions to the crisis cannot be found without facing up to a number of difficult political, structural, and technical problems over the next several decades.

Key questions and policies that must be examined include: What is the record of agrarian capitalism and socialism? Why did the green revolution by-pass Africa? What lessons have been learned from crash food production projects in the Sahel and the development strategies of the 1970s—integrated rural development, helping the poorest of the poor, and the basic needs approach? Are technical packages available for small farmers to step up food production in the 1980s? Can foreign aid assist in the alleviation of the food production crisis and economic stagnation?

OVERVIEW OF AFRICA'S ECONOMY

Despite the fact that Africa is an extremely diverse region, several common features frame the boundaries for addressing its food crisis. First, population densities in Africa are extremely low relative to those in Asia. The Sudan, for example, is two-thirds the size of India, but it has only 18 million people as compared with 670 million in India. Zaire is five times the size of France and has only a small percentage of its arable land under cultivation. But some countries are near their maximum sustainable population densities, given present agricultural technology and available expertise on soil fertility. Much of the arable land in Africa is not farmed because of natural constraints such as low rainfall and tsetse flies, which cause human sleeping sickness and virtually preclude the use of approximately one-third of the continent, including some of the best-watered and most fertile land.[6]

Second, most of the economies are open, heavily dependent on international trade, and small: twenty-four of the forty-five countries have fewer than 5 million people, and only Nigeria has a gross domestic product larger than that of Hong Kong (World Bank 1981, 2). Small countries have special problems in assembling a critical mass of scientific talent and in financing colleges of agriculture and national agricultural research systems.

Third, all but two African states—Ethiopia and Liberia—are former colonies.[7] The colonial legacy is embedded in the top-down orientation of agricultural institutions and the priority given to medicine, law, and the arts rather than agriculture in African universities and partially explains the low priority that African states have assigned to agriculture and to increasing food production over the past twenty-five years.

Fourth, Africa is an agrarian-dominated continent where at least three out of five people work in agriculture and rural off-farm activities. Moreover, since agricultural output accounts for 30–60 percent of the gross domestic product in most countries, the poor performance of the agricultural sector over the past two decades has been a major cause of poverty and economic stagnation.

Fifth, Africa's human resource base is extremely weak relative to those of Asia and Latin America. In most countries, even after twenty years of independence, there are still only small pools of agricultural scientists and managers because of the token priority that colonial governments gave to educating Africans.

PROFILE OF AFRICAN AGRICULTURE

Although there are more than one thousand different ethnic groups in Africa and wide differences in farming and livestock systems by agroecological zones, the following overview pinpoints the major features of African agriculture and some of the differences between it and agriculture in Asia and Latin America.[8] For the most part, land ownership in Africa is remarkably egalitarian as contrasted with that in Latin America. The uniform agrarian structure is partially a function of colonial policies that prohibited foreigners from gaining access to land in some parts of the continent, such as West Africa. But in Zambia and Zimbabwe, colonial policies promoted a dual structure of large and small farms (Blackie 1981).

Empirical research has shown that African farmers, migrants, and traders are responsive to economic opportunities. Although custom, local suspicions, jealousies, ignorance, and fatalism can play a role in inhibiting the introduction of change in a particular situation, these variables do not serve as a general explanation of rural poverty (Jones 1960).

Africa is a region of family-operated small farms, in contrast to Latin America, where land ownership is highly concentrated. The typical smallholder has five to fifteen acres under cultivation in any one year and frequently has as much or more land in fallow in order that soil fertility can be gradually restored. Thus, it is more accurate to describe most African farming systems as land-extensive farming systems rather than land-surplus systems. The typical smallholder gives first priority in terms of land preparation, planting, and weeding to growing staple foods (such as millet, sorghum, yams, cassava, white maize, and beans) to feed his family and second priority to producing cash crops such as coffee, cotton, and groundnuts for the market.

Family labor supplies the bulk of the energy in farming, unlike in Asia, where the main energy source is oxen. The short-handle hoe and the machete are the main implements used in land preparation and weeding. Rural nonfarm activities account for 25–50 percent of the total time worked by male adults in farm households over the course of a year in Africa. Unlike in Asia, there is no landless labor class in most African countries because of the presence of idle land.

Land tenure in Africa can be characterized as a communal tenure system of public ownership and private use rights of land (Cohen 1980). The combination of private use rights and communal control over access to land allows families (*a*) to continue to farm and graze the same land over time and to transfer these use rights to their descendants and (*b*) to have the right to buy and sell rights to trees (such as oil palm and cocoa) through a system of pledging. There is no active rural land market in most countries. Land tenure and land use policy issues will be of strategic importance in the 1980s and 1990s as the frontier phase is exhausted, land markets emerge, irrigation is expanded, and herders shift from nomadic to seminomadic herding and sedentary farming systems that integrate crops and livestock.

Unlike in Asia, where two or three crops are grown sequentially over a twelve-month period, most African farmers produce only one crop during the rainy season and engage in some form of off-farm work during the dry season. Irrigation is a footnote in most countries because farmers can produce food and cash crops more cheaply in rainfed farming systems.

Rural Africa is at a crossroads. Farming and livestock systems are complex, heterogeneous, and changing. African villages are experiencing major changes in response to the penetration of the market economy, drought, explosive rates of population growth, and the oil boom in countries such as Nigeria and Gabon. For example, the oil boom in Nigeria has escalated rural wage rates, induced migration from northern Cameroon and Niger, and provided a market for livestock and food crops from neighboring countries.

The subsistence farmer producing entirely for his family's consumption is hard to find in Africa today except in special cases where inadequate transport, rebellion, or political unrest have forced farmers to withdraw from the market and produce for their subsistence needs (such as in Uganda and Guinea in the 1970s and in Tanzania in the early 1980s). In the 1980s and 1990s, village institutions will be under pressure as rural Africa shifts from extensive to intensive farming and livestock systems in response to the decline in the ratio of land to labor. Inequality between countries—for example, Upper Volta and the Ivory Coast—and within countries—for example, southern and northern Sudan—will likely increase in the coming decades.

UNDERDEVELOPED DATA BASE

Africa has a weak and uneven data base, and there is a need to interpret official statistics with caution. For example, accurate data on acreage under cultivation and yields are available for only a handful of countries. Estimates of land under irrigation vary from 1 percent to 5 percent. Estimates of the size of national livestock herds are notoriously suspect because of cattle tax evasion. Even trade data must be carefully examined. For example, official data on cocoa exports from Togo in the 1970s included a large volume of cocoa from Ghana which was smuggled into Togo. Data on rural income distribution are available for only a few countries. Agricultural statistical agents in most countries rely heavily on guesstimates from extension agents, and they have been known to revise their figures to bring them into line with published estimates from international agencies. The combination of underdeveloped data and the case study nature (village studies, for example) of much of the research in the past decade makes it difficult to generalize about the sources of agricultural output and the causes of poverty, malnutrition, and lagging food production.

There is also a need to beware of the pitfalls of studies that present the results of survey research, such as farm management and nutrition surveys, in terms of averages. For example, data showing that farmers produce enough food to feed each family member an average of two thousand calories a day during a given year are meaningless if some family members do not have enough food to survive during the "hungry season." Moreover, the use of averages promotes the view that there is a homogeneous or classless rural society and that interventions designed to improve the average incomes in an area will automatically improve the incomes of all people, including those on the lower end of the income scale. Numerous researchers have shown that rural inequality is an integral part of Africa's history, that inequality may increase as a result of technical change, and that assistance to particular groups of people will have to be carefully targeted.

In summary, although a few scholars talk glibly about average sorghum yields for a country, the "African case," and uncritically use Africa-wide figures (for example, that women produce at least 80 percent of the food in Africa), serious scholars wisely eschew generalizing about even a subregion such as West Africa—an area as large as the continental United States.

FOOD AND POPULATION TRENDS

Looking at Africa's food production trends, population growth, food imports, and poverty, the overriding pattern emerges clearly: since independence Africa's historical position of self-sufficiency in staple foods has slowly dissipated (FAO 1978). Over the 1960–80 period, aggregate food production in Africa grew very slowly—by about 1.8 percent per year, a rate below the aggregate growth rate of Asia or Latin America. However, the critical numbers are not statistics on total food production but per capita figures. The U.S. Department of Agriculture (1981) statistics show that sub-Saharan Africa is the only region of the world where per capita food production declined in the 1960–78 period. In addition, the average per capita calorie intake was below minimum nutritional levels in most countries.[9]

The per capita figures reflect the fact that Africa is the only region of the world where the rate of growth of population actually increased in the 1970s. The annual population growth rate in Africa was 2.1 percent in the mid 1950s and 2.7 percent in the late 1970s and is projected to increase throughout the 1980s until it levels off at about 3 percent by the 1990s (United Nations 1981). Underlying the upward population trend is a young age structure. The average African woman produces six living children in her reproductive years.

There is little hope for reducing fertility levels in the 1980s because of a complex set of factors, including the economic contribution of children to farming and rural household activities, the pro-fertility cultural environment, the failure of family planning programs to date, the pro-natal policies of some states, such as Mauritania, and the indifference of most African heads of state and intellectuals to population growth in what they consider to be a land-surplus continent. But explosive rates of population growth cannot be ignored much longer. For example, Kenya's annual rate of population growth of more than 4 percent implies a doubling of population in about seventeen years (Kenya 1981). In Senegal, where 95 percent of the population is Muslim and the Muslim leaders have great political power, the government is moving gradually on population intervention as it expands demographic research and quietly opens child and maternal health clinics in urban areas. In sum, for a variety of reasons, it is almost certain that most states will move slowly on population-control policies during this decade. As a result, population growth will press hard on food supplies, forestry reserves, and livestock and wildlife grazing areas throughout the 1980s and beyond.

Food imports are another important dimension of the critical food situation. Many countries that were formerly self-sufficient in food significantly increased their ratio of food imports to total food consumption in the 1960s and 1970s. According to USDA figures, food imports are dominated by grain imports—especially wheat and rice—which have increased from 1.2 million tons a year in 1961–63 to 8 million tons in 1980, at a total cost of $2.1 billion. Significantly, commercial imports of food grain grew more than three times as fast as population over the 1969–79 period. Rising food imports are attributed to many factors: lagging domestic production; structural and sectoral shifts arising from such factors as the oil boom in Nigeria and the increase in minimum wages in Zimbabwe following independence; increasing urbanization; the accompanying shift of consumer tastes from cassava, yams, millet, and sorghum to rice and wheat; availability of food aid on easy terms; and overvalued foreign exchange rates, which often make imported cereals cheaper than domestic supplies. Although data on food aid are imprecise, food aid represented about 20 percent of Africa's total food imports in 1982. Wheat, wheat flour, and rice dominate overall imports.

Given the intimate linkage of hunger and malnutrition to poverty, economists, nutritionists, and food production specialists are coming to agree that food and poverty problems should be tackled together. For if rural and urban incomes are increased, a large increment of the increased income of poor people (50–80 percent) will be spent on food.[10] Unless food production is stepped up, an increase in rural and urban incomes will simply lead to increased food prices and food imports and a hardship on families in absolute poverty. Conversely, while expanded food production should be the centerpiece of food policy in Africa in the 1980s, food policy strategies must go beyond crash food production campaigns to deal with poverty itself because expanded food production by itself will not solve the basic problem of poverty.

Africa's food and poverty problems should not be allowed to overshadow some impressive achievements of the continent over the past twenty-five years. Foremost is the increase in average life expectancy—from an estimated thirty-eight years in 1950 to almost fifty years in 1980. This 30 percent increase is often overlooked by those who are mesmerized by rates of economic growth. Moreover, the achievements in education have been impressive in some countries, and there has been a vast improvement in the capacity of countries such as Nigeria, Kenya, the Ivory Coast, Cameroon, and Malawi to organize, plan, and manage their economies.

HISTORICAL ROOTS OF POVERTY AND THE NEGLECT OF AGRICULTURE

From this overview, one can see that while most Africans are farmers and Africa has enormous physical potential to feed itself, there are substantial barriers to tapping this potential. Experts from academia, donor organizations, and consulting firms emphasize post-independence corruption, mismanagement, repressive pricing of farm commodities, and the urban bias in development strategies. Year after year, African heads of state point to unfavorable weather in their appeal for food aid. In fact, the food production crisis stems from a seamless web of political, technical, and structural constraints which are a product of colonial surplus extraction strategies, misguided development plans and priorities of African states since independence, and faulty advice from many expatriate planning advisers. These complex, deep-rooted constraints can only be understood in historical perspective starting with the precolonial and colonial periods (Eicher and Baker 1982).

The colonial period formally began when the colonial powers met at the Berlin Congress in 1884 and decided how Africa should be partitioned among the main European powers. Until the past decade, much of the literature by economists on the colonial period has been pro-colonial. For example, Bauer boldly asserts that "far from the West having caused the poverty in the Third World, contact with the West has been the principal agent of material progress there" (Bauer 1981, 70). But empirical research over the past two decades has shown that colonial approaches to development created a dual structure of land ownership in some countries and facilitated the production and extraction of surpluses—copper, gold, cocoa, coffee, and so on—for external markets while paying little attention to investments in human capital, research on food crops, and strengthening of internal market linkages. For example, colonial governments gave little attention to the training of agricultural scientists and managers. By the time of independence in the early 1960s, there was only one faculty of agriculture in French-speaking tropical Africa. Between 1952 and 1963, only 4 university graduates in agriculture were trained in Francophone Africa, and 150 in English-speaking Africa (McKelvey 1965). In 1964, 3 African scientists were working in research stations in Kenya, Uganda, and Tanzania (Johnston 1964).

Many colonial regimes focused their research and development programs on export crops and the needs of commercial farmers and managers of plantations. In fact, Evenson (1981) points out that in 1971 cotton was the only crop that enjoyed as much research emphasis in the Third World as in industrial countries.[11] The modest investment in research on food crops could be defended during the colonial period because the rate of population growth was low—1 percent to 2 percent per annum—and surplus land could be "automatically" brought under cultivation by smallholders. But with annual rates of population growth now approaching 3–4 percent in some countries, researchers must devote more attention to food crops and the needs of smallholders and herders. Although the debate on colonialism will continue for decades, we have established the simple but important point that contemporary agricultural problems can only be understood by serious analysis of colonial policies and strategies.

FIVE DEBATES ON FOOD AND AGRICULTURE IN THE POST-INDEPENDENCE PERIOD

Africa's food and poverty problems are also a product of misguided policies, strategies, and priorities over the past two decades. In the post-independence period since 1960, African states have engaged in five key debates on food and agriculture. The first was over the priority to be given to industry and agriculture in development plans and budget allocations. As African nations became independent in the late 1950s and early 1960s, most of them pursued mixed economies with a heavy emphasis on foreign aid, industrial development, education, and economic diversification. For example, the late President Kenyatta promoted capitalism and encouraged investors "to bring prosperity" to Kenya. A small number of countries such as Mali, Ghana, and Guinea shifted abruptly to revolutionary socialism in the early 1960s. But whether political leaders were espousing capitalism or socialism, they generally gave low priority to agriculture. African leaders, like former colonial rulers, thought agricultural development would simply reinforce dependency. They tended to view agriculture as a backward sector that could provide surpluses—in the form of taxes and labor—to finance industrial and urban development. Agricultural policies in many capitalist and socialist countries supported plantations, state farms, land settlement schemes, and the replacement of private traders and moneylenders with government trading corporations, grain boards, and credit agencies. The effects of these policies on agricultural production were typically inhibiting, in some cases highly so.

The second debate was over the relevance of Western neoclassical models versus the "political economy" (stressing dependency and class structure) and radical models of development. As Western economists assumed important roles in helping to prepare development plans and served as policy advisers in the early 1960s, Western modernization and macroeconomic models were introduced into Africa. The dominant neoclassical models emphasized the industrial sector as the driving force of development and the need to transfer rural people to the industrial sector. These models had three major shortcomings. First, they assumed that one discipline—economics—could provide answers on how to slay the dragons of poverty, inequality, and malnutrition. As Hirschman (1981) reminds us, development is a historical, social, political, technical, and organizational process which cannot be understood by means of a single discipline. Second, the cities were unable to provide jobs for the rural exodus because of trade union pressure that elevated minimum wages in government and in industry and capital-intensive techniques in the industrial sector (Byerlee et al. 1983). Third, the neoclassical growth models were unable to provide a convincing micro understanding of the complexity of the agricultural sector—the sector that employs 50–95 percent of the labor force in African states. Although these models were technically elegant, they seem remarkably naive today because they assigned a passive role to the agricultural sector.

The vacuity of the Western neoclassical models of development and their failure to come to grips with the broad social, political, and structural issues, as well as the complexities of the agricultural sector, opened the door for the political economy and dependency models to emerge in the 1960s and gain a large following among African intellectuals.[12] The models that emerged in Africa were greatly influenced by Latin American dependency writers. Samir Amin, an Egyptian economist, has been the preeminent proponent of the dependency and underdevelopment paradigm of development in Africa over the past two decades.[13] The political economy literature attempts to link rural poverty and underdevelopment to historical forces, world capitalism, and surplus extraction. The political economy models have made a valuable contribution in stressing the need to understand development as a long-term historical process, the need to consider the linkages between national economies and the world economic system, and the importance of structural barriers (for example, land tenure in Zimbabwe and Zambia) to development. But there is little empirical support for many of the assertions made by some of the political economy scholars.

The question remains, Can political economy and dependency scholars move beyond their abstract models to develop models based on studies of the behavior of African farmers and herders, on African institutions, and on micro/macro linkages in order to provide policy guidance in a continent in which the majority of the people are farmers?

The third debate—over agrarian capitalism versus socialism—has been one of the most emotional topics over the past thirty years; it will continue to dominate discussions on politics, development strategies, and foreign aid in the 1980s. Even though it is difficult to define African socialism, about one-fourth of the states now espouse socialism as their official ideology. The experiences of Ghana and Tanzania are well documented. Four years after Ghana became independent, President Nkrumah abruptly shifted from capitalism to a socialist strategy that equated modernization with industrialization and large-scale farming and state control over agricultural marketing. Ghana was unable to assemble the technical and managerial skills and incentive structure to operate its vast system of state farms, parastatals, and trading corporations. The failure of agrarian socialism has imposed a heavy toll on the people of Ghana (Nweke 1978; and Killick 1978).

Tanzania's shift to socialism in 1967 produced a voluminous literature, international press coverage, massive financial support from international donors—especially Scandinavian countries and the World Bank—and attention from political leaders and intellectuals throughout Africa.[14] The vision of agrarian socialism is set forth in President Nyerere's essay "Socialism and Rural Development." But after seventeen years of experimentation, it seems fair to examine the balance sheet on socialism in a country where 80 percent of the population live in rural areas. The Tanzanian experiment is floundering in part because of the quantum jump in oil prices in the mid 1970s and the conflict in Uganda but basically because of the sharp decline of peasant crop production[15] and production on government-managed coffee, tea, and sisal estates. One cannot overlook Tanzania's gains in literacy and social services, but one may legitimately worry about their sustainability over the longer term without increased rural incomes or exceptionally heavy foreign aid flows. There are many unanswered questions about Tanzania's experiment with agrarian socialism, such as why President Nyerere authorized the use of coercion to round up farmers living in scattered farmsteads and forced them to live in villages. Many pro-Tanzania scholars avoid this topic. But the failure of Tanzania to feed its people explains why Tanzania is no longer taken seriously as a model which other African countries want to emulate.[16]

Agrarian socialism is now under fire throughout Africa: after twenty years of experimentation, presently no African models are performing well. Even Benin, Mozambique, and Guinea are silently retreating from some of the rigid orthodoxy of socialism. What are the reasons for the failure of agrarian socialism to date? First, and most important, socialist agricultural production requires a vast amount of information and managerial and administrative skills in order to cope with the vagaries of weather, seasonal labor bottlenecks, and the need for on-the-spot decision-making authority. In most African countries, the critical shortage of skills and information is the biggest enemy of agrarian socialism. No amount of socialist ideology can substitute for the lack of soil scientists, managers, bookkeepers, mechanics, and an efficient communication system. Second, many parastatals, state farms, and government-operated grain boards have been plagued with overstaffing, corruption, mismanagement, and high operating costs. Because these constraints cannot be easily overcome, it is unlikely that Africa will make much progress with socialist agriculture in this century.

As the pendulum swings from socialism to private farming and private traders in the 1980s, it is important to stress that to put all or most of the weight on ideology—capitalism or socialism—is to ignore an important lesson learned over the past thirty years in the Third World, namely, that ideology is but one variable influencing the outcome of agricultural development projects. The "correct" choice of ideology cannot in and of itself assure successful development. Examples of failure under both capitalist and socialist models are too numerous to conclude otherwise.

The fourth debate was over the use of pricing and taxation policies to achieve agricultural and food policy objectives. The first issue here is whether Africans are responsive to economic incentives. Empirical research has produced a consensus that African farmers do respond to economic incentives as do farmers in high-income countries but that Africans give priority to producing enough food for their families for the coming one to two years (Helleiner 1975). The next question is whether African states have pursued positive or negative pricing and taxation policies for agriculture.[17] Numerous empirical studies across the continent have provided conclusive evidence that many countries (both capitalist and socialist) are maintaining low official prices for food and livestock in order to placate urban consumers. The impact of these negative policies dampens incentives to produce food and livestock for domestic markets and encourages black market operations and smuggling across borders.

For example, starting in the mid 1960s Tanzania paid farmers throughout the country a uniform price for maize in order to achieve equity objectives. But this policy discouraged regional specialization, increased transportation costs, and encouraged smuggling across borders. In Mali, the government pricing policy for small farmers in a large irrigated rice production scheme in 1979/80 could be labeled "extortion." A meticulous two-year study has shown that it cost farmers 83 Malian francs to produce a kilo of rice but that the government paid farmers only 60 Malian francs per kilo (Kamuanga 1982). Does it seem irrational that farmers smuggled rice across the border into Senegal, Niger, and Upper Volta, where they secured 108–28 Malian francs per kilo?

Not only food crops are subjected to negative pricing policies; export crops are also heavily taxed. In an analysis of pricing and taxation policies for major crops in thirteen countries over the 1971–80 period, the World Bank concluded that, taking the net tax burden and the effect of overvalued currency into account, producers in the thirteen countries received less than half of the real value of their export crops (World Bank 1981, 55). These examples and other studies carried out over the past two decades provide solid evidence that African states are using negative pricing and taxation policies to pump the economic surplus out of agriculture.[18] A simple but powerful conclusion emerges from this experience: African states should overhaul the incentive structure for farmers and livestock owners and adopt increased farm income as an important goal of social policy in the 1980s. Moreover, increasing incentives to farmers and herders is a strategic policy lever for attacking poverty and promoting rural employment.

The fifth debate—about the green revolution and the African farmer—concerns what can be done to increase the low cereal yields in Africa. A dominant cause of rural poverty is the fact that 60–80 percent of the agricultural labor force is producing staple foods at very low levels of productivity. While foodgrain yields in Latin America and Asia have increased since 1965, those of Africa have remained stagnant. Over the past twenty years, the green revolution debate has focused on whether African states could import high-yielding foodgrain varieties directly from International Agricultural Research Centers in Mexico, the Philippines, and other parts of the world or whether improved cereal varieties could be more efficiently developed through investments in regional and national research programs in Africa.

Twenty years ago, foreign advisers were optimistic about transferring green revolution technology to Africa, but after two decades of experimentation the results are disappointing. In fact, the green revolution has barely touched Africa. For example, ICRISAT's transfer of hybrid sorghum varieties from India in the late 1970s to Upper Volta, Niger, and Mali was unsuccessful because of unforeseen problems with disease, variability of rainfall, and poor soils.[19] Moreover, the green revolution crops—wheat and rice—that produced 40–50 percent increases in yields in Asia are not staple foods in most of Africa.[20] Knowledgeable observers agree that African farming systems are extremely complex and that the development of suitable technical packages requires location-specific research by multidisciplinary research teams supported by strong national research programs on the staple foods of each country (Norman 1980).

These five debates illustrate the complex set of problems that have preoccupied African states over the past two decades as they have tried to find a meaningful role for their agricultural sector in national development strategies. Throughout much of the post-independence period, most states have viewed agriculture as a backward and low-priority sector, have perpetuated colonial policies of pumping the economic surplus out of agriculture, and have failed to give priority to achieving a reliable food surplus (food security) as a prerequisite for achieving social and economic goals. The failure of most African states to develop an effective set of agricultural policies to deal with the technical, structural, institutional, and human resource constraints is at the heart of the present food crisis. Part of the failure must be attributed to the colonial legacy and part to the hundreds of foreign economic advisers who have imported inappropriate models and theories of development from the United States, Europe, Asia, and Latin America. In the final analysis, agricultural stagnation in capitalist Zaire and Senegal, socialist Tanzania and Guinea, and many other countries must also be placed before heads of state and planners who have promoted premature industrialization, built government hotels, airlines, and large dams with negative internal rates of return,[21] and spent tens of millions of dollars building villas for heads of state for the annual meetings of the OAU. Moreover, most African political leaders have also exhibited a fundamental misunderstanding of the role of agriculture in national development when 60–80 percent of the people are in farming. Unfortunately, these mistakes in dealing with agriculture over the past twenty years cannot easily be overcome through crash production projects and doubling of aid over the 1980–90 period.

POLICY DIRECTION FOR THE 1980s AND 1990s

Africa's inability to feed itself amid vast amounts of unused land and record levels of foreign aid is, on the surface, one of the major paradoxes in Third World development. What should be done? While the several notable recent reports on Africa's food and economic problems agree on the severity of the food and hunger crisis, each of these assessments underemphasizes the mistakes of African states and in a somewhat self-serving fashion overstresses the need for more foreign aid. Almost all of the reports implicitly assume that capital, rather than human resources, is the most pressing constraint in rural Africa. This preoccupation with capital is understandable because foreign aid institutions such as the International Fund for Agricultural Development (IFAD) and the World Bank have a fixation on capital transfers. Moreover, Third World countries have focused on capital transfers and the need to increase aid in the north/south dialogues, and many donors and African heads of state equate a doubling of aid with an attack on poverty in Africa. The *Lagos Plan of Action,* which was adopted by the heads of state and government in Lagos in April 1980, has little new to say about agricultural development except that food production should be accelerated with the aim of achieving self-sufficiency (OAU 1980). The World Bank's report *Accelerated Development in Sub-Saharan Africa* (1981) correctly singles out domestic policy issues as the heart of the crisis, but it also advances an unsupported appeal for donors to double aid to Africa over the 1980–90 period. Further, while the World Bank report criticizes large-scale irrigation projects, it does not report the Bank's own difficulties (and those of most of the other donors) in designing sound livestock projects. The World Food Council's (1982a) report on the African food problem correctly notes the overemphasis on project-type aid, the excessive number of foreign missions (for example, Upper Volta received 340 official donor missions in 1981), and the small percentage of aid funds for food production projects, but it skirts many of the political and structural barriers to change. The World Food Council's (1982b) report by the African ministers of agriculture avoids the topic of population growth, the empirical record of agrarian socialism, and the disastrous performance of state grain boards. New approaches are needed. The following discussion spells out a comprehensive approach for the 1980s and 1990s.

STEPS TO MEET THE CRISIS

Solutions to Africa's food and poverty problems must, first of all, be long-term. Second, they require a redirection in thinking about agriculture's role in development at this stage of Africa's economic history and about the need for a reliable food surplus as a precondition for national development. Third, there is a need for both African states and donors to admit that the present crisis is not caused by a lack of foreign aid. In fact, in many countries current aid flows cannot be absorbed with integrity. Hence, donors are part of both the problem and the solution. The Berg report underplays these issues in its unsupported case for doubling aid to Africa by the end of the 1980s (World Bank 1981). Fourth, there is a need to recognize that the lack of human resources is an overriding constraint on rural change in Africa. In fact, the human resource constraint severely limits the amount of aid that can be effectively absorbed in the short run. In order to buy time to lay a foundation for long-range solutions, it will be necessary to rely on a number of holding actions. Examples include expanded commercial food imports, food aid, and promoting seasonal and international migration until more land is brought under irrigation and higher rainfall areas can be cleared of tsetse flies and river blindness. But these holding actions must not be allowed to substitute for efforts towards long-range solutions.

Three steps should be taken now to start the process of formulating longer-term approaches. First, African states, donors, and economic advisers should jettison the ambiguous slogans such as "National Food Self-Sufficiency," "Food First," and "Basic Needs."[22] Although these have a powerful emotional and political appeal, they offer little help in answering the key question: What blend of food production, food imports, and export crops should be pursued to achieve both growth and equity objectives? The concept of national food self-sufficiency should be scrapped as a rigid target because it promotes autarchy and ignores the historical and the potential role of trade in food and livestock products between African states. In summary, there is a need to return to the basics of agricultural development: investments in human resources and agricultural research, policy and structural reforms that will help small farmers and herders, revamping the incentive structure, changing the role of the state,[23] and strengthening the administrative capacity to design and implement projects and programs.

The second immediate step should be the phasing out or restructuring of some of the crash food production projects—that is, seed multiplication, irrigated wheat schemes, livestock schemes, and integrated rural development projects—that are floundering. Many of these crash projects were hastily assembled over the past decade without a sound technical package and without being tested in a pilot phase. These unproductive projects consume scarce high-level manpower, perpetuate recurrent cost problems, and create a credibility problem for both African policy makers and international donors. Particularly important is the reassessment of integrated rural development (IRD) projects. The weakness of most IRD projects—their lack of emphasis on food production and income-generating activities—can be corrected by restructuring some of the projects rather than phasing them out. Other projects that have been implemented in advance of a sound knowledge base, like those in livestock, should be either phased out or scaled down and continued as pilot projects for a five-to-ten-year period. A five-to-ten-year pilot phase is unheard of in Africa, but in projects like those in livestock it is a necessary period for solving technical problems and developing appropriate local institutions to solve such key issues as over-stocking.

The third immediate step is to scale down the state bureaucracy, the state payroll, and state control over private farmers and private traders. After twenty years of experience with parastatals, the record is clear: parastatals (public enterprises) are ineffective in producing food, are no more efficient than private traders in foodgrain marketing, are almost all overstaffed,[24] and serve as a sponge for foreign aid. As the number of parastatal employees increases, the pressure intensifies for donors commensurately to increase their contributions to meet the payroll of the expanded bureaucracy. The parastatal disease is well known, but it is not given much attention in the reports cited above, except in the World Bank's *Accelerated Development* report, which should be applauded for its candor on this topic.

The fourth step is to realize that a food policy strategy cannot be pursued in isolation from livestock and export crop policies nor in isolation from policies to deal with rural poverty. A food policy strategy should not rule out the expansion of export crops, because expanded farm income, through food sales, export crops, and off-farm income, and productive rural employment are prerequisites for solving rural poverty problems. Moreover, although food aid can help the rural poor in the short run, the expansion of productive rural employment is fundamental to coping with rural poverty in the long run.

FOOD POLICY STRATEGIES

The starting point for food policy analysis in each country should be the development of a food policy strategy with two goals in mind: achieving a reliable food surplus (based on domestic production, grain storage, and international trade) and reducing rural poverty by focusing on measures to help small farmers produce more food for home consumption and more food, cash crops, and livestock for the market so that they can purchase a better diet.[25] But a word of caution is in order: food policy analysis is every bit as complex and as delicate as family planning.[26] The rice riots in Monrovia, which left more than one hundred dead in 1979, and the sugar riots in Khartoum and other major cities in the Sudan following the doubling of sugar prices in 1981 are reminders of the narrow range of options for policy makers on food policy issues. Consequently, as experiences from the Sudan, Zimbabwe, Nigeria, and Kenya (outlined below) illustrate, most countries will move very slowly on policy reforms unless spurred by famine, a reduction in foreign-exchange earnings from petroleum, or coordinated donor leverage to link long-term food aid with policy reforms.

The Sudan provides a conspicuous example of the difficulty of mobilizing the agricultural sector as an engine of growth and expanded food production. In the mid 1970s the international press frequently asserted that the Sudan could become the "breadbasket of the Middle East" by drawing on several billion dollars of OPEC loans and gifts to develop its vast reserve of idle land. The issue today, however, is not one of exporting food to the Middle East but one of the Sudan's inability to feed its 18 million people. The Sudan was forced to rely heavily on food aid in the early 1980s in order to cope with severe balance-of-payment problems and inflation. Although the Sudan has historically excelled in cotton research, it has devoted only token attention to research on food crops. As long as the Sudan continues to receive food aid and has hopes of striking oil in the southern part of the country, there is little likelihood of policy reforms.

In Zimbabwe, the legacy of the colonial policy of promoting a dual structure of large farms for white farmers and small farms for Africans in poor natural resource regions presents a classic efficiency/equity dilemma for the Mugabe government (Zimbabwe 1981). In the early 1980s Zimbabwe was a significant maize exporter based largely on the surpluses produced by its thirty-five hundred large farmers. But the maize exports were heavily subsidized, and in 1982 the government reconsidered its role as a food security safety net for the southern African region. In 1982 Zimbabwe increased price incentives for soybean oil relative to maize in order to meet the domestic shortage of cooking oil. Although Zimbabwe gains political prestige by exporting maize to black Africa, it realizes that it cannot continue to subsidize maize exports at a time when it is facing large budget deficits.

On the eve of independence in 1960, Nigeria was a net exporter of food, mainly oil palm and groundnuts. But during the 1960s Nigeria pursued import-substituting industrialization, taxed its farmers heavily through export marketing boards, experimented with land settlements, and promoted government plantations. In 1970, ten years after independence, Nigeria was importing food, and by 1981 food imports from the United States alone totaled more than $1 billion. Petroleum exports have enabled Nigeria to pay for food imports and buy time. Although Nigeria is far ahead of most Francophone African countries in trained agricultural manpower, Idachaba (1980) reported that more than 40 percent of the positions for senior agricultural researchers in the eight major research stations were vacant in 1978. The government recently concluded that it will take ten to fifteen years to achieve self-sufficiency in food production. Nigeria has now formed a high-level Green Revolution Committee to address its food problem (Abalu 1982).

Although Kenya is widely regarded as an agricultural success story of the 1960s and 1970s, Kenya was confronted with food shortages in 1980 and 1981 and was forced to import maize, wheat, and milk powder. Although adverse growing conditions contributed to the food shortages of the early 1980s, the National Food Policy paper (Kenya 1981) reveals that other factors were undermining Kenya's capacity to feed itself. These included the unprecedented 4 percent rate of growth of population, the decline in wheat production following the transfer of large farms to smallholders, and a smallholder credit repayment rate of 20 percent. The message of the National Food Policy paper is clear:

Kenya has a major food production constraint that cannot be overcome except through large investments in agricultural research, irrigation, and land reclamation in the 1980s and 1990s. But one wonders why the National Food Policy paper paid lip service to population growth.

These case studies illustrate the complexity of Africa's food problems and the need to analyze each country's problems on a case-by-case basis. Moreover, food policy analysis requires more than the preparation of a National Food Policy strategy paper over a two-to-six-month period. Food policy analysis is an ongoing process that will undoubtedly occupy the attention of policy makers and researchers throughout the 1980s and 1990s.

FOOD AID LEVERAGE

A major issue in achieving policy reforms is whether donor agencies and countries can or should use food aid leverage to promote the required changes. In existence for almost thirty years, food aid is now a topic of growing interest in Africa. Although there is unanimity on using food aid for humanitarian purposes—for example, feeding refugees—food aid for development is more controversial. The opposition to this sort of food aid—where food is sold at concessional terms and extended as grants for food-for-work programs—comes from evidence that food aid (1) can reduce the pressure on recipient countries to carry out policy reforms; (2) can depress farm prices; (3) is unreliable;[27] and (4) can promote an undesirable shift in consumption patterns that will increase rather than reduce dependency or require subsidies (such as wheat production in West Africa) to maintain the Western-acquired consumption pattern.[28]

Food aid programs are firmly institutionalized with donors. Food aid accounted for approximately 40 percent of all U.S. economic assistance to Africa over the 1970–80 period. Even Japan started to dispose of some of its surplus rice in Africa in the early 1980s. To date, there has been little solid academic research on the role of food aid for development purposes in Africa. Moreover, the evaluation of food aid is usually assigned to junior officers in many bilateral agencies. Hence, evaluation studies of food aid by donors should be taken with a grain of salt. The food aid experience in Asia and Latin America, however, shows that the availability of food aid can take the pressure off recipient nations to carry out internal policy reforms.

A compelling case can be made for linking food aid with policy reforms in major food-deficit countries in Africa through the development of food policy reform packages. These reform packages will be useless, however, unless there is an agreement by donors to make three- to five-year forward food aid commitments in exchange for internal policy reforms. Countries such as Mali and the Sudan would be good test cases for linking food aid to tough domestic policy reforms. But unless donors agree to meet minimum forward food aid levels, African states can easily postpone policy reforms and continue to rely on a patchwork of bilateral food aid programs.

AGRICULTURAL RESEARCH

Beyond policy reforms, a long-range solution to food and hunger problems will depend, to a large degree, on achievements in agricultural research. Authorities on food production and livestock projects in the field now commonly bemoan the lack of proven technical packages for small farmers in dry-land farming systems throughout Africa and the uniformly unfavorable technical conditions (low rates of growth, disease) for livestock production. Significant increases are needed over the next twenty years in research expenditures on dryland farming systems with emphasis on food crops (white maize, yams, cassava, millet, and sorghum) and on livestock.

An expanded research program on food and livestock should be viewed in a twenty-year time frame because problems such as low soil fertility and livestock diseases cannot be resolved through a series of short-term, ad hoc research projects. The U.S. experience, wherein forty years (1880–1920) were spent developing a productive system of federal and state research programs, should be heeded by donors who are likely to expect major results in three to five years from new research projects in Africa.

Research on irrigation is particularly important and should be accelerated in the coming decades. The knowledge base for irrigation in Africa is meager. Irrigation has played a minor role in Africa except in large-scale projects in the Sudan and in Madagascar, where there is a history of irrigation by small farmers. The cultivated land under irrigation is probably less than 5 percent in most other countries (as compared with an estimated 30 percent in India). Following the 1968–74 drought in the Sahel, there was considerable optimism about the role of irrigated farming in "drought-proofing" the region. Due to numerous technical and administrative problems and human resource constraints, however, the projected expansion of irrigation in the Sahel is behind schedule, and it is certain that irrigation will not play a significant role in the Sahelian states until early in the next century.

Although research on the economics of irrigation is fragmentary, the limited results provide support for a smallholder irrigation strategy in the 1980s, with priority given to ground-water development with small pumps, land reclamation through drainage and water control, and an increase in small-scale projects that are developed and maintained by groups of farmers with their own family labor. A small-scale irrigation strategy is advocated because the cost of bringing more rainfed land under cultivation is substantially less than the cost of leveling and preparing land for large-scale irrigation. For example, recent irrigation projects in Niger, Mauritania, and northern Nigeria each had costs of more than ten thousand dollars per hectare at 1980 prices (World Bank 1981, 79). On the other hand, farmers in Senegal have cleared and prepared their own land for irrigation, expending several hundred hours of family labor per hectare. Although irrigation will not be a panacea for the recovery of the Sahel nor for feeding Africa in the 1980s and 1990s, a long-term research program on the human, technical, and institutional dimensions of irrigation should be initiated in the immediate future.

It remains to be seen whether donors will have the courage to view research and graduate training within Africa as a long-term investment and whether they will provide guaranteed funding for a minimum of ten years. Another important issue is whether country priorities of bilateral donors will remain stable enough to assure African countries of continuity in funding over a ten-year period. A rule of thumb is that an African country should never embark on a long-term program to upgrade its national agricultural research system with major support from only one bilateral donor. But as we point out below, co-financing by six to eight donors can create as many problems as it solves.

INVESTMENT IN HUMAN RESOURCES

A third essential component of a long-range strategy is massive investments in human capital formation, including graduate training of several thousand agricultural scientists and managers. This is necessary to replace the foreign advisers, researchers, managers, and teachers in African universities and to meet the needs of a science-based agriculture in the next century. Since it takes ten to fifteen years of training and experience beyond high school to develop a research scientist, the investments in human capital will not produce payoffs for Africa until the 1990s.

Building graduate agricultural training programs within Africa necessitates a reexamination of the role of the African university in national development and the relevance of some of the present undergraduate degree programs. For example, in 1982 the Faculty of Law and Economics in the University of Yaoundé in Cameroon was turning out graduates with degrees in law and economics who ended up on the unemployment lines in Yaoundé. The time is propitious for African universities to move from undergraduate to graduate training programs in science and agriculture. Before graduate education is expanded, however, some questions should be raised about priorities in undergraduate education and the relevance of the curriculum. Undergraduate degree programs in agriculture in many universities are still embarrassingly undervalued and underfunded when compared with programs in law, medicine, and history. For example, the University of Dakar in Senegal was formally established in 1957, and in 1960 the Senegalese assumed its administration. In 1982 there were approximately twelve thousand students in the University of Dakar, of whom several thousand specialized in law and economics. Not until 1979 was a National School of Agriculture established at Thies, near Dakar, to offer undergraduate training in agriculture. That university-level teaching of agriculture was not initiated until nineteen years after Senegal's independence reflects an enduring colonial legacy as well as the government's ambivalence about agriculture's role in national development.

Although the structural reforms entailed in redesigning African universities to serve rural Africa will require decades to resolve, it is time for donors to stop merely paying lip service to African universities. Whereas donors embraced African universities in the 1960s, they generally withdrew their support in the

1970s as they promoted crash food production and IRD projects and invested heavily in international agricultural research centers. Money saved ($100 million to $200 million) from phasing out the floundering crash projects cited above can be reallocated to selected African universities with emphasis on faculties of agriculture. Donors should press for long-term structural reform of the curriculum in universities in exchange for long-term aid commitments of ten to twenty years.

In 1982 graduate-level education for African students in the United States cost $1,850 per month, or $39,000–$55,000 for a Master's degree over a twenty- to thirty-month period. Donors should gradually phase out Master's-level training programs in agriculture and related fields in the United States. Instead, U.S. faculty members should be sent to Africa to help develop regional centers of excellence in graduate training in eight to ten African universities over the next ten to fifteen years. In order to achieve this goal, donors will have to give greatly increased priority to aiding African universities, including ten-year authorizations to foreign universities to provide teachers for graduate instruction and research. In the final analysis, the initiative for this second phase—graduate training in agriculture in African universities—will have to come from within Africa.

DEALING WITH RURAL POVERTY

The fourth component of a long-range solution to Africa's food crisis will be an ongoing effort to address the hunger/malnutrition/poverty problem. Rural poverty is potentially a much more difficult problem to solve than the food production gap, but self-sufficiency in food production will be a bogus achievement if the poor do not have access to a decent diet. A society cannot expect to move from a low- to a middle-income stage of development if two-thirds of its population are producing millet, sorghum, maize, and yams at stagnant levels of output. Agricultural research on foodgrain production is a prerequisite to increasing food production. Moreover, since jobs cannot be created in urban areas for all the unemployed, a rural investment strategy should also facilitate the expansion of rural small-scale industries that are labor-intensive and can provide jobs.[29]

IMPLICATIONS FOR FOREIGN AID

The implications of all this for the foreign assistance community flow quite clearly from the foregoing analysis. Currently, forty donors are moving funds and technical assistance through a patchwork of several thousand uncoordinated projects in support of agricultural and rural development throughout Africa. In turn, African states are allocating a high percentage of a scarce resource—trained agricultural professionals—to meet the project reporting requirements of donors, and African governments are asking donors to pay the recurrent costs—salaries, petrol—of the aid-funded projects. In short, both donors and recipients are prisoners of projects and slogans, and they are caught in a vicious circle. Should aid to Africa be doubled in real terms during this decade? The answer depends on whether donors and African states can replace the short-term approaches with long-term investments and address the following in a consistent manner:

1. *Food security policies and strategies.* Donors should urge African policy makers to focus on policies and strategies to achieve a reliable food surplus (food security) based on local production, storage, and international trade. Despite the pleas of international journalists who urge donors to increase the number of food production projects, a single food policy reform in Mali—raising official farm prices—may be more effective than twenty new food production projects. Donors should concentrate their resources on helping local professionals develop an improved micro foundation for food policy analysis that addresses the constraints on achieving a reliable food surplus, with emphasis on food production, storage, and international trade.

2. *Long-term investments.* Emphasis should be placed on reducing the number of tiny projects (such as producing visual aids for the livestock service in a Sahelian country), increasing the lifetime of aid projects, and increasing the volume of aid in program grants that are tied to policy reforms. Long-term investment programs like ten-year research projects, five-to-ten-year pilot livestock projects, twenty-year programs to develop colleges of agriculture, and five-year food aid/policy reform packages should be perceived not as luxuries but rather as prerequisites to solving Africa's technical, structural, and human capital constraints.

3. *Technology generation within Africa.* Professional agriculturalists in most donor agencies privately concede that there is currently an excess of donor funds in search of agricultural production projects supported by agricultural research findings that have been tested and proven on farmers' fields. In short, the international technology transfer model has failed in the direct transfer of foodgrain varieties from Mexico or India to Ghana, Lesotho, and Upper Volta. What can be done? In my judgment, donors should (*a*) admit that the international technology transfer model is not producing the expected results, (*b*) maintain but not increase investments (in real terms) in the four International Agricultural Research Centers (IARCs) in Africa, and (*c*) increase the level of financial assistance to national agricultural research systems and to faculties of agriculture in African universities.

Although the U.S., Mexican, and Indian foodgrain varieties are not directly transferable to Africa, some of the processes these countries used to generate technology appropriate to the needs of their farmers in dry-land areas are applicable in helping to strengthen faculties of agriculture in African universities and national agricultural research services. For example, the U.S. dust bowl crisis in Kansas and Oklahoma in the 1930s gave rise to the U.S. Soil Conservation Service, research on new varieties, irrigation, and other techniques which transformed the dust bowl into a highly productive area of American agriculture over a thirty-year period. In this process, U.S. colleges of agriculture played a strategic role, in cooperation with local and state organizations and with the U.S. Department of Agriculture.

4. *Co-financing.* Co-financing of aid projects by donors is a growing problem in Africa because typically six to eight donors each underwrite a piece of an agricultural project. Co-financing is attractive because it spreads the risk for donors and reduces the dependency of African states on one donor. But co-financing is proving to be a liability for institution-building projects such as research institutes and extension schools. The recipient institutions are caught in a cross fire of imported perspectives from technical advisers, a hodgepodge of buildings, and dubious gifts of equipment from around the world. Moreover, the administrators of these local institutions are overwhelmed by the administrative and reporting requirements of the donors. At most, two donors—one for infrastructure and one for technical assistance and training—should be allowed to assist any one institution. But African states will have trouble getting weaned away from co-financing because they are using this device to pay for part of their recurrent budget deficits and the payroll of the state bureaucracy.[30]

5. *Foreign private investment.* A major topic of debate is whether foreign private investment, especially multinational firms, can contribute to the resolution of Africa's food and poverty problem. A related question is whether bilateral aid should assist foreign private firms in establishing fertilizer plants, processing plants, and in some cases large-scale food production projects. Just as the roles of women in African development cannot be analyzed in isolation from those of men, the role of the private sector can only be analyzed in relation to public investments. The poor record of food and livestock production projects throughout Africa over the past ten years provides ample proof that many of these projects fail because public-sector investments were not made in agricultural research to develop profitable packages for rainfed farming, prevention and control of animal disease, rural roads, and schools to train agricultural managers. Public-sector investments can either facilitate or destroy the conditions for capitalists to function in a market-oriented economy.

In general, inadequate infrastructure, local managerial skills, and technical constraints severely limit the scope for foreign private investment in food production projects and in agroindustries in Africa. Although some foreign firms prospered in colonial periods, when they were given choice land and protected markets, since independence there have been many failures, including the recent efforts of U.S. firms to produce food in Ghana, Liberia, and Senegal. As a rule of thumb, if foreign private firms engaged in food production projects do not receive special subsidies, they cannot compete with African smallholders who have a knowledge of local climate, pests, and soils and are willing to produce food on their own land at rates of return of seventy-five cents to three dollars per day for family labor. Moreover, the large capital-intensive plantations and ranches emphasized by foreign private firms should be questioned on social grounds because they do not produce the badly needed jobs in an area of the world where seasonal unemployment is widespread. Foreign private enterprise, however, can contribute to Africa's food system in countries such as Cameroon, Kenya, the Ivory Coast, and Zimbabwe, which have a good infrastructure and need international managerial skills and capital for investments in food processing plants and in fertilizer and agricultural input industries. But in the final analysis, the focus of foreign aid should be on making public investments in roads, universities, and research stations to help African capitalists—small farmers and herders—produce food for their families and for urban and rural people.

Aid flows to Africa have grown dramatically in recent years: net official aid in 1980 was $13.70 per capita in Africa, compared with an average of $9.60 for all developing countries. In the Sahelian region of West Africa per capita aid was running from $35 to $50 per person in 1982. In many circles in Africa there is a feeling that the continent is already too heavily dependent on aid and foreign transactions relative to the scarcity of African professionals to implement the projects. In fact, in many countries the critical constraint is not land or capital but human resources. This simple fact is overlooked by many donors—including the World Bank. The World Bank, under Robert McNamara, dramatically increased lending in the 1970s, and it has appealed to donors to double lending to Africa in the 1980s. The unsupported case for doubling aid to Africa in the 1980s, in the light of the acute lack of human resources, is, in my judgment, a major flaw in the Berg report (World Bank 1981). If, however, donors take a broad view of the need for massive, long-term public investments in agricultural research, roads, faculties of agriculture in African universities, and land transfer funds (for example, for Zimbabwe) and if African countries change their agricultural development strategies and priorities and introduce policy reforms, then it may be desirable for donors to double aid to Africa in real terms over the 1980–90 period.

SUMMARY

To sum up, agricultural development is a slow and evolutionary process, and it is up to African states and donor agencies to jettison the crash project approach and start now to lay the foundation for long-term investments to solve the food production and poverty problems over a ten-to-twenty-year period. Unless steps are taken in the 1980s to overcome these basic technical, political, structural, and policy constraints, many African states may end up in the 1990s as permanent food-aid clients of the United States, the European Economic Community, and Japan.

NOTES

1. Africa is defined here to include all states in sub-Saharan Africa except the Republic of South Africa.

2. Low and unstable rainfall is a common problem in the Sahelian region of West Africa, parts of the Sudan, Ethiopia, Somalia, Kenya, Tanzania, Zimbabwe, and Botswana. But erratic rainfall, like any other single factor, cannot explain the steady erosion in Africa's capacity to feed itself.

3. Per capita GDP ranges from $120 in Chad to $1,150 in the Ivory Coast. Although per capita income is an imperfect measure that is not well suited to international comparisons, there is no question that rural poverty is a major problem throughout Africa. But because of access to land and the absence of a landless labor class, one does not witness in Africa the grinding poverty that is so pervasive in Haiti, Bangladesh, and India.

4. The average aid flows in the eight Sahelian countries was about $50 per capita in 1982 (USAID 1982, 5). Kenya received $450 million of foreign assistance in 1982, or about $25 per person.

5. Commonly known as the Berg report because Elliot Berg was the study coordinator.

6. Tsetse control is a long-term and costly activity that includes clearing of vegetation that harbors flies, spraying, release of sterile male flies, and human settlement.

7. But Ethiopia was under Italian occupation from 1936 to 1941.

8. For more information see Ruthenberg 1980; and Eicher and Baker 1982.

9. The USDA figures on per capita food production trends in Africa over the past two decades (USDA 1981) should be treated as rough estimates because population and production data for two of the large countries—Nigeria and Ethiopia—are open to question. Since Nigeria and Ethiopia together have about one-third the population of Africa, data distortions in these countries could affect the overall averages for Africa.

10. See Mellor, chapter 10 in this volume.

11. See Evenson, chapter 24 in this volume.

12. For an assessment of the modernization, dependency, and political economy models see Young 1982 and Leys 1982.

13. See the discussion of Amin's work in chapter 1 of this volume.

14. Tanzania received $2.7 billion of Official Development Assistance—a record in Africa—over the ten-year period 1973–82.

15. The sharp decline in real producer prices in the 1970s was undoubtedly an important contributor to the decline in output. Ellis (1982) reports a 35 percent decline in the price- and income-terms of trade of peasant crop producers over the 1970–80 period.

16. Tanzania is slowly dismantling its state control over agriculture following the 1982 Task Force Report (Tanzania 1982) and pressure from donors. The new agricultural policy (Tanzania 1983) has reintroduced cooperatives, turned some government estates over to village cooperatives, and encouraged foreign private investment in tea and sisal production.

17. Positive and negative pricing and taxation policies are shorthand references to the internal terms of trade between agricultural and nonagricultural products. Negative pricing and taxation policies mean that the terms of trade of agriculture are deliberately depressed by government policies (see Krishna, chapter 11 in this volume).

18. The following political constraints are partially responsible for the negative policies towards export crops: need for foreign exchange, politically powerful trade unions and urban groups, the demands of the military, and the absence of alternative ways to tax agriculture when land is not registered and the government does not have enough skilled people to collect land or incomes taxes. The net result of these constraints is that African political leaders have little room to maneuver on

pricing policies for export crops. Hence, the neoclassical economist who argues that "getting prices right" is the core of the development problem is overlooking the imperative of political survival in Africa.

19. The International Crops Research Institute for the Semi-Arid Tropics (ICRISAT) has its headquarters in Hyderabad, India. Recently, ICRISAT made a major policy decision to deemphasize the direct transfer of cereal varieties from Asia to the Sahelian countries and to construct a Sahelian research center on a five-hundred-hectare site near Niamey, Niger. The scientific staff of the Sahelian center will carry out long-term (ten-to-twenty-year) research on cereal production in the Sahel. This is further evidence that agricultural development is a slow and evolutionary process.

20. But wheat and rice consumption are increasing in urban areas throughout Africa.

21. For example, the $900 million Diama and Manantelli dams along the Senegal River are projected to have negative internal rates of return.

22. Although the World Bank was a staunch advocate of basic needs strategies in the late 1970s, it has recently abandoned its support for this dubious concept. Still the International Labour Office continues to confuse African states with recent basic needs missions to Zambia, Tanzania, and Nigeria (ILO 1981).

23. The state should play a less direct role in agricultural production and marketing and emphasize indirect approaches such as agricultural research, extension, credit, and educational programs to help small farmers and herders.

24. Although the government of Senegal dissolved its grain board—ONCAD—in 1980, a large percentage of the employees were transferred to other government agencies.

25. See Timmer, chapter 8 in this volume.

26. Food policy analysis requires a large amount of micro information on production, consumption, nutrition, and the functioning of markets, but this information is not available in most African countries. Although the World Food Council reported that nineteen African countries were preparing national food strategies in 1981, many of these exercises were prepared in capital cities in three to six months, and many of them are likely to be forgotten in three to six months.

27. For example, U.S. food aid to Mozambique was cut off for six months in 1981 (see Anderson 1981).

28. The bulk of U.S. food aid—60 percent to 70 percent—is in the form of wheat and wheat flour even though wheat is not a staple food in most of rural Africa.

29. For empirical support showing that a rural investment strategy for smallholders and small-scale industry can achieve both growth and employment objectives see the results of a nationwide survey in Sierra Leone (Byerlee et al. 1983).

30. For example, the government agency responsible for the development of the Senegal River Valley—SAED—was assisted by thirteen donors in 1982. SAED employed one thousand workers and encountered an $8.5 million recurrent budget deficit in 1982. SAED asked the thirteen donors to pay two-thirds of the cost of the deficit.

REFERENCES

Abalu, G. O. I. 1982. "Solving Africa's Food Problem." *Food Policy* 7(3):247–56.

Amin, S. 1965. *Trois expériences Africaines de développement: le Mali, la Guinée at le Ghana.* Paris: Presses Universitaires de France.

———. 1970. "Development and Structural Change: The African Experience, 1950–1970." *Journal of International Affairs* 24:203–23.

———. 1971. *L'Afrique de l'ouest bloquée: l'économie politique de la colonisation (1880–1970).* Paris: Les Editions de Minuit.

———. 1973. "Transitional Phases in Sub-Saharan Africa." *Monthly Review* 25(5):52–57.

Anderson, David. 1981. "America in Africa, 1981." *Foreign Affairs* 60(3):658–85.

Bauer, P. T. 1981. *Equality, The Third World, and Economic Delusion.* Cambridge: Harvard University Press.

Blackie, Malcolm. 1981. "A Time to Listen: A Perspective on Agricultural Policy in Zimbabwe." Working Paper 5/81. Salisbury: University of Zimbabwe, Department of Land Management. Mimeo.

Byerlee, D.; C. K. Eicher; C. Liedholm; and D. Spencer. 1983. "Employment-Output Conflicts, Factor Price Distortions and Choice of Technique: Empirical Results from Sierra Leone." *Economic Development and Cultural Change* 31(2):315-36.

Cohen, John. 1980. "Land Tenure and Rural Development in Africa." In *Agricultural Development in Africa: Issues of Public Policy,* edited by R. H. Bates and M. F. Lofchie, 349–400. Berkeley and Los Angeles: University of California Press.

Eicher, Carl K., and Doyle C. Baker. 1982. *Research on Agricultural Development in Sub-Saharan Africa: A Critical Survey.* MSU International Development Paper, no. 1. East Lansing: Michigan State University, Department of Agricultural Economics.

Ellis, Frank. 1982. "Agricultural Price Policy in Tanzania." *World Development* 10(4):263–83.

Evenson, R. E. 1981. "Benefits and Obstacles to Appropriate Agricultural Technology." *Annals of the American Academy of Political and Social Science* 458:54–67. Reprinted as chapter 24 in this volume.

Food and Agriculture Organization of the United Nations (FAO). 1978. *Regional Food Plan for Africa.* Rome.

Helleiner, G. K. 1975. "Smallholder Decision Making: Tropical African Evidence." In *Agriculture in Development Theory,* edited by L. G. Reynolds, 27–52. New Haven: Yale University Press.

Hirschman, Albert O. 1981. "The Rise and Decline of Development Economics." In *Essays in Trespassing: Economics to Politics and Beyond,* 1–24, New York: Cambridge University Press.

Idachaba, F. S. 1980. *Agricultural Research Policy in Nigeria.* International Food Policy Research Institute Research Report, no. 17. Washington, D.C.

International Labour Office (ILO). 1981. *First Things First: Meeting the Basic Needs of the People of Nigeria.* Addis Ababa: Jobs and Skills Programme for Africa.

Johnston, B. F. 1964. "The Choice of Measures for Increasing Agricultural Productivity: A Survey of Possibilities in East Africa." *Tropical Agriculture* 40(2):91–113.

Jones, William O. 1960. "Economic Man in Africa." *Food Research Institute Studies* 1:107–34.

Kamuanga, Mulumba. 1982. "Farm Level Study of the Rice Production System at the Office du Niger in Mali: An Economic Analysis." Ph.D. diss., Michigan State University.

Kenya, Republic of. 1981. *Sessional Paper No. 4 of 1981 on National Food Policy.* Nairobi: Government Printer.

Killick, T. 1978. *Development Economics in Action: A Study of Economic Policies in Ghana.* New York: St. Martin's Press.

Leys, Colin. 1982. "African Economic Development in Theory and Practice." *Daedalus* 111(2):99–124.

McKelvey, J. J., Jr. 1965. "Agricultural Research." In *The African World: A Survey of Social Research,* edited by R. A. Lystad, 317–51. New York: Praeger.

Norman, D. W. 1980. *The Farming Systems Approach: Relevancy for the Small Farmer.* MSU Rural Development Paper, no. 5. East Lansing: Michigan State University, Department of Agricultural Economics.

Nweke, F. I. 1978. "Direct Governmental Production in Agriculture in Ghana: Consequences for Food Production and Consumption, 1960–66 and 1967–75." *Food Policy* 3(3):202–8.

Nyerere, Julius K. 1968. "Socialism and Rural Development." In *Ujamaa—Essays on Socialism.* London: Oxford University Press.

Organization of African Unity (OAU). 1980. *Lagos Plan of Action for the Implementation of the Monrovia Strategy for the Economic Development of Africa.* Lagos, Nigeria.

Ruthenberg, H. 1980. *Farming Systems in the Tropics.* 3d ed. London: Oxford University Press.

Tanzania, Ministry of Agriculture, Task Force on National Agricultural Policy. 1982. *The Tanzania National Agricultural Policy (Final Report).* Dar es Salaam: Government Printer.

———, Ministry of Agriculture. 1983. *The Agricultural Policy of Tanzania.* Dar es Salaam: Government Printer.

United Nations. 1981. "World Population Prospects." New York. Mimeo.
United States Agency for International Development (USAID). 1982. *Sahel Development Program: Annual Report to the Congress.* Washington, D.C.
United States Department of Agriculture (USDA). 1981. *Food Problems and Prospects in Sub-Saharan Africa: The Decade of the 1980s.* Washington, D.C.
World Bank. 1981. *Accelerated Development in Sub-Saharan Africa: An Agenda for Action.* Washington, D.C.
———. 1982. *World Development Report 1982.* New York: Oxford University Press for the World Bank.
World Food Council of the United Nations. 1982a. *The African Food Problem and the Role of International Agencies: Report of the Executive Director.* Rome.
———. 1982b. *Nairobi Conclusions and Recommendations of the African Ministers of Food and Agriculture at the World Food Council Regional Consultation for Africa.* Nairobi.
Young, Crawford. 1982. *Ideology and Development in Africa.* New Haven: Yale University Press.
Zimbabwe, Republic of. 1981. *Growth with Equity: An Economic Policy Statement.* Salisbury: Government Printer.

Name Index

Subject Index

484

CARL K. EICHER is professor of agricultural economics at Michigan State University and director of its African Rural Economy Program. He is coeditor of *Agriculture in Economic Development* and *Growth and Development of the Nigerian Economy* and coauthor of *Research on Agricultural Development in Sub-Saharan Africa.*

JOHN M. STAATZ is assistant professor of agricultural economics at Michigan State University. He has been a research associate at the Center for Research on Economic Development, University of Michigan, and the Centre Ivoirien de Recherche Economique et Sociale in Abidjan, Ivory Coast.